Advances in Intelligent Systems and Computing

Volume 1188

D1823923

The series "Advances in Intelligent Systems and Computing" contains publications on theory, applications, and design methods of Intelligent Systems and Intelligent Computing. Virtually all disciplines such as engineering, natural sciences, computer and information science, ICT, economics, business, e-commerce, environment, healthcare, life science are covered. The list of topics spans all the areas of modern intelligent systems and computing such as: computational intelligence, soft computing including neural networks, fuzzy systems, evolutionary computing and the fusion of these paradigms, social intelligence, ambient intelligence, computational neuroscience, artificial life, virtual worlds and society, cognitive science and systems, Perception and Vision, DNA and immune based systems, self-organizing and adaptive systems, e-Learning and teaching, human-centered and human-centric computing, recommender systems, intelligent control, robotics and mechatronics including human-machine teaming, knowledge-based paradigms, learning paradigms, machine ethics, intelligent data analysis, knowledge management, intelligent agents, intelligent decision making and support, intelligent network security, trust management, interactive entertainment, Web intelligence and multimedia.

The publications within "Advances in Intelligent Systems and Computing" are primarily proceedings of important conferences, symposia and congresses. They cover significant recent developments in the field, both of a foundational and applicable character. An important characteristic feature of the series is the short publication time and world-wide distribution. This permits a rapid and broad dissemination of research results.

** Indexing: The books of this series are submitted to ISI Proceedings, EI-Compendex, DBLP, SCOPUS, Google Scholar and Springerlink **

More information about this series at http://www.springer.com/series/11156

Faisal Saeed · Tawfik Al-Hadhrami ·
Fathey Mohammed · Errais Mohammed
Editors

Advances on Smart and Soft Computing

Proceedings of ICACIn 2020

 Springer

Editors
Faisal Saeed
College of Computer Science and
Engineering
Taibah University
Medina, Saudi Arabia

Fathey Mohammed
School of Computing
Universiti Utara Malaysia (UUM)
Sintok, Malaysia

Tawfik Al-Hadhrami
Department of Computing and Technology
Nottingham Trent University
Nottingham, UK

Errais Mohammed
Faculty of science Ain Chock
Hassan II University of Casablanca
Casablanca, Morocco

ISSN 2194-5357   ISSN 2194-5365 (electronic)
Advances in Intelligent Systems and Computing
ISBN 978-981-15-6047-7   ISBN 978-981-15-6048-4 (eBook)
https://doi.org/10.1007/978-981-15-6048-4

This Springer imprint is published by the registered company Springer Nature Singapore Pte Ltd.
The registered company address is: 152 Beach Road, #21-01/04 Gateway East, Singapore 189721,
Singapore

ICACIN'20 Organizing Committee

Honorary Chair

Saddiqi Omar, Dean of the Faculty of Science, Ain Chock, Hassan II University, Casablanca, Morocco

International Advisory Board

Naomie Salim, Universiti Teknologi Malaysia, Malaysia
Rose Alinda Alias, Universiti Teknologi Malaysia, Malaysia
Ahmad Lotfi, Nottingham Trent University, UK
Funminiyi Olajide, Nottingham Trent University, UK
Essa Hezzam, Taibah University, Kingdom of Saudi Arabia
Mohammed Alshargabi, Najran University, Kingdom of Saudi Arabia
Maxim Shcherbakov, Volgograd State Technical University, Russia
Mohammed Al Ashwal, Volgograd State Technical University, Russia
Abdullah Ghareb, University of Saba Region, Yemen

Conference General Chair

Errais Mohammed, Hassan II University, Casablanca, Morocco
Faisal Saeed, President, Yemeni Scientists Research Group (YSRG), Head of Data Science Research Group in Taibah University, Kingdom of Saudi Arabia

Program Committee Co-chairs

Ouail Ouchetto, Hassan II University, Casablanca, Morocco
Raouyane Brahim, Hassan II University, Casablanca, Morocco
Mohammed Al-Sarem, Taibah University, Kingdom of Saudi Arabia

Publications Committee

Fathey Mohammed, Universiti Utara Malaysia (UUM), Malaysia
Tawfik Al-Hadhrami, Nottingham Trent University, UK

Publicity Committee

Abdullah Aysh Dahawi, Universiti Teknologi Malaysia, Malaysia
Mandar Meriem, National School of Applied Science of Berrechid, Morocco

IT Committee

Madrane Nabil, Hassan II University, Casablanca, Morocco
Koulali Rim, Hassan II University, Casablanca, Morocco
Chbihi Louhdi Mohammed Réda, Hassan II University, Casablanca, Morocco
Fuad Abdeljalil Al-shamiri, Universiti Teknologi Malaysia, Malaysia

Secretary

Baddi Youssef, Chouayb Doukkali University, Morocco
Belmakki Mostapha, INPT, Rabat, Morocco
Nadhmi Gazem, Taibah University, Kingdom of Saudi Arabia

Finance Committee

Mohammed Errais, Hassan II University, Casablanca, Morocco
Mohammed Reda Chbihi Louhdi, Hassan II University, Casablanca, Morocco
Wahid Ali, Taibah University, Kingdom of Saudi Arabia

Logistic Committee

Samia Benabdellah, Hassan II University, Casablanca, Morocco
Dehbi Rachid, Hassan II University, Casablanca, Morocco

Registration Committee

Raouyane Brahim, Hassan II University, Casablanca, Morocco
Lahbi Mohammed, Hassan II University, Casablanca, Morocco
Sameer Hasan Albakri, Universiti Teknologi Malaysia, Malaysia

Sponsorship Committee

Jai Andaloussi Said, Hassan II University, Casablanca, Morocco
Rida Mohammed, Hassan II University, Casablanca, Morocco
Fejtah Leila, Hassan II University, Casablanca, Morocco
Tabaa Mohamed, LPRI EMSI, Morocco
Ailane Abdellah, ENSAK, Morocco
Saadaoui Safa, LPRI EMSI, Morocco
Oukaira Aziz, University of Quebec in Outaouais, Canada

Technical Program Committee

Aamre Khalil, Lorraine University, France
Abbas Dandache, Lorraine University, France
Abdallah AlSaeedi, Taibah University, Kingdom of Saudi Arabia
Abdelaziz Ettaoufik, Hassan II University, Casablanca, Morocco
Abdellah Kassem, UND, Lebanon
Abdelltif EL Byed, Hassan II University, Casablanca, Morocco
Abderhmane Jarrou, Lorraine University, France
Abderrahim Maizat, Hassan II University, Casablanca, Morocco
Abderrahim Tragha, Hassan II University, Casablanca, Morocco
Abdullah Ghareb, University of Saba Region, Yemen
Abdulrahman Alsewari, Universiti Malaysia Pahang, Malaysia
Abdulrazak Alhababi, UNIMAS, Malaysia
Adil Sayouti, Ecole Royale Navale, Morocco
Ahmad Al Shami, Higher Colleges of Technology, UAE
Ahmed Lakhssassi, University of Quebec in Outaouais, Canada

Mohammed Rachdi, Hassan II University, Casablanca, Morocco
Mohammed Réda Chbihi Louhdi, Hassan II University, Casablanca, Morocco
Mohammed Rida, Hassan II University, Casablanca, Morocco
Mohammed Talea, Hassan II University, Casablanca, Morocco
Mostafa Bellafkih, INPT, Rabat, Morocco
Mostafa Belmakki, INPT, Rabat, Morocco
Mouhamad Chehaitly, Lorraine University, France
Mounia Abik, ENSIAS, Rabat, Morocco
Nabil Madrane, Hassan II University, Casablanca, Morocco
Nadhmi Gazem, Taibah University, Kingdom of Saudi Arabia
Nadia Saber, EMSI, Morocco
Najib El Kamoun, Chouaib Doukali University, El Jadida, Morocco
Noorminshah Iahad, Universiti Teknologi Malaysia, Malaysia
Noura Aknin, Abdelmalek Essadi University, Tangier, Morocco
Noureddine Abghour, Hassan II University, Casablanca, Morocco
Ouail Ouchetto, Hassan II University, Casablanca, Morocco
Rabab Chakhmoune, Brams, Morocco
Rachid Dehbi, Hassan II University, Casablanca, Morocco
Rachid Ouled Lhaj Thami, ENSIAS, Rabat, Morocco
Rachid Saadane, EHTP, Casablanca, Morocco
Rashiq Rafiq, Taibah University, Kingdom of Saudi Arabia
Redouane Kanazy, EMSI, Morocco
Rim Koulali, Hassan II University, Casablanca, Morocco
Said Jai Andaloussi, Hassan II University, Casablanca, Morocco
Said Ouatik El Alaoui, Ibn Tofeil University, Kenitra, Morocco
Sanaa Elfilali, Hassan II University, Casablanca, Morocco
Slim Lammoun, ENSIT, Tunis University
Sofien Dellagi, Lorraine University, France
Soukaine ELhassnaoui, EMSI, Morocco
Stephen Clark, Nottingham Trent University, UK
Taha Hussein, Universiti Malaysia Pahang, Malaysia
Tarik Nahal, Hassan II University, Casablanca, Morocco
Tawfik Al-Hadhrami, Nottingham Trent University, UK
Wadii Boulila, University of Manouba, Tunisia
Yousef Fazea, Universiti Utara Malaysia (UUM), Malaysia
Youssef Baddi, Chouaib Doukali University, El Jadida, Morocco
Zakaria Sabiri, ENSEA, France
Zineb Besri, Abdelmalek Saadi University, Tangier, Morocco

Preface

Thank you for your participation in the 1st International Conference of Advanced Computing and Informatics (ICACIN 2020) that will be held online in Hassan II University during April 13–14, 2020, and organized by Faculty of Sciences Ain Chock Casablanca, Hassan II University and the Yemeni Scientists Research Group (YSRG) in collaboration with Information Service Systems and Innovation Research Group (ISSIRG) in Universiti Teknologi Malaysia (Malaysia), School of Science and Technology in Nottingham Trent (UK), Data Science Research Group in Taibah University (Kingdom of Saudi Arabia) and Department of Computer-aided Design and Search Design at Volgograd State Technical University (Russia).

ICACIN 2020 is a forum for the presentation of technological advances in the field of computing and informatics. In this conference, 117 papers have been submitted by researchers from 22 countries including Algeria, Bangladesh, Colombia, Egypt, France, Greece, India, Jordon, Latvia, Malaysia, Morocco, Nigeria, Norway, Oman, Peru, Saudi Arabia, Singapore, Tunisia, UAE, UK, USA, Venezuela and Yemen. Of these 117 submissions, 56 submissions (48%) have been selected to be included in this book.

The book presents several research topics which include data science, big data analytics, Internet of things (IoT), smart computing, artificial intelligence, machine learning, information security, intelligent communication systems, health informatics and information systems theories and applications.

We would like to express our appreciation to all authors and keynote speakers for sharing their expertise with us. Special thanks go to the organizing committee for their great efforts in managing the conference. In addition, we would like to thank the technical committee for reviewing all the submitted papers; Prof. Dr. Janusz

Kacprzyk, AISC series editor; Dr. Thomas Ditzinger and Maniarasan Gandhi from Springer.

Finally, we thank all the participants of ICACIN 2020 and hope to see you all again in the next conference.

Medina, Saudi Arabia   Faisal Saeed
Nottingham, UK   Tawfik Al-Hadhrami
Sintok, Malaysia   Fathey Mohammed
Casablanca, Morocco   Errais Mohammed

About This Book

This book presents the papers included in the proceedings of the **1st International Conference of Advanced Computing and Informatics (ICACIN'20)** that was held in Casablanca, Morocco, during April 13–14, 2020. The main theme of the book is *Advances on Smart and Soft Computing*. A total of 117 papers were submitted to the conference, but only 56 papers were accepted and published in this book with an acceptance rate of 48%. The book presents several hot research topics which include artificial intelligence and data science, big data analytics, Internet of things (IoT) and smart cities, information security, cloud computing and networking and computational informatics.

Contents

Artificial Intelligence and Data Science

**Optimization of a Similarity Performance on Bounded Content
of Motion Histogram by Using Distributed Model** 3
El Mehdi Saoudi, Abderrahmane Adoui El Ouadrhiri, Said Jai Andaloussi,
and Ouail Ouchetto

**Automatic Detection of Diabetic Retinopathy Using Custom CNN
and Grad-CAM** 15
Othmane Daanouni, Bouchaib Cherradi, and Amal Tmiri

**A Zero-Knowledge Identification Scheme Based on the Discrete
Logarithm Problem and Elliptic Curves** 27
Salma Ezziri and Omar Khadir

**SMOTE–ENN-Based Data Sampling and Improved Dynamic
Ensemble Selection for Imbalanced Medical Data Classification** 37
Mouna Lamari, Nabiha Azizi, Nacer Eddine Hammami, Assia Boukhamla,
Soraya Cheriguene, Najdette Dendani, and Nacer Eddine Benzebouchi

**Evaluation of Several Artificial Intelligence and Machine Learning
Algorithms for Image Classification on Small Datasets** 51
Rim Koulali, Hajar Zaidani, and Maryeme Zaim

**Yemeni Paper Currency Recognition System Using Deep Learning
Approach** 61
Abdulfattah E. Ba Alawi, Ahmed Y. A. Saeed, Burhan T. Al-Zekri,
Borhan M. N. Radman, Alaa A. Saif, Maged Alshami, Ahmed Alshadadi,
Muneef A. Mohammad, Ala'a M. Mohammed, Osama Y. A. Saeed,
and Adeeb M. Saeed

**Tuning Hyper-Parameters of Machine Learning Methods
for Improving the Detection of Hate Speech** 71
Faisal Saeed, Mohammed Al-Sarem, and Waseem Alromema

Performance Comparison of Machine Learning Techniques in Identifying Dementia from Open Access Clinical Datasets 79
Yunus Miah, Chowdhury Nazia Enam Prima, Sharmeen Jahan Seema, Mufti Mahmud, and M Shamim Kaiser

Big Data Analytics

Clustering Analysis to Improve Web Search Ranking Using PCA and RMSE 93
Mohammed A. Ko'adan, Mohammed A. Bamatraf, and Khalid Q. Shafal

A Parallel Global TFIDF Feature Selection Using Hadoop for Big Data Text Classification 107
Houda Amazal, Mohammed Ramdani, and Mohamed Kissi

Robust and Accurate Method for Textual Information Extraction Over Video Frames 119
Soundes Belkacem and Larbi Guezouli

A Deep Learning Architecture for Profile Enrichment and Content Recommendation 131
Fatiha Sadouki and Samir Kechid

Predictive Data Analysis Model for Employee Satisfaction Using ML Algorithms 143
Madara Pratt, Mohcine Boudhane, and Sarma Cakula

A Novel Approach for Big Data Monetization as a Service 153
Abou Zakaria Faroukhi, Imane El Alaoui, Youssef Gahi, and Aouatif Amine

Ensemble Feature Selection Method Based on Bio-inspired Algorithms for Multi-objective Classification Problem 167
Mohammad Aizat Basir, Mohamed Saifullah Hussin, and Yuhanis Yusof

Automated Binary Classification of Diabetic Retinopathy by Convolutional Neural Networks 177
Ayoub Skouta, Abdelali Elmoufidi, Said Jai-Andaloussi, and Ouail Ochetto

Feature Selection and Classification Using CatBoost Method for Improving the Performance of Predicting Parkinson's Disease 189
Mohammed Al-Sarem, Faisal Saeed, Wadii Boulila, Abdel Hamid Emara, Muhannad Al-Mohaimeed, and Mohammed Errais

Lean 4.0, Six Sigma-Big Data Toward Future Industrial Opportunities and Challenges: A Literature Review 201
Hanane Rifqi, Abdellah Zamma, and Souad Ben Souda

Fog Computing in the Age of Big Healthcare Data:
Powering the Medical Internet of Things 211
Hayat Khaloufi, Karim Abouelmehdi, and Abederrahim Beni-Hssane

Prediction of Diabetes Using Hidden Naïve Bayes:
Comparative Study. 223
Bassam Abdo Al-Hameli, AbdulRahman A. Alsewari,
and Mohammed Alsarem

Toward Data Warehouse Modeling in the Context of Big Data 235
Fatimaez-Zahra Dahaoui, Lamiae Demraoui,
Mohammed Reda Chbihi Louhdi, and Hicham Behja

Internet of Things (IoT) and Smart Cities

Internet of Things Industry 4.0 Ecosystem Based on Zigbee
Protocol 249
Hamza Zemrane, Youssef Baddi, and Abderrahim Hasbi

IOT Search Engines: Study of Data Collection Methods 261
Fatima Zahra Fagroud, El Habib Ben Lahmar, Hicham Toumi,
Khadija Achtaich, and Sanaa El Filali

An Agent-Based Architecture Using Deep Reinforcement Learning
for the Intelligent Internet of Things Applications 273
Dhouha Ben Noureddine, Moez Krichen, Seifeddine Mechti, Tarik Nahhal,
and Wilfried Yves Hamilton Adoni

Toward a Monitoring System Based on IoT Devices for Smart
Buildings 285
Loubna Elhaloui, Sanaa Elfilali, Mohamed Tabaa,
and El Habib Benlahmer

10 Gbps MMW UFMC-Based Backhaul with FiWi Access
Network 295
Abdulaziz Mohammed Al-Hetar and Abdulnasser Abdulgaleel Abdulgabar

Toward a Real-Time Picking Errors Prevention System Based
on RFID Technology 303
El Mehdi Mandar, Wafaa Dachry, and Bahloul Bensassi

UAV Modular Control Architecture. 319
Asmaa Idalene, Khalid Boukhdir, and Hicham Medromi

Trust in IoT Systems: A Vision on the Current Issues, Challenges,
and Recommended Solutions 329
Hanan Aldowah, Shafiq Ul Rehman, and Irfan Umar

Review on Common IoT Communication Technologies for Both
Long-Range Network (LPWAN) and Short-Range Network 341
Abdullah Ahmed Bahashwan, Mohammed Anbar, Nibras Abdullah,
Tawfik Al-Hadhrami, and Sabri M. Hanshi

Information Security

An Overview of Blockchain Consensus Algorithms: Comparison,
Challenges and Future Directions............................. 357
Kebira Azbeg, Ouail Ouchetto, Said Jai Andaloussi, and Laila Fetjah

Critical Information Infrastructure Protection Requirement
for the Malaysian Public Sector 371
Saiful Bahari Mohd Sabtu and Kamaruddin Malik Mohamad

Experimental Evaluation of the Obfuscation Techniques Against
Reverse Engineering.. 383
Mohammed H. Bin Shamlan, Alawi S. Alaidaroos,
Mansoor H. Bin Merdhah, Mohammed A. Bamatraf, and Adnan A. Zain

PassGAN for Honeywords: Evaluating the Defender
and the Attacker Strategies 391
Muhammad Ali Fauzi, Bian Yang, and Edlira Martiri

Measuring the Influence of Methods to Raise the E-Awareness
of Cybersecurity for Medina Region Employees................. 403
Mahmood Abdulghani Alharbe

An Improvement of the Cryptographical Implementation
of Coppersmith's Method.................................... 411
Sohaib Moussaid El Idrissi and Omar Khadir

A Conceptual Model for Dynamic Access Control in Hadoop
Ecosystem.. 421
Hafsa Ait idar, Hicham Belhadaoui, and Reda Filali

Towards a Blockchain and Smart Contracts-Based Privacy
Framework for Decentralized Data Processing................... 431
Marc Gallofré Ocaña, Abdo Ali Al-Wosabi, and Tareq Al-Moslmi

Cloud Computing and Networking

VOIP in MANETs Based on the Routing Protocols OLSR
and TORA 443
Hamza Zemrane, Youssef Baddi, and Abderrahim Hasbi

A Survey on Network on Chip Routing Algorithms Criteria 455
Muhammad Kaleem and Ismail Fauzi Bin Isnin

Modeling and Developing a Geomatics Solution for the Management of Sanitation Networks in Urban Areas: A Case Study of El Fida District of Greater Casablanca 467
Kaoutar El Bennani, Hicham Mouncif, and ELMostafa Bachaoui

Networked Expertise Based on Multi-agent Systems and Ontology Design 481
Yman Chemlal and Anass El Haddadi

Overview of Mobility Management in Autonomous and Centralized Wi-Fi Architecture 495
Hind Sounni, Najib Elkamoun, and Fatima Lakrami

Wireless Technologies and Applications for Industrial Internet of Things: A Review 505
Nisrine Bahri, Safa Saadaoui, Mohamed Tabaa, Mohamed Sadik, and Hicham Medromi

Visual Vehicle Tracking via Deep Learning and Particle Filter 517
Hamd Ait Abdelali, Omar Bourja, Rajae Haouari, Hatim Derrouz, Yahya Zennayi, François Bourzex, and Rachid Oulad Haj Thami

SDNStat-Sec: A Statistical Defense Mechanism Against DDoS Attacks in SDN-Based VANET 527
Faycal Bensalah, Najib Elkamoun, and Youssef Baddi

Smart Expiry Food Tracking System 541
Haneen Almurashi, Bushra Sayed, Ma'ab Khalid, and Rahma Bouaziz

Toward a Real-Time Personal Protective Equipment Compliance Control System Based on RFID Technology 553
El Mehdi Mandar, Wafaa Dachry, and Bahloul Bensassi

Outer Weighted Graph Coloring Strategy to Mitigate the Problem of Pilot Contamination in Massive MIMO Systems 567
Abdelfettah Belhabib, Mohamed Boulouird, and Moha M'Rabet Hassani

A Comprehensive Study of Dissemination and Data Retrieval in Secure VANET-Cloud Environment 577
M. A. Al-Shabi

Toward to Autonomous Broker for Virtual Network Provisioning and Monitoring 587
Mohammed Errais, Mostafa Bellafkih, and Mohammed Al Sarem

Comparative Study on Random Traveling Wave Pulse-Coupled Oscillator Algorithm of Energy-Efficient Wireless Sensor Networks 599
Zeyad Ghaleb Al-Mekhlafi, Jalawi Alshudukhi, and Khalil Almekhlafi

Computational Informatics

**Impact of Inspirational Motivation on Organizational Innovation
(Administrative Innovation, Process Innovation, and Product
Innovation)** 613
Ali Ameen, Sultan Alshamsi, Osama Isaac, Nadhmi A. Gazem,
and Fathey Mohammed

Managing the Smart Grid in Free Market 625
Rim Marah, Inssaf El Guabassi, Sanae Larioui, and Mohammed Abakkali

An Haptic Display for Visually Impaired Pedestrian Navigation 635
Sara Alzighaibi, Rahma Bouaziz, and Slim Kammoun

**GPU Parallelization for Accelerating 3D Primitive Equations
of Ocean Modeling** 643
Abdullah Aysh Dahawi, Norma Binti Alias, and Amidora Idris

Author Index 655

About the Editors

Faisal Saeed has been an Assistant Professor at the Information Systems Department, Taibah University, KSA, since 2017. Previously, he worked as a Senior Lecturer at the Department of Information Systems at the Universiti Teknologi Malaysia (UTM). He holds a B.Sc. in Computers (Information Technology) from Cairo University, Egypt; and an M.Sc. in Information Technology Management and Ph.D. in Computer Science from UTM. His research interests include data mining, information retrieval and machine learning.

Tawfik Al-Hadhrami is a Senior Lecturer at the School of Science and Technology, Nottingham Trent University, UK. He holds a B.Sc. in Control and Computer Engineering from the University of Technology, Baghdad, Iraq; and an M.Sc. in Information Technology, Applied Systems from the School of Engineering and Physical Sciences at Heriot-Watt University, Edinburgh, UK. Since 2009, he has been a Ph.D. student researcher at the University of the West of Scotland, Paisley, UK. His research interests include wireless and mobile networks and video networking, Internet of Things (IoT), 5G/6G, and Internet protocol networks and applications.

Fathey Mohammed received his B.Sc. in Computer Engineering from Isfahan University, Iran, in 2003; his M.Sc. in Information Technology Engineering from the Tarbiat Modares University, Tehran, Iran; and his Ph.D. in Information Systems from the Universiti Teknologi Malaysia (UTM). He is currently a Visiting Senior Lecturer at the School of Computing, Universiti Utara Malaysia (UUM). His research interests include cloud computing, technology innovation adoption, information systems, project management, e-government and e-business.

Errais Mohammed is an Assistant Professor at the Computers Science Department at the Hassan II University, Casablanca, Morocco. He received his Ph.D. in Informatics from Hassan II University, Mohammedia, Morocco, in 2014. His research interests include networking, management networks and virtualization. He has published several research papers and participated in various local/international conferences.

Artificial Intelligence and Data Science

Optimization of a Similarity Performance on Bounded Content of Motion Histogram by Using Distributed Model

El Mehdi Saoudi, Abderrahmane Adoui El Ouadrhiri, Said Jai Andaloussi, and Ouail Ouchetto

Abstract In this paper, a content-based video retrieval (CBVR) system called Bounded Coordinate of Motion Histogram version 2 (BCMH v2) was processed on a distributed computing platform by using Apache Hadoop framework and a real-time distributed storage system using HBase. In fact, the amount of multimedia data is growing exponentially. Most of this data is available in image and video models. Analyzing huge data involves complex algorithms, which leads to challenges in optimizing processing time and data storage capacity. Many content-based video retrieval systems are suitable for processing large video dataset, but they are limited by the computational time and/or storage on a single machine. Thus, this paper presents the effectiveness of the proposed method with the distributed computing platform and its evaluation on the HOLLYWOOD2 video dataset. The experimental results demonstrate the good performances of the proposed approach.

Keywords Distributed system · CBVR · Hadoop · HBase

1 Introduction

Currently, multimedia data is created in various fields such as computer-aided, surveillance systems and Web searches. This data, in particular, images and videos, could be presented through signatures rather than using the entire data in order to

E. M. Saoudi (✉) · A. Adoui El Ouadrhiri · S. Jai Andaloussi · O. Ouchetto
Department of Mathematics and Computer Sciences, Faculty of Science, Computer Science, Modeling Systems and Decision Support Lab, Hassan II University of Casablanca, Casablanca, Morocco
e-mail: elmehdi.saoudi@gmail.com

A. Adoui El Ouadrhiri
e-mail: a.adouielouadrhiri-etu@etude.univcasa.ma

S. Jai Andaloussi
e-mail: said.jaiandaloussi@etude.univcasa.ma

O. Ouchetto
e-mail: ouail.ouchetto@gmail.com

3

F. Saeed et al. (eds.), *Advances on Smart and Soft Computing*, Advances in Intelligent
Systems and Computing 1188, https://doi.org/10.1007/978-981-15-6048-4_1

accelerate the representation, the indexation and the retrieving of videos. Content-based video retrieval (CBVR) techniques propose different combinations of models that showed high performance in terms of accuracy. Küçüktunç et al. [1] have presented a color histogram-based shot-boundary detection algorithm specialized for content-based copy detection applications. Singh and Hemachandran [2] have presented a CBVR system based on color moment and Gabor texture feature. Araujo et al. [3] have presented a retrieval architecture, where the image query can be compared directly based on video descriptors that can be compared directly to image descriptors. Schoeffmann et al. [4] have proposed a video content descriptor called Motion Intensity and Direction Descriptor (MIDD) that can be used to find similar segments in a laparoscopic video database. Münzer et al. [5] have presented a recent survey on content-based processing and analysis of endoscopic images and videos. El Ouadrhiri et al. [6] have implemented a Bounded Content of Motion Histogram (BCMH) using spatiotemporal features (e.g., motion vectors) and fast and adaptive bidimensional empirical mode decomposition (FABEMD) to represent the video frames by signatures. Principle component analysis (PCA), bounded coordinate system (BCS) and K-nearest neighbors (KNN) are used to calculate the similarity rate between videos. Despite these techniques, El Ouadrhiri et al. [6] present compet-itive results, and the processing time remains a critical issue. Jai-Andaloussi et al. [7] have already suggested content-based image recovery (CBIR) using a distributed computer system to take advantage of processing time. Thus, the use of distributed systems in CBVR will play an important role in providing results with a minimum time interval.

In this paper, the main objective is to implement El Ouadrhiri et al. [6] CBVR system under a distributed processing system using Hadoop and a distributed real-time storage system using HBase, for studying its performance in terms of processing time. Distributed processing for video data is one of the dominant research fields over the last decade. Computing cost and execution time required for processing vast datasets are the fundamental reasons for distributed processing. Hadoop MapReduce framework and HBase are the perfect suitable solution for distributed processing and storage. The main fundamental technology of Hadoop is HDFS, YARN and MapReduce. The Hadoop distributed file system (HDFS) is a distributed file system designed to run on commodity hardware. HDFS is highly fault-tolerant and provides high throughput access to application data and is suitable for applications that have large datasets [8]. The fundamental idea of YARN is to split up the functionalities of resource management and job scheduling/monitoring into separate daemons. Thus, the concept is to have a global resource manager (RM) and per-application application master (AM) [9]. Hadoop MapReduce is a software framework for easily writing applications that process vast amounts of data (multi-terabyte datasets) in parallel on large clusters (thousands of nodes) of commodity hardware in a reliable, fault-tolerant manner [10]. Apache HBase is an open-source, distributed, versioned, non-relational database with real-time read/write access to big data [11].

A large number of researchers are working on processing time improvement of video analysis by using Hadoop distributed system and HBase. Liu et al. [12] proposed a method to store massive image data in HBase and process it using

MapReduce. Bukhari et al. [13] proposed a big data handling system under the Hadoop ecosystem to solve the issue of demography large-scale data handling with the integration of HBase. Pandagale et al. [14] proposed a Hadoop-HBase approach for finding association rules using MapReduce algorithm; they use HBase for real-time read/write access to the data. Saoudi et al. [15, 16] proposed a content-based video retrieval system using Hadoop distributed computing. Al-bashiri et al. [17, 18] present a new similarity measurement method.

The rest of this paper is organized as follows. Section 2 presents the use-case CBVR system (BCMH v2). The proposed approach is described in Sect. 3. Experimental results are presented in Sect. 4. Finally, the paper is concluded in Sect. 5.

2   Bounded Coordinate of Motion Histogram Version 2

Bounded Coordinate of Motion Histogram Version 2 (BCMH V2 [6]) is a content-based video retrieval model (CBVR), which is based on two aspects: The first is for creating the features representing the video frames. The second is for classifying the video according to its action class. This classification could help in the video retrieval process. The frame features are based on motion vectors that are the change information between frames. Thus, the first three signatures parameters are "D", "A" and "I", direction, angle and intensity, respectively. The direction means the maximum number of motions vectors, the angle represents the orientation of the dominant direction and the intensity is the mean of the dominant motion vectors. The six further parameters are for presenting more information about the frame changes. In this part, the fast and adaptive bidimensional empirical mode decomposition (FABEMD) algorithm is applied to decompose the frames from high to low frequencies (BIMF frames) [6]. Generalized Gaussian distribution (GGD) is used in this approach to give the best probability density function that represents the frame (BIMF); this density is formulated by two parameters, which are α, the scale factor, and β, the shape factor [6]. Relatively to the computation time, this work is based only on the extraction of the first three BIMFs. The features of this model constitute the signature of the frame with nine parameters. These nine parameters are rich in information, but it is hard to use them for a large dataset. Indeed, the representation of two dimensions is more suitable to interpret the signature. Thus, bounded coordinate system (BCS) based on principle component analysis (PCA) is utilized to represent the corresponding axes of dominating content. To model the comparison between videos features, BCS gives two parameters for each video: the distance between the origins and the rotation of data. For classifying these parameters for each video with K-nearest neighbors (kNN), the non-negative least squares (NNLS) is used to form one coordinate (λ_1, λ_2) of video compared with all videos. Finally, mean average precision (MAP) presents the percentage of similarity of all video's dataset in training and testing phases. Figure 1 presents the overall BCMH v2 process and Fig. 2 shows the classification part.

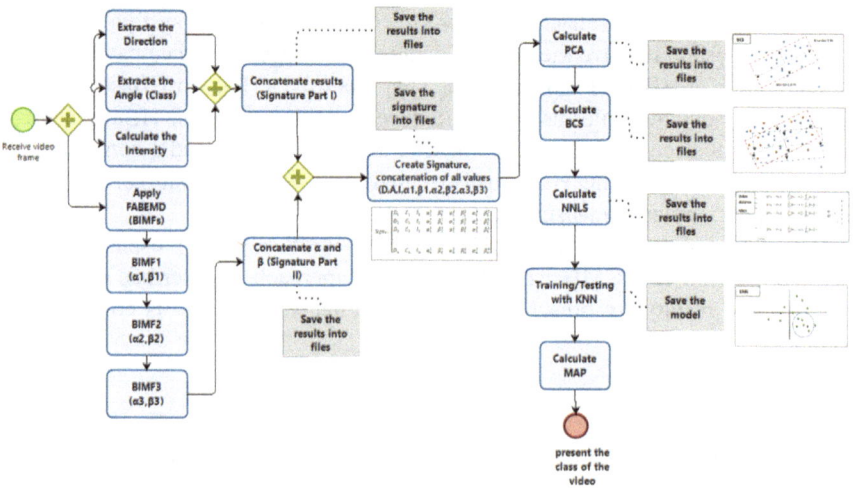

Fig. 1 Overall BCMH version 2 process

$$
Sign_{Vi} = \begin{bmatrix}
D_1 & C_1 & I_1 & \alpha_1^1 & \beta_1^1 & \alpha_1^2 & \beta_1^2 & \alpha_1^3 & \beta_1^3 \\
D_2 & C_2 & I_2 & \alpha_2^1 & \beta_2^1 & \alpha_2^2 & \beta_2^2 & \alpha_2^3 & \beta_2^3 \\
D_3 & C_3 & I_3 & \alpha_3^1 & \beta_3^1 & \alpha_3^2 & \beta_3^2 & \alpha_3^3 & \beta_3^3 \\
\cdot & \cdot & \cdot & \cdot & \cdot & \cdot & \cdot & \cdot & \cdot \\
\cdot & \cdot & \cdot & \cdot & \cdot & \cdot & \cdot & \cdot & \cdot \\
D_n & C_n & I_n & \alpha_n^1 & \beta_n^1 & \alpha_n^2 & \beta_n^2 & \alpha_n^3 & \beta_n^3
\end{bmatrix}
$$

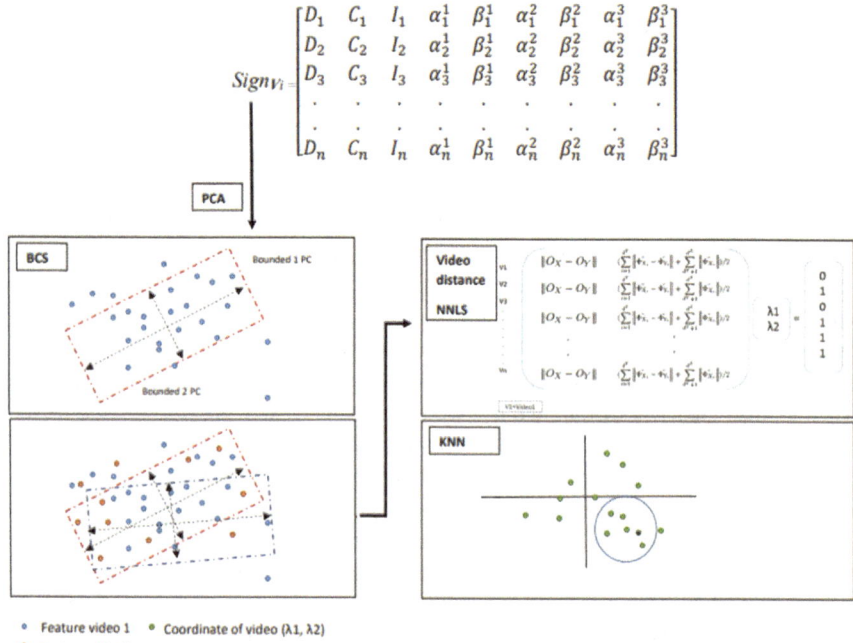

- ○ Feature video 1   ● Coordinate of video ($\lambda 1$, $\lambda 2$)
- ○ Feature video 2   ● Test video coordinate

Fig. 2 BCMH v2 classification process

3   The Proposed Approach

The Bounded Coordinate of Motion Histogram Version 2 [6] has good accuracy in video retrieving, but with the increasing amount of video data, the processing time becomes long, especially that the most crucial objective from this system is the capability to extract large amounts of videos features and process them as fast as possible. To deal with this issue, the contributions of the paper principally include:

- Adopting the Hadoop system to run up all the BCMH v2 processes in a distributed manner using MapReduce jobs.
- To reduce video processing complexity, all the video features will be saved in HBase tables for a real-time read/write query access.

3.1   System Architecture and Implementation

Many CBVR techniques are used in video processing, but the main drawback is that these techniques process data sequentially in a single machine. By providing a huge number of high-resolution videos, the processing will take a long time. Thus, a framework that allows for the distributed processing of large datasets across clusters of computers is recommended. This is guaranteed by Apache Hadoop. Further, Hadoop is a highly scalable storage platform, fault-tolerant, and it accepts a variety of data like images and videos. In order to store a large number of video data, the researchers use the Hadoop distributed file system (HDFS), and for using the extracted features values they utilize the real-time read/write access. This paper proposes a methodology using the NoSQL database HBase. The architecture of this approach is illustrated in Fig. 3. There are principally two components data processing and data storage.

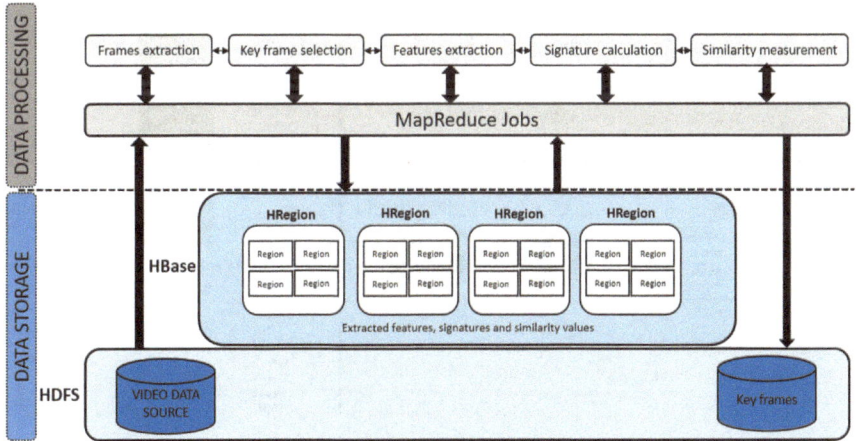

Fig. 3 Architecture of the proposed approach

3.2 Data Processing

The proposed video analysis framework stores video data and extracted features in Hadoop distributed file system and HBase, respectively, and it processes the data using the MapReduce jobs. Chained mappers use metadata for reading the videos from HDFS so that each one will process exactly one video at once. The system process several videos simultaneously, depending on the number of cluster slaves. The sequencing of MapReduce jobs of the approach is demonstrated in Fig. 4. There are mainly six MapReduce jobs:

MapReduce 1: Feature extraction part-1 of video frame. This job extracts vector motions from each video frame stored into HDFS and calculates the first part of the signature (direction, angle and intensity), and then, it stores the results into HBase. Also, this MapReduce creates three BIMF frames using the FABEMD algorithm for each video frame.

MapReduce 2: Feature extraction part-2 of video frame. Extract the second part of the signature from BIMF frames which is (α_1, β_1, α_2, β_2, α_3 and β_3) using generalized Gaussian distribution (GGD) and add results into the appropriate line in HBase. Thus, the signature of each video frame will be represented by (direction, angle, intensity, α_1, β_1, α_2, β_2, α_3, and β_3).

MapReduce 3: Downsizing the signature representation. This job applies the PCA method to represent easier the signature (2 dimensions instead of 9). Also, the process applies the BCS program to calculate the difference between the centers and the rotation angle of the PCA coordinates of the video and add results into HBase.

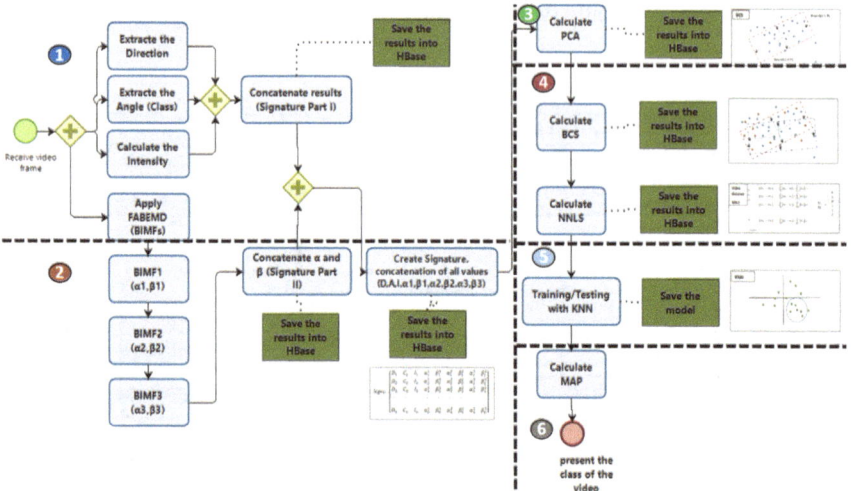

Fig. 4 Sequencing of MapReduce jobs of the proposed approach

MapReduce 4: Representing the video by coordinates. This step calculates the new coordinates of video representation by NNLS using the annotated class file of videos "y", and BCS matrix. Then, it saves results into HBase.

MapReduce 5: Finding the nearest neighbor. The job calculates the coordinate value of kNN for each video in the dataset, with $k = 5$.

MapReduce 6: Calculate the precision "Mean Average Precision (MAP)": The last MapReduce calculates the MAP of the five nearest neighbors' action class and presents the precision and the class label for each video.

3.3 Data Storage

The choice of storage module is crucial for the success of the processing module. In this approach, the training video database, as well as the extracted frames, is stored on an HDFS directory. The videos signatures database is represented as an HBase table titled HSignatures. It defines the video ID as the row key and stores frames signature calculated by the MapReduce job as values in every column family. The structure of the HSignatures table is presented in Table 1. The values of similarity measurement between training videos dataset will be recorded on an HBase table named HSimilarity. It defines the video ID as the row key and stores similarity measurement values in the appropriate column family. The process is executed by a MapReduce job. Table 2 shows the HSimilarity table design.

4 Experimental Results and Analysis

In this part, in subsection, one authors present the dataset. Subsection two introduces the cluster setup. Experimental results are analyzed and discussed in subsection three.

4.1 Dataset

The approach was evaluated on the HOLLYWOOD2 training dataset with 12 classes of human actions (AnswerPhone, DriveCar, Eat, FightPerson, GetOutCar, Handshake, HugPerson, Kiss, Run, SitDown, SitUp, StandUp) dispersed over 823 videos.

Table 1 HSignatures table structure

V_ID	Video_Info		Frame_Signature											
	Classe	HDFS_Link	ID	D	A	I	α_1	β_1	α_2	β_2	α_3	β_3	X_1	X_2
V_1	3	/Link	F_1	...	...	...	...	...	...	...	...	...	...	...
	3	/Link	F_2	...	...	...	...	...	...	...	...	...	...	...
	3	/Link	F_n	...	...	...	...	...	...	...	...	...	...	...
V_n	12	/Link	F_1	...	...	...	...	...	...	...	...	...	...	...
...	...	...	...	...	...	...	...	...	...	...	...	...	...	...

Table 2 HSimilarity table structure

V_ID	Similarity_measurement_variables					
	Video_In_Comparison		BSC X	BCS Y	NNLS X	NNLS Y
V_1	V_2		...	...	...	...
	V_3		...	...		
V_2	V_3		...	...	...	...
V_2	V_4	... ...				

4.2 Cluster Setup

The experimental tests were executed on a Hadoop-HBase cluster including one master with 2 cores of Intel(R) Core (TM) T9900 @ 3.06 GHz 4 GB RAM and four slaves with Intel 2 cores, 4 threads @ 2.6 GHz 4 GB RAM. All nodes of the Hadoop-HBase cluster are running under a 64-bit Ubuntu Linux and connected by the same network segment. The latest stable release of Hadoop and HBase that are Hadoop 3.2.1 and HBase 2.2.2 was installed.

4.3 Experimental Results

The proposed distributed approach has been compared with a local single-node system which is detailed in [6] and with a similar method named BCMHv1 which is presented in [16]. The time is considered as a principal parameter for measuring the efficiency of the approach. The experiments were executed on a Hadoop-HBase cluster with one, two and four slave nodes, respectively. For each cluster size, researchers rerun the experiment three times and keep average execution time. The purpose of the first experiment was to verify the effect of the BCMH v2 based on Hadoop-HBase, and the time-processing comparisons in a various cluster and video data sizes are shown in Fig. 5. It can be concluded that the processing time on a single node grows linearly with increasing video files. But when tests are implemented on a four slave node, the processing time is reduced to 83.19% of that of a Hadoop single machine and to 81.13% of that of a local machine. Also, the graph indicates that the proposed approach is more accurate than [16] on Hadoop cluster. So, it is clear that the Hadoop-HBase approach is relevant for large video datasets. The processing time can be more reduced with more machines added.

In order to have a clear and valid performance comparison between extracted features storage in Hadoop-text files and Hadoop-HBase, researchers implemented the two techniques and compare performances. The experimental results are shown in Fig. 6, the amount of data is limited, and the difference between time processing with Hadoop-HDFS using text files and Hadoop-HBase is undistinguished, although, with the increase of data, the time processing becomes significant. The HBase operates better with a massive amount of data.

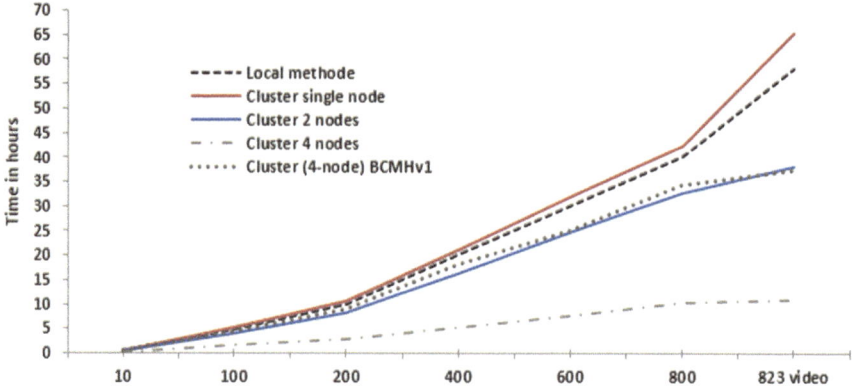

Fig. 5 Time-processing comparisons in a various cluster and video data sizes

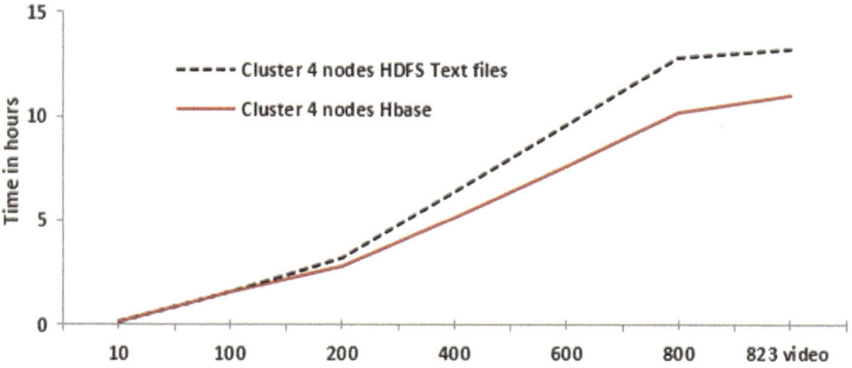

Fig. 6 Time-processing comparisons using text file storage and HBase

5 Conclusion and Future Works

This paper has proposed an approach to process huge amounts of video data on a distributed system using HBase and Hadoop MapReduce. To implement the approach, authors present a new architecture which has six chained MapReduce jobs in various phases and procedure of the use case (BCMH v2) and an HBase storage distributed system for real-time access to the data. The results show that the proposed approach reduces the processing time to 83.19% of that of a Hadoop single machine and to 81.13% of that of a local machine. Thus, the expected results have been achieved by reducing the processing time. However, the present approach is still not complete. In future works, a similarity study with other huge data processing systems is an important issue. Furthermore, it is planned to add more machines to the cluster for more evaluations. Also, an improvement in the time complexity of the BCMH v2 algorithm can affect the performance greatly.

References

1. Küçüktunç, O., Güdükbay, U., Ulusoy, Ö.: Fuzzy color histogram-based video segmentation. Comput. Vis. Image Underst. **114**(1), 125–134 (2010)
2. Singh, S.M., Hemachandran, K.: Content-based image retrieval using color moment and Gabor texture feature. Int. J. Comput. Sci. Issues (IJCSI) **9**(5), 299 (2012)
3. Araujo, A., Girod, B.: Large-scale video retrieval using image queries. IEEE Trans. Circuits Syst. Video Technol. **28**(6), 1406–1420 (2017)
4. Schoeffmann, K., Husslein, H., Kletz, S., Petscharnig, S., Muenzer, B., Beecks, C.: Video retrieval in laparoscopic video recordings with dynamic content descriptors. Multimedia Tools Appl. **77**(13), 16813–16832 (2018)
5. Münzer, B., Schoeffmann, K., Böszörmenyi, L.: Content-based processing and analysis of endoscopic images and videos: a survey. Multimedia Tools Appl **77**(1), 1323–1362 (2018)
6. El Ouadrhiri, A. A., Andaloussi, S. J., Saoudi, E. M., Ouchetto, O., Sekkaki, A.: Similarity performance of keyframes extraction on bounded content of motion histogram. In: International Conference on Big Data, Cloud and Applications, pp. 475–486. Springer, Cham, Apr 2018
7. Jai-Andaloussi, S., Elabdouli, A., Chaffai, A., Madrane, N., Sekkaki, A.: Medical content based image retrieval by using the Hadoop framework. In: 2013 20th International Conference on Telecommunications (ICT), pp. 1–5. IEEE (2013)
8. https://hadoop.apache.org/docs/stable/hadoop-project-dist/hadoop-hdfs/HdfsDesign.html
9. https://hadoop.apache.org/docs/current/hadoop-yarn/hadoop-yarn-site/YARN.html
10. https://hadoop.apache.org/docs/current/hadoop-mapreduce-client/hadoop-mapreduce-client-core/MapReduceTutorial.html
11. https://hbase.apache.org/
12. Liu, Y., Chen, B., He, W., Fang, Y.: Massive image data management using HBase and MapReduce. In: 2013 21st international conference on geoinformatics (pp. 1–5). IEEE, June 2013
13. Bukhari, S.S., Park, J., Shin, D.R.: Hadoop based Demography big data management system. In: 2018 19th IEEE/ACIS International Conference on Software Engineering, Artificial Intelligence, Networking and Parallel/Distributed Computing (SNPD), pp. 93–98. IEEE, June 2018
14. Pandagale, A.A., Surve, A. R.: Hadoop-HBase for finding association rules using Apriori MapReduce algorithm. In: 2016 IEEE International Conference on Recent Trends in Electronics, Information & Communication Technology (Rteict) (pp. 795–798). IEEE, May 2016
15. Saoudi, E.M., El Ouadrhiri, A.A., Andaloussi, S.J., El Warrak, O., Sekkaki, A.: Content based video retrieval by using distributed real-time system based on storm. Int. J. Embedded Real-Time Commun. Syst. (IJERTCS) **10**(4), 60–80 (2019)
16. Saoudi, E.M., El Ouadrhiri, A.A., El Warrak, O., Andaloussi, S.J., Sekkaki, A.: Improving content based video retrieval performance by using Hadoop-MapReduce model. In: 2018 23rd Conference of Open Innovations Association (FRUCT), pp. 1–6. IEEE, Nov 2018
17. Al-bashiri, H., Abdulgabber, M. A., Romli, A., Hujainah, F.: Collaborative filtering similarity measures: revisiting. In Proceedings of the International Conference on Advances in Image Processing, pp. 195–200, Aug 2017
18. Al-Bashiri, H., Abdulgabber, M.A., Romli, A., Salehudin, N.: A Developed collaborative filtering similarity method to improve the accuracy of recommendations under data sparsity. Int. J. Adv. Comput. Sci. Appl. (IJACSA) **9**(4), 135–142 (2018)

Automatic Detection of Diabetic Retinopathy Using Custom CNN and Grad-CAM

Othmane Daanouni, Bouchaib Cherradi, and Amal Tmiri

Abstract Diabetic retinopathy (DR) is a complication that affects eyes and is one of the common causes of blindness in the developed world. With the adoption of unhealthy lifestyles and the number of diabetic patients is rising more rapidly, there is a growing need for an automated system for early diagnosis and treatment to avoid blindness. With the development of different technologies [e.g., smart devices, cloud computing and the Internet of Things (IoT)], remote diagnosis and treatment of patients have become a feasible option for delivering health care to large populations which will minimize the need for hospitalization. In our study, we propose a lightweight customized convolutional neural network (CNN) architecture for the diagnosis of DR using optical coherence tomography (OCT) images to explore potential application in the IoT environment. We have used pre-trained CNN models, i.e., MobileNet with transfer learning approach to retrain a custom CNN for robust OCT classification. In the interest of demonstrating the feasibility of our method, we used Gradient-weighted Class Activation Mapping (Grad-CAM) to highlight important regions in the images used for prediction. For the computational task, we used an eight virtual CPUs Intel(R) Xeon(R) CPU @ 2.20 GHz with $2 \times$ NVIDIA Tesla K80 GPUs. Different metrics are used to evaluate the proposed model performance such as accuracy, recall, precision and ROC Area. The proposed architecture achieved 80% accuracy, 85% precision and 80.5% recall.

Keywords Diabetic retinopathy (DR) · Convolutional neural network (CNN) · Transfer learning · Grad-CAM · Clinical diagnosis

O. Daanouni (✉) · B. Cherradi · A. Tmiri (✉)
LaROSERI Laboratory, Chouaib Doukkali University, El Jadida, Morocco
e-mail: daanouni34@gmail.com

A. Tmiri
e-mail: b_tmiri@yahoo.fr

B. Cherradi
e-mail: bouchaib.cherradi@gmail.com

B. Cherradi
STIE Team, CRMEF Casablanca-Settat, Provincial Section of El Jadida, Casablanca, Morocco

© The Editor(s) (if applicable) and The Author(s), under exclusive license to Springer Nature Singapore Pte Ltd. 2021
F. Saeed et al. (eds.), *Advances on Smart and Soft Computing*, Advances in Intelligent Systems and Computing 1188, https://doi.org/10.1007/978-981-15-6048-4_2

1   Introduction

Diabetic retinopathy is a health condition and one of a common form of eye disease that damages blood vessels due to changes in blood glucose levels of the light-sensitive tissue at the back of the eye (retina). Diabetic retinopathy distorts vision due to the leakage of fluid in the retinal blood vessels and forms lesions in the retina. It is the most common cause of blindness in adults with diabetes [1]. Diabetic retinopathy is considered asymptomatic or causes noticeable mild vision problems, but eventually, it can cause blindness especially for individuals that have type 1 or type 2 diabetes [2, 3]. Early clinical diagnosis and detection of diabetic retinopathy are vital in order to protect patients from losing their vision.

One of the most popular diagnosis methods of DR is called optical coherence tomography (OCT). OCT was developed for retinal pathology to obtain high-resolution images of biological tissues of the eye retina [4, 5]. The increasing number of diabetic patients and limited access to optical heath experts have caused manual diagnosis of diabetic retinopathy to be a difficult task and also impractical [6]. One way to overcome this limitation is with the emergence of deep learning (DL) techniques; specifically, CNN is widely used in the field of diabetic retinopathy detection [7, 8]. This technique largely outperforms the previous image recognition methodologies [8]. However, implementing a convolution neural network model from scratch is time-consuming and requires a large amount of data. Transfer learning technique was effective for overcoming those problems, and it is a machine learning technique that uses robust pre-trained models that have been used for another task and are then adapted to solve new problems, generally related to the first task. Transfer learning reveals promising results in computer vision and especially in the field of medical image recognition.

In this work, we develop a custom CNN to solve the problem of DR detection on OCT images. In addition, we used Gradient-weighted Class Activation Mapping (Grad-CAM) to highlight important regions (regions of interest) in the OCT images used for prediction.

The paper is organized as follows: In Sect. 2, we present the latest research conducted in DR prediction. In Sect. 3, we present an overview of the materials and methods used. In particular, we introduce the OCT dataset and some rudiments about the DL techniques. In Sect. 4, we present a brief description of our proposed custom CNN adapted to automatic detection of DR diseases. Implementation setup, some results and discussion are given in Sect. 5. In Sect. 6, we conclude this paper and present some perspectives of this work.

2   Related Work

Pedro Costae et al. [9] proposed a weakly supervised method for detection of DR in eye fundus image with regional localization that contains a lesion, while the model is trained on images' labels only. The model was trained and validated through Messidor dataset and tested using E-ophtha MA without retraining of the model. The study achieved high performance with 95.8% area under the receiver operating curve.

Arun Das et al. [10] introduce a cloud-based teleophthalmology architecture for diagnosis and progression of age-related macular degeneration (AMD) via the Internet of Medical Things (IoMT). To predict AMD, a convolution neural network based on ResNet with 152 layers is trained with age-related eye disease study (AREDS) images, and for prediction of the progression of AMD, they used a long-short term memory (LSTM) deep neural network. The architecture achieved promising results with specificity $98.32 \pm 0.1\%$ and sensitivity $94.97 \pm 0.5\%$.

Liang et al. [11] proposed a multi-self-attention network structure based on a three-architecture inception V3 model to generate a feature map and feed it into a multi-self-attention network to calculate different self-attention features and finally used a convolution network to predict DR. The experiment implemented with a Tensorflow framework, and the results show that the classification result is the best when the self-attention network is used twice ($n = 2$) and achieved 87.7%.

3   Materials and Methods

This section aims to present and analyze different methodologies used to perform classification of diabetic retinopathy such as CNN, transfer learning, data collection and others. However, first, we offer in the following sub-section a brief overview of the dataset used.

3.1   OCT Dataset

The repository dataset [8] holds 207,130 OCT images. These images are split into two sets: (1) 108,312 training images (37,206 images with choroidal neovascularization, 11,349 images with diabetic macular edema, 8617 images with drusen and 51,140 normal images) taken from 4,686 patients, (2) 1000 testing images (250 images from each category) taken from 633 patients. Images are split into four directories: CNV, DME, drusen and normal and labeled as follow:

(disease)-(randomized patient ID) − (image number for this patient).

Figure 1 presents an illustration of different DR captured using OCT images.

3.2   Convolutional Neural Network (CNN)

A convolutional neural network (CNN) is a form of supervised learning technique that consists of convolution layers for automatic feature extraction and neural networks for

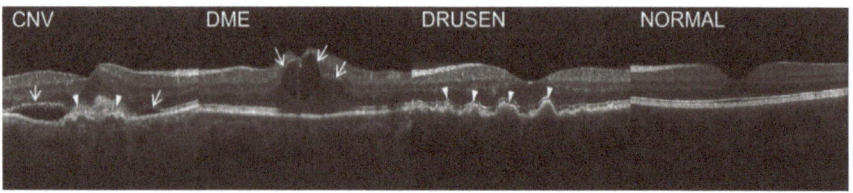

Fig. 1 Different categories of diabetic retinopathy

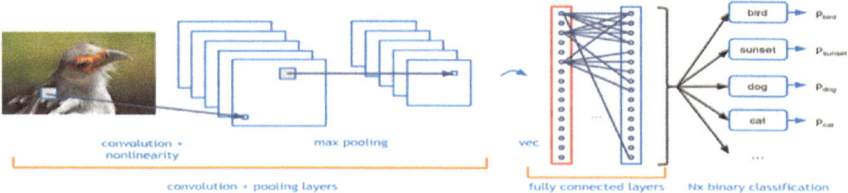

Fig. 2 Example of convolution neural network

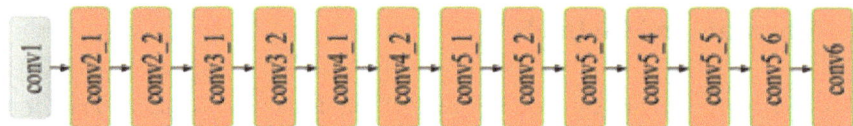

Fig. 3 MobileNet architecture

classification. A CNN model is a feed-forward neural network, and it is inspired by an animal's visual cortex. It is constructed with different layers such as conv2D, pooling layers and feature maps. These are successfully applied in recognition tasks such as handwriting [12] and others [13]. In order to detect required shapes, colors and other patterns, the convolution layer utilizes kernel sizes such as 3×3 or 5×5, convolved across the image. Robust pre-trained models of CNN are already available namely MobileNet, AlexNet, VggNet and others. Figure 2 presents a typical architecture of CNN.

3.3 MobileNet Architecture

MobileNet[1] is a CNN architecture suitable for mobile- and embedded-based vision applications; because of its lightweight architecture, there is no need for high computing power and it has low latency. The MobileNet was introduced by Google (Fig. 3).

---

[1]https://arxiv.org/abs/1704.04861.

3.4 Transfer Learning (TL)

Transfer learning (TL)[2] is a training method aiming to adopt the weights of a pre-trained CNN and appropriately re-train the CNN to optimize the weights for a specific task, i.e., AI classification of retinal image. TL is not only an efficient optimization procedure but improves classification. Early convolution layers learn to recognize generic features, edges, patterns or textures although deeper layers focus on specific aspects of the new considered image task, such as tumor segmentation or blood vessels. Actually, TL is widely used in computer vision in terms of efficiency gains, reducing a major problem in the artificial intelligence (AI) field represented by overfitting [14] (Fig. 4).

4 Proposed Custom CNN for Automatic Detection of DR

In this section, we present a brief description of our proposed CNN model-based transfer learning technique. Indeed, this custom CNN architecture used for the classification of fundus images is a modification of MobileNet architecture with the transfer learning approach described above. To achieve this, we truncated the last fully connected layers and replaced them with three fully connected layers and a new softmax layer that is relevant to our problem (classification of OCT images). In this manner, we achieved 3,228,864 non-trainable parameters in the customized CNN from 5,854,916 parameters. Then, we have to train only 2,626,052 trainable parameters. Figure 5 shows the partial architecture of our custom CNN.

In order to adjust the input to the correct shape for the dense layer, we used the GlobalAveragePooling2D function. We used a softmax activation function with four classes (drusen, DME, CNV and normal) in order to compute the probability of each class.

5 Results and Discussion

In this section, we conduct a performance analysis of MobileNet architecture with transfer learning methodology in order to predict diabetic retinopathy.

For this purpose, different steps in transfer learning are used. Firstly, the weights of MobileNet architecture are set to be not trainable and we replace the last fully connected layer with a new fully connected layer on the top to address our problem. After that, we train the network with OCT images in order to learn new weights and new features extraction.

---

[2]https://machinelearningmastery.com/transfer-learning-for-deep-learning/.

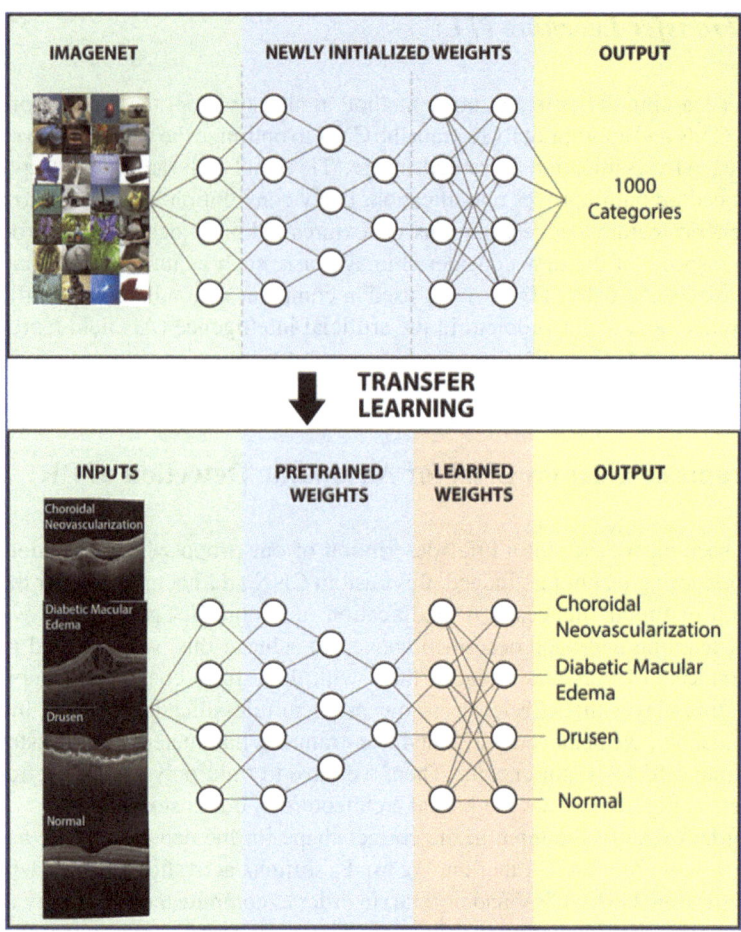

Fig. 4 Transfer learning technique

5.1 Experiment Setting

The experiment set conducted in this research used Python 3 as the programming language with Keras 2.3.1 and TensorFlow 2.0 as backend.

For the activation function, we used ReLu activation and we chose the categorical cross-entropy with Adam optimization with a learning rate of 0.001. In order to prevent our model from overfitting, we defined an early stop to monitor validation loss with min delta equal to zero and number of patience = 2 as 2 number of epochs with no improvement after which training would be stopped.

For resolving the problem of local minimum, we used a reduce on a plateau to monitor validation loss with a factor equal to 0.2 and minimum delta = 0.00001 with two patience.

lobal_average_pooling2d_2 (	(None, 1024)	0
ense_5 (Dense)	(None, 1024)	1049600
ense_6 (Dense)	(None, 1024)	1049600
ense_7 (Dense)	(None, 512)	524800
ropout_1 (Dropout)	(None, 512)	0
ense_8 (Dense)	(None, 4)	2052

```
=================================================================
otal params: 5,854,916
rainable params: 2,626,052
on-trainable params: 3,228,864
```

Fig. 5 New fully connected architecture

In order to make the model more robust, we used real-time data augmentation on images on CPU in parallel to training the model on GPU.

5.2 Performance Evaluation

The performance metrics will have a serious impact on retinopathy diabetics prediction. Therefore, the accuracy, precision and recall are mean indexes considered in the evaluation. The confusion matrix and ROC curve are also used in order to give a valuable evaluation of the classification performance of the model.

$$\text{Recall} = \frac{\text{TP}}{\text{TP} + \text{FN}} \tag{1}$$

$$\text{Precision} = \frac{\text{TP}}{\text{FP} + \text{TP}} \tag{2}$$

$$\text{Accuracy} = \frac{\text{TP} + \text{TN}}{\text{TP} + \text{TN} + \text{FP} + \text{FN}} \tag{3}$$

The training and validation accuracy and loss curves of custom MobileNet over nine epochs are shown in Figs. 6 and 7. The training process is stopped to avoid overfitting of models when accuracy no longer improves and the training loss converges to a minimum. Additionally, the training process took approximately 2.79 h.

It can be seen from Figs. 6 and 7 that the training loss of the custom MobileNet model quickly converges, and the accuracy no longer improves. After the training, we evaluated the network model on the testing set with 1000 OCT images divided into 250 images per class. The testing scores are shown in Table 1, and their ROC curve [15] is shown in Fig. 9.

Fig. 6 Training and validation accuracy after nine epochs

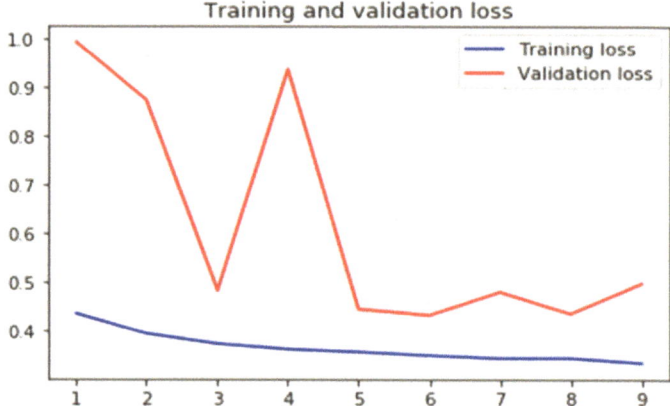

Fig. 7 Training and validation loss after nine epochs

Table 1 Evaluation metrics of custom CNN

	Precision	Recall	Support
CNV	0.89	0.92	250
DME	0.97	0.44	250
DRUSEN	0.91	0.86	250
NORMAL	0.63	1	250
Accuracy	0.80		1000

The micro- and macro-average area values achieved by custom MobileNet are equal to 0.87, and the highest area among the four classes is achieved by the CNV class (area = 0.94); however, the lowest area is achieved by the DME class (area = 0.72).

Figure 8 presents the four class prediction results on a confusion matrix.

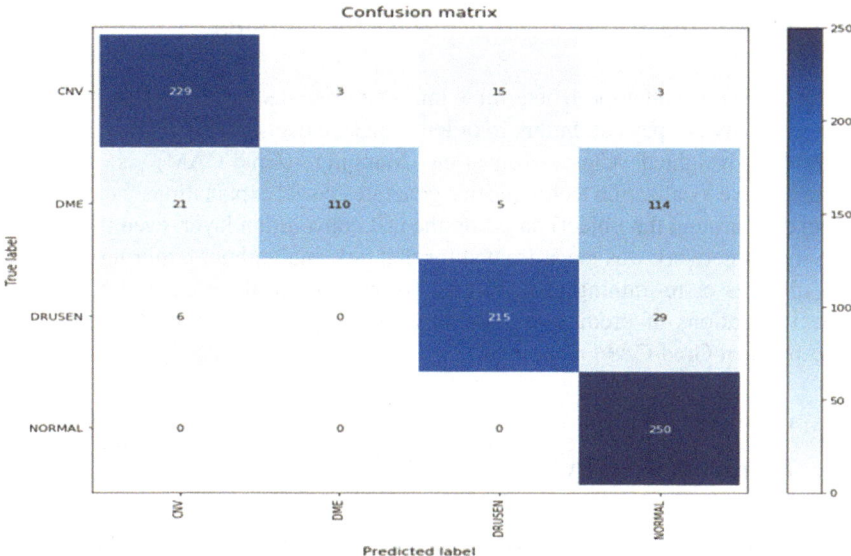

Fig. 8 Confusion matrix of 1000 OCT images classified in four classes

Figure 9 presents other performance evaluation results in terms of receiver operating characteristic (ROC) variation.

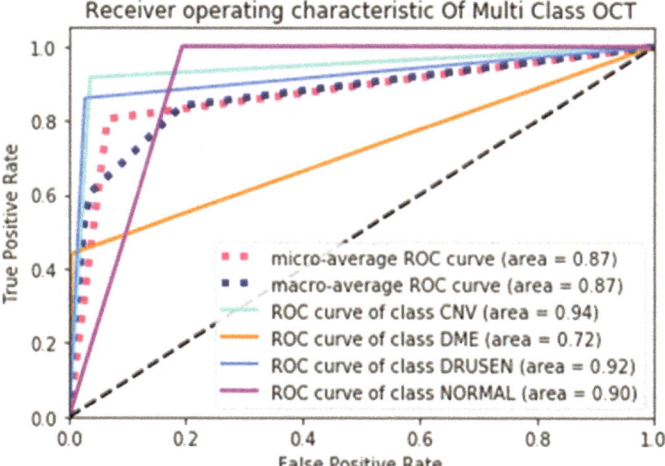

Fig. 9 Receiver operating characteristic performance

5.3   *Visualizing Neural Network Decision-Making*

In this section, we attempt to determine important regions in OCT images to which the neural network pays attention, in order to make a decision.

Gradient-weighted Class Activation Mapping (Grad-CAM) is a class-discriminative localization technique that generates visual explanations (i.e., drawing a bounding around the object) based on the last convolution layer, even though the CNN-based network was never trained for that task and without requiring architectural changes or re-training [16]. In Fig. 10, we present three examples showing some illustrations of prediction results by the proposed custom CNN and their correspondent Grad-CAM mapping.

5.4   *Threats to Validity*

The major focus of this study is to test capability for a lightweight customized CNN, i.e., MobileNet that demonstrates its effectiveness in different tasks with feasible accuracy to reduce size and latency [17] that is required for the IoT environment. MobileNet architecture has not yet been explored deeply for prediction of diabetic retinopathy. The result of this study is to identify the specific key element for the prediction of diabetes disease and to enhance the model in the future work.

6   Conclusion and Perspectives

Accurate and efficient prediction of diabetic retinopathy using computer vision and smart mobile devices can be the adequate solution to the DR problem.

In this paper, an automated approach using customized CNN architecture is suggested to detect diabetic retinopathy using OCT images. This approach is based on transfer learning from a MobileNet model and replaces the last fully connected layer with three fully connected layers with a softmax layer. The customized model is trained on 207.130 OCT images to classify OCT into four classes. The Grad-CAM technique is used to show a localization and generates visual explanations of the region of interest directly responsible for DR disease. The customized model shows more promising results within nine epochs; the model achieves 0.80 accuracy within 0.87 ROC Area.

This work is extended, to focus more on enhancing the accuracy of MobileNet model and reduce complexity and latency by combining the model with a shallow neural network and a pre-trained feature map to achieve high accuracy in prediction DR.

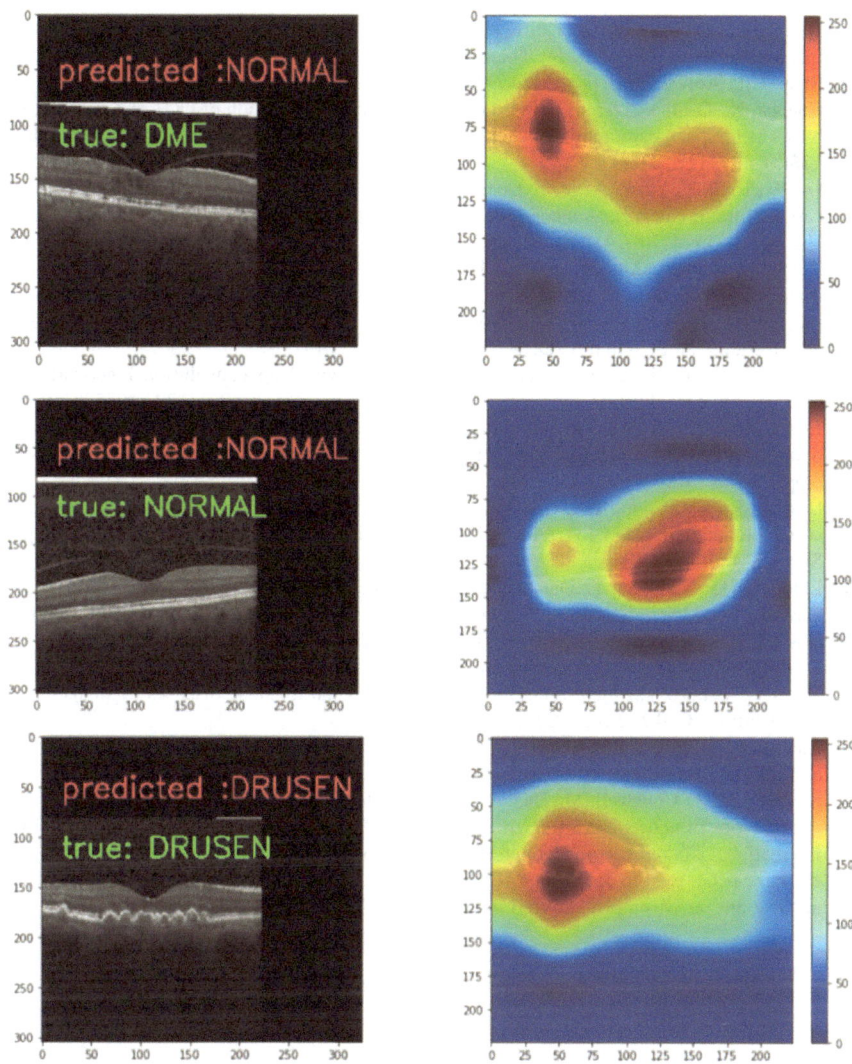

Fig. 10 Visualization of Grad-CAM tested on OCT images

References

1. Faust, O., Acharya, R., Ng, E.Y.K., Ng, K.H., Suri, J.S.: Algorithms for the automated detection of diabetic retinopathy using digital fundus images: a review. J. Med. Syst. **36**(1), 145–157 (2012)
2. Daanouni, O., Cherradi, B., Tmiri, A.: Predicting diabetes diseases using mixed data and supervised machine learning algorithms. In: Proceedings of the 4th International Conference on Smart City Applications, p. 85 (2019)

3. Daanouni, O., Cherradi, B., Tmiri, A.: Type 2 diabetes mellitus prediction model based on machine learning approach. In: The Proceedings of the Third International Conference on Smart City Applications, pp. 454–469 (2019)
4. Cuenca, N., Ortuño-Lizarán, I., Pinilla, I.: Cellular characterization of OCT and outer retinal bands using specific immunohistochemistry markers and clinical implications. Ophthalmology 125(3), 407–422 (2018)
5. Lu, W., Tong, Y., Yu, Y., Xing, Y., Chen, C., Shen, Y.: Deep learning-based automated classification of multi-categorical abnormalities from optical coherence tomography images. Transl. Vis. Sci. Technol. 7(6), 41–41 (2018)
6. Asiri, N., Hussain, M., Abualsamh, H.A.: Deep learning based computer-aided diagnosis systems for diabetic retinopathy: a survey. arXiv preprint arXiv:1811.01238 (2018)
7. Mateen, M., Wen, J., Song, S., Huang, Z.: Fundus image classification using VGG-19 architecture with PCA and SVD. Symmetry 11(1), 1 (2019)
8. Wang, Z., Yang, J.: Diabetic retinopathy detection via deep convolutional networks for discriminative localization and visual explanation. arXiv preprint arXiv:1703.10757 (2017).
9. Costa, P., et al.: EyeWes: weakly supervised pre-trained convolutional neural networks for diabetic retinopathy detection. In: 2019 16th International Conference on Machine Vision Applications (MVA), pp. 1–6 (2019)
10. Das, A., Rad, P., Choo, K.K.R., Nouhi, B., Lish, J., Martel, J.: Distributed machine learning cloud teleophthalmology IoT for predicting AMD disease progression. Fut. Gener. Comput. Syst. 93, 486–498 (2019)
11. Liang, Q., Li, X., Deng, Y.: Diabetic retinopathy detection based on deep learning (2019)
12. Cireşan, D., Meier, U., Schmidhuber, J.: Multi-column deep neural networks for image classification. arXiv preprint arXiv:1202.2745 (2012)
13. Sayamov, S.: Weakly supervised learning for retinal lesion detection (2019)
14. Kermany, D.S., et al.: Identifying medical diagnoses and treatable diseases by image-based deep learning. Cell 172(5), 1122–1131. e9 (2018)
15. Rice, M.E., Harris, G.T.: Comparing effect sizes in follow-up studies: ROC Area, Cohen's D, and R. Law Hum. Behav. 29(5), 615–620 (2005)
16. Selvaraju, R.R., Cogswell, M., Das, A., Vedantam, R., Parikh, D., Batra, D.: Grad-cam: visual explanations from deep networks via gradient-based localization. In: Proceedings of the IEEE International Conference on Computer Vision, pp. 618–626 (2017)
17. Howard, A.G., et al.: Mobilenets: efficient convolutional neural networks for mobile vision applications. arXiv preprint arXiv:1704.04861 (2017)

A Zero-Knowledge Identification Scheme Based on the Discrete Logarithm Problem and Elliptic Curves

Salma Ezziri and Omar Khadir

Abstract In cryptography, zero-knowledge proof has become frequently used in several types of transactions, for example, the authorization to access a server, the communication between a card and an automatic teller, and the verification of customers identity by banks. The idea of cryptographic identification protocols is that one entity, the claimant, proves its identity to another party, the verifier, by demonstrating the knowledge of a secret, of course without revealing the private key itself to the verifier during the protocol. In this work, a new identification method is presented, which bases its security on the discrete logarithm problem and partially on the Schnorr protocol. The security analysis and the complexity will be examined. An application to elliptic curves will be also given.

Keywords Zero-knowledge proof · Identification · Discrete logarithm problem: DLP · Schnorr's protocol · Elliptic curves

1 Introduction

Identification protocols are an important issue in public key cryptography. They are used in various situations, due to their simplicity and security. These schemes are usually used in microprocessor-based devices, access to software applications, local or remote access to computer accounts.

The idea of cryptographic zero-knowledge protocols is that: one entity, the claimant, proves its identity to another entity, the verifier, by demonstrating the knowledge of a secret but without revealing the secret itself to the verifier during the protocol.

S. Ezziri (✉) · O. Khadir
Laboratory of Mathematics, Cryptography, Mechanics and Numerical Analysis, FSTM,
University Hassan II of Casablanca, Casablanca, Morocco
e-mail: Salma.ezziri@gmail.com

O. Khadir
e-mail: khadir@hotmail.com

27
F. Saeed et al. (eds.), *Advances on Smart and Soft Computing*, Advances in Intelligent Systems and Computing 1188, https://doi.org/10.1007/978-981-15-6048-4_3

The method is generally based on developing the solutions of hard mathematical problems, like factoring, discrete logarithm issues and computing the square root modulo a large composite number.

The identification is carried out by supplying an answer to a time-variant challenge, where the response relies on both the prover's secret and the challenge.

The challenge is a number chosen by the verifier, randomly and privately. The verifier, Bob, checks if the answer given by the prover, Alice, is valid or not. It is very difficult for anybody other than Alice to imitate her identification.

A zero-knowledge proof must satisfy three properties:

Completeness: If the statement is valid, then the honest verifier, Bob, will be convinced of this fact by an honest prover. Consequently Bob will accept Alice's identification with probability one.

Soundness: If the statement is invalid, no cheating prover can convince the honest verifier that it is true. Consequently the verifier, Bob, will reject the proof with probability of at least $1/2$.

Zero-knowledge: If the statement is true, no cheating verifier can discover anything new other than this information.

In 1985, Goldwasser et al. [1] introduced the idea of an interactive proof system, thus creating an important branch of cryptography and computational complexity theory zero-knowledge proof. In the same year, Koblitz [2] proposed the use of elliptic curves in cryptography. He proved that there exist groups more complex than the conventional multiplicative group $((Z/pZ)^*)$, where Z is the set of all integers and p is a prime number. These structures are practical in public-key cryptography. In 1986, Fiat and Shamir [3, 4] developed a remarkable scheme of identification, based on the factoring and computing the square root modulo a large composite number. In 1988, Guillou and Quisquater [5, 6] published a paper in which they proposed an extension of the Fiat-Shamir protocol. Their method was based on the famous RSA algorithm [7]. It permits a reduction in the memory required for the user's secrets and the count of messages exchanged. In the same year, another variant of the zero-knowledge scheme was introduced by Blum et al. [8], named non-interactive proofs. They showed that a common random string shared between the verifier and the prover is sufficient to accomplish computational zero-knowledge without requiring interaction. In 1989, Schnorr [9], see also Stinson [10, pp. 371–374], used the discrete logarithm problem to propose a new identification technique. In 1993, Okamoto [11], Stinson [10, pp. 378–383], put forward a scheme which is provably secure against active attacks under the discrete logarithm assumption. In 2005, Bangerter et al. [12] presented an efficient zero-knowledge proof for exponentiation and multi-exponentiation based on a discrete logarithm. In 2018, a new variant of Guillou–Quisquater zero-knowledge protocol was suggested [13], and the security of this method relies simultaneously on RSA and on Rabin's cryptosystem. Recently, in 2020, Gonzlez et al. [14] present a zero-knowledge proof scheme resistible against quantum attacks. Its security was based on two solid problems: the isomorphism of polynomials problem and the graph isomorphism problem.

However, with the development of rapid computer computation and quantum computers, most of the algorithms cited above have been broken. This leads to the necessity of inventing new alternative communication protocols in order to protect sensitive data.

In this work, a new identification protocol will be presented. In order to improve the security of the existing Schnorr scheme, the proposed scheme will be based simultaneously on the discrete logarithm problem and elliptic curves. Finally, a theoretical generalization will be presented.

The paper is organized as follows: Sect. 2 outlines the basic Schnorr's identification scheme. Section 3 presents the main results, while Sect. 4 summarizes the work and presents its conclusions.

In all the following, let N and Z be the set of natural numbers and relative integers, respectively. In this paper, the notations in Schnorr's paper [7] will be respected.

First the classical Schnorr's scheme is described.

2 Methods

2.1 Description of the Classical Schnorr's Protocol

In mathematics, the discrete logarithm problem is how to compute the exponent x that verifies the modular equation $a^x \equiv b[p]$, where a and b are given elements in the finite group (Z/pZ).

For large prime numbers p, calculating the discrete logarithm for an element of the multiplicative group (Z/pZ) is believed to be a difficult problem. So far, no efficient general method for solving discrete logarithms on conventional computers is known. Several important data protection algorithms in public-key cryptography base their security on this assumption.

In 1989, Schnorr suggested one of the simplest identification protocols based on the discrete logarithm problem. The protocol is defined for a cyclic group Z/pZ where p is a prime integer.

In order to prove the knowledge of a number, s, that verifies $g^s \equiv v[p]$, where p, g and v are given, Alice, the prover, interacts with the verifier, Bob, as follows:

1. In the first round, Alice commits herself to a random number r. She sends to Bob the message $R \equiv g^r[p]$.
2. Bob replies by a challenge c chosen at random and less than p.
3. After receiving c, Alice sends the third and last message $a \equiv r - cs[p - 1]$.

   Bob accepts the identification if and only if $R \equiv g^a v^c [p]$.

2.2 Elliptical Curves Version of Schnorr's Protocol

In this section, a summary of the basic protocol of Schnorr over elliptic curves is presented.

Let E be an elliptic curve defined over Z/pZ, p prime, and P, $Q \in E$. In addition, P is a generator.

The elliptical curve discrete logarithm problem is how to find the integer s such that $Q = sP$.

In fact, nobody knows how to solve this equation. No efficient algorithm is known. It is however considered to be more complicated to resolve than the conventional discrete logarithm problem.

Because the elliptical curve discrete logarithm problem tells us that for $Q = s.P$, the secret integer s is very difficult to find, let Q and P be the public keys of Alice and s be her private key. In order to prove the knowledge of the solution of a discrete logarithm problem, the prover, Alice, interacts with the verifier, Bob, as follows:

1. Alice commits herself to a random integer r. She sends to Bob the message $R = rP$.
2. Bob replies with a challenge number c chosen at random.
3. After receiving c, Alice sends the third and last message: the response $a = r + cs$.

The verifier accepts the identification, if and only if $aP = R + cQ$. In the following section, the main work of this paper is presented.

3   Results

3.1 Description of the Proposed Identification Scheme

The protocol is defined in a cyclic group Z/pZ of order p with a generator g. In order to prove the knowledge of s such that $g^s \equiv v[p]$, where p, g and v are given, Alice interacts with the verifier Bob as follows:

1. Alice chooses a random number r and computes $R \equiv g^r[p]$. She sends the result R to the verifier Bob.
2. Bob selects two random numbers c and d, less than p, and sends them to Alice.
3. Alice computes $a \equiv r - cs - ds^2 \ [p-1]$, and $w \equiv g^{s^2}[p]$, then sends the results to Bob.

Bob accepts the identification if and only if $R \equiv g^a v^c w^d \ [p]$.

Example 1 Let $p = 4933$. Alice's secret key is $s = 2587$ and her public keys are $g = 2$, $v \equiv g^s \equiv 2^{2587} \equiv 3575[p]$. She interacts with the verifier Bob by following this procedure:

1. Alice chooses a random number $r = 1982$ and computes $R \equiv g^r \equiv 2^{1982} \equiv 3020[p]$. She sends the result R to the verifier Bob.
2. Bob selects two random numbers $c = 4818$ and $d = 1609$ and sends them to Alice.
3. Alice replies by sending $a \equiv r - cs - ds^2 \equiv 1982 - 4818.2587 - 1609.2587^2 \equiv 3775[p - 1]$, and $w \equiv g^{s2} \equiv 3065[p]$.

Bob checks that $R \equiv g^a v^c w^d \equiv 2^{3775} 3575^{4818} 3065^{1609} \equiv 3020[p]$. Here, Bob accepts the identification of Alice.

Security Analysis

Theorem 1 *If an attacker breaks the Schnorr protocol, the previous identification method remains valid. Then, it is an alternative of Schnorr's zero-knowledge scheme, hence the effectiveness of the presented result.*

Proof Suppose that Oscar finds a way to break the protocol proposed by Schnorr, in other words he manages to solve the equation $a \equiv r - cs[p - 1]$ and find Alice's secret key. However, he will not be able to solve both the equations: $a \equiv r - cs - ds^2[p - 1]$ and $w \equiv g^{s^2}[p]$, hence the efficiency of the suggested scheme.

Theorem 2 *If an attacker is able to break the proposed identification protocol, then he also can break Schnorr's scheme.*

Proof Suppose that the attacker, Oscar, finds a way to break the suggested protocol. So, for every two parameters, c and d, sent by the verifier Bob, Oscar can find a number a that verifies the identification equation. In particular, that is the case if Bob chooses $d = 0$. And that is exactly the procedure in the Schnorr's identification. Therefore, the attacker can identify himself as Alice in the Schnorr identification method.

Note that, reciprocally, if Oscar breaks the protocol proposed by Schnorr, there is no evidence that he can break the proposed scheme. So, from a point of security, the result presented is stronger than that suggested by Schnorr.

Now, let us discuss some possible and realistic attacks. Suppose that Oscar is an Alice adversary.

Attack 1: Knowing Alice's public keys, if the attacker intercepts the value of a and w, he will not be able to find Alice's secret keys. Instead, he will be blocked by a mathematical equation with too many unknown parameters.

Attack 2 (man in the middle): Suppose that Oscar intercepts the values of R, a and w. If he tries to imitate the identification of Alice he cannot succeed. Indeed, he will be blocked by the challenge numbers c and d which are changeable with every trial of identification.

Attack 3: If the attacker intercepts the value of R, he will not be able to find r. Here, he will be blocked by a discrete logarithm problem.

Attack 4: If the enemy finds a way to solve the equation of identification proposed by Schnorr: $R \equiv g^a v^c \ [p]$, he will not be able to solve the equation: $R \equiv g^a v^c w^d \ [p]$, as the equations are different. So breaking the Schnorr scheme does not necessarily lead to breaking the proposed protocol.

Complexity Let T_{exp} and T_{mult} be the time required to calculate an exponentiation and a multiplication, respectively. The time used for modular subtractions, additions, and comparisons is neglected.

Following the protocol Alice needs to perform one modular exponentiation to generate her public keys. In the identification protocol, she must execute two modular exponentiations and four modular multiplications. Finally, the verifier, Bob, calculates three modular exponentiations and two modular multiplications to check the identification process. Thus, there are six modular exponentiations and six multiplications. The total time required to execute all the identification's operations is as follows:

$$
\begin{aligned}
T_{tot} = 6T_{mult} + 6T_{exp} &= 6O\big((\log p)^2\big) + 6O\big((\log p)^3\big) \\
&= O\big((\log p)^2\big) + O\big((\log p)^3\big) \\
&= O\big((\log p)^2 + (\log p)^3\big)
\end{aligned}
$$

Finally, the proposed identification system works on a polylogarithmic time.

3.2 Elliptic Curves Version of the Proposed Identification Scheme

Now, let us view the proposed protocol over elliptic curves.

Let E be an elliptic curve defined over Z/pZ and two points $P, Q \in E$, such that $Q = sP$.

Let Q and P be the public keys of Alice and s be her private key.

In the identification protocol, Alice proves that she knows the parameter s that verifies $Q = sP$. Alice interacts with the verifier Bob as follows:

1. Alice chooses a random number, r, and computes $R = rP$. She sends the result, R, to the verifier Bob.
2. Bob chooses two random numbers c and d and sends them to Alice.
3. Alice replies by sending $a = r + cs + ds^2$, and $W = s^2 P$.

   Bob accepts the identification if and only if $aP = R + cQ + dW$.

Example 2 Let $p = 609,359$ and consider the curve: E: $Y^2 = X^3 + 1013X + 893$. Alice's secret key is $s = 8435$. Her public keys are $P = (354,066,257,183)$ and $Q = sP = 8435P = (251,658,23,824)$.

Alice needs to prove that she knows a number s that verifies $Q = sP$. To be carried out, the identification protocol works as follows:

1. Alice chooses a random number $r = 24{,}835$ and computes $R = rP = 24835P = (108{,}146{,}104{,}514)$. She sends the result R to the verifier Bob.
2. Bob chooses two random numbers $c = 9541$ and $d = 3826$ and sends them to Alice.
3. Alice replies by sending $a = r + cs + ds^2 = 24{,}835 + 9541.8435 + 3826.8435^2 = 272{,}297{,}438{,}020$, and $W = s^2P = 71149225P = (407{,}826{,}271{,}721)$ to Bob.

Bob checks that: $aP = R + cQ + dW = (356{,}131{,}44{,}382)$.
Then, he accepts the identification of Alice.

Security Analysis

Theorem 3 *If an attacker breaks the Schnorr protocol over elliptic curves, the proposed method of identification remains valid, then it can be used as an alternative of Schnorr's zero-knowledge scheme, hence the effectiveness of the proposed result.*

Proof Suppose that Oscar finds a way to break the Schnorr protocol over elliptic curves, in other words, he has managed to solve the equation $a = r + cs$ and find Alice's secret key. However, he will not be able to solve both equations: $a = r + cs + ds^2$ and $W = s^2P$, hence the efficiency of the proposed scheme.

Theorem 4 *If an attacker is able to break the suggested identification protocol, then he also can break Schnorr's scheme over elliptic curves.*

Proof This is similar to the proof of Theorem 2. Here also, from the point of view of security, the proposed result is stronger than the Schnorr protocol over elliptic curves.

Now, let us discuss some possible and realistic attacks.

Attack 1: Knowing Alice's public keys, if the attacker intercepts the value of a and W, he will not be able to find Alice's secret keys. Indeed, he will be blocked by a mathematical equation with too many unknown parameters.

Attack 2 (man in the middle): Suppose that Oscar intercepts the values of R, a and W. If he tries to imitate the identification of Alice, then, he will be blocked by the challenge numbers c and d which are changeable with every trial of identification.

Attack 3: If the attacker intercepts the value of R, he will not be able to find r. Indeed, he will be blocked by a discrete logarithm problem.

Theoretical generalization For its theoretical interest, a generalization of the previous method over elliptic curves is suggested in this section.

Let E be an elliptic curve defined over F_p and two points $P, Q \in E$.

In the identification protocol, Alice proves that she knows a secret number, s, that verifies $Q = sP$, where p, P and Q are given. Alice interacts with the verifier, Bob, as follows:

1. Alice chooses a random number r and computes $R = rP$. She sends the result, R, to the verifier Bob.
2. Bob chooses n random numbers $(c_1, c_2, ..., c_n)$ and sends them to Alice.
3. Alice replies by sending (a, W_i) such that:

$$
\begin{cases}
a = r + \sum_{i=1}^{n} s^i c_i \\
W_i = s^i P / i = \{2, 3, \ldots, n\}
\end{cases}
$$

Bob validates the identification if and only if:

$$
a \cdot P = R + c_1 Q + \overset{n}{\underset{i=2}{P}} c_i W_i
$$

4 Conclusion

In order to protect privacy of data, cryptography is commonly used in everyday life without users being aware of its use, for example, in Internet communications, payment by credit card, electronic money, passports, file encryption, mobile phones, and electronic signatures. However, with the development of quantum computers, the cryptanalysis of existing cryptographic protocols has become much easier. This has increased the need to develop new alternatives to secure sensitive data.

In this work a novel, alternative zero-knowledge identification protocol has been proposed whose security relies on the difficulty of solving the discrete logarithm problem over elliptic curves. It has also been proved that the proposed scheme is more difficult to break compared to the original one. Finally, a theoretical generalization has been presented. This proposed protocol can be eventually used in authentication systems where the prover wants to prove his identity to the verifier via a secret item of information (called a password) without revealing any useful information about this secret. This is a special kind of zero-knowledge identification scheme, where the size of passwords is limited.

References

1. Goldwasser, S., Micali, S., Rackoff, C.: The knowledge complexity of interactive proof systems. SIAM J. Comput. **18**(1), 186–208 (1989)
2. Koblitz, N.: Elliptic curve cryptosystems. Mathematics of computation **48**(177), 203–209 (1987)

 3. Fiat, A., Shamir, A.: How to prove yourself: practical solutions to identification and signature problems. In: Conference on the Theory and Application of Cryptographic Techniques, pp. 186–194. Springer, Berlin (1986)
 4. Shamir, A.: Identity-based cryptosystems and signature schemes. In: Workshop on the Theory and Application of Cryptographic Techniques, pp. 47–53. Springer, Berlin (1984)
 5. Guillou, L.C., Quisquater, J.-J.: A practical zero-knowledge protocol fitted to security microprocessor minimizing both transmission and memory. In Workshop on the Theory and Application of Cryptographic Techniques, pp. 123–128. Springer, Berlin (1988)
 6. Quisquater, J.-J., Quisquater, M., Quisquater, M., Quisquater, M., Guillou, L., Guillou, M. A., Guillou, G., Guillou, A., Guillou, G., Guillou, S.: How to explain zero-knowledge protocols to your children. In: Conference on the Theory and Application of Cryptology, pp. 628–631. Springer, Berlin (1989)
 7. Rivest, R.L., Shamir, A., Adleman, L.: A method for obtaining digital signatures and public-key cryptosystems. Commun ACM **21**(2), 120–126 (1978)
 8. Blum, M., Feldman, P., Micali, S.: Non-interactive zero-knowledge and its applications. In: Providing Sound Foundations for Cryptography: On the Work of Shafi Goldwasser and Silvio Micali, pp. 329–349 (2019)
 9. Schnorr, C.-P.: Efficient identification and signatures for smart cards. In: Conference on the Theory and Application of Cryptology, pp. 239–252. Springer, Berlin (1989)
10. Stinson, D.R.: Cryptography: Theory and Practice. Chapman and Hall/CRC (2005)
11. Okamoto, T.: Provably secure and practical identification schemes and corresponding signature schemes. In: Annual International Cryptology Conference, pp. 31–53. Springer, Berlin (1992)
12. Bangerter, E., Camenisch, J., Maurer, U.: Efficient proofs of knowledge of discrete logarithms and representations in groups with hidden order. In: International Workshop on Public Key Cryptography, pp. 154–171. Springer, Berlin (2005)
13. S Ezziri and O Khadir. Variant of guillou-quisquater zero-knowledge scheme. Int. J. Open Problems Compt. Math. **10**(2) (2018)
14. Fernández, E.G., Morales-Luna, G., Sagols, F.: A zero knowledge proof system with algebraic geometry techniques. Appl. Sci. **10**(2):465 (2020)

SMOTE–ENN-Based Data Sampling and Improved Dynamic Ensemble Selection for Imbalanced Medical Data Classification

Mouna Lamari, Nabiha Azizi, Nacer Eddine Hammami, Assia Boukhamla, Soraya Cheriguene, Najdette Dendani, and Nacer Eddine Benzebouchi

Abstract During the last few years, the classification of imbalanced datasets is one of the crucial issues in medical diagnosis since it is related to the distribution of normal and abnormal cases which can potentially affect the performance of the diagnosis system. For solving this problem, various techniques have been designed in order to achieve acceptable quality. Ensemble systems are one of those techniques, and they have proven their ability to be more accurate than single classifier models. Classifier selection is related to the choice of an optimal subset within a pool of classifiers. Selection of classifier can be broadly split into two classes: static and dynamic. This paper proposes a novel set selection scheme for the classification of imbalanced medical datasets. The suggested approach is based on the combination of an improved dynamic ensemble selection called META-DES framework combined with a hybrid sampling method called SMOTE–ENN. The experimental results prove the superiority of the proposed ensemble learning system using three UCI datasets.

M. Lamari (✉)
LRI Laboratory, Badji Mokhtar University of Annaba, Annaba, Algeria
e-mail: mouna.lam23@gmail.com

N. Azizi (✉) · A. Boukhamla · S. Cheriguene · N. Dendani · N. E. Benzebouchi
Labged Laboratory, Badji Mokhtar University of Annaba, Annaba, Algeria
e-mail: azizi@labged.net

A. Boukhamla
e-mail: assiaboukhamla@gmail.com

S. Cheriguene
e-mail: soraya_cheriguene@yahoo.fr

N. Dendani
e-mail: n_dendani@YAHOO.FR

N. E. Benzebouchi
e-mail: NASrobenz@hotmail.fr

N. E. Hammami
Faculty of Computer and Information Sciences, Jouf University, Sakaka, Kingdom of Saudi Arabia
e-mail: nacereddine.hammami@gmail.com

© The Editor(s) (if applicable) and The Author(s), under exclusive license to Springer Nature Singapore Pte Ltd. 2021
F. Saeed et al. (eds.), *Advances on Smart and Soft Computing*, Advances in Intelligent Systems and Computing 1188, https://doi.org/10.1007/978-981-15-6048-4_4

37

Keywords Imbalanced dataset · Medical data classification · Ensemble learning ·
Smote–ENN · META-DES · Dynamic ensemble selection

1 Introduction and Related Works

One of the areas where researchers have been paying attention in recent years is
automatic disease diagnosis. The classification of imbalanced medical datasets is
a very significant problem in this area of search as it is linked to the distribution
of normal and abnormal cases which can potentially affect the performance of the
diagnosis system [1, 2]. Particularly, in a binary classification problem, the learning
procedure of the most classification paradigm is frequently biased toward the majority
class examples (called the "negative"), and thus, there is a higher classification error
rate in the minority ones (known as the "positive" class).

Various techniques have been designed for solving the imbalanced problems in
order to obtain an equitable distribution for both classes. These techniques can be
classified into three main parts [3, 4]: data resampling [5–7], algorithmic modification
[8–10] and ensemble learning (EL) methods [11–13]. In data resampling approaches,
training instances are modified to generate a more balanced class distribution which
can be classified into three main groups [3]: the majority class undersampling, the
minority class oversampling and hybridizations of undersampling and oversampling.

In contrast, algorithmic techniques are oriented toward the adaptation of basic
learning methods in order to be more in tune with the problems of class imbalance
[14, 15]. The ensemble classifier approach is another family of techniques that have
been proposed in the literature. The principle of these techniques makes it possible
to make any type of algorithm sensitive to asymmetry, in particular by boosting
or bagging methods. Since EL becomes a famed resolution to tackle imbalanced
learning problems, this article aims to focus on building an ensemble model.

Classifier selection is one of the major and complex problems in setting up a multi-
classifier system (MCS) [16, 17]. This work is linked to the selection of an optimal
subset from within a pool of classifiers for reducing the number of ensemble members
as well as the computational time. Classifier selection can be broadly divided into two
classes: static and dynamic [16]. In the static classifier selection (SCS), an optimal
classifier subset is chosen once during the training phase and used to classify objects
of the test set. During dynamic selection, one or more classifiers are identified for
each unknown sample based on forming performances and also diverse parameters
of the real sample to be categorized. Presently, dynamic selection is divided into
two concepts: dynamic classifier selection (DCS) and dynamic ensemble selection
(DES) [17–19].

Usually, the selection process in DCS is based on the skills of individual classifiers,
which are calculated during classification operation. The skill of each classifier will
be assessed on a region of the characteristic space defined as the vicinity of the
test model on a validation set. Several different types of DES have been proposed
in the literature so far. *META-DES* is an interesting dynamic selection (DS) scheme

proposed in [20] and has been considered as it has surpassed various DS techniques in several classification benchmarks. This method uses various sources of information to estimate the skill level of a basic classifier, to generate the right class for an unknown sample.

The present work aims to enhance the classification performance of ensemble learning for imbalanced medical datasets. The proposed MCS is based on *META-DES* combined with a hybrid sampling method called *SMOTE–ENN*. SMOTE–ENN is a preprocessing algorithm proposed in [21] which rebalances class distribution by resampling the data space. This algorithm combines the undersampling of the majority class with the oversampling of the minority class in order to overcome shortcomings associated with applying each of them separately.

The rest upcoming sections of this work are structured as follows: In Sect. 2, dynamic ensemble selection method used is presented. In Sect. 3, the main stages of the suggested approach are described. In Sect. 4, used datasets and experimental results are presented and discussed.

2   The Improved Dynamic Ensemble Selection META-DES

Given a new test sample, dynamic techniques aim to select the most competent classifiers of which the local region of the feature space, the test sample is located. Only classifiers who achieve a certain level of competence, according to selection criteria, are chosen to classify this sample.

In **META-DES** technique, the dynamic ensemble selection problem can be considered as a meta-problem:

- meta-classes are "competent" or "not competent" to classify the sample x_j.
- meta-features f_i corresponding to a different criteria for measuring the competence of a base classifier are encoded into a meta-features vector $v_{i;j}$.
- A meta-classifier λ is trained based on the meta-features $v_{i;j}$ to predict whether a base classifier c_i is competent enough to classify x_j.

Figure 1 illustrates the different steps of META-DES framework.

In **Overproduction step**: In this stage, an initial set of classifier is created and trained using the initial dataset T. The used learning models can be homogenous or heterogeneous.

In **Meta-Training step**: The dataset will be used to extract the meta-features which will be considered as entries to train a meta-classifier. It will be able to predict the robustness of the initial base of classifiers to classify an input sample. The region of competence is defined using the K-nearest neighbor algorithm, and sets of meta-feature are calculated [20].

In **Generalization step**: To predict an input sample, the region of competence θ_j and the set of output profiles φ_j are calculated using dynamic selection dataset. The meta-features vector $v_{i,j}$ is constructed all individual classifier c_i and transmitted as

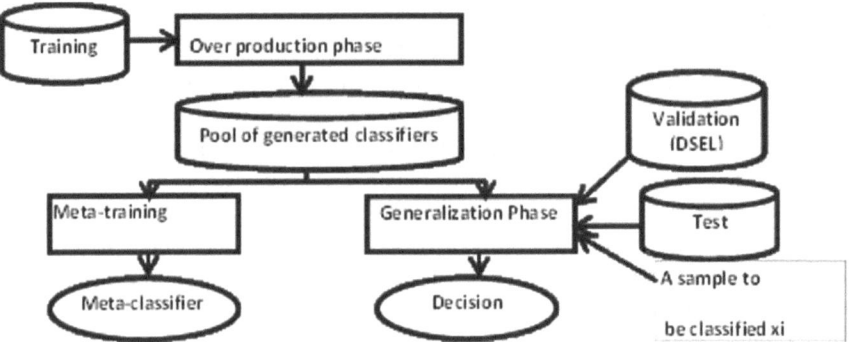

Fig. 1 Main steps of Meta_DES technique

input to the meta-classifier λ. The meta-classifier will be considered as supervisor to estimate the competence of the base of classifiers choosing in the first stage.

3   The DES–SMOTE–ENN Proposed Approach

The proposed system is composed of three main phases: (a) data balancing using a hybrid sampling approach SMOTE–ENN, (b) the generation of the initial set of heterogeneous classifiers C and (c) the dynamic selection of a subset of classifiers C^* $\subset C$ to classify each test sample x_i. The two steps of balancing and generalization are performed during the training phase, while the selection step is performed in the test phase. Outputs of selected classifiers subset fusion are done in the dynamic selection step by using majority voting as aggregation method to obtain the final decision.

Figure 2 summarizes the main steps of our *DES–SMOTE–ENN* approach.

3.1   The Dataset Balancing Stage

To redefine class distribution, we used SMOTE–ENN hybrid sampling approach. In this study we aim to analyze the impact of this method on the efficiency of the final decision using dynamic selection of the set of classifiers. The SMOTE–ENN approach combines the two methods SMOTE and ENN, the first technique is an oversampling method, and the second approach is an edited nearest neighbor undersampling method (ENN) [21].

Figure 3 shows a comparison between the original dataset, the data after application of the SMOTE method and the data after use of the SMOTE–ENN hybrid approach.

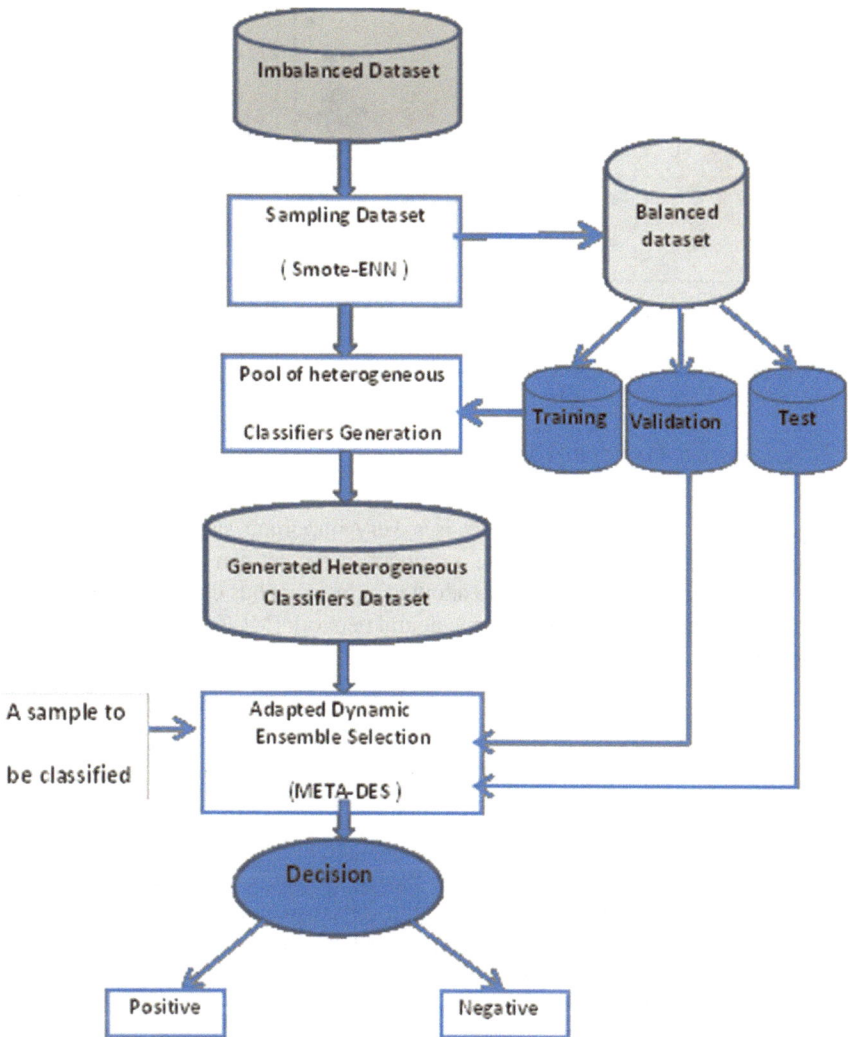

Fig.2 Main stages of proposed DES–SMOTE–ENN approach

Once dataset is balanced, it will be divided into three parts: training, validation (DSEL) and testing. The training examples are used to build the initial set of classifiers. This step is the same for all the selection methods. The validation data is used in the dynamic selection method to select classifiers. The test data is used to measure the performance classifiers.

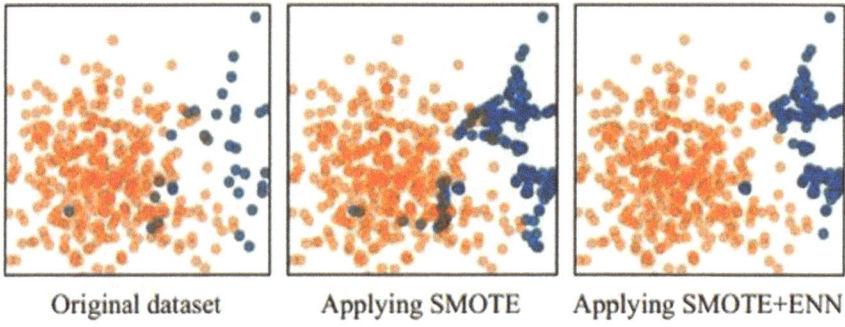

Original dataset   Applying SMOTE   Applying SMOTE+ENN

Fig. 3 Dataset distribution using SMOTE plus ENN technique

3.2   Classifier Generation

The generation of a set of classifiers is a very important step in the process of creating multi-classifier systems, the members of the set can be homogeneous or heterogeneous, in this work, we have adopted heterogeneous classifiers set of three different types classification algorithms: neural network (MLP), (SVM) and (DT for decision tree).

3.3   The Dynamic Selection of a Subset of Classifiers

Once the pool of classifiers is created and trained individually, the dynamic selection step proceeds. In our case, an adapted META-DES technique explained in Sect. 2 is applied. The main advantage of this algorithm is the use of a meta-classifier that has the ability to choose the best subset of classifier for each unknown sample.

In order to generate the final decision of the suggested system, the majority vote technique is employed.

4   Experimental Results and Discussion

In order to validate the proposed classification system, three imbalanced medical binary class datasets are selected from the UCI Machine Learning Repository [22]. Their description is provided in Table 1.

Table 1 Characteristics of used datasets

Dataset	Instances	Positive	Negative	Features
Appendicitis	106	21	85	8
Diabetes	768	268	500	8
Parkinson's	195	48	147	22

4.1 Experimental Parameters and Measures

In the generalization phase, two main criteria must be verified between the different classifiers: complementarily and diversity.

There are many ways to generate different experts. In our experiments, the diversity is guaranteed by using three different topologies of learning algorithms as bases of classifiers: ten multi-layer perceptions (MLP), three support vector machines (SVM) and ten decision trees (DTs).

To better evaluate the performance of the dynamic selection method on imbalanced data, the experiments are divided into two main parts: before and after database balancing. In each step, the results of the base of classifiers tested individually; the results of the selection method based on heterogeneous sets of classifiers are presented.

4.2 Imbalanced Medical Data Classification Using META-DES

The performance of initial pool of classifiers is presented in Table 2. The following measures are used to estimate general performance of the proposed approach: (sen: Sensitivity, SPE: specificity; AUC; Air under curve based on ROC curve, F1: weighted average of precision and recall; Gm harmonic mean); these measures are defined as follows:

Sensitivity or Recall (Sen) = TP/(TP+FN)
Specificity (Spe) = TN/(TN+FP)
F-measure or Fscore (F1) = (2*ACC*Sen)/(ACC+Sen)
G-measure (Gm) = $\sqrt{}$(Sens+Spe)
Area Under roc Curve (AUC) = (Sens+Spe)/2

where,

Accuracy (ACC) = (TP+TN)/(TP+TN+FP+FN)
True Positives (TP) describe the samples of sick people with a positive test.
True Negatives (TN) describe the samples of non-sick people with a negative test.
False Positives (FP) represent the samples of non-sick people with a positive test.
False negatives (FN) represent the samples of sick people with a negative test [23–25].

Table 3 presents the main results obtained by META-DES dynamic selection methods applied to an initial pool of classifiers.

4.3   Integration of SMOTE–ENN with META-DES for Imbalanced Medical Data Classification

By using SMOTE–ENN hybrid resampling technique to balance each dataset, the same classification process is used. It refers to the generation of the pool of individual classifiers and the application of the improved dynamic ensemble selection representing by META-DES approach. Table 4 shows the performance of each used topology of classification after data balancing.

It is noticeable that the performance of the individual classifiers applied to the balanced data outperforms the performance of the same ones trained by original dataset. In the case of imbalanced databases, the average of AUC of the three classifiers is: 0.80 for (MLP) classifier, 0.73 for (DT) and 0.81 for (SVM). After the generation of synthetic data to balance the dataset, the average of AUC increases to 0.81 for (MLP) 0.88 for (DT) and 0.95 for (SVM) classifiers. This implies the importance of the balancing phase of the imbalanced bases prior to data classification. It can be concluded that the boundaries between classes were well defined by balancing the databases.

4.4   Dynamic Ensemble Selection Results

To analyze the behavior of META-DES dynamic selection method using balanced datasets, Table 5 illustrates the obtained performance measures.

As shown in Table 5, the META-DES method gives encouraging results on most databases with an average AUC equal to 0.996.

Table 2 Base of classifiers performance

Datasets	Sen	Spe	F1	Gm	AUC	Sen	Spe	F1	Gm	AUC	Sen	Spe	F1	Gm	AUC
Appendicitis	0.77	0.69	0.79	0.73	0.72	**0.83**	**0.7**	**0.84**	**0.76**	**0.76**	0.8	0.56	0.81	0.66	0.74
Diabetes	0.69	**0.68**	0.7	0.68	0.72	0.63	0.59	0.64	0.61	0.61	**0.72**	0.67	**0.72**	**0.69**	**0.78**
Parkinson's	**0.89**	**0.87**	**0.9**	**0.88**	**0.97**	0.86	0.77	0.86	0.81	0.82	0.83	0.44	0.79	0.52	0.92
Average	0.78	0.75	0.80	0.76	0.80	0.77	0.69	0.78	0.73	0.73	0.78	0.56	0.77	0.62	0.81

Table 3 Applying META-DES for imbalanced medical datasets

Datasets	META-DES				
	Sen	Spe	F1	Gm	AUC
Appendicitis	**0.89**	**0.58**	**0.88**	**0.69**	**0.77**
Diabetes	**0.77**	0.68	0.76	0.71	**0.79**
Parkinson's	0.74	0.62	0.73	0.66	0.74
Average	0.80	0.63	0.79	0.69	0.77

4.5 Discussion

The comparison between Tables 2 and 4 illustrates that the basic classifiers trained on balanced datasets gives better results compared to those trained by imbalanced bases. In a comparison between Tables 3 and 5, the level of AUC increased in a good way from 0.77 to 0.996. We can affirm then that the combination of the *SMOTE–ENN* method for data balancing and the use of the improved dynamic ensemble selection *META-DES* method for selecting the best subset of classifiers for each example gave better results. It can be concluded that dynamic selection of classifiers is a very efficient technique for improving classification performance because it selects the best subset of classifiers that fits with each given test sample. It can also be seen that the step of data balancing using a resampling method is a very important for imbalanced data since the imbalance between classes can influence the performance of the basic classifiers and consequently of the final chosen set by the dynamic selection methods.

Table 4 Base of classifiers performances using balanced data

Datasets	MLP					DT					SVM				
	Sen	Spe	F1	Gm	Auc	Sen	Spe	F1	Gm	AUC	Sen	Spe	F1	Gm	AUC
Appendicitis	0.90	0.90	0.90	0.90	**0.95**	**0.93**	**0.93**	**0.93**	**0.93**	0.93	0.88	0.90	0.87	0.88	0.91
Diabetes	0.62	0.38	0.47	0.30	0.50	**0.89**	**0.89**	**0.89**	**0.89**	**0.89**	0.88	**0.89**	0.88	**0.89**	**0.97**
Parkinson's	**0.93**	**0.93**	**0.93**	**0.93**	**0.98**	0.85	0.82	0.84	0.83	0.83	0.90	0.90	0.90	0.90	0.96
Average	0.82	0.74	0.77	0.71	0.81	0.89	0.88	0.89	0.88	0.88	0.89	0.90	0.88	0.89	0.95

Table 5 Results of Proposed DES–SMOTE–ENN System

Dataset	META-DES				
	Sen	Spe	F1	Gm	AUC
Appendicits	0.97	0.98	0.98	0.98	**1.00**
Diabetes	0.93	0.92	0.93	0.92	**0.99**
Parkinson's	**0.96**	**0.97**	**0.96**	**0.96**	1.00
Average	0.82	0.96	0.96	0.95	0.996

5 Conclusion

The imbalanced data is an open problem in medical diagnostic assistance (MDA) systems based on classification. In this paper, a new scheme of ensemble learning is proposed to classify medical imbalanced data by balancing the original datasets with a hybrid sampling method called SMOTE–ENN combining oversampling and undersampling. Three different topologies have permitted to create a pool of heterogeneous classification algorithms. After the validation of the proposed approach on the three imbalanced medical datasets: appendicitis, diabetes, Parkinson's, the obtained results are very promising. Future works on this topic will involve the evaluation of other schemes of dynamic ensemble selection like KNORA-U and KNORA-E with other medical imbalanced datasets. Homogenous classifiers generation can be also suggested to be studied our proposed system to more understanding the integration of resample technique with general dynamic ensemble algorithms.

References

1. Mena, L.M., Gonzalez, J.A.: Machine learning for imbalanced datasets: application in medical diagnostic In: Proceedings of the 19th International FLAIRS Conference, pp 11–13, Melbourne Beach, Florida (2006)
2. Zemmal, N., Azizi, N., Sellami, M., Dey N.: Automated Classification of mammographic abnormalities using transductive semi supervised learning algorithm. In: El Oualkadi, A., Choubani, F., El Moussati, A. (eds.), Proceedings of the Mediterranean Conference on Information & Communication Technologies 2015. Lecture Notes in Electrical Engineering, vol. 381. Springer, Cham (2015)
3. Fernández, A., García, S., Galar, M., Prati, R.C., Krawczyk, B., Herrera, F.: Foundations on imbalanced classification. In: Learning from Imbalanced Data Sets. Springer, Cham (2018)
4. Galar, M., Fernandez, A., Berrenechea, A., Bustince, H., Herrera, F.: A Review on ensembles for the class imbalance problem: bagging, boosting, and hybrid-based approaches. IEE Trans. Syst. Mans Cybern. Part C Appl. Rev. **42**(4). IEEE (2012)
5. Shilaskar, S., Ghatol, A.: Diagnosis system for imbalanced multi-minority medical dataset. Soft Comput. **23**, 4789–4799 (2019)
6. Mustafa, N., Memon, E.R.A., LI, J.P., Omer, M.Z.: A classification model for imbalanced medical data based on PCA and farther distance based synthetic minority oversampling technique: (IJACSA). Int. J. Adv. Comput. Sci. Appl. **8**(1) (2017)

7. Elhassan, A.T., Aljourf, M., Al-Mohanna, F., Shoukri, M.: Classification of imbalance data using Tomek Link (T-link) combined with random undersampling (RUS) as a data reduction method. Global J. Technol. Optim. S1 (2017)
8. Geng, Y., Luo, X.: Cost-sensitive convolutional neural networks for imbalanced time series classification: Intell. Data Anal. **23**(2), 357–370 (2019)
9. Jia, F., Lei, Y., Lu, N., Xing, S.: Deep normalized convolutional neural network for imbalanced fault classification of machinery and its understanding via visualization. In: Mechanical Systems and Signal Processing, vol. 110, pp. 349–367. Elsevier (2018)
10. Gan, D., Shen, J., An, B., Xu , M., Liu, N.: Integrating TANBN with cost sensitive classification algorithm for imbalanced data in medical diagnosis. Comput. Ind. Eng. (2020)
11. Sun, J., Lang, J., Fujita, H., Li, H.: Imbalanced enterprise credit evaluation with DTE-SBD: Decision tree ensemble based on SMOTE and bagging with differentiated sampling rates. Inf. Sci. **425**, 76–91 (2018)
12. Gong, J., Kim, H.: RHSBoost: improving classification performance in imbalance data. Comput. Stat. Data Anal. **111**, 1–13 (2017)
13. Han, S., Choi, H., Choi, S., et al.: Fault diagnosis of planetary gear carrier packs: a class imbalance and multiclass classification problem. Int. J. Precis. Eng. Manuf. **20**(2), 167–179 (2019)
14. Azizi, N., Farah, N., Sellami, M.: Off-line handwritten word recognition using ensemble of classifier selection and features fusion. J. Theor. Appl. Inf. Technol. **14**(2), 141–150 (2010)
15. Azizi, N., Farah, N., Sellami, M.: Ensemble classifier construction for Arabic handwritten recognition. In: 7th International Workshop on Systems, Signal Processing and their Applications, WoSSPA, pp. 271–274 (2011)
16. Azizi, N., Farah, N.: From static to dynamic ensemble of classifiers selection: application to Arabic handwritten recognition. Int. J. Knowl. Based Intell. Eng. Syst. **16**(4), 279–288 (2012)
17. Alceu, S., Britto, J.A., Sabourin, R., Oliveira, L.E.S.: Dynamic selection of classifiers—a comprehensive review. Pattern Recogn. **47**(11), 3665–3680 (2014)
18. Cruz, R.M.O., Souza, M.A., Sabourin, R., Cavalcanti, G.D.C.: On dynamic ensemble selection and data preprocessing for multi-class imbalance learning in international. J. Pattern Recogn. Artif. Intell. **33** (11) (2018)
19. Cruz, R.M.O., Sabourin, R., Cavalcanti, G.D.C.: Dynamic classifier selection: recent advances and perspectives. Inf. Fusion **41**, 195–216 (2018)
20. Cruz, R.M.O., Sabourin, R., Cavalcanti, G.D.C., Ren, T.I.: META-DES: a dynamic ensemble selection framework using meta-learning. Pattern Recogn. **48**(5), 1925–1935 (2014)
21. Batista, G.E., Bazzan, A.L., Monard, M.C.: Balancing training data for automated annotation of keywords: a case study. In: WOB, pp. 10–18 (2003)
22. https://archive.ics.uci.edu/ml/index.php
23. Berbaum, K.S., Dorfman, D.D., Franken, E.A.: Measuring observer performance by ROC analysis: indications and complications. Invest. Radiol. **24**, 228–233 (1989)
24. Fawcett, T.: An introduction to ROC analysis. Pattern Recogn. Lett. **27**, 861–874 (2006)
25. Benzebouchi, N.E., Azizi, N., Ashour, A.S., Dey, N., Sherratt, R.S.: Multi-modal classifier fusion with feature cooperation for glaucoma diagnosis. J. Exp. Theor. Artif. Intell. **31**(6), 841–874 (2019)

Evaluation of Several Artificial Intelligence and Machine Learning Algorithms for Image Classification on Small Datasets

Rim Koulali, Hajar Zaidani, and Maryeme Zaim

Abstract In this paper, a benchmark of machine learning (ML) algorithms for single-label image classification is proposed and evaluated on a small dataset. The dataset is obtained through a mobile application allowing citizens to upload images related to water and electricity distribution infrastructure problems. The collected dataset is preprocessed, organized and used to train and evaluate classical supervised ML algorithms (SVM, NB, DT, KNN and MLP) along with deep learning methods (CNN and transfer learning). Data augmentation and fine-tuning techniques are explored to handle the overfitting problem. Conducted experiment results show the effectiveness of transfer learning with data augmentation and fine-tuning using the VGG16 network as the precision reaches 89%.

Keywords Image classification · Artificial intelligence · Classical supervised machine learning · Deep learning · Transfer learning

1 Introduction

Computer vision is a challenging research field that aims to provide machines with the capacity to see while overcoming the diversity and complexity of images and videos [1, 2]. Many computer vision systems are developed using machine learning algorithms, due to their high accuracy in comparison with handcrafted programs [3].

Image classification (single label) is a core task within computer vision that aims to assign to each image a unique content specific label. A model is trained using a labeled dataset of images. Then, the performances of the obtained model are evaluated on a

R. Koulali (✉) · H. Zaidani · M. Zaim
Mathematics and Informatics Department, LIMSAD Laboratory, Ain Chok Sciences College, Hassan II University, Casablanca, Morocco
e-mail: rim.koulali@gmail.com

H. Zaidani
e-mail: zaidaniihajar01@gmail.com

M. Zaim
e-mail: meryem.Zaim@gmail.com

© The Editor(s) (if applicable) and The Author(s), under exclusive license to Springer   51
Nature Singapore Pte Ltd. 2021
F. Saeed et al. (eds.), *Advances on Smart and Soft Computing*, Advances in Intelligent
Systems and Computing 1188, https://doi.org/10.1007/978-981-15-6048-4_5

pre-labeled test dataset. Many methods have been proposed as image classification systems which can be classified into two categories: traditional methods and deep learning methods.

The implementation of image classification systems based on traditional methods is in general centered around handcrafted features and range from the study of texture and color [4, 5] to using classical machine learning algorithms including support vector machine (SVM) [6], random forests [7, 8] and artificial neural networks [9]. Nowadays, Deep learning methods have been widely implemented due to the automatic learning of features using big datasets and the decrease of deployed resources and processing time. Studies on deep learning explore using many techniques including convolutional neural network (CNN) [10, 11], transfer learning [12], data augmentation [13] and deep feature fusion network [14].

The current study is part of an application dedicated to improving the maintenance of water and electricity conduction and infrastructure services. Citizens use the application to upload images related to a deficient system (water leak, lightning problem, water color change …). Our contribution consists of developing an efficient image classification system to be integrated into the application. Therefore, the uploaded images will be classified one after another by assigning to each image an appropriate problem-dependent label. Furthermore, the reported problem will be reported automatically to the corresponding department, and a decision will be made to tackle the problem. Consequently, our contribution will help find the exact defect, reduce the claims processing time and enhance the quality of the services.

For the proposed image classification system, we begin by collecting and organizing the data. The obtained benchmark dataset is qualified as a small one, with a total of 2000 images, which is a challenging constraint as image classification systems are trained on big datasets [15, 16]. Afterward, we conducted a benchmark study of several supervised machine learning algorithms, including support vector machine (SVM), Naive Bayes, multi-layer perceptron (MLP), decision tree and K-nearest neighbors (KNN). We also trained and tested deep learning methods such as CNN and transfer learning using the VGG16 model. To tackle the problem of overfitting, we used data augmentation and fine-tuning techniques.

The remainder of this paper is structured as follows: Section 2 provides details about the benchmark dataset. Machine learning methods are described in Sect. 3. Experimental results are presented in Sect. 4. Finally, we conclude the paper and announce our future works in Sect. 5.

2   Dataset

In this section, we describe the used benchmark dataset and explain the process of data collection, organization and augmentation.

Fig. 1 Examples of images of each one of the nine classes

2.1 Data Collection

The dataset was created from the uploaded images through a mobile application that allows citizens to report problems related to water/electricity distribution, management of street lighting and liquid sanitation. The dataset is composed of 52,207 images belonging to nine classes, namely electricity supply, subsidence degradation, overflows, public lighting, water leakage, lack of water, faulty electrical equipment, electricity and quality of water. Figure 1 illustrates a group of images belonging each one to each class.

2.2 Data Organization

The images have various sizes and belong to the following types: "jpg" and "png." Besides, the resolution of the images varies from sharp images to gradually blurred ones. Moreover, since the uploaded images were stocked with no prior verification, there were many images unrelated to the nine classes. We discarded these images as they are not significant to the defined classification system. We also discarded images related to more than one class. Then, we generated our dataset of 2000 images with a relatively balanced distribution of images over different classes.

2.3 Data Augmentation

Data augmentation is an efficient method aiming to enrich the image dataset by creating new training images from the existing dataset which will be attributed with same class labels [17]. Data augmentation is performed using a range of image manipulation operations such as shifting, flipping and zooming. We used the Image-DataGenerator class from the Keras deep learning library [18], which offers the possibility to use automatic data augmentation when generating a model. Among the main methods of data augmentation, we used horizontal and vertical random image shifting, image flipping, image rotation and image zooming. Hence, it is expected that data augmentation should help prevent overfitting which is a common problem with small datasets. Overfitting takes place when the model, exposed to too few samples, learns patterns that do not generalize to new data. Therefore, we rely on this technique to improve the developed model's ability to generalize.

3   Image Classification Methods

In this section, we detail the state-of-the-art machine learning algorithms that were evaluated on our image dataset, including classical supervised learning algorithms and deep learning methods.

3.1   Classical Supervised Machine Learning Algorithms

Machine learning is an important field of research related to artificial intelligence. Machine learning enables the generation of models to solve problems that are impossible to handle with explicit algorithms especially image classification. Supervised machine learning algorithms learn models based on labeled training sets. The obtained models will be used to associate each new data with its appropriate class label. Supervised machine learning algorithms (SMLM) tested in this study are given below:

- Support vector machine (SVM) builds one or more hyperplanes mapped in space [19]. An appropriate kernel is used to map each data type. The final purpose is to find the optimal hyperplane which will attribute a class to each data point while having the largest distance to the nearest training data point of any class.
- Decision tree (DT) is a hierarchical classifier that repeatedly partitions a given dataset into equal subsets to calculate class membership. As a result, in each repetition, a class label can either be accepted or rejected. The classifier ends by finding terminal nodes and allocating class labels to each one of them [20].
- K-nearest neighbor (KNNs) is an intuitive classifier as the input data is labeled using a distance-based similarity calculus. Hence, for each new observation, KNN searches for the k-nearest labeled training data by using the distance metric and attributes the label which appears the most to the new observation [21]. In our study, we use Euclidean distance as a distance metric.
- Multi-layer perceptron (MLP) is a feedforward artificial neural network (ANN) algorithm generating a set of output data using a nonlinear transformation. MLP is represented as a directed graph composed of several multiple layers of nodes where

every two consecutive layers are fully connected. MLP uses backpropagation for training the network to find nonlinearly separable data [22].

- Naive Bayes (NB) is qualified as a simple probabilistic classifier. NB uses Bayes theorem with the hypothesis of naive independence between the observed data. Therefore, NB labels data by calculating probabilities based on a given set of characteristics [23].

3.2   Deep Learning (DL) Techniques

Deep learning is a subfield of machine learning and artificial intelligence. DL uses artificial neural network models to learn data representations with several levels of abstraction [24]. Deep learning models can build their characteristics while learning from a hierarchical data structure. Deep learning methods (DLM) tested in this study are given below:

- Convolutions neural network (CNN) is a specific type of ANN that uses layers with convolving filters applied to local characteristics [24]. CNN was invented as a solution to computer vision problems such as image classification. We implemented our CNN training script using Keras. We used a sequential model that allows us to build a model layer by layer.
- Transfer learning allows the re-use of pre-generated knowledge by using models and generated data from the previous tasks for training and testing other tasks [25]. Therefore, data for training and testing can be different. For our study, we exploit the pre-trained neural network using the VGG16 model which is known to achieve great performances in image recognition and classification [10]. The VGG16 model is composed of 16 layers belonging to three different types of layers. We also apply fine-tuning to the VGG16 network to ameliorate the classifier performances.

4   Experiments and Results

After generating the dataset following the steps detailed in Sect. 2, we trained and tested the ML algorithms listed in Sect. 3 for image classification. We report in this section the classification performance of the benchmarked algorithms along with their execution time.

4.1   Evaluation Metrics

We use precision, recall and accuracy of the generated classes as evaluation metrics. The precision is the ratio of the number of correctly annotated images and the number of generated labels; the recall is the fraction of the number of properly annotated images and the number of ground-truth labels. Accuracy is the ratio of the number of correct images labels predictions and the total number of annotated images.

Besides, we use macro-precision and macro-recall scores, where the average is taken over all classes and all test samples, respectively, to calculate F1-score.

4.2   Classical Supervised Learning Machine Algorithms

The performances of supervised learning machine algorithms are depicted in Fig. 2. The classes are represented in the following order: subsidence degradation (A), electric supply (B), overflows (C), public lighting (D), water leakage (E), lack of water (F), faulty electrical equipment (G), electricity (H) and quality of water (I). Images belonging to class I and D have the highest precision for all the tested models. This is mainly due to the quality of images as the objects are easily detected. However, they have a poor performance in terms of recall because images belonging to these two classes can be attributed to other classes such as electricity, supply electricity, overflows, ...

Among the tested algorithms, SVM exhibits an overall better performance for each class of images. Almost all the models deliver low results for the classes as shown in Table 1 which presents the performances of each model using macro-precision, recall and $F1$-measure scores. The obtained accuracy scores are low, whereas SVM

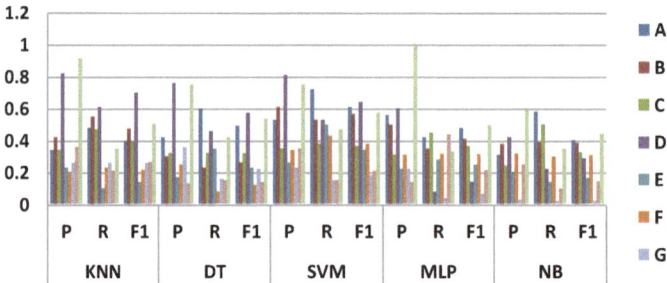

Fig. 2 Models performances in terms of precision (P), recall (R) and $F1$-measure ($F1$) per class

Table 1 Image classification performances using supervised learning machine algorithms

Models	Macro-precision (%)	Macro-recall (%)	Macro-$F1$-measure (%)
NB	28	28	28
SVM	46	44	43
KNN	41	39	38
MLP	39	29	29
DT	38	34	33

gives a maximum accuracy of 46%. We believe the number of images used per class contributed to this a poor performance as a large amount of data is needed to get better accuracy. For example, for a single class, we need at least 500–1000 images which is a time-consuming task.

4.3 Deep Learning Algorithms

As illustrated in Table 2, the classification using the CNN, despite its high accuracy of 84% and a turnaround time of 10 h, had shortcomings in classifying some images even though they were clear. The model based on transfer learning is trained with the VGG16 model outperformed the other models in terms of accuracy and precision, since it reached up to 89%, with an execution time of no more than 9 h. According to the tests carried out, after a prediction of 6000 images, errors were almost rare.

In comparison with the classical ML algorithms, the deep learning methods gave an enhanced performance with more than 100%. This is mostly due to the time and the storage space consumed as every image of the test dataset is compared to all the stored training images. However, although deep learning methods take a lot of time during the training, the classification of new data is done in a brief time.

As depicted in Fig. 3, the CNN model needs nearly 30 epochs to reach a good precision with a minimal loss. However, there exists strong overfitting since training

Table 2 Image classification system accuracy based on CNN and VGG16 models

Models	Accuracy (%)	Execution time (%)
Deep learning (CNN)	84	10 h
Transfer learning (VGG16)	89	9 h

Fig. 3 Training and loss accuracy for CNN model

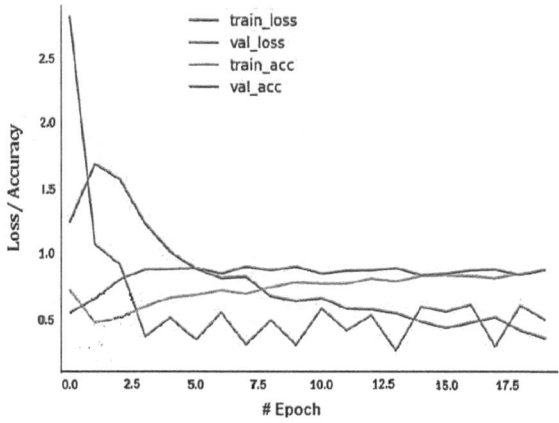

Fig. 4 Training and loss
accuracy for VGG16 model

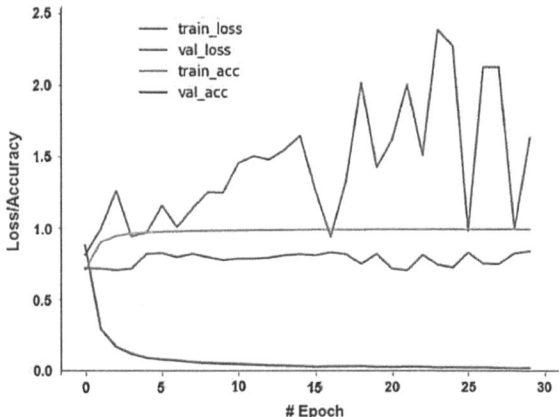

accuracy is better than validation accuracy and validation loss value is greater than training loss one. The overfitting is due to the small size of the used dataset. We tackled this problem while using transfer learning VGG16 using the pre-trained neural network model.

Figure 4 depicts the accuracy and loss of evaluation of training and testing for developed networks. The performance of image classification improves a lot compared to the CNN model performance. We were able to generate a model with good accuracy and low loss after 20 epochs for the VGG16 model. As we can see, validation accuracy is better than training accuracy and training loss value is greater than the validation loss one. Therefore, we can conclude that using fine-tuning and data augmentation helped to reduce the overfitting problem using a small dataset. Figure 5 shows some images which have been correctly classified with almost 100% in term of precision

5   Conclusion and Future Works

In this paper, we presented a detailed comparison of the reported water and electricity image classification performance of machine learning algorithms including supervised learning and deep learning algorithms. The benchmark dataset is a small one composed of 6000 images.

Supervised learning algorithms gave low results. Indeed, the most performing algorithm was SVM with 44% in terms of $F1$-measure. Using deep learning methods enhanced the classification results with more than 100%. Transfer learning yielded the best performance results by using data augmentation and fine-tuning as accuracy reaches 89% and time execution is the lowest with a total of 9 h. Hence, we were able to obtain high-performance results for image classification while using a small dataset.

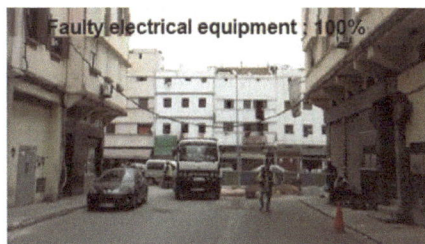
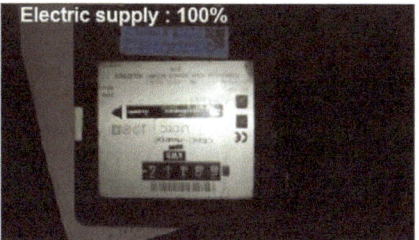

Fig. 5 Examples of VGG16 model classification

To improve the image classification performance, we intend to implement a multi-label image classification with transfer learning methods.

References

1. Szeliski, R.: Computer Vision: Algorithms and Applications. Springer Science and Business Media (2010)
2. Steger, C., Ulrich, M., Wiedemann, C.: Machine Vision Algorithms and Applications. Wiley, New York (2018)
3. Bishop, C.M.: Pattern Recognition and Machine Learning. Springer, Berlin (2006)
4. Riaz, F., Hassan, A., Nisar, R., Dinis-Ribeiro, M., Coimbra, M.T.: Content-adaptive region-based color texture descriptors for medical images. IEEE J. Biomed. Health Inf. **21**(1), 162–171 (2015)
5. Akbarizadeh, G., Rahmani, M.: Efficient combination of texture and color features in a new spectral clustering method for PolSAR image segmentation. Natl. Acad. Sci. Lett. **40**(2), 117–120 (2017)
6. Wang, M., Wan, Y., Ye, Z., Lai, X.: Remote sensing image classification based on the optimal support vector machine and modified binary coded ant colony optimization algorithm. Inf. Sci. **402**, 50–68 (2017)
7. Belgiu, M., Drăguț, L.: Random forest in remote sensing: a review of applications and future directions. ISPRS J. Photogramm. Remote Sens. **114**, 24–31 (2016)
8. Xia, J., Ghamisi, P., Yokoya, N., Iwasaki, A.: Random forest ensembles and extended multi-extinction profiles for hyperspectral image classification. IEEE Trans. Geosci. Remote Sens. **56**(1), 202–216 (2018)
9. Kaymak, S., Helwan, A., Uzun, D.: Breast cancer image classification using artificial neural networks. Procedia Comput. Sci. **120**, 126–131 (2017)

10. Simonyan, K., Zisserman, A.: Very deep convolutional networks for large-scale image recognition. arXiv preprint arXiv, 1409.1556 (2014)
11. Szegedy, C., Liu, W., Jia, Y., Sermanet, P., Reed, S., Anguelov, D., Rabinovich, A., et al.: Going deeper with convolutions. In: Proceedings of the IEEE Conference on Computer Vision and Pattern Recognition, pp. 1–9 (2015)
12. Liu, S., Tian, G., Xu, Y.: A novel scene classification model combining ResNet based transfer learning and data augmentation with a filter. Neurocomputing **338**, 191–206 (2019)
13. Perez, L., & Wang, J.: The effectiveness of data augmentation in image classification using deep learning. arXiv preprint arXiv:1712.04621 (2017)
14. Chaib, S., Liu, H., Gu, Y., Yao, H.: Deep feature fusion for VHR remote sensing scene classification. IEEE Trans. Geosci. Remote Sens. **55**(8), 4775–4784 (2017)
15. Krizhevsky, A., Sutskever, I., Hinton, G. E.: Imagenet classification with deep convolutional neural networks. In Advances in neural information processing systems, pp. 1097–1105 (2012)
16. Deng, J., Dong, W., Socher, R., Li, L.J., Li, K., Fei-Fei, L.: Imagenet: a large-scale hierarchical image database. In 2009 IEEE Conference on Computer Vision and Pattern Recognition, pp. 248–255. IEEE (2009)
17. Wong, S.C., Gatt, A., Stamatescu, V., McDonnell, M.D.: Understanding data augmentation for classification: when to warp? In 2016 International Conference on Digital Image Computing: Techniques and Applications (DICTA), pp. 1–6. IEEE (2016)
18. Keras Image Preprocessing. https://keras.io/preprocessing/image/
19. Cortes, C., Vapnik, V.: Support-vector networks. Machine Learn. **20**(3), 273–297 (1995)
20. Quinlan, J.R.: Induction of decision trees. Machine Learn. **1**(1), 81–106 (1986)
21. Kantardzic, M.: Data mining: concepts, models, methods, and algorithms. Wiley, New York (2011)
22. Gurney, K.: An Introduction to Neural Networks. CRC Press (2014)
23. Cooper, G.F., Herskovits, E.: A Bayesian method for the induction of probabilistic networks from data. Machine Learn. **9**(4), 309–347 (1992)
24. LeCun, Y., Bengio, Y.: Convolutional Networks for Images, Speech, and Time Series. The Handbook of Brain Theory and Neural Networks **3361**(10) (1995)
25. Torrey, L., Shavlik, J.: Transfer learning. In: Handbook of research on machine learning applications and trends: algorithms, methods, and techniques. IGI Global 242–264 (2010)

Yemeni Paper Currency Recognition System Using Deep Learning Approach

Abdulfattah E. Ba Alawi, Ahmed Y. A. Saeed, Burhan T. Al-Zekri, Borhan M. N. Radman, Alaa A. Saif, Maged Alshami, Ahmed Alshadadi, Muneef A. Mohammad, Ala'a M. Mohammed, Osama Y. A. Saeed, and Adeeb M. Saeed

Abstract The recognition of paper currency with various designs is a difficult task for automated banking machines and people with low visions. Due to the wide use of automated system such as ATM, for getting cash, safe and precise methods for recognizing paper currency are highly demanded. Recently, the use of an efficient automatic classifying system becomes one of the most important needs of current

A. E. Ba Alawi (✉) · A. Y. A. Saeed (✉) · B. T. Al-Zekri · B. M. N. Radman · A. A. Saif ·
M. Alshami · A. Alshadadi · A. M. Mohammed · A. M. Saeed
Department of Software Engineering, Al-Saeed Faculty of Engineering & IT, Taiz University,
Taiz, Yemen
e-mail: baalawi.abdulfattah@gmail.com

A. Y. A. Saeed
e-mail: ahmed12.12f@gmail.com

B. T. Al-Zekri
e-mail: burhanalzekri77@gmail.com

B. M. N. Radman
e-mail: borhanpfaloubidy91@gmail.com

A. A. Saif
e-mail: aalaaameenmohammedsaif@gmail.com

M. Alshami
e-mail: magedalshami333@gmail.com

A. Alshadadi
e-mail: ahmednomanalshadadi2018@gmail.com

A. M. Mohammed
e-mail: alaashaka0@gmail.com

A. M. Saeed
e-mail: adeebalshameeri@gmail.com

M. A. Mohammad · O. Y. A. Saeed
Department of Communication and Computer Engineering, Al-Saeed Faculty of Engineering &
IT, Taiz University, Taiz, Yemen
e-mail: mneefbasha@gmail.com

O. Y. A. Saeed
e-mail: osama12.12.12c@gmail.com

61

F. Saeed et al. (eds.), *Advances on Smart and Soft Computing*, Advances in Intelligent
Systems and Computing 1188, https://doi.org/10.1007/978-981-15-6048-4_6

banking services. The main aim of this paper is to introduce an intelligent model that can distinguish between different types of Yemeni paper currencies using convolution neural networks (CNN). The system is based on image processing methods to classify different types of currencies efficiently. Moreover, deep learning techniques are used to perform the classification process effectively. The obtained results in the actual implementation of this model are encouraging. The accuracy rate reaches 92.7% in the testing phases.

Keywords Paper currency · Recognition · ResNet · Currency classification · Convolution neural network (CNN)

1 Introduction

There are over 250 types of paper currency in the globe. Each country has special currency designs. The differences in the texture, size, or length of the currency bills are still insufficient to be identified accurately. Due to the increased improvement in backing services that require currency recognition, this field becomes a highly demanded area for researchers who have made plenty of attempts to resolve the issues in this domain. Paper currency recognition system is used for many purposes such as visually impaired people, automated teller machines (ATMs), and automatic goods seller machines [1, 2]. Implementing accurate currency recognition systems will facilitate many banking services.

Since 1994, some researchers have been risen. Takeda et al. [3, 4] designed paper currency recognition using a small size neural network with optimized masks [3]. The neural networks used were to be extended in the next year (1995) to build a high-speed recognition for paper currency [4].

At the beginning of the twenty-first century, many researchers used neural networks to design paper currency recognizers [3–10] such as Euro currency recognition system. Ahmadi et al. [5] have implemented PCA algorithm to classify paper currency. He et al. [6] designed and implemented a model for paper currency characteristic acquisition. In 2010, Junfang et al. [9] introduced a reliable system for paper currency recognition based on LBP. In 2011, Hasanuzzaman et al. [11] brought in an effective component-based banknote recognition that is based on SURF features. In 2012, Althafiri et al. [12] discussed the plan of a system that can recognize Bahraini paper currency using image processing techniques. In 2014, Sargano et al. [13] used a fuzzy system to design an intelligent model for recognizing paper currency. Currency classification is based on image processing [13–18] with intelligent techniques. The paper [19] introduced a system to Naira currency recognition. Ascribed to this field of study, recent researches have been constructed to recognize Ethiopian [20], Bosnian [21], Iraqi [22], or Bangla [23] currency.

The proposed scheme is implemented using deep learning approach. It can be used for sorting paper currency tasks for banks. In addition to the main objective of this research stated above, this scheme can be used to help visually impaired people with voice assistant pronouncing the obtained result or the recognized type when needed. Moreover, this work is a stepping toward a comprehensive national currency recognizer system with the different values of the global paper currency.

This paper has been structured as follows: Sect. 2 introduces Yemeni paper currency (YPC) recognition methodology. Section 3 presents and describes in detail the techniques used; they are CNN, ResNet-50 pre-trained model, and transfer learning. Section 4 describes the obtained results of the proposed system. Section 5 is the conclusion of this paper, which is an attempt to present the findings as a whole.

2 Methodology

This paper attempts to build the case of a system that can classify Yemeni paper currency by using a number of methods starting with data collection. They can be summarized as follows:

2.1 Data Collection Phase

Data is collected from scratch starting with taking photographs of Yemeni paper currencies in different environments. The collected images reach 300 in number related to six classes. Fifty images are collected for each class. About 220 images are used during the training process. 60 images are used for validation processes. The rest of the images (20 in number) are used in the testing phase.

2.2 Training Phase

In this phase, four steps are performed as follows:

Input the YPC dataset
This task is to prepare the images in a particular folder in order to start a training process.

Image preprocessing
During this phase, each image is resized or rescaled to 224 * 224 to fit the input layer of the pre-trained model (ResNet-50).

Pre-trained model
Transfer learning is used for transferring the knowledge of the pre-trained model (ResNet) to perform new tasks of classification with the dataset of YPC. In this task,

we use the model which we trained in a large dataset to be trained in the task of classifying paper currency.

Obtaining the expertise model

The output of the previous steps is a trained model that is capable of recognizing different types of Yemeni paper currencies. These steps are performed by using Python with Pycharm environment. The model has the extension (.pt) and saved in a specific file. Figure 1 summarizes the previous steps of training stage.

Fig. 1 Training Steps

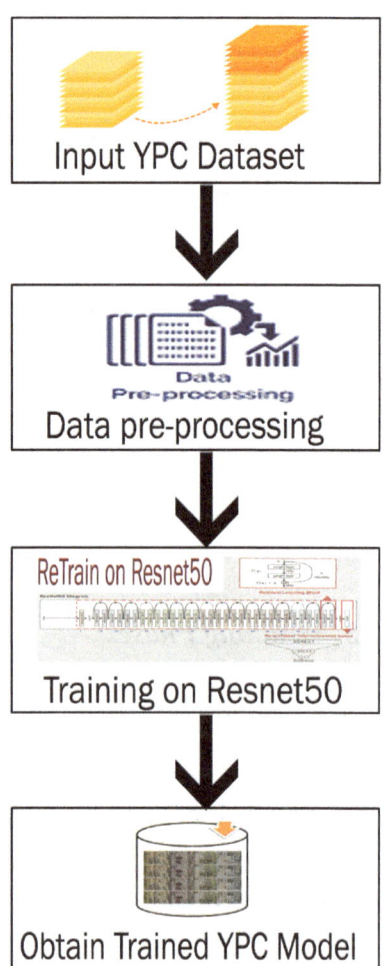

2.3 Testing Phase

This stage is an elementary phase for measuring the validity and the effectiveness of the proposed scheme. It includes the following steps:

Input a test image In this phase, we input the image we desire to test and recognize to which class it belongs.

Preprocess the input image In this phase, image is to be resized to 224 * 224.

Classify the input image In this stage, the input image is processed using the obtained expertise model to recognize and specify type.

Show predicted type After recognizing the type of the input image, the label of the class is printed. Figure 2 is a brief show of the testing steps.

3 Techniques Used

Recent techniques have been used during the implementation of this intelligent model; they are:

Fig. 2 Testing steps

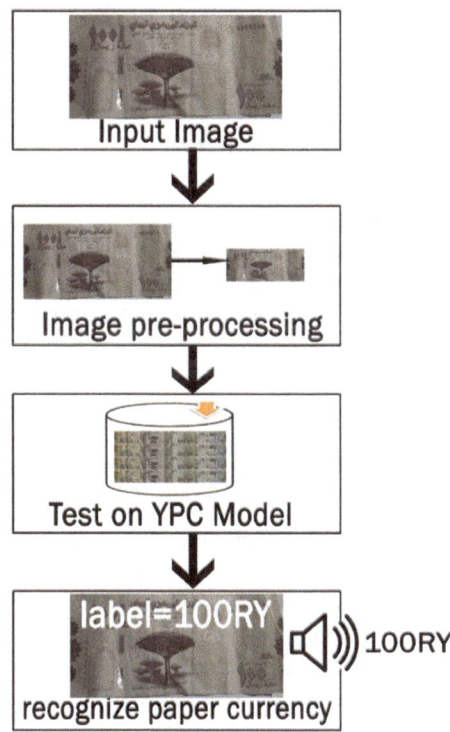

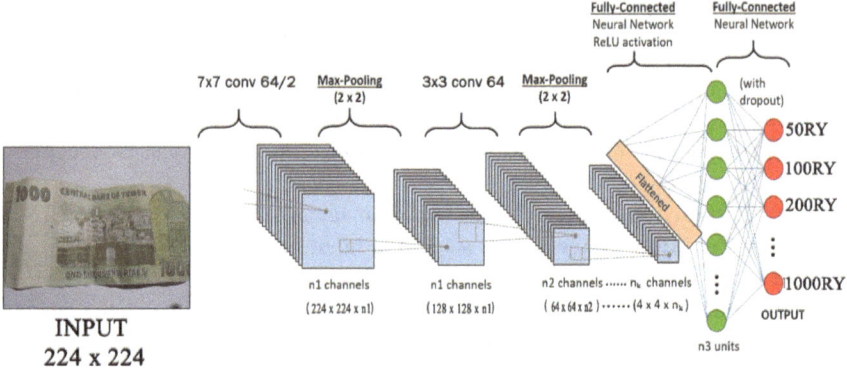

Fig. 3 Convolution neural networks

3.1 Deep Learning Approach

Convolution neural networks (CNN) algorithm is used here due to its vital role in image classification tasks. The power of CNN is in the hidden layers, which are located between the input layer and the output layer. The classification tasks by CNN show high performance results. Figure 3 illustrates the architecture of CNN.

3.2 Transfer Learning

Transfer learning is a new technique that has been used recently. The most important advantage of this technique is reducing the required time and resource for training. Instead of training from scratch that takes more time and GPU resource and large dataset of images, the pre-trained model ResNet-50 is used to transfer the knowledge and perform the task.

4 Results and Observations

During the training phase on the collected dataset, training accuracy and validation accuracy are the loss to each phase which is used as performance metrics. With the use of 32-batch size and 25 epochs, the following performance statistics is obtained. Table 1 is a summary of the performance metrics:

Vividly, the experimental findings show the feasibility of the proposed method. The findings are in the initial stages, and they can be enhanced with the use of large datasets.

Table 1 Performance
metrics results

Metrics	Value
Training accuracy	95%
Validation accuracy	92.7%
Training loss	0.108
Validation loss	0.163

To prove the system's good performance, an image of Yemeni paper currency has been tested. The result showed a 100 Yemeni Rails exactly as supposed. The output is shown in Fig. 4:

Figure 4 shows a YR 100 with a green color bellow left corner.

The main differences between this work and the related done works are the currency type and the techniques used. Here, deep learning approach is used with (ResNet-50) pre-trained model. In addition, this work has used a Yemeni paper currency that is the first step toward recognizing other Yemeni paper currencies.

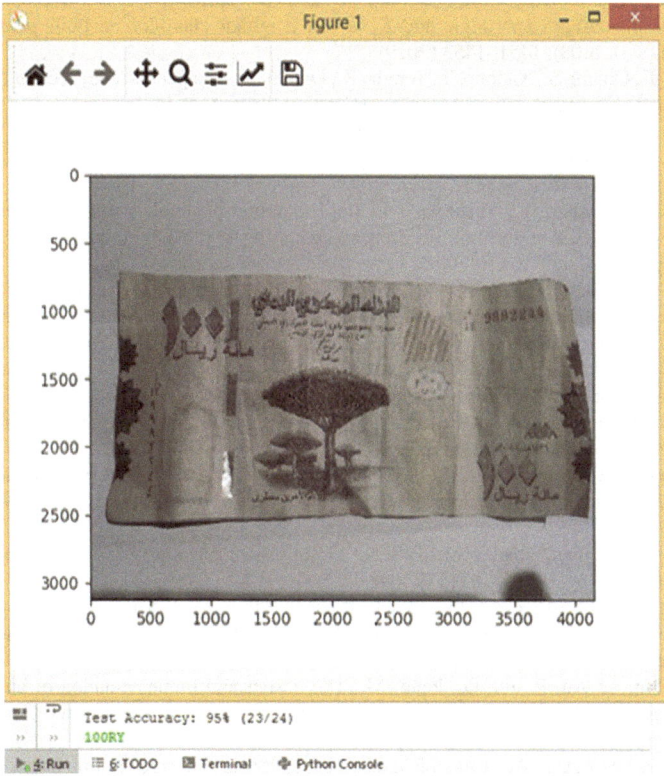

Fig. 4 Output of tested image of a YR 100

5 Conclusion and Future Work

This paper has discussed an efficient method to accurately recognize Yemeni paper currency using deep learning approach. The proposed system includes six papers of Yemeni bank notes; they are 50, 100, 200, 250, 500, and 1000. Though it is a successful proposed method, there are some points that should be recommended any future improvement of the present work. Enhancement of the processing speed and expansion of the system to include identification of currencies from different countries are major points to be taken for consideration.

References

1. Raval, V., Shah, A.: A Comparative analysis on currency recognition systems for iCu₹ e and an improved grab-cut for android devices. In: Emerging Technologies in Data Mining and Information Security, pp. 279–286. Springer (2019)
2. Bae, R.E.T., Arboleda, E.R., Andilab, A., Dellosa, R.M.J.I.J.S.T.R.: Implementation of template matching, fuzzy logic and K nearest neighbor classifier on Philippine banknote. Recogn. Syst. **8**(08), 1451–1453 (2019)
3. Takeda, F., Onami, S., Kadono, T., Terada, K., Omatu, S.: A paper currency recognition method by a small size neural network with optimized masks by GA. In: Proceedings of 1994 IEEE International Conference on Neural Networks (ICNN'94), pp. 4243–4246. IEEE (1994)
4. Takeda, F., Omatu, S.J.I.T.o.N.N.: High speed paper currency recognition by neural networks. IEEE Trans. Neural Netw. **6**(1), 73–77 (1995)
5. Ahmadi, A., Omatu, S., Yoshioka, M.: Implementing a reliable neuro-classifier for paper currency using PCA algorithm. In: Proceedings of the 41st SICE Annual Conference. SICE 2002, pp. 2466–2468. IEEE (2002)
6. Zhang, E.-H., Jiang, B., Duan, J.-H., Bian, Z.-Z.: Research on paper currency recognition by neural networks. In: Proceedings of the 2003 International Conference on Machine Learning and Cybernetics (IEEE Cat. No. 03EX693) 2003, pp. 2193–2197. IEEE (2003)
7. Takeda, F., Nishikage, T.: Multiple kinds of paper currency recognition using neural network and application for Euro currency. In: Proceedings of the IEEE-INNS-ENNS International Joint Conference on Neural Networks. IJCNN 2000. Neural Computing: New Challenges and Perspectives for the New Millennium 2000, pp. 143–147. IEEE (2000)
8. Hassanpour, H., Yaseri, A., Ardeshiri, G.: Feature extraction for paper currency recognition. In: 2007 9th International Symposium on Signal Processing and Its Applications 2007, pp. 1–4. IEEE (2007)
9. Hassanpour, H., Farahabadi, P.M.J.E.S.w.A.: Using Hidden Markov Models for paper currency recognition. **36**(6), 10105–10111 (2009)
10. Nishikage, T., Takeda, F.: Multiple kinds of paper currency recognition using neural network and application for Euro Currency. Neural Networks, IEEE-INNS-ENNS International Joint Conference on **2000**, 2143 (2000)
11. He, J., Hu, Z., Xu, P., Jin, O., Peng, M.: The design and implementation of an embedded paper currency characteristic data acquisition system. In: 2008 International Conference on Information and Automation 2008, pp. 1021–1024. IEEE (2008)
12. Guo, J., Zhao, Y., Cai, A.: A reliable method for paper currency recognition based on LBP. In: The 2nd IEEE International Conference on Network Infrastructure and Digital Content 2010, pp. 359–363. IEEE (2010)
13. Sargano, A.B., Sarfraz, M., Haq, N.J.J.o.I., Systems, F.: An intelligent system for paper currency recognition with robust features. **27**(4), 1905–1913 (2014)

14. Althafiri, E., Sarfraz, M., Alfarras, M.J.J.o.A.C.S., Research, T.: Bahraini paper currency recognition. **2**(2), 104–115 (2012)
15. Rajebhosale, S., Gujarathi, D., Nikam, S., Gogte, P., Bahiram, N.J.I.R.J.o.E., Technology: currency recognition system using image processing. **4**(3) (2017)
16. Abburu, V., Gupta, S., Rimitha, S., Mulimani, M., Koolagudi, S.G.: Currency recognition system using image processing. In: 2017 Tenth International Conference on Contemporary Computing (IC3) 2017, pp. 1–6. IEEE (2017)
17. Sarfraz, M.J.P.C.S.: An intelligent paper currency recognition system. **65**, 538–545 (2015)
18. Dalvi, M., Palve, S., Pangare, P., Modani, L., Shukla, R.J.i.: Intelligent currency recognition system. **6**(4) (2017)
19. Almu, A., Muhammad, A.B.J.A.: Image-based processing of Naira currency recognition. **15**(1) (2017)
20. Devi, O.R., Padma, Y., Kavitha, J.S.D., Sindhura, K.J.I.J.o.C.S., Security, I.: A novel technique for paper currency recognition system. **15**(4) (2017)
21. Almisreb, A.A., Turaev, S.: Bosnian currency recognition using region convolutional neural networks. In: Book of Proceedings 2019, p. 46 (2019)
22. Abbas, A.A.J.A.-N.J.o.S.: An image processor bill acceptor for Iraqi Currency. **22**(2), 78–86 (2019)
23. Murad, H., Tripto, N.I., Ali, M.E.: Developing a bangla currency recognizer for visually impaired people. Proceedings of the Tenth International Conference on Information and Communication Technologies and Development **2019**, 1–5 (2019)

Tuning Hyper-Parameters of Machine Learning Methods for Improving the Detection of Hate Speech

Faisal Saeed, Mohammed Al-Sarem, and Waseem Alromema

Abstract Social media platforms have a main role in hate crimes worldwide. Detecting hate speech from social media is a big challenge. Many studies utilized machine learning methods for classifying the text as hate speech. However, the performance of machine learning method differs when using different parameters settings. Selecting the best values of parameters for machine learning method yields directly in the performance of the method. It is very time consuming for methods to find the best values manually. In this paper, grid search and random search hyper-parameters (HP) tuning methods were used with several machine learning methods in order to enhance the performance of detecting hate speech. The experimental results showed great improvements when HP methods were applied.

Keywords Hate speech · Hyper-parameters · Machine learning · Parameters tuning

1 Introduction

Hate speech is a critical issue that seriously affects the dynamics and usefulness of online social communities. Recently, detecting hate speech automatically become hot research topic due to the explosion in the contents on the social media that have a great impact on the public opinion and the outcome of democratic processes in different countries [1, 2]. Therefore, the huge Internet-based life suppliers, for example, Facebook and Twitter, have components for clients to report crime discourse. However, this approach needs efficient automatic methods for the evaluation of the contents such as in [3–5]. Therefore, hate speech detection is a vital problem that is affecting the usefulness of online social media [1].

F. Saeed (✉) · M. Al-Sarem
College of Computer Science and Engineering, Taibah University, Medina, Saudi Arabia
e-mail: fsaeed@taibahu.edu.sa

W. Alromema
Computer Science and Information Department, Khayper Community College, Taibah University, Medina, Saudi Arabia

F. Saeed et al. (eds.), *Advances on Smart and Soft Computing*, Advances in Intelligent Systems and Computing 1188, https://doi.org/10.1007/978-981-15-6048-4_7

Several machine learning (ML) methods have been used for hate speech detection such as support vector machine approach, Naïve Bayes and logistic regression. The good way to report the best machine learning model for a specific research problem is by comparing different ML algorithms and choosing the one that performs better. Therefore, hyper-parameters (HP) of the machine learning algorithm is a critical aspect of the model training process that is considered the best practice for obtaining a successful ML applications [6]. HP tuning can affect the predictive performance of ML algorithms [7]. In addition, HP tuning is an essential task in deep learning, which can make significant changes in network performance.

Moreover, finding good values for parameters of machine learning method during the training stage is a very time-consuming process. The trial and error could be used to configure hyper-parameters of machine learning method, and the best obtained configuration can lead to improve the performance of the ML method. However, search methods can be a good alterative HP method. Many classifiers used HP methods, which include decision trees [8], fast Bayesian optimization [9], deep semi-supervised learning [10], deep learning and support vector machine.

This study investigated the effect of two HP methods: grid search and random search, to improve the performance of machine learning methods for detecting hate speech. The paper is organized as a following: Sect. 2 shows the related works. Section 3 presents the methods used in this study and the experimental design. Section 4 discusses the obtained results. Finally, Sect. 5 concludes the whole paper.

2   Related Works

In [11], an experimental statistical analysis was proposed to study the effect of the factor size of datasets used in the hyper-parameter optimization (HPO) for the performance of several ML algorithms. Also, the effect of using different partitions of a dataset in the HPO phase was studied over the efficiency of a machine learning algorithm.

The recent methods of hate-speech detection on twitter were analyzed in [1]. In addition, the experimental methodologies used in the previous works and their generalizability to other datasets were studied. The study indicated that decision-theoretic approaches are based on searching different combinations of hyper-parameters, calculating the obtained accuracy, and selecting the best combination. The grid search HP method can be used if the experiment is conducted on a fixed domain of hyper-parameters values. In addition, the Random Search (RS) was considered more effective than other methods such as grid search and Bayesian optimization techniques [1].

Different methods have been compared to evaluate the effect of several hyper-parameters optimization methods on the efficiency and performance of machine learning methods. When applying HPO on a dataset, there are different ways to tune the parameters including the number of iterations and type of optimization [3, 8, 12].

In [13], the authors investigated the sensitivity of decision tree to a HP optimization process. To find the best values of J48 parameters, four hyper-parameters methods were applied, which are Random Search (RS), Distribution Algorithm (EDA), Particle Swarm Optimization (PSO) and Genetic Algorithm (GA). To evaluate the impact of this parameters tuning, 102 diverse datasets were used in the conducted experiments. It was concluded that the change was statistically significant in most instances, even though there was a low average enhancement on all datasets.

Several techniques for HP tuning of classification algorithms have been suggested, such as [14, 15]. For some studies, grid search (GS) was used [16].

Random search starts with a simple HP configuration, which is extended by a randomly generated HP configuration at each iteration. Normally, after a given number of iterations, the cycle stops. In optimizing deep learning (DL) algorithms, RS has obtained efficient results [1, 15]. A successful HP tuning technique in ML has also emerged as sequential model-based optimization (SMBO) [17].

In [18], a generative model named FABOLAS was proposed, which is a Bayesian optimization method based on entropy search that mimics human experts in evaluating algorithms on subsets of the data quickly gather information about good hyper-parameter settings. High-quality solutions were found 10–100 times faster than other state-of-the-art Bayesian optimization methods.

In literature, many studies worked on detecting hate speech. For instance, an approach to classify tweets on Twitter was proposed in [19]. The approach used several machine learning methods to classify the tweets into three classes: hateful, offensive and clean. In addition, the hate speech detection was done using deep learning method [20]. As opposed to more traditional methods, deep learning methods are able to automatically learn representations of the input data that can be used as features to classify it. This paper extends the works done in detecting hate speech in social media. The hyper-parameters tuning methods will be applied on several machine learning methods in order to improve the effectiveness of hate speech detection.

3 Materials and Methods

In this section, we present the details of the followed methodology. As shown in Fig. 1, the dataset was cleaned and transformed into the appropriate format. Since the data are in textual format, the traditional text cleansing was conducted including stop word removal and lemmatization. Then, the TF/IDF tokenization was used to extract the main features set. The dataset, next, was split into training set (67%) and testing set (33%). At training phase, we train several individual machine learning classifiers namely, support vector machine (SVM) with radial basis function kernel, linear regression (LR), Bernoulli Naïve Bayes (BNB), decision tree (J48) and two ensemble classifiers: random forest (RF) and stochastic gradient descent (SGD). The default parameters for these methods were used at beginning.

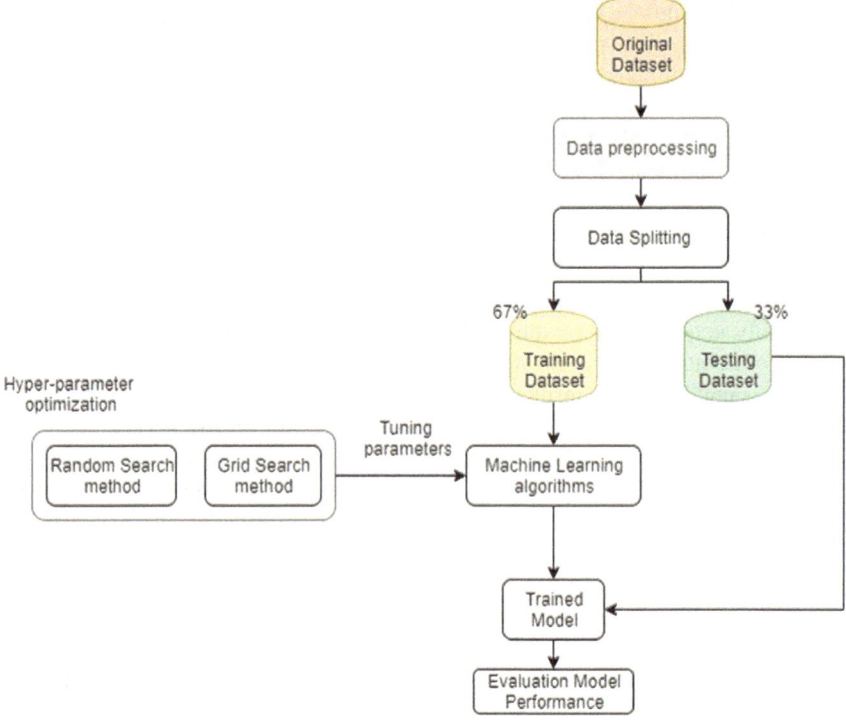

Fig. 1 Research methodology

To ensure getting better results from the aforementioned classifiers, tuning the hyper-parameters of each classifier is time-consuming and tedious labor intensive task. Thus, we employed two well-known search optimization algorithms, grid search and random search. The optimization search yields set of parameters that fed back to classifier for building the final trained model. The recall measure was used during the optimization process of the two hyper-parameters methods. At the end, the performance of the trained model was measured using the testing set, and the results are presented in terms of accuracy, precision, recall and F-measure.

3.1 Description of Data

The dataset used in this study is available online at Kaggle Repository,[1] which consists of a label column followed with date and text of tweets. The dataset is designed to solve a single-class classification problem where the label column is either 0 meaning neutral comment, or 1 meaning an insulting comment.

---

[1] https:// "www.kaggle.com/c/detecting-insults-in-social-commentary/data".

3.2 Experimental Setup

The experimental protocol was designed for finding those parameters that yield the highest performance of detecting hate speech. The experiments were carried out on Jupyter Notebook (Anaconda3) with Python version 3.8 and 64-bit Windows 10 Pro operating system with 8 RAM and Intel® CoreTM i7-4600U CPU @2.10 GHz 2.70 GHz.

4 Results and Discussion

The experiments were conducted on the dataset from Kaggle Repository that included labeled tweets (neutral and insult). Several machine learning have been applied on this dataset. Tables 1, 2, 3 and 4 show the results of the precision, recall, F-measure and accuracy for the machine learning methods with the default parameters and with hyper-parameters methods: grid search and random search. The value of the best performance was bolded in each table.

Tables 1 and 2 show the performance of the machine learning methods in terms of precision and recall. It is showed that the performances of SVM, LR and SGD methods were improved when grid search and random search methods were applied. The best precision and recall obtained were (0.84) by SGD method. The other methods (NB, RF and J48) showed no improvements obtained by using HP methods.

Table 1 Precision of machine learning methods before and after using hyper-parameters methods

Classifier	Default settings	Grid search	Random search
SVM	0.55	0.83	0.83
LR	0.82	0.83	0.83
Bernoulli NB	0.81	0.73	0.74
RF	0.80	0.55	0.55
SGD	0.83	**0.84**	**0.84**
J48	0.78	0.78	0.78

Table 2 Recall of machine learning methods before and after using hyper-parameters methods

Classifier	Default settings	Grid search	Random search
SVM	0.74	0.83	0.83
LR	0.82	0.83	0.83
Bernoulli NB	0.74	0.72	0.73
RF	0.81	0.74	0.74
SGD	0.83	**0.84**	**0.84**
J48	0.76	0.75	0.75

Table 3 F-measure of machine learning methods before and after using hyper-parameters methods

Classifier	Default settings	Grid search	Random search
SVM	0.63	0.83	0.83
LR	0.80	0.83	0.83
Bernoulli NB	0.63	0.73	0.74
RF	0.77	0.63	0.63
SGD	0.83	**0.84**	**0.84**
J48	0.75	0.76	0.76

Table 4 Accuracy of machine learning methods before and after using hyper-parameters methods

Classifier	Default settings (%)	Grid search (%)	Random search (%)
SVM	74.18	83.04	82.78
LR	82.28	83.12	83.12
Bernoulli NB	74.26	72.24	75.44
RF	80.51	74.18	74.18
SGD	82.53	**84.22**	**84.22**
J48	74.51	75.44	75.53

In addition, Tables 3 and 4 showed the performance of the machine learning methods in terms of F-measure and accuracy. It is showed that the performances of SVM, LR, NB and SGD methods were improved when grid search and random search methods were applied. The best F-measure and accuracy were obtained also by SGD which achieved 0.84. However, the performance of NB was improved using random search method only; while no improvements were reported for RF with HP methods.

The best values for SGD parameters using grid search methods were alpha: 0.0001, loss: "log", max_iter: 1000, n_jobs: -1 and penalty:l2. While the best values for SGD parameters using random search methods were penalty: l2, n_jobs: -1, max_iter: 1000, loss: "log", alpha: 0.0001.

As the main task in this study is to detect the hate speech, more investigation was carried on the precision, recall and F-measure values for the insult class. As shown in Tables 1, 2, 3 and 4, the weighted average values of precision, recall and F-measure for SVM (default parameters) were 0.55, 0.74 and 0.63, respectively. However, Table 5 showed that the precision, recall and F-measure values for the insult class using SVM (default parameters) were 0. That means the classifier cannot detect any tweet from insult class. However, using HP methods, the precision, recall and F-measure values for insult class using SVM (using HP methods) were totally improved. The classifier is able to detect the hate speech with a better performance. Similarly, the values of precision, recall and F-measure for insult class using NB and SGD were improved when using HP methods.

Table 5 Precision, recall and F-measure of machine learning methods for insult class

Classifier		Precision	Recall	F-measure
SVM	Default parameters	0.00	0.00	0.00
	Grid search	0.66	0.71	0.68
	Random search	0.66	0.68	0.67
NB	Default parameters	1.00	0.00	0.01
	Grid search	0.47	0.53	0.50
	Random search	0.48	0.52	0.50
SGD	Default parameters	0.65	0.70	0.67
	Grid search	0.73	0.61	0.67
	Random search	0.73	0.61	0.67

5 Conclusion

Several machine learning methods have been used in this study, which included support vector machine, linear regression, Bernoulli Naïve Bayes, decision tree and two ensemble classifiers: random forest and stochastic gradient descent. The experimental results showed that applying these methods with default parameters for detecting hate speech yields poor results. For instance, the precision, recall and F-measure for detecting hate speech (the insult class) were 0. Finding the best HP values for the machine learning methods helped to improve the rate of detecting hate speech. In the conducted experiments, the grid search and random search hyper-parameters methods showed a great effect on the performance of machine learning methods. The best performance was obtained by SGD method, which achieved 84% using all evaluation measures including the accuracy. In future work, more hyper-parameters tuning and optimization methods could be investigated. In addition, more machine learning and ensemble methods could be applied.

References

1. Arango, A., Pérez, J., Poblete, B.: Hate speech detection is not as easy as you may think: a closer look at model validation. In: Proceedings of the 42nd International ACM SIGIR Conference on Research and Development in Information Retrieval, pp. 45–54. ACM (2019)
2. Goodfellow, I., Bengio, Y., Courville A.: Deep Learning. MIT Press (2016)
3. Konen, W., Koch, P., Flasch, O., Bartz-Beielstein, T., Friese, M., Naujoks, B.: Tuned data mining: a benchmark study on different tuners. In: Proceedings of the 13th Annual Conference on Genetic and Evolutionary Computation, pp. 1995–2002 (2011).
4. MacAvaney, S., Yao, H.R., Yang, E., Russell, K., Goharian, N., Frieder, O.: Hate speech detection: challenges and solutions. PLoS ONE **14**(8), 1–16 (2019). https://doi.org/10.1371/journal.pone.0221152
5. Djuric, N., Zhou, J., Morris, R., Grbovic, M., Radosavljevic, V., & Bhamidipati, N.: Hate speech detection with comment embeddings. In Proceedings of the 24th international conference on world wide web, pp. 29–30, (2015).

6. Thornton, C., Hutter, F., Hoos, H.H., Leyton-Brown, K.: Auto-WEKA: combined selection and hyperparameter optimization of classification algorithms. In: Proceedings of the 19th ACM SIGKDD International Conference on Knowledge Discovery and Data Mining KDD 2013, pp. 847–855 (2013)
7. Feurer, M., Springenberg, T., Hutter, F.: Initializing bayesian hyperparameter optimization via meta-learning. In: Proceedings of the Twenty-Ninth AAAI Conference on Artificial Intelligence (2015)
8. Koch, P., Wujek, B., Golovidov, O., Gardner, S.: Automated hyperparameter tuning for effective machine learning. In: Proceedings of the SAS Global Forum 2017 Conference, SAS Institute Inc., Cary (2017)
9. Klein, A., Falkner, S., Bartels, S., Hennig, P., Hutter, F.: Fast bayesian optimization of machine learning hyperparameters on large datasets (2016). arXiv preprint arXiv:1605.07079
10. Random search for hyper-parameter optimization: J. Bergstra and Y. Bengio. J. Mach. Learn. Res. **13**, 281–305 (2012)
11. DeCastro-García, N., Muñoz Castañeda, Á.L., Escudero García, D., Carriegos, M.V.: Effect of the Sampling of a Dataset in the Hyperparameter Optimization Phase over the Efficiency of a Machine Learning Algorithm. Complexity- special issue vol-2019. Hindawi Publisher doi.org/10.1155/2019/6278908 (2019)
12. Eggensperger, K., Hutter, F., Hoos, H.H., Leyton-Brown, K.: Efficient benchmarking of hyperparameter optimizers via surrogates. In: Proceedings of the Twenty-Ninth Association for the Advancement of Artificial Intelligence, pp. 1114–1120 (2015)
13. Mantovani, R.G., Horváth, T., Cerri, R., Vanschoren, J., de Carvalho, A.C.: Hyper-parameter tuning of a decision tree induction algorithm. In: 2016 5th Brazilian Conference on Intelligent Systems (BRACIS), pp. 37–42. IEEE (2016)
14. Bergstra, J., Bardenet, R., Bengio, Y., K´elg, B.: Algorithms for hyper-parameter optimization. In: Proceedings of Neural Information Processing Systems 24 (NIPS2011), pp. 2546–2554 (2011)
15. Bardenet, R., Brendel, M., K´egl, B., Sebag, M.: Collaborative hyperparameter tuning. In: Dasgupta, S., Mcallester, D. (Eds.), Proceedings of the 30th International Conference on Machine Learning (ICML-13), vol. 28(2). JMLR Workshop and Conference Proceedings, pp. 199–207 (2013)
16. Braga, I., do Carmo, L.P., Benatti, C.C., Monard, M.C.: A note on parameter selection for support vector machines. In: Castro, F., Gelbukh, A., Gonz´alez, M. (Eds.), Advances in Soft Computing and Its Applications, ser, vol. 8266, pp. 233–244. LNCC, Springer Berlin Heidelberg (2013)
17. Brochu, E., Cora, V.M., de Freitas, N.: A tutorial on bayesian optimization of expensive cost functions, with application to active user modeling and hierarchical reinforcement learning. CoRR, vol. abs/1012.2599 (2010)
18. Guo, X.C., Yang, J.H., Wu, C.G., Wang, C.Y., Liang, Y.C.: A novel LS-SVMs hyper-parameter selection based on particle swarm optimization. Neurocomputing **71**(16–18), 3211–3215 (2008)
19. Gaydhani, A., Doma, V., Kendre, S., & Bhagwat, L.: Detecting hate speech and offensive language on twitter using machine learning: an n-gram and tfidf based approach. In: IEEE International Advance Computing Conference Computation and Language (2018). arXiv preprint arXiv:1809.08651
20. Agrawal, S., Awekar, A.: Deep learning for detecting cyberbullying across multiple social media platforms. In: Advances in Information Retrieval—40th European Conference on IR Research, ECIR 2018, Grenoble, France, March 26–29, Proceedings, pp. 141–153 (2018)

Performance Comparison of Machine Learning Techniques in Identifying Dementia from Open Access Clinical Datasets

Yunus Miah, Chowdhury Nazia Enam Prima, Sharmeen Jahan Seema, Mufti Mahmud, and M Shamim Kaiser

Abstract Identified mainly by memory loss and social inability, dementia may result from several different diseases. In the world with ever growing elderly population, the problem of dementia is rising. Despite being one of the prevalent mental health conditions in the community, it is not timely identified, reported and even understood completely. With the massive improvement in the computational power, researchers have developed machine learning (ML) techniques to diagnose and detect neurodegenerative diseases. This current work reports a comparative study of performance of several ML techniques, including support vector machine, logistic regression, artificial neural network, Naive Bayes, decision tree, random forest and K-nearest neighbor, when they are employed in identifying dementia from clinical datasets. It has been found that support vector machine and random forest perform better on datasets coming from open access repositories such as open access series of imaging studies, Alzheimer's disease neuroimaging initiative and dementia bank datasets.

Keywords Dementia · Magnetic resonance imaging (MRI) · Memory · Mild cognitive impairment (MCI) · Neurodegenerative diseases

Y. Miah · C. N. E. Prima · S. J. Seema
Department of ICE, Bangladesh University of Professionals, Dhaka 1216, Bangladesh
e-mail: 16511042@student.bup.edu.bd

C. N. E. Prima
e-mail: 16511006@student.bup.edu.bd

S. J. Seema
e-mail: seema@yahoo.com

M. Mahmud (✉)
Department of Computing and Technology, Nottingham Trent University, Clifton, Nottingham NG11 8NS, UK
e-mail: muftimahmud@gmail.com; mufti.mahmud@ntu.ac.uk

M. Shamim Kaiser (✉)
Institute of Information Technology, Jahangirnagar University, Savar, Dhaka 1342, Bangladesh
e-mail: mskaiser@juniv.edu

79

F. Saeed et al. (eds.), *Advances on Smart and Soft Computing*, Advances in Intelligent Systems and Computing 1188, https://doi.org/10.1007/978-981-15-6048-4_8

1 Introduction

Dementia is not a particular disease. The word itself describes a degenerative disease which is a combination of a cluster of symptoms influencing deterioration in memory, communication, language, thinking, reasoning and judgement, perception and the ability to perform daily activities (Fig. 1).

It is mainly caused by permanent damages in brain cells which disrupt brain's normal functioning. Though dementia can affect a person at any age, but it is most routinely diagnosed in people over the age of 65 years. Experts have estimated that about 47.5 million is living with dementia worldwide.

According to recent statistics, it has been discovered that one new case of dementia is diagnosed in every 4 s [1]. There is no specific test to diagnose dementia. Doctors usually identify it based on patient's medical history, doing laboratory tests, asking questions, tracking mood and behavioral changes and inspecting neuroimages (e.g., magnetic resonance imaging (MRI), computed tomography (CT), positron emission tomography (PET), etc.).

These days machine learning (ML) techniques have been applied to many fields where recognizing patterns in data plays a vital role. Major contributions have been made in biological data mining [2], medical image analysis [3], disease detection [4], anomaly detection in smart homes for the elderly [5], financial forecasting [6], natural language processing [7] and strategic game playing [8].

The application of ML techniques in detecting dementia is increasing and to understand how they are contributing to dementia detection, the recent literature published during the last decade (from 2011 to 2019; extracted from the Scopus database by searching for keywords) was investigated, and a pool of papers were created. As

Fig. 1 Cognitive functions and declines of the human brain

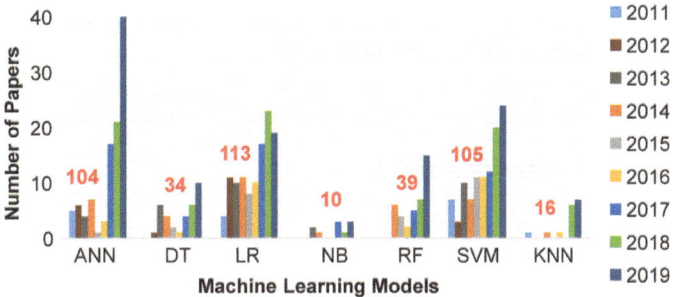

Fig. 2 Yearly published research work from 2011 to 2019 involving ML models to detect dementia from clinical datasets. Total number of papers are shown in red bold

shown in Fig. 2, various techniques, including artificial neural network (ANN), decision tree (DT), logistic regression (LR), Naive Bayes (NB), random forest (RF) and support vector machine (SVM), have been employed for this purpose. Also, the number of papers has increased steadily from 16 in 2011 to 111 in 2019, and the reported works have used mainly three open access clinical data sources, namely open access series of imaging studies (OASIS), Alzheimer's disease neuroimaging initiative (ADNI) and dementia bank (DBANK). In this paper, we evaluated performances of seven ML techniques on three datasets to detect dementia. The performances reported on this paper were obtained from randomly selected papers available in this pool to give a fair representation of the techniques and the data sources.

Here, Sect. 2 presents a brief introduction to the major dementia types. Section 3 describes the materials and methods in two sections (available open access datasets in Sect. 3.1 and different ML techniques which have been applied in dementia detection in Sect. 3.2). Whereas Sect. 4 describes the results with discussions on the pipeline in Sect. 4.1 and performance analysis in Sect. 4.2 which compares the performance of different ML techniques used in detecting dementia.

2 Types of Dementia

There are many variations of dementia, and the major ones are introduced below.

2.1 Alzheimer's Disease (AD)

AD is an irreversible and neurodegenerative brain disorder in which memory and thinking skills are destroyed slowly [9]. From recent survey [10], it has been estimated that people aged above 85 years will have the probability to suffer from AD by 2050.

Microscopic hallmarks of AD are neuritic plaques comprising amyloid-β peptide (Aβ42) [11, 12] and neurofibrillary tangles [13].

2.2 Vascular Dementia (VaD)

VaD is a syndrome of acquired intellectual deficit which is caused by cerebrovascular disease. Different types of vascular dementia include stroke-related dementia, post-stroke dementia, subcortical dementia, single-infarct and multi-infarct dementia [14]. Diagnosis of vascular dementia can be done by visiting specialist and having a brain scan. Lipid-lowering agents are also important in the secondary prevention of stroke and may play a vital role to protect against VaD.

2.3 Lewy Bodies Dementia (LBD)

LBD, most common type of dementia after AD, can be distinguished by the accumulation of amalgamated Alpha synuclein protein in LB and Lewy neurites in the neuronal processes [15]. Single-photon emission computed tomography (SPECT) or PET images are used to detect LBD. This particular type of image uses gamma rays and a ligand which binds to presynaptic dopamine transporters for its detection process.

2.4 Parkinson's Disease (PD)

PD is a prevalent progressive neurodegenerative disease affecting people over the age of 55 [4]. PD is predicted to be doubled by the year 2030 [16]. For producing sleek and regulatory movements, dopamine acts as a carrier between two brain areas namely the substantia nigra and the corpus striatum. Due to the low level of dopamine, difficulties in movement associated with the Parkinson's disease are appeared.

3 Materials and Methods

3.1 Open Access Datasets for Dementia Research

Table 1 outlines the datasets with their features (subject types and numbers) along with image type or modality. The datasets which have been used so far in detecting and predicting dementia include OASIS, ADNI, DBANK, clinical data rating (CDR),

Table 1 Open access data repositories for dementia research

References	Source	Data category (No. of subjects)	Modality/Type
[17]	Diagnosis	AD (55), MCI (101), CHOA (179)	Neuropsychological, demographical
	(SS)-/CDR	ND (174), VMD (100), MD (27)	
[18]	D-group	Probable AD (293), possible AD (21), VD (5), MCI (43), memory problem (3), unidentified (4)	Verbal utterances
	Control group	Probable AD (189), possible AD (8), VD (4), MCI (37), memory prob (3), unidentified (1)	
[19]	OASIS	NC (97), MCI (57), AD (24)	MRI, SES, EC, MMSE, ASF, CDR, nWBV, eTIV
[20]	OASIS	NC (49), VMD (49)	MRI
[21]	ADNI	Stable MCI (230), progressive MCI (43)	PET/MRI
[22]	RECOUH & NIST	HC (15), D-group (14)	Demographical
[23]	ADNI	NC (208), MCI-NC (281), MCI-C (69), AD (160)	MRI
[24]	OASIS	CS brain MRI scan (416)	MRI 3D
[25]	OASIS	CS (416), longitudinal (373)	MRI

Abbreviations *CHOA* Cognitively healthy older adult, *ND* no dementia, *VMD* Very mild dementia, *MD* mild dementia, *NC* no controls, *SES* socio economic status, *EC* education codes, *ASF* Atlas scaling factor, *nWBF* normalized whole brain volume, *eTIV* estimated total intracranial volume, *GGCDS* Gangbuk-Gu Center for Dementia in Seoul, *CERAD-K* consortium to establish a registry for Alzheimer's disease-K, *HC* healthy controls, *D* Dementia, *CS* cross sectional

semi supervised clinical data rating (SS-CDR), mini-mental state examination (MMSE) and RECOUH & NIST (The Research Ethics Committees of Osaka University Hospital and Nara Institute of Science and Technology). These databases provide open access to a number of datasets which covers many experimental conditions including AD, mild cognitive impairment and healthy control.

3.2 Machine Learning Methods for Dementia Detection

This section briefly introduces the various ML techniques that have been found instrumental in the literature and shown in Fig. 2.

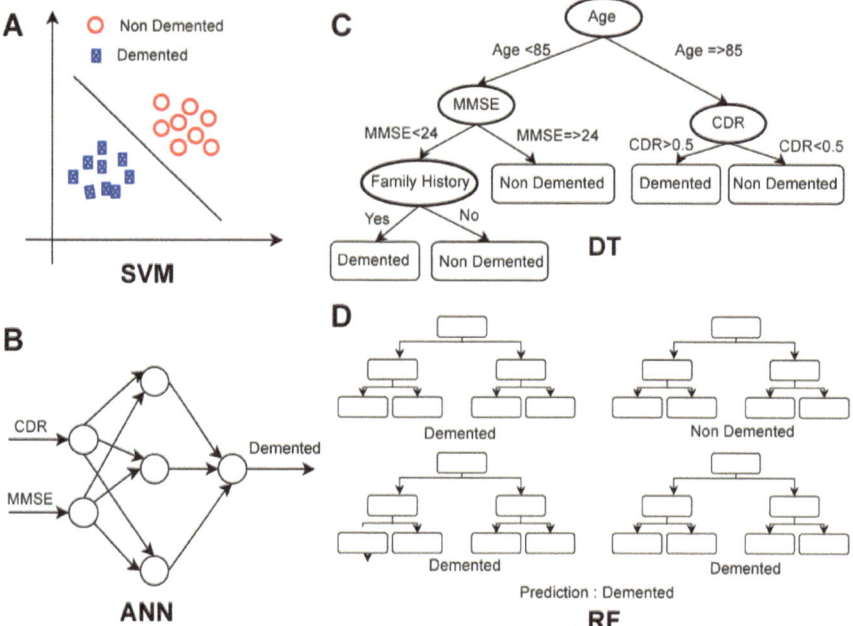

Fig. 3 Various ML models (A. SVM, B. ANN, C. DT, D. RF) which can be employed in classifying neurological disorders, e.g., demented versus non demented

Support Vector Machine (SVM) SVM is a popular ML technique which classifies data by computing optimal hyperplanes (decision boundaries) in an N-dimensional space (N represents the number of features) (see Fig. 3a) [26]. The classification is being performed by mapping the data points (usually known as support vector) into a feature-based space. Then, we have to choose an optimal hyper plane which has maximum margin distance [27]. To deal with nonlinear separable data, SVM has introduced a function named Kernel which provides a higher dimension where the data is separable [20].

Logistic Regression (LR) LR is one of the most popular ML techniques which is used in modeling the probability of a certain class due to its less computational complexity. To predict a model, the input variables need not to be scaled and by transforming the output using any function, e.g., sigmoid, LR will return a probability value. For each real value as input, the function will map a probability value over a range between 0 and 1 [28].

Artificial Neural Network (ANN) An ML technique is considered as a nonlinear statistical data modeling tool that contains several interconnected nodes in a multi-layer architecture (see Fig. 3b). The architecture has three layers—an input layer, one or more hidden layers and an output layer. The edge refers to a weight. For predicting the output, we have to adjust the weights incrementally to the known outputs [29].

Naive Bayes (NB) A probabilistic-based classification technique follows the Bayes' theorem by estimating parameter using maximum likelihood estimation. The reason of calling itself as 'Naive' because this algorithm does not take into consideration the relationships between variables [28].

Decision Tree (DT) This recursive approach has earned a significant value by splitting a dataset based on different conditions for both classification and regression. It uses a tree like structure which consists of previously learned knowledge and performs classification through translating them to if–then rules (see Fig. 3c). The deeper the tree, the more complex the tree is [28].

Random Forest (RF) As the name suggests, the RF is made up of many DTs. Therefore, it is a simple technique that consists of a large number of individual DTs which are relatively uncorrelated models (trees) and can operate as an ensemble. Collecting the models, it chooses the most predictive model [21]. The classification and prediction are usually decided based on the mode and mean of the ensemble results (see Fig. 3d).

K-Nearest Neighbor (KNN) Among all the shallow machine learning algorithms, the KNN is user-friendly but effective supervised algorithm. Based on certain criteria (e.g., Euclidean distance), it performs classification or regression of data points. In KNN, the function approximation happens locally, and the computation happens during the evaluation. For this reason, it is known as instance-based or lazy learning method [28].

4 Results and Discussion

4.1 Disease Detection Pipeline

Figure 4 shows a block diagrammatic representation for identifying dementia. As it is shown in that diagram, based on the type and severity of the disease, patients undergo specific types of diagnostic methods (e.g., MRI, CT/PET scan, MMSE/CDR, etc.) to acquire data suitable to that type of symptoms. Specific regions of interest are then located from these data, which are then fed to appropriate ML techniques to identify dementia.

4.2 Performance Analysis

The OASIS database provides neuroimages, processed images, cognitive and genetic spectrum datasets for communities to use. It is an easily accessible platform for clinical, cognitive and neuroimaging research. Using this dataset, Zhang et al. [19]

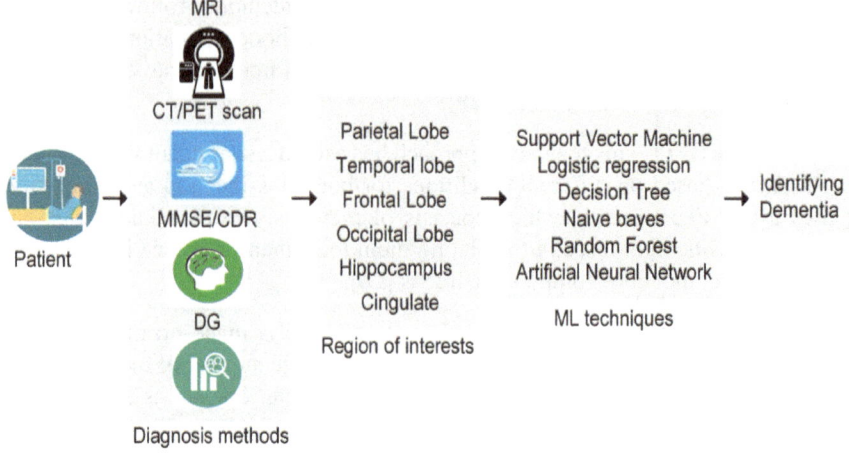

Fig. 4 Block diagrammatic presentation for identifying dementia

analyzed features like images, demographical data, clinical data, MMSE and got 77% accuracy based on SVM and DT technique. Aruna et al. [20] achieved an accuracy of 86.03% by analyzing texture and intensity feature using SVM technique. Bansal et al. [25] got an accuracy of (99.28, 99.52%) for NB, (99.52, 99.2%) for DT, (99.55, 99.08%) for RF and (96.88, 97.53%) for ANN by analyzing individual predictive ability (Fig. 5). Similarly, Farhan et al. [30] evaluated white matter, gray matter and cerebrospinal fluid using SVM, DT, ANN techniques and secured an accuracy of 93.75% for each individual method. Kamathe et al. [31] applied KNN (acc-92.31%) taking six features from 26 AD images.

On DBANK, which contains communications data from dementia patients, Rudzicz et al. [32] applied NB and SVM on all 200 features and obtained 63.1 and

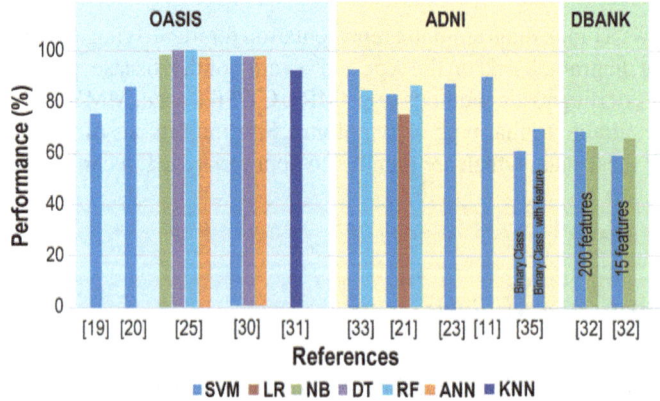

Fig. 5 Performance of different ML models on OASIS, ADNI and DBANK datasets

68.4% accuracies, respectively. When they employed only 15 features, the accuracy increased to 67.0% for NB and decreased to 59.2% for SVM.

ADNI defines the progression of AD by gathering, validating and utilizing the data. Lebedev et al. [33] employed SVM (acc 93%) and RF (acc 84.224% with ensemble and accuracy 83.294% without Ensemble) on ADNI dataset using features, e.g., cortical thickness, sulcal depth, Jacobian, volumes and thickness. Mathotaarachchi et al. [21] implemented SVM, LR and RF on this ADNI dataset. Using PET SUVR, age, gender, APOE4 genotype status as features, accuracy of 83.4%, 75.5% and 86.5% was obtained for SVM, LR and RF, respectively, 76.08% accuracy was achieved using linear SVM classifier along with the features—volume of cortical and subcortical regions, average of cortical thickness, standard deviation of thickness and surface area. Kim et al. [23] implemented SVM and achieved accuracy of 86.78% for structural T1 MRI data. Long et al. [34] utilized SVM and achieved an accuracy of 96.5% in differentiating mild AD from healthy elderly with the whole-brain gray matter or temporal lobe (see Fig. 5). Lama et al. [35] found that the accuracy of binary classification (NC vs AD) using SVM, IVM and RELM classifiers on ADNI dataset was 60.10, 59.50 and 77.30% (tenfold cross validation or CV) versus 78.01, 73.36 and 75.66% (leave-one-out or LOO CV), respectively. And the accuracy of binary classification on ADNI dataset with feature selection using SVM, IVM and RELM was 75.33%, 60.20% and 76.61% (tenfold CV) versus 80.32%, 74.47% and 77.88% (LOO CV), respectively.

So, based on the results reported in the considered papers, when the techniques were compared, the SVM and RF provided the highest accuracy in detecting dementia.

5 Conclusion

With the massive development of the computational power, better understanding of ML techniques and improved healthcare delivery frameworks [36], dementia can be detected at an early stage. Performance comparison of several ML approaches for dementia detection using clinical datasets is the core objective of the current study. The ML techniques such as SVM, LR, ANN, NB, DT, RF and KNN were chosen by a statistical analysis based on the relationship among the number of features, feature extraction method and classifier over the time period of 2011 to 2019. This study shows that among the investigated techniques, the SVM and RF perform better for on datasets provided by the OASIS, ADNI and DBANK repositories.

References

1. WHO, ADI.: Dementia: A Public Health Priority. WHO, Geneva (2012)
2. Mahmud, M., Kaiser, M.S., Hussain, A., Vassanelli, S.: Applications of deep learning and reinforcement learning to biological data. IEEE Trans. Neural Netw. Learn. Syst. **29**(6), 2063–2079 (2018)
3. Ali, H.M., Kaiser, M.S., Mahmud, M.: Application of convolutional neural network in segmenting brain regions from MRI data. In: Liang, P., Goel, V., Shan, C. (eds.) Brain Informatics, pp. 136–146. Springer, Cham (2019)
4. Noor, M.B.T., et al.: Detecting neurodegenerative disease from MRI: a brief review on a deep learning perspective. In: Liang, P., Goel, V., Shan, C. (eds.) Brain Informatics, pp. 115–125. Springer International Publishing, Cham (2019)
5. Yahaya, S.W., Lotfi, A., Mahmud, M.: A consensus novelty detection ensemble approach for anomaly detection in activities of daily living. Appl. Soft Comput. **83**, 105613 (2019)
6. Orojo, O., Tepper, J., McGinnity, T.M., Mahmud, M.: A multi-recurrent network for crude oil price prediction. In: Proc. IEEE SSCI, pp. 2953–2958 (2019)
7. Rabby, G., et al.: Teket: a tree-based unsupervised keyphrase extraction technique. Cogn. Comput. (2020), https://doi.org/10.1007/s12559-019-09706-3, [epub ahead of print].
8. Silver, D., et al.: Mastering the game of go with deep neural networks and tree search. Nature **529**(7587), 484 (2016)
9. Akhund, et al.: Adeptness: Alzheimer's disease patient management system using pervasive sensors—early prototype and preliminary results. In: Wang, S. (ed.) Brain Informatics, pp. 413–422. Springer International Publishing, Cham (2018)
10. Association, A.: 2016 Alzheimer's disease facts and figures. Alzheimer's Dementia **12**(4), 459–509 (2016)
11. Fontana, R., et al.: Early hippocampal hyperexcitability in ps2a pp mice: role of mutant ps2 and app. Neurobiol. Aging **50**, 64–76 (2017)
12. Leparulo, A., et al.: Dampened slow oscillation connectivity anticipates amyloid deposition in the ps2a pp mouse model of Alzheimer's disease. Cells **9**(1), 54 (2020)
13. Singh, S.K., et al.: Overview of Alzheimer's disease and some therapeutic approaches targeting aβ by using several synthetic and herbal compounds. Oxidative Med. Cell. Longev. **2016** (2016)
14. Roman, G.C., Erkinjuntti, T., Wallin, A., Pantoni, L., Chui, H.C.: Subcortical ischaemic vascular dementia. Lancet Neurol. **1**(7), 426–436 (2002)
15. Spillantini, M., et al.: α-synuclein in lewy bodies. Nature **388**(6645), 839–840 (1997)
16. Tsoulos, I., et al.: Application of machine learning in a Parkinson's disease digital biomarker dataset using neural network construction (nnc) methodology discriminates patient motor status. Front. ICT **6**, 10 (2019)
17. Williams, J.A., et al.: Machine learning techniques for diagnostic differentiation of mild cognitive impairment and dementia. In: 27 AAAI Conference AI, pp. 71–76 (2013)
18. Orimaye, S.O., et al.: Learning predictive linguistic features for Alzheimer's disease and related dementias using verbal utterances. In: Proceedings of Workshop Computing Linguistic Clinical Psychology: Linguistic Signal Clinical Reality, pp. 78–87 (2014)
19. Zhang, Y.D., Wang, S., Dong, Z.: Classification of AD based on structural MRI by kernel SVM decision tree. Prog. Electromagn. Res. **144**, 171–184 (2014)
20. Aruna, S., Chitra, S.: Machine learning approach for identifying dementia from MRI images. WASET Int. J. Comput. Inf. Eng. **9**(3), 881–888 (2016)
21. Mathotaarachchi, S., et al.: Identifying incipient dementia individuals using machine learning and amyloid imaging. Neurobiol. Aging **59**, 80–90 (2017)
22. Tanaka, H., et al.: Detecting dementia through interactive computer avatars. IEEE J. Translation. Eng. Health Med. **5**, 1–11 (2017)
23. Kim, J., Lee, B.: Automated discrimination of dementia spectrum disorders using extreme learning machine and structural t1 MRI features. In: Proceedings of EMBC, pp. 1990–1993 (2017)

24. Ullah, H.T., et al.: Alzheimer's disease and dementia detection from 3d brain mri data using deep convolutional neural networks. In: Proceedings of I2CT, pp. 1–3 (2018)
25. Bansal, D., et al.: Comparative analysis of various machine learning algorithms for detecting dementia. Proc. Comput. Sci. **132**, 1497–1502 (2018)
26. Battineni, G., et al.: Machine learning in medicine: performance calculation of dementia prediction by SVM. Inform. Med. Unlocked **16**, 100200 (2019)
27. Cortes, C., Vapnik, V.: Support-vector networks. Mach. Learn. **20**(3), 273–297 (1995)
28. Hastie, T., Tibshirani, R., Friedman, J.: The Elements of Statistical Learning: Data Mining, Inference, and Prediction. Springer Science & Business Media (2009)
29. Montan˜o, J., Palmer, A.: Artificial neural networks, opening the black box. Metodolog´ıa de las Ciencias del Comportamiento **4**(1), 77–93 (2002)
30. Farhan, S., Fahiem, M.A., Tauseef, H.: An ensemble-of-classifiers based approach for early diagnosis of Alzheimer's disease: classification using structural features of brain images. Comput. Math. Methods Med. **2014** (2014)
31. Kamathe, R.S., Joshi, K.R.: A robust optimized feature set based automatic classification of Alzheimer's disease using k-nn and adaboost. ICTACT J. Image Video Process. **8**(3) (2018)
32. Rudzicz, F., et al.: Automatically identifying trouble-indicating speech behaviors in Alzheimer's disease. In: Proceedings of ACM SIGACCESS, pp. 241–242 (2014)
33. Lebedev, A., et al.: RF ensembles for detection and prediction of Alzheimer's disease with a good between-cohort robustness. NeuroImage: Clin. **6**, 115–125 (2014)
34. Long, X., et al.: Prediction and classification of Alzheimer disease based on quantification of MRI deformation. PloS One **12**(3) (2017)
35. Lama, R.J., et al.: Diagnosis of Alzheimer's disease based on structural MRI images using a regularized extreme learning machine and PCA features. J. Healthc. Eng. **2017** (2017)
36. Asif-Ur-Rahman, M., et al.: Toward a heterogeneous mist, fog, and cloud-based framework for the internet of healthcare things. IEEE Internet Things J. **6**(3), 4049–4062 (2019)

Big Data Analytics

Clustering Analysis to Improve Web Search Ranking Using PCA and RMSE

Mohammed A. Ko'adan, Mohammed A. Bamatraf, and Khalid Q. Shafal

Abstract Classification of web pages is the first step of web page ranking (or we can call it indexing), one of the most common ways to achieve indexing process is clustering that pages into groups as per the similarity, whenever the misclassification is less, the result will be perfect. Moreover, clustering is a collection of algorithms that dive the data into groups related to each other. Thus, we chose Microsoft learn to rank dataset, to achieve the analysis and model building on it, this dataset is specially designed for researches in this field, and it has huge and different information about ranking process. Because of the quantity of the information, we chose randomly 16,015 observations only from MSLR-WEB30K_2 _ fold 1, in this study according to the ability of our hardware, and the algorithms of analysis, some of algorithms which were used in analysis (determine the optimal number of clusters) cannot handle the huge quantity of observations. Hence, in this paper, we are going to use clustering analysis to improve the web search ranking using principle component analysis (PCA) with root main square error as a feature reduction technique to compute the errors rate and the accuracy of the model result to get the best number of attributes; this process was achieved with cross-validation approach using extreme gradient boost algorithm as a training model to estimate the sum of errors during training operation.

Keywords Clustering analysis · PCA · RMSE · Web page ranking · XGBoost · Feature reduction

M. A. Ko'adan (✉)
Alrayan University, Mukalla, Hadhramout, Yemen
e-mail: m.kowdan@alrayan-university.edu.ye

M. A. Bamatraf
Hadhramout University of Science and Technology, Mukalla District, Hadhramout, Yemen
e-mail: mbamatraf@hu.edu.ye

K. Q. Shafal
Aden University, Aden, Yemen
e-mail: khalidshafal@yahoo.com

© The Editor(s) (if applicable) and The Author(s), under exclusive license to Springer   93
Nature Singapore Pte Ltd. 2021
F. Saeed et al. (eds.), *Advances on Smart and Soft Computing*, Advances in Intelligent
Systems and Computing 1188, https://doi.org/10.1007/978-981-15-6048-4_9

1 Introduction

Data mining is extremely helpful in information pre-processing and integration of databases. Data processing permits the researchers to spot co-occurring sequences and the correlation between any activities. Data visualization and visual data processing facilitate the data scientist with a transparent read of the data [1].

Each search engine has to implement three stages to be optimized (1) indexing, (2) sorting, (3) storing. The indexing stage is considered our concern during this study to find the optimum result in search method; in web search engines, restoration process of data consists of the three important activities that made the search engine, which is cowling, dataset and the algorithm of search program. As short explanation about how this process happened? We can see that it happens by calling the sites, which related to the words or phrase that the user enters it into searching interface. The extremely trick half is that the result ranking. The ranking is additionally what spend all time and effort attempting to affect [2].

Nowadays, the researches target to boost this field by decreasing the loss within the data to present higher results. Therefore, we decided to choose cluster analysis because it is a statistical technique in which collections of objects or points with similar characteristics are jointed together along in clusters. It encompasses a variety of various algorithms and strategies that are used for grouping objects of comparable types into various classes [3, 4].

2 Feature Selection

Feature selection is a popular technique for processing steps used for pattern recognition, classification and compression schemes. In many data analytics problems, data redundancy is a challenge must be avoided, so reducing dimensionality is an essential step before achieving any data analysis. The general criterion for reducing the attributes is the goal to present most of the related information of the original data accordingly to some optimality criteria. In some applications, it might be necessary to pick a subset of the features instead of finding a mapping that uses all of the features. The benefits of using this subset of features could reduce the usage of resources during computation of unnecessary features and costs of sensors (in physical measurements systems). Lu et al. [5] As an unsupervised technique principle component analysis is a feature reduction algorithm for projection of high-dimensional data into a new subset of low dimension, which represents the data as possible with minimum reconstruction error. PCA is a quantitatively rigorous method to perform this facilitation. The mechanism of this algorithm is based on creating a new subset of features called principle components, all of them are linear combination to the original features, and all of them are orthogonal to each other. This means that there is no redundancy in dataset [6].

2.1 Principle Component Analysis

In conventional supervised FS evaluation methods working with various feature subsets using an evaluation function or metrics to choose only the features that have relation to the decision classes of the data. However, in many data mining fields, decision class labels always unknown or incomplete, this indicates the meaning of unsupervised features selection, in unsupervised learning, decision class labels are unknown always [7].

Keeping the efficiency and accuracy in model building is a big challenge and too difficult especially with big number of features, so here we need to use PCA as an efficient feature reduction technique. Therefore, we are looking to linear transformation of a random vector with zero mean and variance matrix x to a lower dimension random vector

$X \in K^n$ With zero mean and covariance matrix Σ_x
x to a lower dimension random vector $Y \in K^q$, $q < n$

$$Y = AqTX$$

With $A_q^T A_q = I_q$

where I_q is the $q \times q$ identity matrix

In PCA, A_q is a $n \times q$ matrix whose columns are the q orthonormal eigenvectors corresponding to the first q largest eigenvalues of the covariance matrix $\sum x$. It has been many perfect characteristics of the linear transformation; one of them is maximization the distribution of points between the axes in the graph which represents the avoiding the linearity of the data. Another important characteristic is minimization the mean square error between the prediction data to the original data [5].

Here, we recognize how to implement PCA in the real models, I wish through recognizing of PCA will provide an essential base for relativizing the fields of data mining and dimensional reduction [8].

Please note that the first paragraph of a section or subsection is not indented. The first paragraphs that follow a table, figure, equation etc. do not have an indent, either.

Subsequent paragraphs, however, are indented.

Algorithm: PCA

Input: Data Matrix
Output: Reduce set of features.
Step 1: X ← Create N x d data Matrix with one row vector Xn per data point.
Step 2: subtract mean x from each row vector xn in X.
Step 3: Σ ← covariance matrix of X.

Step 4: Find eigenvectors and eigen values of Σ.
Step 5: PC's $\leftarrow$ the M eigenvectors with largest eigen values.
Step 6: Output PCs.

2.2 Root Mean Square Error (RMSE)

Root mean square error is the standard way to measure the error of a model in predicting quantitative data; formally, it is defined as:

$$\text{RMSE} = \sqrt{\sum_{i=1}^{n} \frac{(\hat{y}_i - y_i)^2}{n}}$$

where

$y_1, y_2, \ldots . y_n$ is predicted value.
$\hat{y}_1, \hat{y}_2, \ldots \hat{y}_n$ is observed value.
n number of observations [7].

To understand why this measure of error makes sense for a mathematical perspective, we will ignore the division by n under square root; we can notice that there is a similarity to the Euclidean distance formula.

$$\text{distance}(\mathbf{x}, \mathbf{y}) = \sqrt{\sum_{i=1}^{n} (x_i - y_i)^2}$$

If we keep, n fixed and rescale the Euclidean distance by a factor of $\sqrt{1/n}$.

Moreover, consider our observed value is determined by adding random error to all predicted value:

$$y_i = \hat{y}_i + \epsilon_i \quad \text{for } i = 1, \ldots, n$$

where: $\epsilon_1, \ldots, \epsilon_n$ independent, identically distributed error.

Those errors, as random variables, might have Gaussian distribution with mean μ and standard deviation σ.

The mean of distribution error is μ; when we want to estimate the standard deviation, we can see that:
where

- $E[..]$ is the expectation.
- Var (..) is the variance [9, 10].

Moreover, because ε_i is already variable with the same ε, so we can replace it to $E\left[\varepsilon_i^2\right]$ that is the average of the expectations.

In addition, to be remembered, we will suppose our error distributed with $\mu = 0$ and by putting this in equation and add the square root of both sides

$$\mathbb{E}\left[\frac{\sum_{i=1}^{n}\left(\hat{y}_i - y_i\right)^2}{n}\right]$$

$$= \mathbb{E}\left[\frac{\sum_{i=1}^{n}\epsilon_i^2}{n}\right]$$

$$= \frac{1}{n}\sum_{i=1}^{n}\mathbb{E}\left[\epsilon_i^2\right]$$

$$= \mathbb{E}\left[\epsilon^2\right]$$

$$= \text{Var}(\epsilon) + \mathbb{E}[\epsilon]^2$$

$$= \sigma^2 + \mu^2$$

then the output:

$$\sqrt{\mathbb{E}\left[\frac{\sum_{i=1}^{n}\left(\hat{y}_i - y_i\right)^2}{n}\right]} = \sqrt{\sigma^2 + 0^2} = \sigma$$

In addition, here, we can observe the left side looks familiar, if we ignore the practice $E[\ldots]$ from the square root; it is definitely the formula of RMSE [8].

2.3 Extreme Gradient Boost (XGBoost)

Extreme gradient boost is a decision-tree-based ensemble data-mining algorithm that uses a gradient boosting framework.

It includes efficient linear model solver and tree learning algorithm, and applied in predicting of unstructured data problems (Image, text ...etc.), the algorithm differentiates in the:

- It can be employed in regression, classification, ranking, and so.
- It is a general portable algorithm that can run smoothly with all famous operating systems.
- Support AWS, Azure and yarn cluster.
- Can handle various types of input data.
- Support customized objective function and evaluation function.
- has higher performance on various datasets.
- Fast [11].

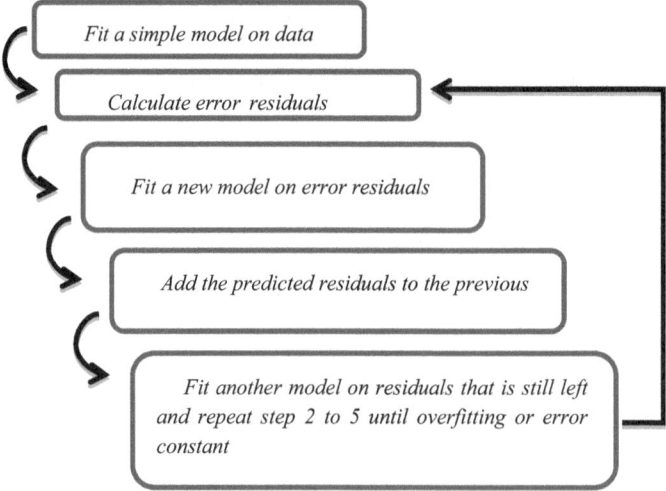

Fig. 1 The XGBoost algorithm diagram

The power of this algorithm represents in its scalability, which can achieve fast learning through parallel, distributed computing, and presents efficient memory usage. Therefore, to understand however, it is operating, this will explain the steps of the algorithm:

1. Fit a simple linear regression or decision tree on data.
2. Computes error residuals. Actual aim value, minus predicted target value.
3. Fit a new model on error residuals as the aim variable with the same input variables.
4. Add the predicted residuals to the previous predictions.
5. Fit another model on residuals that is still left. And repeat steps 2–5, until it starts overfitting or the sum of residuals becomes constant.

Overfitting can be controlled by systematically checking accuracy on validation data [12] (Fig. 1).

2.4 Data Description

Microsoft learn to rank dataset is our concern in this study; this data have the matched document for at most 30,000 queries, in fivefolds separated for fivefold cross-validation. The dataset contains of 136 variables; here, in this research work, we aim to use 136 variables with only random 16,015 observations from fold 1 in MSLR-WEB30K-2 folder and use it in analysis with PCA and building model with K-means algorithm to prove our hypotheses [13, 14].

3 Methodology

3.1 Feature Reduction Process

Here, we are aiming to clean the dataset from the variables, which are zero variance predictors, which means we have to eliminate those variables, which have very few unique values according to the number of samples, and the ratio of the frequency of the most common value against to the frequency of the second most common value is large. After that, we tend to transform the data via principle component analysis, to prepare the data for modeling. Actually, we use PCA here for two reasons: (1) to solve multi-collinearity problem, which is shown in variables, which have high correlation between each other, (2) to reduce the features (feature selection process). After that, we will use extreme gradient boost (which we call it XGBoost function in *R*) to train the model after the previous two steps (cleaning and transforming data) using XGBoost to computing the error rate meanwhile we use number of PCA components to find the proper number of these components which give stable results, instead of using it all, this step to avoid the outlier problem that happened because of irregular objects that make the model unstable. So initially we will Error = 1, and PC from 1 to 5, during running the model, the model will run with 100 round each time to calculate the sum of root mean square error (RMSE), and gradually, we will increase the PCA component 5 each time as below:

As Fig. 2 shown, we initiate the PCA component fromPC1 to PC5, to fit the XGBoost model for first time modeling, and to compute the error rate of 100 round, second time we will increase the PCA components another 5 components (range will be from PC1 to PC10), and run the model to compute the error rate again, and so on.

Each time we will have an increment of PCA components gradually 5 components to fit the model and compute the error rate, until we reach to final number of PCA components. As a result of these steps, we get Table 1.

Table 1 describes the result of using XGBoost algorithm with PCA to reduce the features with RMSE as metric to determine the suitable number of features to fit in clustering model to check is our hypotheses is valid or not? so the next step is to experiment the result to build *K*-means algorithm model and check the stability of clusters and the distance between an within clusters.

3.2 Clustering Model

Clustering model is the final step to test our hypotheses empirically to evaluate the results; here, we will use *k*-means algorithm to build the model.

First, we have to determine the optimal number of clusters, finding the optimal should be is one of *k*-means algorithm's parameters, and this parameter is responsible of stability of clusters and the whole model.

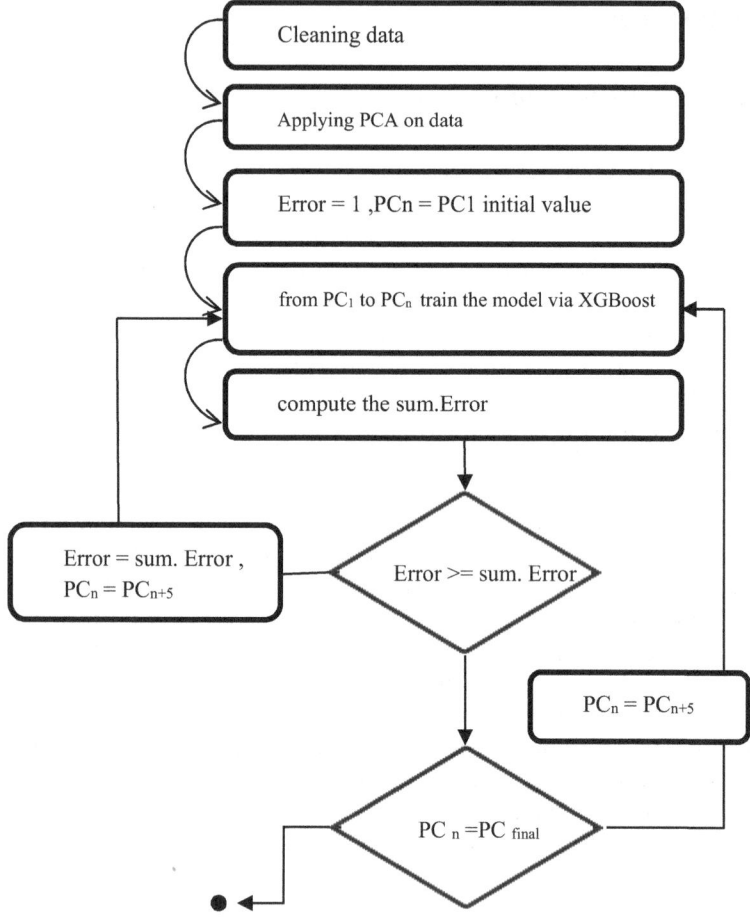

Fig. 2 Feature reduction algorithm

Therefore, to achieve this step, there are many metrics to find the optimal clusters number, what we used is as follows:

1. Elbow method: it defines the clusters as that the total intra-cluster variation or total within-cluster sum of square (WSS) is minimized; it measures the compactness of the clustering and we want it to be as small as possible [15].

In Fig. 3, we can see that the optimal number which is given by elbow method is three where is the knee of graph is indicated to three in addition of above this method sometimes can give us the exactly number of k, so it is not the perfect method to depend only in clustering analysis, as example of this we can see here else the graph can shows the k may be 5 or even 13 and so.

2. NbClust: it is a package provides 30 indices for determining the optimal number of clusters, it proposes the best clustering by variant methods to determine the

Table 1 Error estimation for XGBoost model table

RMSE	PCA comp
0.0703	PC5
0.055	PC10
0.0501	PC15
0.0503	PC20
0.0449	PC25
0.0418	PC30
0.0412	PC35
0.0366	PC40
0.0374	PC45
0.0373	PC50
0.038	PC55
0.038	PC60
0.0381	PC65
0.0398	PC70
0.0399	PC75
0.0394	PC80
0.0398	PC85
0.0399	PC90
0.038	PC95
0.0372	PC100
0.0367	PC105
0.0368	PC110
0.0369	PC115
0.0368	PC120

Fig. 3 The result of elbow method

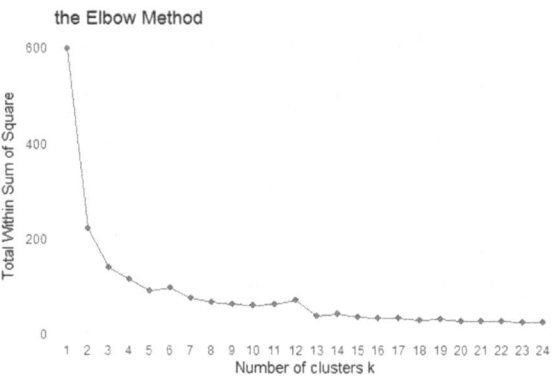

Fig. 4 NbClust method

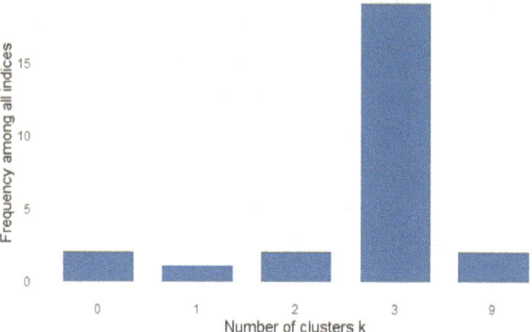

best number, all used different assumed numbers of clusters, distance measures and clustering methods [16].

Figure 4 shows us that number 3 is the optimal number of clusters by most of NbClust package measures.

3. Gap statistics: it uses the comparative between the total intra-cluster variation and their expected values under null reference distribution of the data, the optimal value will be that maximize the gap [15].

Here in Fig. 5 clearly, we can see gap statistics method suggests the optimal number of clusters is two.

According to previous methods, we recommend to choose number 3 as the optimal number of clusters because it is the estimated number of two methods of determining optimal number of clustering and that is enough for building model.

Fig. 5 Gap method results

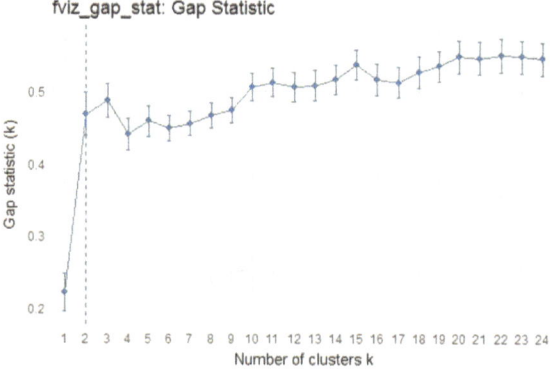

3.3 K-Means Algorithm Model

It is an unsupervised algorithm and it is an easy way to classify any dataset through the number of clusters, firstly needs a number of clusters as an urgent parameter to divide the dataset to groups of related objects according to the distance between those objects [17].

Number of clusters we can call it k, after choosing k, we will represent the k initial cluster center; the next step is to keep all point related to the given dataset and aggregate it to the nearest center according to closeness of the object with cluster center using distance metric (which is Euclidean distance in this study). If all objects are distributed, we will recalculate a new k cluster center. The algorithm will keep running until there is no change in k cluster centers [18].

So after we determine the optimal number of clusters, and to make the comparative more clearly, we will use the k-means algorithm two times, once with all components of PCA and second with only the proper number, which we got it from xgboost model with low error estimated.

As seen in Fig. 6, we can see the effect of PCA on the model, which shows the three clusters separated from each other according the distance, and showing the objects in each cluster according to the similarity between them, this figure shows only using all PCA components.

Now after the last step, we will fit the K-means algorithm model with feature reduction of PCA results (only 40 PCA components) and we will see the effect of using the error computing in the model.

We can see clearly how the distance between the clusters in Fig. 7 becomes far as possible, and how the clusters still stable and the outliers objects is joints to the closest clusters without any conflict. This is the point of using feature reduction, using the minimum of features without and defect in result.

Fig. 6 The use of all PCs results diagram

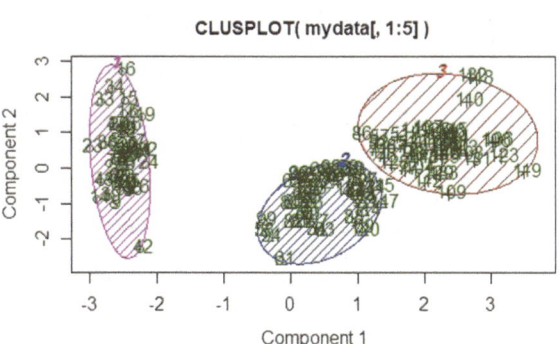

CLUSPLOT(mydata[, 1:5])

Component 1
These two components explain 94.9 % of the point variability.

Fig. 7 PCA with RMSE
results diagram

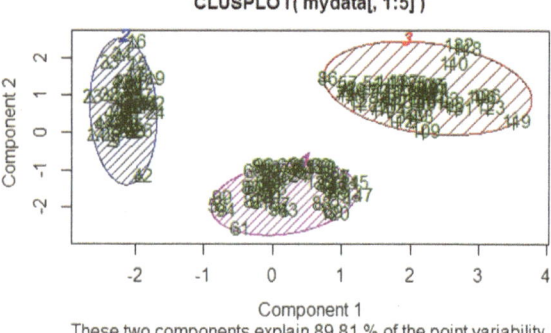

4   Conclusion

As implementation of RMSE with PCA, we found out that model can be stable and can provide new approach to reduce the attributes of dataset for cluster analysis, especially with huge data; in this experiment, we prove that we can use RMSE to evaluate the model to fit it with low number of attributes, according to low error rate to ensure the quality of results. Using of PCA here to avoid multicollinearity because high rate of correlation between some of main attributes in this dataset.

References

1. Mohan, A.: An Empirical Analysis on Point-wise Machine Learning Techniques using Regression Trees for Web-search Ranking. Master Search in Washington University in St, Louis (2010)
2. Ledford, J.L.: Search Engine Optimization Bible, 2nd edn. Wiley Publishing Inc.
3. Chawla, S.: A novel approach of cluster based optimal ranking of clicked URLs using genetic algorithm for effective personalized web search. Appl. Soft Comput. **46**, 90–103 (2016). https://doi.org/10.1016/j.asoc.2016.04.042
4. Young, M.: The Technical Writer's Handbook. University Science, Mill Valley (1989)
5. Lu, Y., Cohen, I., Zhou, X.S., Tian, Q.: Feature selection using principal feature analysis. In: Proceedings of the 15th International Conference on Multimedia, pp. 301–304. ACM (2007). ISBN: 978-1-59593-702-5
6. Ahmad, I., Amin, F.-e.: Towards feature subset selection in intrusion detection. In: 2014 IEEE 7th Joint International Information Technology and Artificial Intelligence Conference, ITAIC 2014, pp. 68–73 (2015). https://doi.org/10.1109/itaic.2014.7065007
7. Jolliffe, I.T.: Principal Component Analysis‖. Springer, New York (2002)
8. Masaeli, M., Yan, Y., Cui, Y., Fung, G., Dy, J.: Convex principal feature selection. In: Proceedings of the 10th SIAM International Conference on Data Mining, SDM2010, pp 619–628 (2010). https://doi.org/10.1137/1.9781611972801.54
9. Chai, T., Draxler, R.R.: Root mean square error (RMSE) or mean absolute error (MAE)? Arguments against avoiding RMSE in the literature. Geosci. Model Dev. **7**, 1247–1250 (2014). https://doi.org/10.5194/gmd-7-1247-2014

10. Madsen, H., Pinson, P., Kariniotakis, G., Nielsen, H.A., Nielsen, T.S.: Standardizing the performance evaluation of short-term wind power prediction models. Wind Eng. **29**(6), 475–489 (2005)
11. Friedman, J., Hastie, T., Tibshirani, R., et al.: Additive logistic regression: a statistical view of boosting (with discussion and a rejoinder by the authors). Ann. Stat. **28**(2), 337–407 (2000)
12. Friedman, J.H.: Greedy function approximation: a gradient boosting machine. Ann. Stat. 1189–1232 (2001)
13. Qin, T., Liu, T.-Y.: Introducing letor 4.0 datasets. CoRR, abs/1306.2597 (2013)
14. Qin, T., Liu, T.-Y., Xu, J., Li, H.: Letor: a benchmark collection for research on learning to rank for information retrieval. Inf. Retr. **13**(4), 346–374 (2010)
15. Kassambara, A.: Practical Guide to Cluster Analysis in R, Unsupervised Machine Learning. Published by STHDA (http//www.sthda.com),copy right © 2017 by alboukadel kassambara
16. Charrad, M., Ghazzali, Boiteau, V., Niknafs, A.: NbClust: an R package for determining the relevant number of clusters in a data set. J. Stat. Softw. **61**(6), 1–36 (2014). doi: http://dx.doi.org/10.18637/jss.v061.i06
17. Bachem, I., Lucic, M., Hamed Hassani, S., Krause, A.: Fast and provably good seedings for k-means. In: Advances in Neural Information Processing Systems (NIPS), b (2016)
18. Dhanachandra∗, N., Manglem, K., Chanu, Y.J.: Image segmentation using K-means clustering algorithm and subtractive clustering algorithm. National Institute of Technology, Manipur 795 001, India. In: Eleventh International Multi-Conference on Information Processing (IMCIP-2015) (2015)
19. Yorozu, Y., Hirano, M., Oka, K., Tagawa, Y.: Electron spectroscopy studies on magneto-optical media and plastic substrate interface. In: IEEE Translate Journal Magnetic Japan, vol. 2, pp. 740–741 (Digests 9th Annual Conference Magnetics Japan, pp. 301, 1982) (1987)
20. Ilango, V., Subramanian, R., Vasudevan, V.: Cluster analysis research design model, problems, issues, challenges, trends and tools. Int. J. Comput. Sci. Eng. **3**(8), 3064–3070 (2011)
21. Parveen, A.N., Inbarani, H.H., Kumar, E.N.S.: Performance analysis of unsupervised feature selection methods. In: 2012 International Conference on Computing, Communication and Applications, Dindigul, Tamil Nadu, 2012, pp. 1–7 (2012). https://doi.org/10.1109/iccca.2012.6179181
22. Jachner, S., van den Boogaart, K., Petzoldt, T.: Statistical methods for the qualitative assessment of dynamic models with time delay (R package qualV). J. Stat. Softw. **22**, 1–31 (2007)

A Parallel Global TFIDF Feature Selection Using Hadoop for Big Data Text Classification

Houda Amazal, Mohammed Ramdani, and Mohamed Kissi

Abstract Feature selection (FS) is a fundamental process in machine learning. This process aims to remove irrelevant features, reduce the dimensionality of the feature space, and increase the accuracy of classification algorithms. For this, many feature selection techniques were proposed through several studies. However, as big data is still coming, the classical feature selection methods are enabled to store or process efficiently large datasets. Therefore, this work aims to improve the well-known term weighting method term frequency inverse document frequency (TFIDF) using MapReduce programming model on Hadoop framework. Our proposed approach, we called global TFIDF (GTFIDF), aims to introduce a simple and efficient parallel feature selection method which addresses the drawbacks of the classical TFIDF. The proposed approach includes two weighting schemes based on averaging and maximizing the GTFIDF. Experimental results on two well-known datasets prove that GTFIDF improves the performance of classification in term of the Micro-$F1$ and Macro-$F1$ measures.

Keywords Big data · Hadoop · Feature selection · Text classification

1 Introduction

With the quick development of communication technologies, huge amount of electronic documents are generated permanently. Hence, to make the content of these documents useful, it becomes fundamental to organize them. This situation increases the role of text classification which aims to designate documents to their categories

H. Amazal (✉) · M. Ramdani · M. Kissi
Computer Science Laboratory, Faculty of Sciences and Technologies, Hassan II University of Casablanca, BP 146, 20650 Mohammedia, Morocco
e-mail: houda.kamouss@gmail.com

M. Ramdani
e-mail: ramdani@fstm.ac.ma

M. Kissi
e-mail: mohamed.kissi@univh2c.ma

F. Saeed et al. (eds.), *Advances on Smart and Soft Computing*, Advances in Intelligent Systems and Computing 1188, https://doi.org/10.1007/978-981-15-6048-4_10

based on their textual content [1]. Actually, many text classification approaches were introduced in several domains such as spam e-mail filtering [2], author identification [3], sentiment analysis [4], Web page classification [5], and documents classification [6]. However, with a large amount of documents, the space dimensionality can reach thousands of features (words). Consequently, it is a major challenge from scarce and rare data, not relevant features and overfitting to text classification [7]. Hence, reducing dimensionality is a very important domain of research.

The main purpose of dimensionality reduction is to reduce the amount of selected features while maintaining or improving the performance of the system. Feature selection (FS) is an effective approach to dimensional reduction [8]. It removes irrelevant features in order to obtain the best features from the whole dataset and it provides efficiency in terms of time and space complexity. One of the FS strengths is that the dimensionality reduction process does not modify the interpretation of the important features of the original set.

FS approaches are generally divided into two main types: "wrapper methods" and "filter methods." Wrapper methods enable to choose a subset of features which are the most appropriate based on a particular algorithm of classification [9]. On the other hand, the advantage of filtering methods is that they do not depend on any classification algorithm. Filter methods are based on a function that makes it possible to consider the relevance of a specific feature in the process of classification [10]. So, relevant features are choosing using the "evaluation function" which classify values of features. For this, filter methods are frequently used in the classification of documents.

The TFIDF is a widely used filter method which is based on the utilization of term frequency technique (TF) and inverse document frequency technique (IDF) [11]. TF is the rate at which a word occurs in a specific document, and IDF denotes how frequent a term is found in all documents of a specific category. Therefore, the TFIDF value of a given term must be higher when this term occurs frequently in a given document, but it must also occur rarely in the whole set of documents related to a category. For this, TFIDF is considered as a local feature selection since it gives weight to a feature according to its relevancy for a document for a particular category.

Another challenge related to the large dimensionality is that regarding the currently limited computing power, most existing feature selection methods are not specifically designed to address large textual dataset, and such methods may not proven effective performance for the development of text mining and text classification for large datasets. Therefore, it becomes necessary to use parallel feature selection platforms.

To deal with all aforementioned challenges, this paper introduces an improved version of the classical TFIDF using MapReduce for parallel programming. In the proposal, unlike the traditional use of TFIDF based on document frequencies, the importance of a term is evaluated according to its importance for both document and category. Actually, terms are weighted using TFIDF but to select the most relevant ones, the weight of a single term is considered within all documents in the same category, introducing by this the average weight and the maximum weight for each term. Thus, this work presents a category-based method for TFIDF which transforms

TFIDF to a global feature selection that we call global TFIDF (GTFIDF). To assess the efficiency of our method, we use Hadoop framework to run experimental verification on two datasets. For this, we use multinomial Naive Bayes classifier (MNB) which is considered as one of the most effective used algorithms for classification [12].

The next part of this paper is ordered in four sections: in Sect. 2, we give an overview of the related work concerning FS method and MapReduce; in Sect. 3, we introduce our GTFIDF FS method based on category frequency; in Sect. 4, we detail the used datasets and describe the results obtained during the experiments; in the final section, we present the conclusion and future work.

2 State of the Art

Filter-based feature selection techniques are generally based on feature ranking. Therefore, all features are evaluated using a metric that computes their importance and then features with the highest scores are selected. Because of its simplicity and efficiency, the filter FS methods, such as document frequency (DF) [13], chi-square (CHI) [14], mutual information (MI) [15], and information gain (IG) [16], have been widely used in text classification.

2.1 Document Frequency Method

Document frequency (DF) is one of the most widely used filter feature selection techniques. It is a simple and efficient method which makes it comparable to chi-square method. The document frequency of a term t_i in a category c_k, that we note $DF(t_i, c_k)$, represents the number of documents in category c_k where t_i occurs. A term t_i is considered relevant when it occurs frequently in a category c_k and rarely in other categories [17]. The $DF(t_i, c_k)$ can be defined as follows:

$$DF(t_i, c_k) = \frac{a_i}{n_k}, \tag{1}$$

where a_i is the number of documents containing word t_i and belonging to category c_k, and n_k represents the total number of documents belonging to the category c_k.

2.2 Term Frequency Inverse Document Frequency (TFIDF)

The weight of a term can be computed using the term frequency (TF). However, term weighting using TF may assign large weights to the common words with poor text distinctive power. To overcome this weakness, the inverse document frequency (IDF) is introduced in TFIDF. Indeed, for a term with high frequency in a document, the lower its document frequency in the category is, the higher its relevance for the specific document is. Thus, TFIDF is more reasonable than both TF and IDF schemes and has been widely used. For a term t_i, its TFIDF weight, $\text{tfidf}_{i,j}$, in a text document d_j is commonly represented in text classification as follows:

$$\text{tfidf}_{i,j} = tf_{i,j} \times \log\left(\frac{n}{df_i}\right) \tag{2}$$

where n is the number of all documents in the dataset, $\text{tf}_{i,j}$ represents the frequency that the term t_i occurs in the document d_j, and df_i is the document frequency of t_i, i.e., the total occurrences of documents which contain t_i. However, TFIDF does not take into account the class label of training documents while weighting a term, so the computed weight cannot reliably give the importance of a term in text classification. For example, assume that there were two terms with the same document frequencies, one occurring in several classes of text document while another appearing in only one class of text document. Apparently, the second term has a stronger class discriminating power than the first, but their overall weighting factors, IDF, are the same, which is not normal.

2.3 *Hadoop and MapReduce*

Hadoop is a distributed open source framework for storing and analyzing big data in systems using master–slave architecture [18]. It has two key components, namely Hadoop Distributed File System (HDFS) and MapReduce. HDFS is designed to process large-scale data in a distributed manner under frameworks such as MapReduce and is optimized for high throughput. Besides, it automatically replicates data blocks on nodes, manages fault tolerance, load balance and data distribution. MapReduce is a data processing model that has the advantage of easy scaling of data processing over a cluster of multiple computing nodes. Its storage and computing capacities increase with the addition of other slave nodes to a Hadoop cluster and can reach very large volume sizes. The MapReduce program processes data using key-value pairs for each map and reduce phase as shown in Fig. 1.

3   Global TFIDF (GTFIDF)

Our proposed method is presented in this section on two steps. The first step introduces the preprocessing phase we adopted to prepare the datasets for the feature selection. The second step presents our proposed algorithm GTFIDF to select features as well its parallelization using MapReduce.

Fig. 1 MapReduce scheme

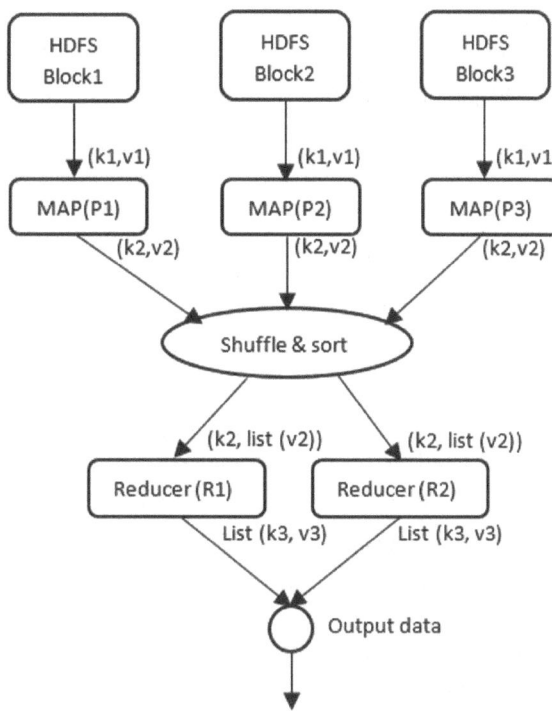

3.1 Preprocessing

Preprocessing documents are a very important step in text classification which can significantly affect the performance of classification algorithms. There are many cleaning techniques which can be performed in the preprocessing step [19]. In this work, we applied the widely used cleaning techniques. First, documents are tokenized and normalized to lower case, and then, stop words, special character, and numbers are removed. Next, the Porter's stemming algorithm is used to reduce words to their stems. Thus, each document can be represented using the vector space model (VSM) [20].

3.2 GTFIDF Using MapReduce

The task is to complete the four values of the statistics: The frequency of feature words in each document, the document frequency of feature words appearing in a document, the TFIDF weight of each feature, and the global TFIDF weight of each feature. The input is the training preprocessed text documents which are labeled using notation $d_j\_c_k$, where d_j is the name of the document and c_k is its category. The offset of the document is the key, following by the line content as the value.

1. The job of calculating feature occurrences is performed as follows:

   - The map function: counts the occurrences of each feature t_i which occurs in document $d_j\_c_k$; output the intermediate results of word count. Its key is constituted by feature t_i and $d_j\_c_k$, which is emit: $\langle(t_i, d_j\_c_k), 1\rangle$.
   - The reduce function: After the shuffling phase, the sum is calculated over the $\langle(t_i, d_j\_c_k), 1\rangle$ set, since $(t_i, d_j\_c_k)$ is the same, and get the number of words in a document.

2. The job of calculating feature frequencies:

   - The map function: obtains the total number of features containing in one document and outputs the intermediate results $\langle d_j\_c_k, (t_i, \text{count}\_t_i)\rangle$.
   - The reduce function: As to the $\langle d_j\_c_k, (t_i, \text{count}\_t_i)\rangle$ set, if $d_j\_c_k$ is the same, then sum count_t_i and get the frequency of a each feature in each document.

3. The job of calculating the IDF value as well the local TFIDF:

   - The map function: obtains the frequency of a feature in a document and outputs the intermediate results $\langle t_i, (d_j\_ck, tf\_t_i)\rangle$.
   - The reduce function: As to the $\langle t_i, (d_j\_ck, tf\_t_i)\rangle$ set, if t_i is the same, then count (d_j_ck) where $tf\_t_i \neq 0$, this gives the DF. Then TFIDF weight is computed using Eq. 2.

4. The job of calculating the average and the maximum values of TFIDF:

   - The map function: obtains the TFIDF of a feature for each document and output the intermediate results $\langle c_k, (t_i, d_j, \text{tfidf}\_t_i)\rangle$.
   - The reduce function: In order to overcome the weakness of standard TFIDF method, as the $\langle c_k, (t_i, d_j, \text{tfidf}\_t_i)\rangle$ set, if category c_k is the same, then determine $\overline{\text{tfidf}\_t_{i,k}}$ and $\text{tfidf}\_t_{i_{max},k}$ by the following:

$$\overline{\text{tfidf}\_t_{i,k}} = \frac{1}{N_k} \sum_{j=1}^{N_k} \text{tfidf}_{i,j} \tag{3}$$

and

$$\text{tfidf}\_t_{i_{max},k} = \max_{j=1}^{N_k}\{\text{tfidf}_{i,j}\} \tag{4}$$

where $\text{tfidf}_{i,j}$ is the tfidf_t_i in the document d_j.

Finally, in our method, we generate two different weighting schemes. In the first scheme, in the final results, only words with tfidf_t_i superior to the average value are retained as relevant for category, whereas in the second scheme, each feature is assigned to document where its weight is maximum.

4 Experiments and Results

4.1 Datasets

In order to study the efficiency of our proposed method, we use two datasets in the experiments: namely Reuters-21578 and 20-Newsgroup [21]. The Reuters corpus is a standard benchmark containing 135 categories collected from the Reuters newswire. In this work, only the top-10 categories of Reuters were used including around 11,083 words. Also, the 20-Newsgroup corpus is a benchmark which contains around 20,000 documents collected from the Usenet newsgroup dataset including around 46,962 words. All documents were evenly spread over 20 different categories.

4.2 Performance Measures

To evaluate the performance for classification, we use the Macro-F1 and Micro-F1 measures [22]. The macro-average gives equally weight to all categories, ignoring the number of documents belong to the category, whereas the micro-average gives to all documents an equally weight. Concretely, Micro-$F1$ and Macro-$F1$ are computed as following:

$$Macro - F1 = \frac{\Sigma_{k=1}^{c} F_k}{C}, F_k = \frac{2 \times p_k \times r_k}{p_k + r_k}, \tag{5}$$

where pair of (p_k, r_k) designates, respectively, precision and recall values of the class c_k, and C is the total number of categories.

$$Micro - F1 = \frac{2 \times r \times p}{r + p}, \tag{6}$$

where p and r designates, respectively, precision and recall values over the entire classification decisions of all classes in the dataset.

4.3 Results

Tables 1, 2, 3, 4 show the experimental results obtained by the two schemes proposed in this paper, namely the Av-GTFIDF and Max-GTFIDF. The MNB algorithm was used for classification. Indeed, it is simple and competitive in text classification compared to different classical algorithms such as support vector machines (SVM) [23].

Table 1 Reuters-21587 Macro-$F1$

Nb. of features	500	1000	1500	2000	2500	3000	3500	4000
Max-GTFIDF	0.4942	0.5803	0.5292	0.7122	0.6713	0.7259	0.7501	0.7650
Av-GTFIDF	0.4918	0.5679	0.5267	0.7026	0.6510	0.7068	0.7364	0.7473
TFIDF	0.4104	0.4663	0.4248	0.6180	0.6084	0.5941	0.6444	0.6440
CHI	0.4919	0.5740	0.5158	0.6985	0.6452	0.6902	0.7218	0.7260

Table 2 20-Newsgroup Macro-$F1$

Nb. of features	500	1000	1500	2000	2500	3000	3500	4000
Max-GTFIDF	0.5869	0.6795	0.7791	0.8031	0.8086	0.7917	0.8276	0.8378
Av-GTFIDF	0.5663	0.6428	0.7679	0.7720	0.7880	0.7666	0.7834	0.7876
TFIDF	0.5475	0.6007	0.7092	0.7276	0.7087	0.5761	0.5984	0.6030
CHI	0.5820	0.6557	0.7602	0.7813	0.7810	0.6604	0.7401	0.7176

Table 3 Reuters-21587 Micro-$F1$

Nb. of features	500	1000	1500	2000	2500	3000	3500	4000
Max-GTFIDF	0.4865	0.5753	0.5373	0.6951	0.6713	0.7158	0.7325	0.7621
Av-GTFIDF	0.4841	0.5511	0.5131	0.6859	0.6453	0.7058	0.7219	0.7461
TFIDF	0.3827	0.4333	0.3686	0.5903	0.6079	0.5821	0.6167	0.6152
CHI	0.4758	0.5474	0.4908	0.6774	0.6232	0.6802	0.7148	0.7233

Table 4 20-Newsgroup Micro-$F1$

Nb. of features	500	1000	1500	2000	2500	3000	3500	4000
Max-GTFIDF	0.5894	0.6659	0.7758	0.8139	0.7792	0.7877	0.8174	0.8279
Av-GTFIDF	0.5768	0.6543	0.7638	0.8052	0.7540	0.7487	0.7716	0.7947
TFIDF	0.5457	0.5872	0.7093	0.7236	0.6970	0.5655	0.5885	0.5836
CHI	0.5817	0.6481	0.7569	0.7827	0.7742	0.6585	0.7256	0.7206

To examine the reliability of the proposal, the Av-GTFIDF and Max-GTFIDF are compared to the classical TFIDF and CHI methods. Tables 1 and 2 show that both schemes Av-GTFIDF and Max-GTFIDF give the best results in terms of Macro-$F1$ measure when compared to the classical TFIDF and CHI. Compared to Av-GTFIDF, the Max-GTFIDF method gives, generally, more important results. It can also be noted that for small number of selected features (500), the Macro-$F1$ measure achieves the lowest performance for all methods.

Tables 3 and 4 show the Micro-$F1$ measure for all methods. It can be noted that Max-GTFIDF performs, generally, better than Av-TFIDF, classical TFIDF and CHI. But, for a small number of selected terms, all methods perform almost identically.

While the proposed schemes achieve their best performance for high number of selected features (4000), the performance of the classical TFIDF start to decrease.

A reason behind the efficiency of Av-GTFIDF and Max-GTFIDF schemes is that both are using a frequency-based intra-category condition to evaluate features. In addition, the superiority of Max-GTFIDF is due to using the inter-category condition in the computation of weights of terms.

Figures 2and3 give the $F1$-measure of two datasets with Max-GTFIDF, Av-GTFIDF, TFIDF, and CHI at different numbers of features such making it possible to observe performance of filter feature selection methods.

More precisely, Av-GTFIDF and Max-GTFIDF show the effectiveness with large number of features of Reuters-21578 and 20-Newsgroup datasets when compared to TFIDF and CHI. It comes down to the fact in case of TFIDF the weight of a word is computed based on its relevancy for documents without taking into account its relation with the category, while in the case of CHI the score of a term is computed without taking under consideration its frequency but only the existence of this term in a document.

Max-GTFIDF is actually an improved version of Av-GTFIDF to select more discriminative features. As analyzed in this paper, in a particular category, if a term has the highest weight, then that term is discriminant for that category. Therefore, the performance of Max-GTFIDF is superior to that of Av-GTFIDF.

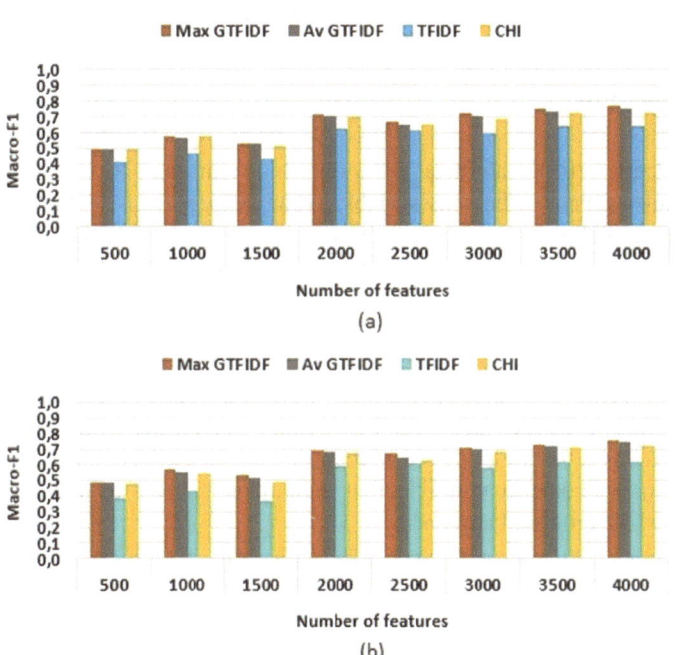

Fig. 2 Reuters-21578 dataset Macro-$F1$ (**a**) and Micro-$F1$ (**b**)

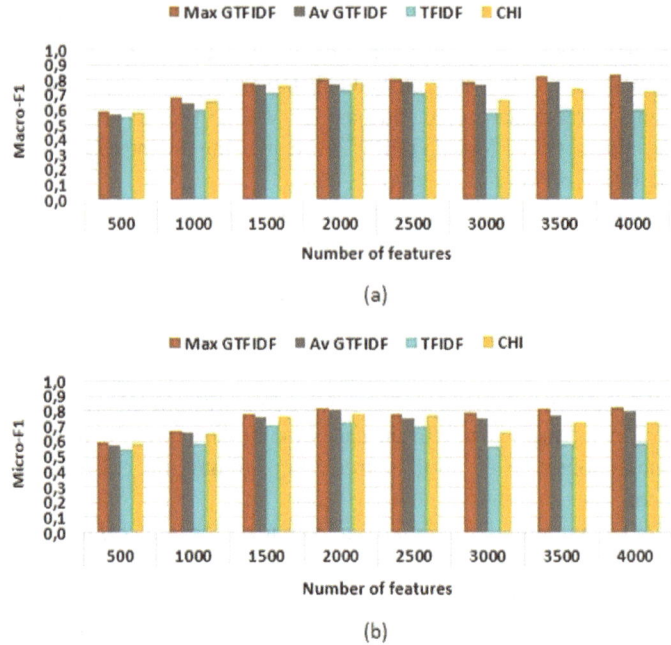

Fig. 3. 20-Newsgroup dataset Macro-$F1$ (**a**) and Micro-$F1$ (**b**)

5 Conclusion

Feature selection (FS) is a crucial task in machine learning. Obviously, the current FS techniques are not sufficiently scalable to handle large datasets. In this paper, we propose a parallel global feature selection method based on TFIDF. The proposal includes two weighting schemes, namely AV-GTFIDF and Max-GTFIDF. The algorithm was implemented using Apache Hadoop and tested on two different large datasets through the MNB classifier. The results obtained during the experiments approve the effectiveness of our approach, according to Micro-$F1$ and Macro-$F1$ measures, which proves the reliability of the global feature selection techniques. As future work, the experiments can be extended to a cluster with more machines, which increases computing power and reduces time costs. Also, other classification algorithms as well other feature selection techniques will be used to better study the reliability of the proposed approach.

References

1. Zhao, S., Yao, H., Zhao, S., Jiang, X., Jiang, X.: Multi-modal microblog classification via multi-task learning. Multimedia Tools Appl. **75**(15), 8921–8938 (2016)
2. Idris, I., Selamat, A.: Improved email spam detection model with negative selection algorithm and particle swarm optimization. Appl. Soft Comput. **22**, 11–27 (2014)
3. Zhang, C., Wu, X., Niu, Z., Ding, W.: Authorship identification from unstructured texts. Knowl.-Based Syst. **66**, 99–111 (2014)
4. Wang, Y., Youn, H.: Feature weighting based on inter-category and intra-category strength for twitter sentiment analysis. Appl. Sci. **9**(1), 92 (2019)
5. Saraç, E., Özel, S.A.: An ant colony optimization based feature selection for web page classification. Sci. World J. (2014)
6. Jiang, L., Li, C., Wang, S., Zhang, L.: Deep feature weighting for Naive Bayes and its application to text classification. Eng. Appl. Artif. Intell. **52**, 26–39 (2016)
7. Liu, H., Yu, L.: Toward integrating feature selection algorithms for classification and clustering. IEEE Trans. Knowl. Data Eng. **4**, 491–502 (2005)
8. Labani, M., Moradi, P., Ahmadizar, F., Jalili, M.: A novel multivariate filter method for feature selection in text classification problems. Eng. Appl. Artif. Intell. **70**, 25–37 (2018)
9. Dy, J.G., Brodley, C.E.: Feature selection for unsupervised learning. J. Mach. Learn. Res. 5, 845–889 (2004)
10. Song, Q., Ni, J., Wang, G.: A fast clustering-based feature subset selection algorithm for high-dimensional data. IEEE Trans. Knowl. Data Eng. **25**(1), 1–14 (2011)
11. Zhang, L., Jiang, L., Li, C., Kong, G.: Two feature weighting approaches for naive bayes text classifiers. Knowl.-Based Syst. **100**, 137–144 (2016)
12. Abbas, M., Memon, K.A., Jamali, A.A., Memon, S., Ahmed, A.: Multinomial Naive Bayes classification model for sentiment analysis. IJCSNS **19**(3), 62 (2019)
13. Azam, N., Yao, J.: Comparison of term frequency and document frequency based feature selection metrics in text categorization. Expert Syst. Appl. **39**(5), 4760–4768 (2012)
14. Yang, J., Liu, Y., Zhu, X., Liu, Z., Zhang, X.: A new feature selection based on comprehensive measurement both in inter-category and intra-category for text categorization. Inf. Process. Manage. **48**(4), 741–754 (2012)
15. Vergara, J.R., Estévez, P.A.: A review of feature selection methods based on mutual information. Neural Comput. Appl. **24**(1), 175–186 (2014)
16. Uğuz, H.: A two-stage feature selection method for text categorization by using information gain, principal component analysis and genetic algorithm. Knowl.-Based Syst. **24**(7), 1024–1032 (2011)
17. Tang, B., Kay, S., He, H.: Toward optimal feature selection in Naive Bayes for text categorization. IEEE Trans. Knowl. Data Eng. **28**(9), 2508–2521 (2016)
18. Tang, Z., Xiao, W., Lu, B., Zuo, Y., Zhou, Y., Li, K.: A parallel algorithm for bayesian text classification based on noise elimination and dimension reduction in spark computing environment. In: International Conference on Cloud Computing, pp. 222–239. Springer (2019)
19. Pradhan, L., Taneja, N.A., Dixit, C., Suhag, M.: Comparison of text classifiers on news articles. Int. Res. J. Eng. Technol. **4**(3), 2513–2517 (2017)
20. Zhang, B.: Analysis and Research on Feature Selection Algorithm for Text Classification. University of Science and Technology of China, Anhui (2010)
21. Agnihotri, D., Verma, K., Tripathi, P.: An automatic classification of text documents based on correlative association of words. J. Intell. Inf. Syst. **50**(3), 549–572 (2018)
22. Uysal, A.K., Gunal, S.: A novel probabilistic feature selection method for text classification. Knowl.-Based Syst. **36**, 226–235 (2012)
23. Hadi, W., Al-Radaideh, Q.A., Alhawari, S.: Integrating associative rule-based classification with Naive Bayes for text classification. Appl. Soft Comput. **69**, 344–356 (2018)

Robust and Accurate Method for Textual Information Extraction Over Video Frames

Soundes Belkacem and Larbi Guezouli

Abstract This paper presents a robust and accurate method for textual information extraction over video frames. Text has to be located and successfully discriminated from his background and other scene objects. Extraction is a challenging task due to varieties degradations related to environmental changes, acquisition conditions in addition to characteristics related to video content. Method consists of a three-step process: Slide region detection, text detection, and non-text filtering; text detection use region invariants features extraction based on moment functions to label frame pixels for text/non-text. Experimental results show high accuracy for text/non-text discrimination, and this is mainly related to orthogonal moment functions expressivity and invariants properties.

Keywords Text detection · Text localization · Lecture videos · Moment functions · Feature extraction

1 Introduction

Recently, lecture videos receive an increasing interest and are widely for distance learning [1, 2]. In fact, courses and different presentations are recorded and made online available by educational institutions and universities. Hence, the amount of available online and archived lecture videos is increasing explosively and their contents must be easy to access and search within.

For efficient, automatic, and effective access and search content-based access and retrieval systems are required. Performances are strongly related to content description efficiency. The used feature must be robust and expressive enough for accurate contents characteristics extraction and description, but also, must be invariant to

S. Belkacem (✉)
Department of Mathematics and Computer Science, University of Medea, Medea, Algeria
e-mail: soundes.belkacem@gmail.com

L. Guezouli
LaSTIC Laboratory, Department of Computer Science, University of Batna 2, Batna, Algeria
e-mail: larbi.guezouli@univ-batna2.dz

© The Editor(s) (if applicable) and The Author(s), under exclusive license to Springer   119
Nature Singapore Pte Ltd. 2021
F. Saeed et al. (eds.), *Advances on Smart and Soft Computing*, Advances in Intelligent
Systems and Computing 1188, https://doi.org/10.1007/978-981-15-6048-4_11

degradations and noises. Moreover, the selected feature must be easy to extract and to use.

Textual information within slides presentation is the most important visual content for lecture video understanding and can be used as outline for video navigation and indexing.

Textual information extraction systems (TIE) from images and videos have been an active research area in the last decade [2–7]. The main purpose is to segment text from others scene objects. However, till now problem remains unresolved and no global solution is given. In fact, text detection is a challenging task due to varieties of text characteristics such as alignments, fonts, and size; also, frames show significant degradations caused mainly by acquisition conditions resulting on blurred frames with low resolution, uneven illumination, view distortion, and textured complex background.

Moreover, traditional methods for text detection and localization method cannot be applied directly to lecture videos because of lecture videos characteristics:

1. Camera motion;
2. Environmental changes;
3. Homogenous scene composition.

Making text segmentation more difficult and challenging task.

Among visual features, shape is a powerful feature incorporated on almost every content-based image retrieval system [8]. Shape descriptors are generally divided into two categories: contour-based and region-based. Contour-based approaches consider only contour information while region-based approaches consider both contour and interior information. Moments functions are the most common shape descriptors, In fact, due to their computational efficiency, moments are widely used in image analysis and pattern recognition applications [9].

Moments functions differ per their basic function. Hence, various moments can be obtained regarding used polynomial. They are generally categorized into two categories: *orthogonal* (Zernike moments ZMs [10] and pseudo-Zernike moment's PZMs [11]) and *non-orthogonal* moments [12]. Author in [9] shows that orthogonal moment's with orthogonal complex basis functions outperform commonly used moments. In fact, they are rotation, scale, and translation invariants with high expressivity and minimum information redundancy.

In this paper, we propose a robust and accurate approach for text detection and localization from lecture video frames. Text region properties are extracted using moments functions. The computed feature is used to label frame pixels into text/non-text regions. Both moments categories are used: the orthogonal moments: Zernike moment ZM and pseudo-Zernike moment PZM; also, non-orthogonal one: Hu moment.

The rest of this paper is organized as follows: Sect. 2 gives an overview of the related work on text detection and localization. Section 3 gives an overview of used moment functions. Section 4 describes the proposed method. Experimental results and comparisons are given in Sect. 5. Section 6 concludes the paper.

2 Related Works

Text detection methods are generally categorized into two categories [2–7, 13]: texture-based and region-based methods.

The texture-based methods analyzed image/frame based on texture analysis techniques such as wavelet decomposition, discrete cosine transform (DCT), local binary pattern (LBP), and Fourier transform. Texture-based methods present good results over noisy images, but feature extraction requires to be on several levels. Hence, high computation time is required. Moreover, texture detection is a delicate problem and introduced solution based on DCT or LBP which are sensitive to alignment and perspective distortion.

Region-based methods extract text region properties: color, edge, or stroke to group text pixels into characters and words. Region-based methods have the advantage of producing results that can be directly processed by character recognition systems OCR with lower execution time comparing to texture-based methods. Text region properties are used for connected components, edges, and strokes for text detection.

Hybrid approaches combining two or more methods from both categories allow taking advantages of one and overcoming the limitations of the other. This combination seems to be an effective solution, as is the case with the recently proposed methods.

Preprocessing or improvement (blur removing and filtering) step is usually necessary to prepare images for the detection stage for higher accuracy and thereby increasing TIE system performances.

3 Moments Functions

3.1 Hu Moments

Moments of order p and repetition q over image with $f(x, y)$ intensity function is given by equation:

$$M_{p,q} = \iint x^p x^q f(x, y) \mathrm{d}x \mathrm{d}y_D, \quad p, q = 0, 1, 2 \ldots \tag{1}$$

In 1962, Hu introduced a set of invariants moments (HU moment's) to represent image. Thus, seven moments are invariants to directions, scales, and orientations. Using the following equation:

$$I_1 = \eta_{20} + \eta_{02} \tag{2}$$

$$I_2 = (\eta_{20} - \eta_{02})^2 + 4\eta_{11}^2 \tag{3}$$

$$I_3 = (\eta_{30} - 3\eta_{12})^2 + (\eta_{21} - 3\eta_{03})^2 \tag{4}$$

$$I_4 = (\eta_{30} - \eta_{12})^2 + (3\eta_{21} - \eta_{03})^2 \tag{5}$$

$$I_5 = (\eta_{30} - 3\eta_{12})(\eta_{30} - \eta_{12})\left[(\eta_{30} - \eta_{12})^2 - 3(\eta_{21} - \eta_{03})^2\right]$$
$$+ (3\eta_{21} - \eta_{03})(\eta_{21} - \eta_{03})\left[3(\eta_{30} - \eta_{12})^2 - (\eta_{21} - \eta_{03})^2\right] \tag{6}$$

$$I_6 = (\eta_{20} - \eta_{012})\left[(\eta_{30} - \eta_{12})^2 - (\eta_{21} - \eta_{03})^2 + 4\eta_{11}(\eta_{30} - \eta_{12})(\eta_{21} - \eta_{03})\right] \tag{7}$$

$$I_7 = (3\eta_{21} - \eta_{03})(\eta_{30} - \eta_{12})\left[(\eta_{30} - \eta_{12})^2 - 3(\eta_{21} - \eta_{03})^2\right]$$
$$+ (\eta_{30} - 3\eta_{012})(\eta_{21} - \eta_{03})\left[3(\eta_{30} - \eta_{12})^2 - (\eta_{21} - \eta_{03})^2\right] \tag{8}$$

3.2   Zernike Moments

For image of size $(N * N)$ with intensity function $f(r, \theta)$, computation of Zernike moments (ZM) of order p and repetition q is given by equation:

$$\text{ZM}_{p,q} = \frac{p+1}{\pi} \iint\limits_{x^2+y^2 \leq 1} V_{p,q}^*(r, \theta) f(r, \theta) dxdy \tag{9}$$

where $V_{p,q}^*(x, y)$ is the complex conjugate of the complex of Zernike $V_{p,q}$ polynomials (x, y), which can be separated into two functions:

$$V_{p,q}(x, y) = R_{p,q}(r)e^{jq\theta} \tag{10}$$

where

- $R_{p,q}(r)$: Radial polynomial over polar coordinates (r, θ). $R_{p,q}(r)$ is given by:

$$R_{p,q}(r) = \sum_{s=0}^{\frac{p-|q|}{2}} (-1)^s \frac{(p - s)!}{s!\left(\frac{p+|q|}{2} - s\right)!\left(\frac{p-|q|}{2} - s\right)!} r^{p-2s} \tag{11}$$

- $e^{jq\theta}$: Angular function,

    where $e^{jq\theta} = (\cos\theta + j\sin\theta)^q$.

- p: Moment order, $0 \leq p$.
- q: Moment repetitions, integer $0 \leq |q| \leq p$. For computation, only positive values are used. Negatives values can be computed using complex conjugate: $ZM_{p,-q} = ZM^*_{p,q}$.
- j: Imaginary number $j = \sqrt{-1}$.
- θ: angle between the vector r and axis X. $\theta = \tan^{-1}(x/y)$ and $\theta \in [0, 2\pi]$.
- r: Length of the vector from the origin $(\bar{x}, \bar{y})$ to pixel (x, y). $r = \sqrt{x^2 + y^2}$.

Zernike moments are defined over polar coordinates in a unit circle; then the pixels of square image must be normalized to the interval $[0, 1]$ where $x^2 + y^2 \leq 1$. Linear transformation is required to normalize pixel coordinates into polar system.

3.3 Pseudo-Zernike Moments

Pseudo-Zernike moments (PZM) use pseudo-Zernike polynomial as a basis function. They are defined over a polar coordinate within a united circle. Pseudo-Zernike moment of order p and repetition q for image of size $(N * N)$ with intensity function $f(r, \theta)$ is computed with equation:

$$PZM_{p,q} = \frac{p+1}{\pi} \iint\limits_{x^2+y^2 \leq 1} V^*_{p,q}(x, y) f(x, y) dx dy \tag{12}$$

where

- $V^*_{p,q}(x, y)$: the complex conjugate of the complex pseudo-Zernike $V_{p,q}$ polynomials (x, y):

$$V_{p,q}(x, y) = R_{p,q}(r) e^{jq\theta} \tag{13}$$

- $R_{p,q}(r)$: Radial polynomial over polar coordinates (r, θ).
- $e^{jq\theta}$: Angular function, where $e^{jq\theta} = (\cos \theta + j \sin \theta)^q$.
- p: Moments order, $0 \leq p$.
- q: Moments repetitions, integer $0 \leq |q| \leq p$. Only $0 \leq q$ values are considered for computation. $0 \geq q$ values are computed with: $PZM_{p,-q} = PZM^*_{p,q}$.
- j: Imaginary number $j = \sqrt{-1}$.
- θ: angle between axis X and vector r. $\theta = \tan^{-1}(x/y)$ and $\theta \in [0, 2\pi]$.
- r: Length of the vector from the origin $(\bar{x}, \bar{y})$ to pixel (x, y). $r = \sqrt{x^2 + y^2}$.
- $R_{p,q}(r)$ is given by:

$$R_{p,q}(r) = \sum_{s=0}^{p-|q|} (-1)^s \frac{(2p+1-s)!}{s!(p+|q|+1-s)!(p-|q|-s)!} r^{p-s} \tag{14}$$

Fig. 1 Overview of
proposed method

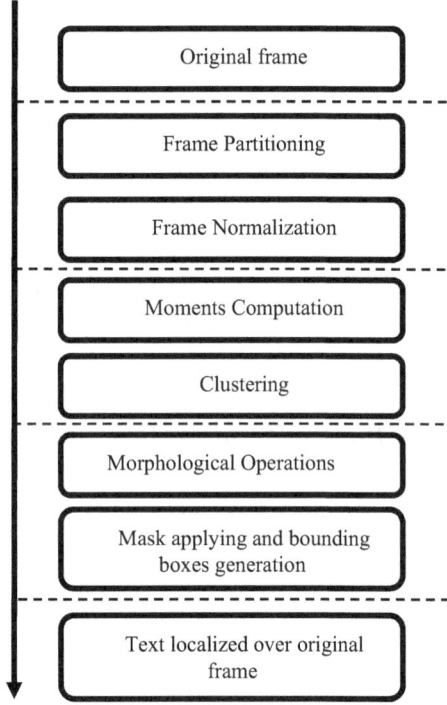

In order to compute PZM for 2D images, pixels coordinate need to be normalized into interval [0, 1] over polar coordinates where $x^2 + y^2 \leq 1$.

4   Proposed Method

4.1   Overview

Proposed method consists mainly of three-step process: Slide region detection and segmentation and feature extraction and non-text filtering. Figure 1 gives the flowchart of the proposed method.

4.2   Slide Region Detection

In lecture videos, slide projection region contains most reveals features for content-based indexing. Hence, video frames are firstly segmented into slide/non-slide region. Non-slide region is considered as background, and it includes other scene objects:

speaker and/or audience. Slide region detection avoids useless feature extraction and similarity computation. Frame regions are classified with k-means algorithm into slide/non-slide region. Result is a binary image containing slide and background.

4.3 Feature Extraction

1. *Color space transformation*
   Once slide region is segmented and extracted, obtained frame is converted into HSV color space. V component will be used for further computation.
2. *Normalization*
   Once slide region is segmented and extracted, obtained frame is converted into HSV color space. V component will be used for further computation. For PZM computation, pixel coordinates are normalized into polar system.
3. *Moments computation*
   Moment functions are used for feature computation. For each sampled frame, local descriptor is computed using moment functions. ZM, PZM, and Hu moments are computed within sliding window over frame regions using equations given in Sect. 3.
   Frame descriptor is given by

$$V = \left\{ \cup D_{x,y} | x = 0 \dots \overset{\scriptscriptstyle\backprime}{N}, y = 0 \dots \overset{\scriptscriptstyle\backprime}{M} \right\} \tag{15}$$

   where
   $D_{x,y} = \left\{ \underset{P_{max}}{\Leftarrow} \text{PZM}'(p, q) \right\}$ is window descriptor.
4. *Clustering*
   Descriptors are used to text discrimination from other objects. Two classes are obtained for $k = 2$, and text is well-separated from its background. K-means is unsupervised classifier, and no training dataset is needed.

4.4 Non-text Filtering

To filter out non-text, morphological operations are used to smooth and extract text edges. Than contours are approximated, and largest quadruplet shape surrounding edges. If more than 45% of quadruplet area is filled, then it is considering as containing text. Other constraints on region size are also applied to filter out false insertions. Figure 2 shows sample detection results.

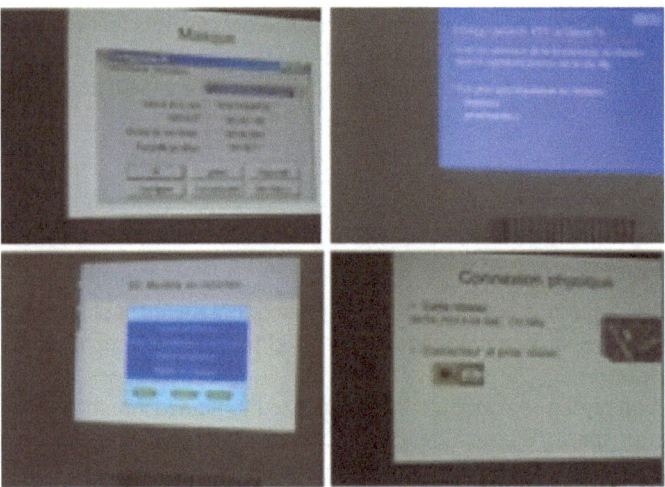

Fig. 2 Sample frames from used dataset

5 Experiments and Results

5.1 Experimental Setup

Proposed method is implemented in C++ using the OpenCV library, and the experiments are performed on a 2.50 GHz CPU with 4 GB of RAM, under Windows 8.

Dataset consists of 50 keyframes extracted from lecture videos with average size 800 * 480. Frames are very low resolution, high luminance variation and contain slides of different perspectives with uniform, colored, or textured background. Slide region contains text and/or images. Text has different: size, font, shape, orientation, perspective distortion and incorporates mathematical symbols. Camera and speaker motion are also allowed resulting in slide region occlusion. Some sample frames from used dataset are shown in Fig. 2.

5.2 Evaluation Metrics

Method accuracy is evaluated using recall R, precision P, and F-measure.

$$P = \frac{\text{Correctly Detected Text Area}}{\text{Totally Detected Text Area}} \tag{16}$$

$$R = \frac{\text{Correctly Detected Text Area}}{\text{Total Text Area}} \tag{17}$$

$$F_1 = 2 * \frac{R * P}{R + P} \tag{18}$$

Correctly Detected Text Area: Text regions with ratio $\geq 90\%$ are considered correctly recognized. *Ratio* is computed by:

$$ratio = \frac{interarea}{minArea} \tag{19}$$

where

InterArea: area of intersection between text bounding box and ground truth bounding box;

MinArea: area of minimum bounding box containing both of them (Fig. 3).

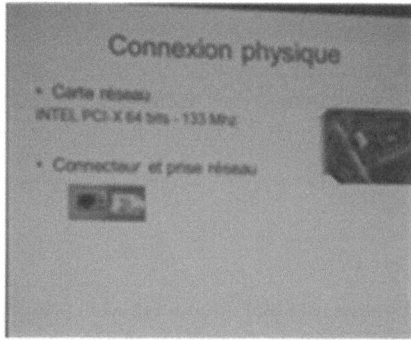

(1)

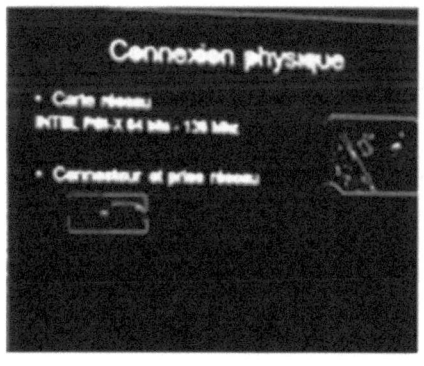

(2)

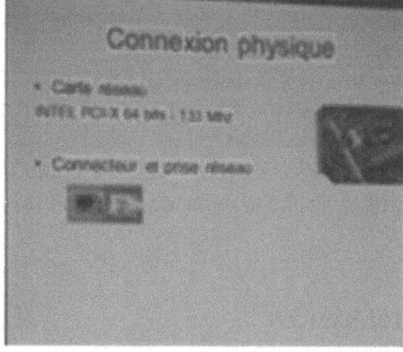

(3)

Fig. 3 Text detection example (1) slide frame (2) clustering results (3) detected text over original frames

Table 1 Detection results using different moments

	Precision	Recall	$F1$	Average time per frame	
				Feature extraction (ms)	Text detection (ms)
PZM	0.861	0.723	0.78	0.10	0.31
HU	0.709	0.680	0.69	0.28	0.46
ZM	0.834	0.720	0.77	0.33	0.52

5.3 Proposed Method Evaluation

Experiments results show that (Table 1):

- The use of moment function allows text/background discrimination;
- Text regions within frames are detected with no prior information on it presence, and no preprocessing step is performed;
- Orthogonal moments show the highest accuracy comparing to non-orthogonal ones;
- PZM gives better results in comparison with ZM and HU moments;
- PZM and ZM are almost similar; however, PZM has less processing time;
- Case of significant light variation HU moments shows poor results.

6 Conclusion

In this paper, accurate and robust text detection method is presented. Moment functions are used for robust and accurate feature extraction over sliding window to group pixels into text/non-text regions. And then, a binary mask is obtained using k-means algorithm. Finally, morphological operations are used to locate detected text by bounding boxes. Experiments show that moment functions give good performances for text detection. However, orthogonal property involves more local features and shows less sensitivity to luminance variations which results on high accuracy regarding to the non-orthogonal moments.

In future works, (i) we intend to use obtained results for optical characters recognition steps. (ii) Moreover, we could combine regions features with other feature such as stroke to enhance detection results.

References

1. Pedrotti, M., Nistor, N.: Online lecture videos in higher education: acceptance and motivation effects on students' system use. In: 2014 IEEE 14th International Conference on Advanced Learning Technologies, pp. 477–479. IEEE (2014)
2. Furini, M.: On introducing timed tag-clouds in video lectures indexing. Multimed. Tools Appl. **77**(1), 967–984 (2018)

3. Basu, S., Yu, Y., Zimmermann, R.: Fuzzy clustering of lecture videos based on topic modeling. In: 2016 14th International Workshop on Content-Based Multimedia Indexing (CBMI), pp. 1–6. IEEE (2016)
4. Yang, H., Meinel, C.: Content based lecture video retrieval using speech and video text information. IEEE Trans. Learn. Technol. 7(2), 142–154 (2014)
5. Yang, H., Siebert, M., Luhne, P., et al.: Lecture video indexing and analysis using video OCR technology. In: 2011 Seventh International Conference on Signal Image Technology & Internet-Based Systems, pp. 54–61. IEEE (2011)
6. Balasubramanian, V., Doraisamy, S.G., Kanakarajan, N.K.: A multimodal approach for extracting content descriptive metadata from lecture videos. J. Intell. Inf. Syst. 46(1), 121–145 (2016)
7. Yin, X.-C., Zuo, Z.-Y., Tian, S., et al.: Text detection, tracking and recognition in video: a comprehensive survey. IEEE Trans. Image Process. 25(6), 2752–2773 (2016)
8. Wang, F., Ngo, C.W., Pong, T.C.: Structuring low-quality videotaped lectures for cross-reference browsing by video text analysis. Pattern Recogn. 41(10), 3257–3269 (2008)
9. Kanan, H.R., Faez, K., Gao, Y.: Face recognition using adaptively weighted patch PZM array from a single exemplar image per person. Pattern Recogn. 41(12), 3799–3812 (2008)
10. Teague, M.R.: Image analysis via the general theory of moments. JOSA 70(8), 920–930 (1980)
11. Teh, C.-H., Chin, R.T.: On image analysis by the methods of moments. IEEE Trans. Pattern Anal. Mach. Intell. 10(4), 496–513 (1988)
12. Hu, M.K.: Visual pattern recognition by moment invariants. IEEE Trans. Inf. Theory 8(2), 179–187 (1962)
13. Amato, F., Greco, L., Persia, F., et al.: Content-based multimedia retrieval. In: Data Management in Pervasive Systems, pp. 291–310. Springer, Cham (2015)

A Deep Learning Architecture for Profile Enrichment and Content Recommendation

Fatiha Sadouki and Samir Kechid

Abstract The objective of our work is to propose a personalized recommender system based on user–resource interactions and user/resource characteristics in order to suggest the most appropriate resources to users. Recommender systems are commonly used in connection with artificial intelligence because machine learning techniques are frequently used to create recommendation algorithms. We suggest using neural networks to improve resource suggestions. Neural networks can quickly perform complex tasks and easily handle massive data. Our new approach is based on an auto-encoder used as a data extractor to improve user performance and a deep neural network (DNN) for item recommendation. For this, we use an information extraction technique based on the reduction of dimensionality by the use of an auto-encoder to compress the features (the proposed tags). We want to create some semantics between the data in order to enrich the profile of the user. Then based on the history of user ratings as well as data on users and items (useful information such as genre, year, tag and rating), we develop a framework for deep learning to learn a similarity function between users and predict item ratings. We evaluate the effectiveness of the proposed approach through an experimental study on a real-world dataset such as the MovieLens film recommender system, according to a number of properties. Experimental validation combines both the accuracy of the recommendation system and a set of quality metrics.

Keywords Recommender system · Machine learning · Neural networks · Auto-encoder · Deep learning · Deep network

F. Sadouki (✉)
Department of Mathematics and Computer Science, Faculty of Sciences, University of Medea, Medea, Algeria
e-mail: sadouki.fatiha@univ-medea.dz

S. Kechid
LRIA Laboratory, Department of Computer Science, Faculty of Electronic and Sciences, USTHB, Algiers, Algeria
e-mail: skechi@usthb.dz

F. Saeed et al. (eds.), *Advances on Smart and Soft Computing*, Advances in Intelligent Systems and Computing 1188, https://doi.org/10.1007/978-981-15-6048-4_12

131

1 Introduction

Recommender systems (SR) have been studied in various fields such as the web, e-commerce and many others. A recommender system is a software tool designed to generate and deliver items or content that a specific user wishes to use. Recommendations are based on predicting the score a user will give to content. These predictions are made from a user profile template which generally summarizes the user's previous behaviour.

Machine learning (ML) is an area in continuous development and a powerful new tool dedicated to solving various problems, and it differs from traditional approaches. The recent appearance of recommender systems based on learning algorithms has shown that they can obtain interesting results and thus improve the quality of the recommendations offered to users. In this paper, we propose a recommendation model based on DNNs that do need information from users, items and the interaction between users and items. First, we use the dataset to obtain the features of the users and items, which we will discuss in Sect. 3. Then, we regard these features as the input of the neural network. In the output layer, we will obtain some probability values that represent the probabilities of the scores that the user might give. Finally, the score with the highest probability will be used as the prediction result. By comparing with some commonly used algorithms on public dataset, it is proved that the proposed model can effectively improve the recommendation accuracy. Our model combines two deep neural networks. First, it uses social information to enrich the user's profile; then the data will be submitted to the proposed deep neural network model which is used to predict the assessment scores.

This paper is divided into three sections. In the first section, we present some works on SR using ML and approaches based on semantic relationships between tags. The second consists of the representation of the proposed model, and the last section presents the results obtained.

2 Related Works

Machine learning is widely applied in various fields. As our paper focuses on recommendation systems based on deep learning, we, respectively, introduce the use of tags and the application of deep learning in recommender systems.

2.1 Tags in Recommendation Systems

Many traditional methods have used tags. Nakamoto et al. [1] introduced a contextual collaborative filtering (CF) model which takes into consideration users' context based upon tagging information when calculating the user similarity between users and predicting ratings for the user. Another work of Karen et al. [2] who proposed a generic method that allows tags to incorporate to standard CF methods, by reducing three-dimensional correlations to three two-dimensional correlations, then by applying a fusion method to re-associate these correlations.

For the semantic relationship between tags, Cattuto [3] and Markines et al. [4] have proposed different ways to measure the similarity between tags and resources in a folksonomy: simple co-occurrence measure (for each pair of tags, the number of annotations that contain the two tags), distributional distance of aggregation and cosine (the idea is to project the tripartite model of folksonomy into bipartite by aggregating data according to a given context), distance between two tags (these distances can be used to match written words with syntactic errors with the correct words) and the classification of tags (Specia and Motta [5] applied the technique of grouping tags into several groups according to similarity measures in Tag-Tag context).

To resolve the problem of redundancy and ambiguity caused by tags, Szomszor et al. [6] constructed a semantic knowledge base enriched by movies folksonomy with descriptions and categorizations of movie titles, and user interests and opinions. Wetzker et al. proposed an algorithm [7] to exploit the semantic contribution of tags. Rendle et al. [8] proposed a tensor factorization algorithm ranking with tensor factorization (RTF) directly optimizes the factorization model for the best-personalized ranking. Zhang et al. [9] proposed a recommendation algorithm based on an integrated of tagging information on user–item–tag tripartite graphs.

2.2 Recommender Systems Based on Deep Learning

Salakhutdinov et al. proposed a hybrid recommendation approach (CDR) with implicit feedback. Specifically, CDR employs a stacked denoising auto-encoder (SDAE) to extract deep feature representation from side information and then integrates into pair-wise ranking model for alleviating sparsity reduction [10]. To solve the sparsity problem of ratings, Kim et al. proposed a novel context-aware recommendation model that integrates convolutional neural network into probabilistic matrix factorization to enhance the prediction accuracy [11].

Van den Oord et al. [12] proposed a model that applies convolutional neural networks in music recommendations to predict latent factors from music audio when they cannot be obtained from usage data. Moreover, Google proposed a general framework of wide and deep learning—jointly trained wide linear models and deep neural networks—that combines the benefits of memorization and generalization for recommender systems [13]. The model DeepFM [14] combined the power of factorization machines for recommendation and deep learning for feature learning in new neural network architecture. Xu et al. [15] proposed a tag-aware personalized recommendation system using a deep-semantic similarity model, DSPR, to extract recommendation-oriented representations for social tags, to address the uncontrolled vocabulary problem and to achieve superior personalized recommendations. Liang et al. [16] used a forward neural network to simulate matrix factorization to explore the feasibility of predicting ratings via two vectors from different deep feature space. A model that combines a collaborative filtering recommendation algorithm and a deep learning technology has been proposed by Zhang et al. [17]. The model first uses a feature representation method based on a quadric polynomial regression model; the second part is input data of the proposed deep neural network model that is used to predict the evaluation scores. Zhuang et al. proposed a new representation learning framework called recommendation via dual-autoencoder (ReDa) [18]. In this framework, they simultaneously learn the new hidden representations of users and items using auto-encoders and minimize the deviations of training data by the learnt representations of users and items.

3   Proposed Model

This section presents the model of our recommender system, as well as its steps; in the same way, we detail each part with its role. First, we will reconcile the tags using a dimensionality reduction method using an auto-encoder. The middle layer (the centre layer) of the latter is used to enrich the user profile in order to provide additional information to improve the recommendation result. The second step is to prepare (preprocess) the data, which consists of combining this data to obtain a single file ready for processing. The next step is to train the model using a well-known and best-performing machine learning algorithm which is the deep neural network. The last step is to make the recommendation with the new proposed approach which is different from the traditional approaches, in order to recommend a list of resources with scores to users (Fig. 1).

3.1   Enrichment

The representation of the profile is a procedure which consists in producing a vector of characteristics describing the user. This operation involves different algorithms categorized into two classes, the technique of extracting information based on the reduction of dimensionality (reducing the number of tags offered) by using an auto-encoder, and the technique of enriching the profile which consists in using the central layer of the auto-encoder.

By providing a list of tags as inputs and comparing the output with inputs, the network can learn by itself the rule for reducing features. The data will be the input to transition from the input layer to the hidden layer which is called the coding step which is a great way to create some semantics between the data. Then the decoding step is the transition from the hidden layer to the output layer.

The activation of these layers is done with the most common activation function; it is the ReLU nonlinear function.

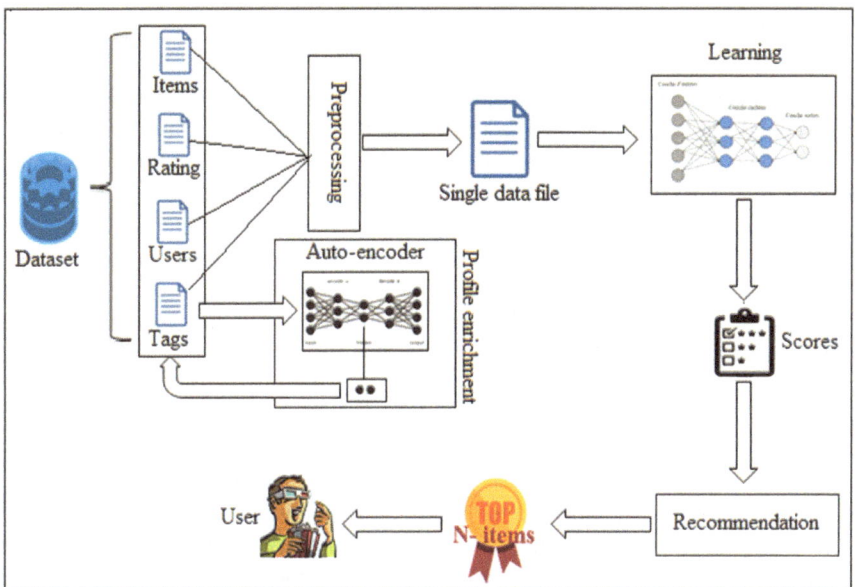

Fig. 1 Overview of the proposed model

$$\text{ReLU} = \text{Max}(0, x) \tag{1}$$

where x is the input value and the output is the maximum between 0 and the input value. Finally, the intermediate part of the auto-encoder is used to enrich the user profile.

3.2 Preprocessing

The objective of this part is the analysis and the normalization of the data; it thus consists in cleaning the data and in removing noise.

First of all, it is a question of extracting the essential attributes (characteristics) and eliminating those which are not essential or those which are painful. Then, it is necessary to convert this data into numerical values (numerical valuation). And finally, join the data files that we have selected in a single file, in order to present a new method which takes into account several attributes and not just one as in traditional recommendation systems.

3.3 Learning

Before proceeding to the learning of the model, we will normalize the values of the attributes of our dataset, except the attribute of the score (vote). This operation will normalize the values between 0 and 1. The reason for normalization is that the neural network does not take the big values in its training of the model.

The deep neural network model takes the features of users and items as the inputs and uses the forward propagation algorithm to predict the rating scores.

According to our model, in the input layer, the input vector x_0 is concatenated by the features of users and items. When x_0 passes through the first hidden layer, the output of the first hidden layer is obtained by the following equation:

$$x_1 = \text{activation}(W_1 x_0 + b_1) \tag{2}$$

where W_1 is the weight matrix between the input layer and the first hidden layer, b_1 is the bias vector, the activation() indicates the activation function, which is designed to make the neural network model nonlinear and multilayer neural networks become meaningful. We choose ReLU as the activation function for our model; the reason for choosing this function is that ReLU is economical in computation and capable of accelerating the speed of training of neural networks, as well we want an output value between 1 and 5.

By Eq. (2) and the discussion above, we can obtain the output at the lth hidden layer:

$$x_l = \text{ReLU}(W_l x_{l-1} + b_l) \tag{3}$$

Finally, we use a mean squared error method to evaluate the difference between the predicted result $\hat{y}_j$ and the true value y_j:

$$\text{MSE} = \frac{1}{n} \sum_{j}^{n} (y_j - \hat{y}_j)^2 \tag{4}$$

In the output layer, our training predicts the ith user's rating score on the jth item $\hat{R}_{ij}$ (Eq. 5).

For optimization, algorithms Adam (adaptive moment estimation (Adam) is an optimization algorithm used to update iterative network weights based on training data) and RMSProp (for calculating gradients and updating parameters) are used.

3.4 Recommendation

Once the model is well-trained and memorized, we can use it to predict a user's rating score on the items that have not been rated by the user. The items with the highest predicted score are recommended to the user.

4 Experimental Evaluation

4.1 Data Description

In order to test the performance of our method, we used a MovieLens dataset, ml-latest-small, which contains 100,836 ratings and 3683 tag applications across 9742 movies. These data were created by 610 users. Each note is between 1 and 5 and always takes an integer value. Each user has voted at least 20 films. In this dataset, only 1572 films were tagged by 58 users.

4.2 Experimental Environment

Hardware We use in this experiment, CPU Intel Core i7-8560U, RAM 16 GB, 512 GB SSD, execution in Google Colab (is a free cloud service with a free GPU).

Software The operating system used in this experiment is Windows 10, and we use Python language 3.7 and popular framework to implement the deep learning modules—Tensorflow 2.0.

Evaluation Measures We adopt the mean absolute error (MAE) method [19], which is commonly used in the field of recommendation systems to evaluate the performance of algorithms. The expression is as follows:

$$\text{MAE} = \frac{1}{N} \sum_{i,j}^{N} \left| R_{ij} - \hat{R}_{ij} \right| \tag{5}$$

where N is the number of testing data samples and $\hat{R}_{ij}$ represents the prediction score.

We also verify the accuracy of our model, and we use the classic measure of precision for the X first films recommended (*precision @ X*). This is the proportion of relevant recommendations in the X first recommendations offered to each user.

Table 1 Comparison of the MAE values

Algorithm	Decision tree	Linear regression	Deep learning
With enrichment	1.2409	0.7760	0.8017
Without enrichment	0.9788	0.9264	0.9100

Table 2 Accuracy results

Algorithm	Decision tree	Linear regression	Deep learning
With enrichment	0.77	0.79	0.78
Without enrichment	0.75	0.73	0.74

4.3 Experimental Results and Discussions

We randomly selected 80% of the data of the dataset as the training set and the remaining 20% as the test set. We used the MAE measure method in this experiment (Tables 1 and 4).

In order to verify the efficiency and the accuracy of our model (Tables 1 and 2), we choose two recommendation algorithms to compare with ours, which are the linear regression and the decision tree. In addition, we test the model with and without enrichment of the profile.

By observing the experimental results of Tables 1 and 2, we notice that the results are different according to the algorithms; the linear regression gave the best result with profile enrichment and the decision tree without profile enrichment. The neural network model gave average results between the two previously mentioned algorithms. We observe that our results are different compared to the algorithms used, but are acceptable and the enrichment using the auto-encoder slightly improved the results of our model.

We retain the precisions of the first 1, 5, 10, 15 and 20 films rated P@1, P@5, P@10, P@15 and P@20, respectively.

As described in Table 3, our model achieves an accuracy of up to 84.48%.

Our SR generates 6364 recommendations for 58 users with 4889 relevant recommendations; a satisfaction rate of 76.82% of users is moderately important for an experiment.

Table 3 Precision@X results

Algorithm	P@1	P@5	P@10	P@15	P@20
With enrichment	84.48%	78.96%	72.93%	67.81%	63.70%

Table 4 Comparison of MAE values for the dataset

Model	MAE
ItemCF	1.091
UserCF	1.237
TAG-CF	1.042
ReDa	0.733
Zang-Alg	0.698
TRSDL	0.658
Our	0.802

To show the advantage of the proposed model, we compare it with six baselines. The baselines are: traditional algorithms based on collaborative filtering [16]: (i.e. ItemCF, UserCF and TAG-CF), models based on deep learning for recommendation (i.e. ReDA and Zang-alg), and tag-aware recommender based on deep learning [16] (i.e. TRSDL). We first describe the competing methods as follows:

- ItemCF: Item collaborative filtering.
- UserCF: user collaborative filtering.
- TAG-CF: This method integrates tag information to the collaborative filtering.
- ReDa [18]: Recommendation via dual-autoencoder.
- Zang-alg [17]: combines a collaborative filtering recommendation algorithm and a deep learning technology.
- TRSDL: a hybrid deep learning model for tag-aware recommender systems.

Table 4 depicts in detail prediction performances of our model and other six baselines on the MovieLens dataset in terms of MAE. Observing the experimental results, we have the following findings and discussions.

Table 4 demonstrates that, when comparing our model to ItemCF, UserCF and TAG-CF models, TAG-CF and our model perform better than ItemCF and UserCF in terms of indicator MAE. TAG-CF and our model utilize additional tag information in recommender systems, which can be regarded as the good reflections of users' preferences and items' characteristics. With the other models, our model does not give good results. The MAE of TRSDL is the best of 0.658, followed by Zang-alg, with an MAE of 0.693, then that of ReDA with 0.733. Our model integrated user tags but did not use collaborative filtering as in the other models. However, the results are not far behind and can be improved.

5 Conclusion

We have proposed a new method of recommending content. The method uses two deep neural networks, one to enrich the user profile and the other to predict the user score.

We have shown that using the tag as a new feature improves accuracy results.

Finally, the experimental results on the real dataset show that the proposed method can slightly improve the accuracy of the predictions. We believe that the lack of content (tags that the user uses) is the cause of the poor results.

As perspectives, we can improve our recommendation system by using a larger dataset that contains more tags and attributes. This development will lead to large file size of the training data, which demands the increase of the network depth and width. Also, we can increase the performance of the results by hybridizing this model with a recommendation approach like collaborative filtering.

References

1. Nakamoto, R., Nakajima, S., Miyazaki, J., Uemura, S.: Tag-based contextual collaborative filtering. IAENG Int. J. Comput. Sci. **34**, 214–219 (2007)
2. Karen, H.L., Tso-Sutter, K.H., Marinho, L.B., Schmidt-Thieme, L.: Tag-aware recommender systems by fusion of collaborative filtering algorithms. In: Proceedings of the 2008 ACM Symposium on Applied Computing, Fortaleza, Ceara, Brazil, 16–20, pp. 1995–1999 (2008)
3. Cattuto, C.: Semiotic dynamics and collaborative tagging. Proc. Natl. Acad. Sci. **104**, 1461–1464 (2007)
4. Markines, B., Cattuto, C., Menczer, F., Benz, D., Hotho, A., Stumme, G.: Evaluating similarity measures for emergent semantics of social tagging. In: 18th International World Wide Web Conference, pp. 641–641 (2009)
5. Specia, L., Motta, E.: Integrating folksonomies with the semantic web. In: 4th European Semantic Web Conference (2007)
6. Szomszor, M., Cattuto, C., Alani, H., O'Hara, K., Baldassarri, A., Loreto, V., Servedio, V.D.: Folksonomies, the semantic web, and movie recommendation. In: Proceedings of the 4th European Semantic Web Conference, Innsbruck, Australia, 3–7, pp. 25–29 (2007)
7. Wetzker, R., Umbrath, W., Said, A.: A hybrid approach to item recommendation in folksonomies. In: Proceedings of the WSDM'09 Workshop on Exploiting Semantic Annotations in Information Retrieval, Barcelona, Spain, 9 Feb 2009, pp. 25–29
8. Rendle, S., Balby Marinho, L., Nanopoulos, A., Schmidt-Thieme, L.: Learning optimal ranking with tensor factorization for tag recommendation. In: Proceedings of the 15th ACM SIGKDD International Conference on Knowledge Discovery and Data Mining, Paris, France, pp. 727–736 (2009)
9. Zhang, Z.K., Zhou, T., Zhang, Y.C.: Personalized recommendation via integrated diffusion on user–item–tag tripartite graphs. Phys. A Stat. Mech. Appl. **389**, 179–186 (2010)
10. Salakhutdinov, R., Mnih, A., Hinton, G.: Restricted Boltzmann machines for collaborative filtering. In: Proceedings of the 24th International Conference on Machine Learning, Corvalis, OR, USA, pp. 791–798 (2007)
11. Kim, D., Park, C., Oh, J., Lee, S., Yu, H.: Convolutional matrix factorization for document context-aware recommendation. In: Proceedings of the 10th ACM Conference on Recommender Systems, Boston, MA, USA, pp. 233–240 (2016)
12. Van den Oord, A., Dieleman, S., Schrauwen, B.: Deep content-based music recommendation. In: Proceedings of the Advances in Neural Information Processing Systems, Lake Tahoe, Nevada, pp. 2643–2651 (2013)
13. Cheng, H.T., Koc, L., Harmsen, J., Shaked, T., Chandra, T., Aradhye, H., Anderson, G., Corrado, G., Chai, W., Ispir, M., et al.: Wide & deep learning for recommender systems. In: Proceedings of the 1st Workshop on Deep Learning for Recommender Systems, Boston, MA, USA, pp. 7–10 (2016)

14. Guo, H., Tang, R., Ye, Y., Li, Z., He, X.: DeepFM: a factorization-machine based neural network for CTR prediction. arXiv 2017. arXiv:1703.04247 (2018)
15. Xu, Z., Chen, C., Lukasiewicz, T., Miao, Y., Meng, X.: Tag-aware personalized recommendation using a deep-semantic similarity model with negative sampling. In: Proceedings of the 25th ACM International on Conference on Information and Knowledge Management, Indianapolis, IN, USA (2016)
16. Liang, N., Zheng, H.-T., Chen, J.-Y., Sangaiah, A.K., Zhao, C.-Z.: TRSDL: tag-aware recommender system based on deep learning—intelligent computing systems (2018)
17. Zhang, L., Luo, T., Zhang, F., Wu, Y.: A recommendation model based on deep neural network. https://doi.org/10.1109/ACCESS.2018.2789866 (2018)
18. Zhuang, F., Zhang, Z., Qian, M., Shi, C., Xie, X., He, Q.: Representation learning via dual-autoencoder for recommendation. Neural Netw. **90**, 83–89 (2017)
19. Herlocker, J.L., Konstan, J.A., Terveen, L.G., Riedl, J.T.: Evaluating collaborative filtering recommender systems. ACM Trans. Inf. Syst. (2004)

Predictive Data Analysis Model for Employee Satisfaction Using ML Algorithms

Madara Pratt, Mohcine Boudhane, and Sarma Cakula

Abstract The digitalization has changed the way people accomplish their daily tasks, communicate with each other, and work. There are many benefits like greater flexibility in terms of place and time, and increase of speed the tasks can be performed. At the same time, technology-based communication can limit social relationships within the company. These changes require company leaders to be cautious about their employee wellbeing and satisfaction. It is not an easy task, and there is little research on how new technologies are affecting employees. The purpose of this research is to create an algorithm to predict the behavior of employees. Dataset of 102 people was used for the analysis. An algorithm was designed to help employers discover in which areas and ways the company should improve their work. The algorithm can predict which employees are likely to leave the company and for which reasons. At the same time, it can be used as a guide for technology developers to improve the quality of their communication technologies.

Keywords Machine learning algorithms · Employee satisfaction · Information and communications technologies (ICT) · Data prediction and analysis

1 Introduction

The access of information and communications technologies (ICT) has been increasing past decades. The International Telecommunication Union (ITU) statistics data shows that at the end of 2019, 56.6% of the global population is using Internet and 93% of the world population lives within a reach if a mobile Internet service [1]. Between 2005 and 2019, the number of Internet users has been growing on

M. Pratt (✉) · M. Boudhane · S. Cakula
Vidzeme University of Applied Sciences, Valmiera, Latvia
e-mail: madara.pratt@va.lv

M. Boudhane
e-mail: mohcine.boudhane@va.lv

S. Cakula
e-mail: sarma.cakula@va.lv

143

F. Saeed et al. (eds.), *Advances on Smart and Soft Computing*, Advances in Intelligent Systems and Computing 1188, https://doi.org/10.1007/978-981-15-6048-4_13

average 10% per year [1]. Technological development has changed the way people think, interact, and complete their tasks [2]. The speed of technology development has been incredible. This technological development and connectedness have created 24/7 work culture [3]. Although there is a lot of research on employee motivation and satisfaction, social human behavior science has not kept pace with the development of ICT. There is a lack of new models and solutions for guiding managerial practices in this new digitalized era [4, 5]. Nowadays since many companies excel in technology, it has been predicted that competitive advantage will be gained not only from quality of technologies but also from quality of their human resources [6].

Employee satisfaction is closely related with production and turnover intent [7–9]. Within this research, turnover intent is describing an employee's intention to leave the company in near future. The cost of hiring and training new employees can be very high. Satisfaction is the outcome of employee motivation, which is very complex construct. Employees are not motivated only by pay and benefits. These factors rather contribute for avoiding dissatisfaction. For this research, Herzberg's two-factor theory is used [10]. The hygiene (extrinsic) factors are assuring employees to feel normal. If they are under the desired level, it causes dissatisfaction. These factors include physical and working conditions, supervision, salary, job security, company policies, benefits, and others. These factors are expected to be optimal. Controversially, motivators (intrinsic) are factors that are the ones bringing actual satisfaction. Some of these factors are recognized for achievement, responsibility, work itself, advancement opportunities, and recognition for achievements. The levels of motivators and hygiene factors are predictors of satisfaction and in turns can be used in predicting turnover intent [11, 12].

The importance of these aspects and complexity of work-related problems within the new digital work setting require new solutions for solving these problems. Therefore, the authors have created an algorithm for predicting employee satisfaction and the possibility of leaving the company. Also, this algorithm is an aid to find the problem areas within the company giving guidance for company leaders. Another important use of the algorithm is for technology developers. It is highly recommended to acknowledge these findings when developing new ICTs.

2   General Context

According to ITU, between 2005 and 2019 the number of Internet users grew on average by 10% every year [1]. Nowadays, Internet is accessible almost everywhere. 93% of world's population lives within a reach of mobile Internet service and 53.6% of world's population use the Internet. The access of ICT has drastically transformed society [13]. The way we communicate and interact with each other has changed.

The integration of ICT in organization functions is necessary for increased efficiency, cost-effectiveness, and competitiveness [14, 15]. The satisfaction and well-being of employees are the responsibility of company management, but it has never been as complex as it is nowadays. The lack of understanding of how new communication formats are affecting social relations within the company can cause problems as miscommunication, dissatisfaction, lack of interpersonal relations, anonymity, and misunderstanding [16–20]. On controversy, companies who will have quality of both—technology and human resources—will have competitive advantage in future [4, 6].

The main attribute for evaluating the wellbeing of employees is satisfaction rate. Satisfaction is influenced by a complex array of personal and situational circumstances [21]. Positive workplace relationships and atmosphere of teamwork are important contributors for positive satisfaction measures [21]. The result of dissatisfaction can be lower productivity and in the worst-case higher increased turnover rate—employees leaving the company [7–9]. The cost of hiring and training employees can be very high, and it can result in other issues. Work satisfaction is a result of motivation and performance [22].

Motivation can be described as a person's internal psychological forces which move them to work [23]. Salary and benefits are not only motivator [10, 24]. Furthermore, it is not even motivation employees, but only maintaining employee neutral feeling. It has been discovered that employees are being motivated by being singled out, made feel important, appreciated, responsibility, development of self-respect, and respect of other [10, 22, 24]. Herzberg's two-factor theory divided motivational factors into motivators (intrinsic) and hygiene (extrinsic) factors [25]. Motivators are the factors, which bring actual satisfaction. These factors include recognition (REC), work itself (WRK), opportunity for advancement (ADV), professional growth opportunities (GTH), responsibility, good feeling about organization (GFO), and clarity of the mission (MIS). Hygiene factors are the ones, which are necessary for employees to dissatisfaction–avoidance. These factors are effective senior management (SMG), effective supervisor (SPV), good relationships with co-workers (CWR), satisfaction with salary (SAL), satisfaction with benefits (BEN), and presence of core values (VAL). Only when those factors are neutral and conditions are optimal, it is possible to get motivated by intrinsic factors [10, 25].

Nowadays, in the business world, the use of artificial intelligence (AI) technologies is essential. The potential of this technology could be used by organizations to solve many problems which were a bottleneck a few years ago. AI serves to proceed with automating tasks, to analyze and even predict futuristic facts. Efficiency and better management are the main benefits of the use of such technology.

Nowadays, companies and organizations pay on average 1/5 of a total employee's salary to replace that employee. The situation could be more difficult in case of a specific employee. For example, in case of high-paid employees, the cost is much higher than 1/5. The use of ML techniques to estimate employee's satisfaction and predict employee's turnover becomes a necessity.

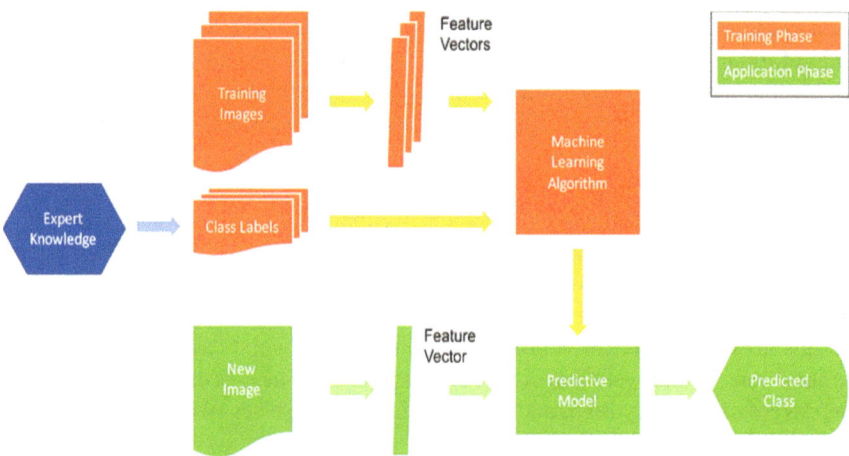

Fig. 1 Proposed prediction model process

3 Data Modeling Process

Prediction models allow businesses to make highly accurate predictions of some issues using previous or historical data (Fig. 1). The reason behind the proposed model in Fig. 1 is significant. The results could uncover factors that most contribute to those outcomes. In other words, it will allow us to find the most relevant features that affect strongly on the dataset. In our case, the proposed model aims to determine whether or not employees are satisfied, so this model can use prediction explanations to gain insight into employees' turnover.

The system is divided into two phases: training phase and application phase (Fig. 1). The first is made to train our data and define the relation between features according to historical data. The second is to predict if an employee (or a group of employees) would quit his/her (their) job. In this case, we used 20% of the dataset for this phase to check the algorithm accuracy. With that view, we develop a model that easily explain model outcomes to stakeholders and identify high-impact factors to help focus their business strategies and support and improve their decisions.

4 Experiments

Data used in this paper was provided by an anonymous company. The results acquired from a questionnaire were collected from their employees. Data should not define any sign that could disclose employee's identity. Besides the questions about demographics and the use of technology, the feature used in the dataset could be split into three groups:

- Motivation:
  - Intrinsic factors (REC, WRK, ADV, GTH, GFO, and MIS);
  - Extrinsic factors (SMG, SPV, CRW, SAL, BEN, and VAL);
- Satisfaction (SAT);
- Turnover intent (TRN).

In this phase, we exploit the data extracted by analyzing the satisfaction rate of each employee. For that, we will use machine learning algorithms in order to predict employee's motivation. The process is divided into three big parts:

1. Data processing;
2. Data training;
3. Data testing and prediction.

4.1 Data Processing

In this phase, we try to explore sort and analyze a vast amount of data extracted from the raw dataset. We are also selecting the most relevant data that could improve the prediction accuracy. In this phase, we will be able to:

- Deal with the missing values.
- Compare the different variables to each other, test correlation hypotheses (see Fig. 2). Figure 2 presents for illustration purposes only. It consists of an immense amount of data with many different findings.
- Propose several hypotheses about the causes underlying the generation of the dataset.
- Propose, if necessary, new sources of data that would help to understand the phenomenon better.

4.2 Data Training

We have used different statistical modeling approaches to train our data and solve the considered problem. In this phase, we propose four ML algorithms—support vector machine, logistic regression, random forest, and k-nearest neighborhoods.

Support Vector Machine (SVM) These large margin separators are a set of supervised learning techniques intended to solve problems of discrimination and regression. SVMs are a generalization of linear classifiers. SVM has been applied to a large number of fields. According to the data, the performance of support vector machines is in the same order, or even better, than that of a neural network or a Gaussian mixture model [26].

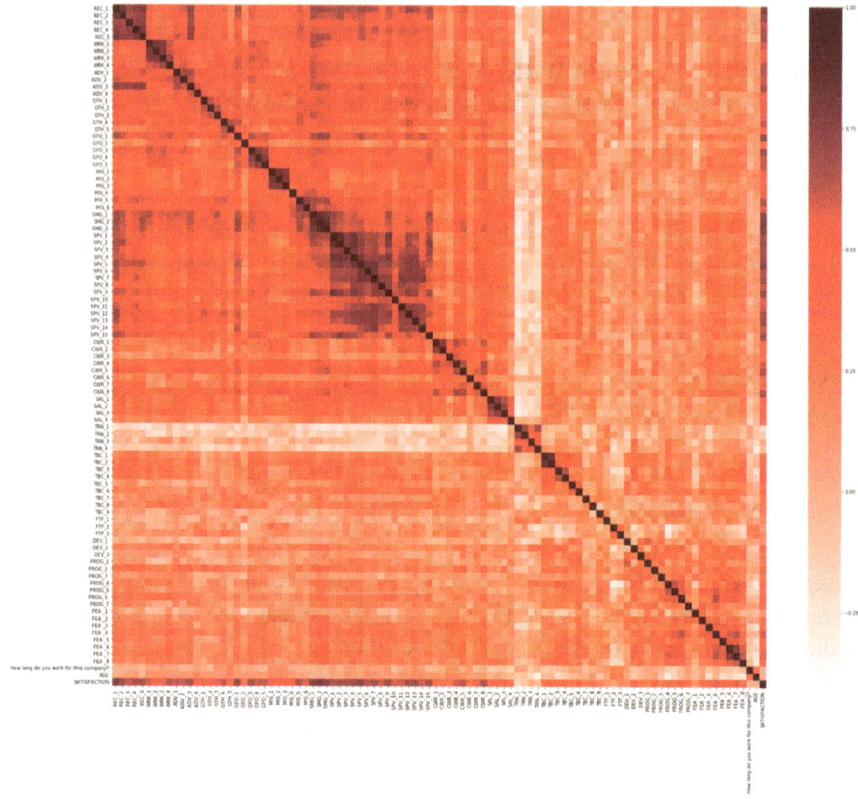

Fig. 2 Correlation between all attributes presented on the dataset. In white: the attributes that are not correlated. In red: the degree of correlation with the other attributes

Logistic Regression It is one of the most commonly used multivariate analysis models. It is a predictive technique. It aims to build a model allowing to predict/explain the values taken by a qualitative target variable (most often binary, we then speak of binary logistic regression; if it has more than two modalities, we talk about polyatomic logistic regression) from a set of quantitative or qualitative explanatory variables [27].

Random Forest It is a part of machine learning techniques. This algorithm combines the concepts of random subspaces and bagging. The decision tree forest algorithm performs training on multiple decision trees trained on slightly different subsets of data [28].

K-Nearest Neighborhoods (KNN) In artificial intelligence, the k-nearest neighbor method is a supervised learning method: abbreviated kNN or KNN. In this context, we have a learning database made up of N "input–output" pairs. To estimate the output associated with a new input x, the KNN method takes into account (identically) the

k training samples whose input is closest to the new input x, according to a distance to be defined [29].

4.3 Data Testing and Prediction

Figure 3 presents a comparative study of the different ML algorithms used in this work. Based on this comparison analysis, the logistic regression and random forest algorithms show the highest mean AUC scores. To analyze those algorithm performances, we will compare the confusion matrix provided by both methods. This last describes not only satisfaction prediction, but it will also extract the relevant features that are responsible for employee's turnover.

We have made a comparative study between the confusion matrixes obtained in the test phase (see Fig. 4): on one left—the random forest algorithm with an accuracy of 95.24% and on the other right, the logistic regression classifier on the test set with an accuracy of 80.95.

Random forest fits a number of decision tree classifiers on various subsamples of the dataset and uses averaging to improve the predictive accuracy and control over-fitting. Random forest can handle a large number of features and is helpful for estimating which of your variables are important in the underlying data being modeled.

As a result, we conclude the most relevant features that are responsible to the employee's satisfaction according to our dataset. Some of the most important are visualized in Fig. 5 by their order of importance (to the left—more influential). In the x-axis: the features by order of importance, in the y-axis: the degree of importance (/0.1).

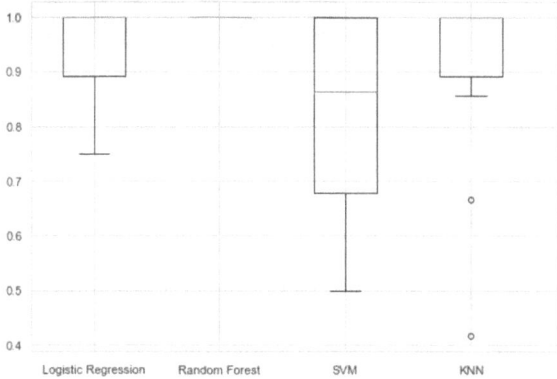

Fig. 3 Box plot of the processed data. As shown in the figure random forest and logistic regression had the best performance

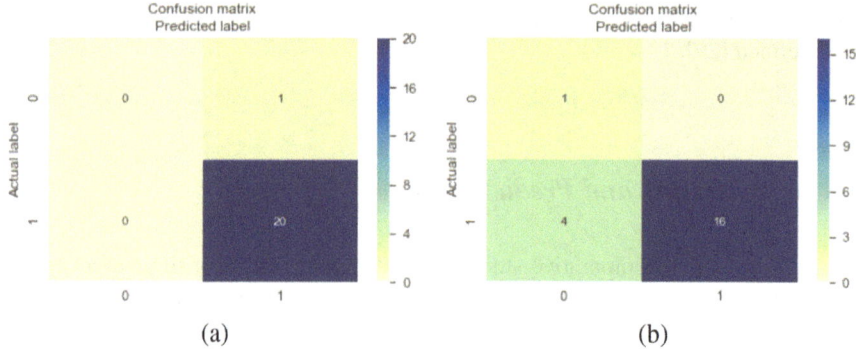

Fig. 4 Confusion matrix prediction label concerning the machine learning algorithms **a** random forest, and **b** logistic regression

Fig. 5 Some of the most relevant features presented by the random forest algorithm

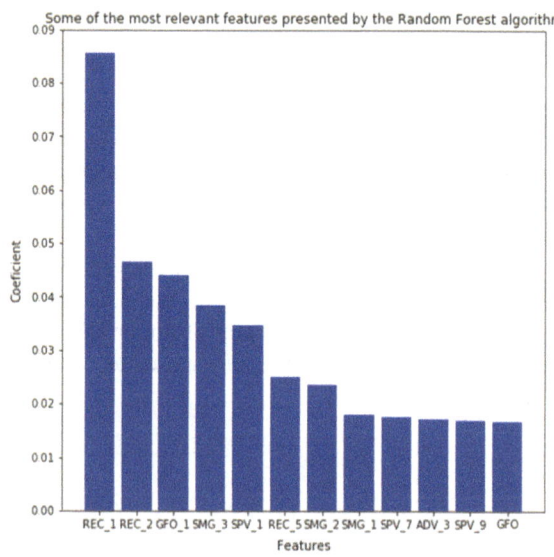

As a summary of Fig. 5, we distinguish the top ten relevant features in our datasets in Table 1.

The most relevant features for employee satisfaction extracted using random forest algorithm are associated with recognition (REC), good feelings about organization (GFO), effective senior management (SMG), and effective supervisor (SPV). Interestingly, all of these are intrinsic factors confirming that pay and benefits are not enough to have satisfied and long-lasting employees.

Table 1 Most relevant features extracted by random forest algorithm

Feature	Coefficient
REC_1	0.085641
REC_2	0.046565
GFO_1	0.044058
SMG_3	0.038517
SPV_1	0.034761
REC_5	0.024978
SMG_2	0.023529
SMG_1	0.018062
SPV_7	0.017657
ADV_3	0.017084

5 Conclusions

In this paper, we presented a predictive data analysis model for employee satisfaction using ML algorithms. It illustrates the groundwork in order to create a predictive system that can automatically estimate employees that are likely to quit their jobs. The algorithm can also determine several parameters responsible for the employee's satisfaction. For this purpose, we have proposed a comparative study to different machine learning algorithms in order to select the best estimator. The data was collected from a company and includes more than 100 employees. Experimental results have shown that random forest was the best estimator. The obtained maximal accuracy of the prediction of using 20% of the data is 95.24%.

Along with the accumulation of data and experience, getting ample experiment data and finding more effective data-patching algorithm, we believe the algorithm will be more precise. It may make sense in minimizing the total risk of job abandonment of the employees. For future work, we will apply the algorithm on other databases taking into consideration more motivational factors.

Acknowledgements Support for effective participation of Vidzeme University of Applied Sciences within the international scientific circles (ViA-Int), project number 1.1.1.5/18/I/005.

References

1. ITU: Measuring Digital Development (2019)
2. Marius, M.: 6 ways technology is changing the way we do business. Bus. Caribbean Comput. ICT/Tech. List. Netw. (2012)
3. Piazza, C.F.: 24/7 workplace connectivity: a hidden ethical dilemma. Bus. Organ. Ethics Partnersh. Markkula Cent. Appl. Ethics, St. Clara Univ., St. Clara, CA. **21**, 2014 (2007)
4. Steers, M.L.N., Wickham, R.E., Acitelli, L.K.: Seeing everyone else's highlight reels: how Facebook usage is linked to depressive symptoms. J. Soc. Clin. Psychol. **33**(8), 701–731 (2014)

5. De Wet, W., Koekemoer, E., Nel, J.A.: Exploring the impact of information and communication technology on employees' work and personal lives. SA J. Ind. Psychol. **42**(1), 1–11 (2016)
6. Thurow, L.C.: Who owns the twenty-first century? Sloan Manage. Rev. **33**(3) (1992)
7. Coomber, B., Louise Barriball, K.: Impact of job satisfaction components on intent to leave and turnover for hospital-based nurses: a review of the research literature. Int. J. Nurs. Stud. **44**(2), 297–314 (2007)
8. Rusbult, C.E., Farrell, D.: A longitudinal test of the investment model: the impact on job satisfaction, job commitment, and turnover of variations in rewards, costs, alternatives, and investments. J. Appl. Psychol. **68**(3), 429–438 (1983)
9. Yücel, İ.: Examining the relationships among job satisfaction, organizational commitment, and turnover intention: an empirical study. Int. J. Bus. Manag. **7**(20) (2012)
10. Herzberg, F., Mausner, B., Barbara, S.: The Motivation to Work. Wiley, New York (1959)
11. Smerek, R.E., Peterson, M.: Examining Herzberg's theory: improving job satisfaction among non-academic employees at a university. Res. High. Educ. **48**(2), 229–250 (2006)
12. Ahmed, S., Taskin, N., Pauleen, D.J., Parker, J.: Motivating information technology professionals: the case of New Zealand. Australas. J. Inf. Syst. **21** (2017)
13. Schmidt, A., Pfleging, B., Alt, F., Sahami, A., Fitzpatrick, G.: Interacting with 21st-century computers. IEEE Pervas. Comput. **11**(1), 22–30 (2012)
14. Tusubira, F.F., Mulira, N.: Integration of ICT in organisations: challenges and best practice recommendations based on the experience of Makerere University and other organizations. Presented to Conference on Universities: Taking a Leading Role in ICT Enabled Human Development (2004)
15. Saleem, A., Sadik Batcha, M., Tabusum, S.S.Z.: Impact of ICT in information access pattern among professional and non-professional students in Chennai: a case study (2015)
16. Hartman, K., et al.: Patterns of social interaction and learning to write: some effects of network technologies. Writ. Commun. **8**(1), 79–113 (1991)
17. Levi, D.: Group Dynamics for Teams, 5th edn. Sage (2017)
18. Driskell, J.E., Radtke, P.H., Salas, E.: Virtual teams: effects of technological mediation on team performance. Gr. Dyn. Theory Res. Pract. **7**(4), 297–323 (2003)
19. Levi, D.: Group Dynamics for Teams. Sage (2014)
20. Vignovic, J.A., Thompson, L.F.: Computer-mediated cross-cultural collaboration: attributing communication errors to the person versus the situation. J. Appl. Psychol. **95**(2), 265 (2010)
21. Volkwein, J.F., Zhou, Y.: Testing a model of administrative job satisfaction. Res. High. Educ. **44**(2), 149–171 (2003)
22. Whetten, D.A., Cameron, K.S.: Developing Management Skills, pp. 377–441. Prentice Hall (2011)
23. Renge, V.: Organizāciju psihologija. Birznieka SIA "Kamene", Riga (1999)
24. Wren, D.A., Bedeian, A.G.: The Evolution of Management Thought. Wiley (1994)
25. Herzberg, F.: One more time: how do you motivate employees? Harv. Bus. Rev. **81**(1), 87–96 (2003)
26. Yang, Y., Wang, J., Yang, Y.: Improving SVM classifier with prior knowledge in microcalcification detection. In: 2012 19th IEEE International Conference on Image Processing, pp. 2837–2840 (2012)
27. Gutiérrez, P.A., Hervás-Martínez, C., Martínez-Estudillo, F.J.: Logistic regression by means of evolutionary radial basis function neural networks. IEEE Trans. Neural Netw. **22**(2), 246–263 (2010)
28. Liu, Y., Liu, L., Gao, Y., Yang, L.: An improved random forest algorithm based on attribute compatibility. In: 2019 IEEE 3rd Information Technology, Networking, Electronic and Automation Control Conference (ITNEC), pp. 2558–2561 (2019)
29. Lu, S., Tong, W., Chen, Z.: Implementation of the KNN algorithm based on Hadoop (2015)

A Novel Approach for Big Data Monetization as a Service

Abou Zakaria Faroukhi, Imane El Alaoui, Youssef Gahi, and Aouatif Amine

Abstract As a result of the digital transformation that has led by different production sectors, the organizations were flooded by vast, various, and complex data, called Big Data. To manage their data assets, the adoption of the Big Data value chain (BDVC) was suitable for a value realization as well as a data-intensive decision-making. The manipulation and exploitation of these data gave rise to the concept of Big Data monetization. The marriage between the BDVC and data monetization remains an own and recent approach. This combination allows the organization's process to become entirely data-driven and expand the scope of its exchanges. Besides, advanced deployment models such as cloud computing will enable this combination to become more resilient by providing large amounts of computational and networking capabilities. Furthermore, it allows sharing data between different BDVCs to build an expandable ecosystem. In this contribution, we review the literature regarding the association between BDVC, data monetization, and cloud computing. Then, we propose a Big Data monetization-driven value chain model that relies on cloud computing and analytical capabilities. It allows monetizing both data and insight as a service under a collaborative concept.

Keywords Big Data value chain · Big Data monetization as a service · Big Data ecosystem · Cloud computing

A. Z. Faroukhi · Y. Gahi (✉) · A. Amine
Laboratoire Génie Des Systèmes, Ecole Nationale Des Sciences Appliquées, Ibn Tofail University, Kenitra, Morocco
e-mail: gahi.youssef@uit.ac.ma

A. Z. Faroukhi
e-mail: zakaria.faroukhi@gmail.com

A. Amine
e-mail: amine.aouatif@uit.ac.ma

I. El Alaoui
Laboratoire Des Systèmes de Télécommunications Et Ingénierie de La Décision, Ibn Tofail University, Kenitra, Morocco
e-mail: elalaoui91@gmail.com

© The Editor(s) (if applicable) and The Author(s), under exclusive license to Springer Nature Singapore Pte Ltd. 2021
F. Saeed et al. (eds.), *Advances on Smart and Soft Computing*, Advances in Intelligent Systems and Computing 1188, https://doi.org/10.1007/978-981-15-6048-4_14

153

1   Introduction

The digital transformation in the world has led to an unprecedented flood of data that organizations were not ready and able to manage or benefit from it. Massive data streams coming from various systems and sources are vast, varied, and complex (e.g., social media and Internet of things (IoT)), making them difficult to handle using conventional analysis methods. Also, data processing infrastructure capabilities and traditional database systems have become outdated [1], which led to the emergence of Big Data.

Big Data has been defined originally as high-volume, high-velocity, and high-variety of data, commonly known as 3V features, later extended to 7Vs (volume, variety, velocity, veracity, variability, visualization, and value), widely described in the literature [2, 3].

These Big Data characteristics bring many challenges to organizations, as they had to change the way the data are analyzed and how they manage processes. They have brought positive sides as well, whereas organizations with a large amount of data gain access to a better understanding of insights, discover new values, improve processes, and stimulate innovation [4].

Thus, the use of Big Data value chain (BDVC) is recommended, even necessary, to quickly go through unified data-driven processes that accomplish an end-to-end digital organization. Beyond the fact that value chains and data value chains are the basic models for data-driven processes, the BDVC enables to process information flows by considering all Big Data aspects such as the 7Vs, the specific architecture as well as the nature of the deployed use cases. BDVC consists of a succession of steps that generate meaningful values aiming to improve organizations' data assets [5–7].

The BDVC outputs are a goldmine of knowledge that deserves to be monetized to maximize business revenue, optimize the quality of processes, and make them entirely data-driven. The BDVC enables data monetization in various ways, directly (factual data), by selling it to third parties and partners, or indirectly (intangible data) through some exchange policies.

This way of dealing with data requires access to remote and distributed resources over the Internet, the adoption of specific services and platforms, as well as aggregation of data through powerful computational capabilities. This need arises the concept of building novel data management strategies relying on distributed architectures such as cloud computing. This kind of architectures is an efficient model for the deployment of data monetization as a service. It also enables BDVC to increase its processing efficiency and extend its influence by building an expandable ecosystem.

Although business process management has become almost full data-driven in the era of Big Data, the monetization of this data is not yet natively integrated into the value generation process. The interest of our contribution is to propose a model that includes data monetization within Big Data value chain using cloud computing capabilities. It should be mentioned that, to the best of our knowledge, there are no contributions that address this association.

The rest of this paper is organized as follows. The next section surveys the related work regarding the association between BDVCs, cloud computing, and data monetization. In Sect. 3, we explore BDVC, including the data monetization context. Section 4 shows an adapted configuration to monetize data as a service throughout BDVC based on cloud computing architecture. Finally, Sect. 5 concludes the paper and shows some research outlooks.

2 Related Work

In this section, we review the latest research contributions regarding the association of BDVC, cloud computing, and data monetization.

Although the combination of BDVC and data monetization is a suitable approach to build data-driven entities, few contributions have been oriented to this research area. Even less, those that considered the cloud to handle an end-to-end workflow. Next, we explore some contributions that focused on BDVC and data monetization.

According to Miller and Mork, the BDVC is a model based on continuity of services through discovery, integration, and data exploitation to achieve decision-making. All stakeholders in this chain collaborate to create value [6]. Furthermore, authors in [5] have proposed a BDVC model to enable value creation and generate useful information from Big Data systems. The proposed model is used to manage the flow of information's activity within Big Data ecosystems. For a higher IT business value, the authors in [8] have suggested that the adopted Big Data models should be designed and mainly developed around data since it is the essential source for generating business value. Businesses must first identify the Big Data strategy at every step throughout the knowledge value chain. The authors in [9] have proposed another BDVC concept; they have presented an approach to managing quality in a hybrid way based on data-driven and process-oriented strategies. Data-driven quality aims to improve data management, especially storage and processing, while process-driven quality seeks to provide preprocessing quality by using filtering and cleaning techniques. The approach also defines requirements for evaluating the specifications' quality, process, and metrics. Visconti et al. [7] have considered that the Big Data-driven value chain is the pillar of value co-creation. Their chain includes a series of sequential, agile, and flexible steps that lead to a final monetization step. They also designed a BDVC with many participants contributing to create an added as part of a global strategy.

Increasingly assimilated to productivity and agility, the cloud has become a real alternative to the traditional approach of value chain built and managed internally. Thus, the cloud constitutes support to promote the adoption of BDVC models. However, few contributions have considered this aspect in the BDVC context.

In [10], Kasim et al. have proposed a framework for data value chain as a service (DVCaaS) to manage the acquisition, integration, management, and sharing related to specific business flow. The framework integrates components to manage data in a service-oriented way by providing elastic and scalable cloud assets. According to the authors in [1], a Big Data pipeline process requires establishing communication tunnels via business entities to reduce the complexity inherent in large amounts of data and a variety of morphologies. This integration through a Big Data framework based on cloud services allows firms to reach and achieve their strategic objectives. Also, the authors in [4] have proposed an architecture for Big Data ecosystem that is based on cloud computing. It ensures the deployment and management of infrastructure and analytics. The model employs a collaborative on-demand infrastructure service. It also supports a workflow for data collection, storage, processing, and visualization. Following the same path, Sharma has inspected in [11] the intrusion of cloud computing technologies into the Big Data ecosystem. He has presented a classification named XCLOUDX that can support Big Data models. He also listed a broad list of emerging as a service platforms, hosted by cloud computing.

We note that most of the available BDVC models focus on data processes improvement and value creation. However, data monetization has not been treated sufficiently. In what follows, we show some works that addressed data monetization.

By tracing the journey of an American retailer, Najjar and Kettinger have advocated that organizations seeking to monetize data must first check that is aligned with the objectives. The authors propose that every enterprise should think about sharing their information with their suppliers to improve their analytical capabilities. They also have suggested several ways to monetize data to enhance analytical skills [12]. The authors in [13] have provided a unified analysis data monetization framework for telecommunications. This framework relies on several functions that process and analyze various data extracted from customer databases, location-based services, and social interactions. It allows deducing similar user communities that have the same characteristics, which could be a potential target of marketing strategies. The contribution in [14] has exposed a personal data monetization platform via a broker that manages the economic data asset and shares transaction activities. Suppliers and consumers can either produce or consume transactions that offer fair pricing, depending on various criteria (e.g., data types, quality, economic). In [15], authors have discussed the data monetization as value creation and its related data management capabilities. According to the authors, businesses could generate additional benefits either via external insight selling or via internal operations improvement. In this regard, they have proposed a framework to improve data assets and identify data monetization opportunities.

It is worth mentioning that Big Data monetization has not attracted the attention of researchers and academics despite its obvious importance.

In the next section, we present the global environment of data monetization to highlight a BDVC model dedicated to this purpose.

3 BDVC Toward Data Monetization

With the global digital transformation, organizations' processes have adapted always facing so abundant data coming from outside rather than inside. To face these data-led transformations, organizations must effectively manage and monetize their valuable data assets following suitable business models. In this regard, this section will propose an adequate framework for BDVC to host this monetization.

3.1 Big Data Monetization

Data monetization is not a new trend. With the expansion of the Internet and IT platforms in the early 2000s, firms using business intelligence (BI) and their related processes were able to monetize data, but in old fashion ways. It was relying on internal data providing intangible value products (e.g., statistics, analytical results, indicators, and trends). With the emergence of Big Data, monetization has inherited the complexity but also the richness of Big Data characteristics. This volume and variety of data offer numerous possibilities for processing and transformation to finely analyze the performance of the company. It also provides an opportunity to launch new activities and generate incomes by considering several data formats such as raw, processed, aggregated, or at the most excellent state. It also includes information enhanced with indicators from BI processes and forecasts from data intelligence and machine learning models. A large number of companies position themselves as data provider, data broker, or even data consumer. Data exchanges, data sales, and data purchases are now a concrete way to use valuable datasets.

It is thus necessary to adopt a data-driven pattern to follow up the data collection, data processing, and data sharing by carefully considering Big Data monetization in each step separately [15–18]. Data monetization is often characterized by (i) data consumers who are entities that rely on shared data such as customers' activities, customer relationship management platforms, and products' feedback. Based on that data, they build their business models and enable specific data-driven decision-making. (ii) Data suppliers: refer to data brokerage firms that are engaged in data collection and data sharing, which are either specialized entities in this field or partners belonging to the same group. (iii) Data aggregators are data mining systems that spread aggregated business information. They collect and share business data with a multitude of sources, including search engines. (iv) Data facilitators: correspond to technology platforms that use various tools for data collection, preprocessing, storage, and analysis, to realize more informed decisions. (v) Data regulators: refer to organizations that recommend and ensure the privacy and ethical use of data. Also, they define standard technologies and standardization of data transfers.

It is essential to mention that an excellent Big Data monetization mindset relies on the implementation of a data-driven strategy within the company. This strategy mainly depends on a reliable BDVC, allowing them to go through every step in the data processes.

Thus, BDVC framework adoption remains adequate to Big Data monetization, primarily that it can support business processes, becoming more and more digital and completely data-oriented, and able to integrate external actors like customers.

3.2 Big Data Monetization-Driven Value Chains

On the one hand, the value chain natively enables organizations to model and plan their activities to create value and achieve strategic goals. Form the other side, the Big Data monetization-driven value chain refers to the lifecycle and sequence of successive steps that manipulate data relating to a set of activities. Several phases, starting from raw data, gradually accumulate the value shared with different contributors to co-create tangible or intangible products to monetize.

Our proposed model enables to handle and process data in different chain levels, according to a certain degree of data quality and maturity. These data can be monetized either directly (raw data or information goods) or indirectly (value of data, result or conclusion) monetization [15, 16, 18].

The model, as illustrated in Fig. 1, presents various steps that form a global BDVC and could provide monetized data. These steps are data generation, data acquisition, data preprocessing, data storage, data analysis, data visualization, data sharing, insight sharing, and data monetization. Next, we describe every step:

Data Generation Refers to all data generated by internal and external sources such as internal databases, CRM, social media, IoT sensors, mobile device applications, and business applications. It is either structured, semi-structured, or unstructured.

Data Acquisition Consists of identifying and collecting varied data from many sources with different formats. These data are then transferred and stored into datacenters as raw as it is while respecting specific layers such as temporary, share, internal, and confidential. These layers are subjects of a preprocessing phase.

Data Preprocessing Data are often incomplete, lacking attribute values, lacking specific attributes of interest, or containing only aggregate data. Furthermore, data are most of the time either noisy, including errors and outliers, or inconsistent with discrepancies in specific values. For these reasons, data preprocessing consists of cleaning, completing, and sometimes transforming incomplete and noisy raw data, having to provide understandable content.

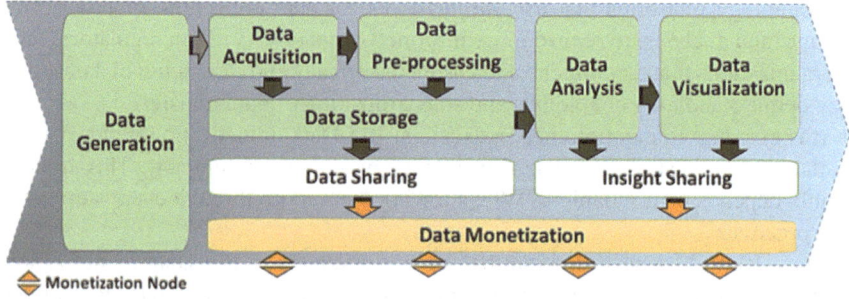

Fig. 1 BDVC model based on various steps to provide monetized data

Data Storage It is essential to underline that the data storage phase consists of storing both raw data, which is gathered during the acquisition phase, but also the preprocessed output. It aims to persist data as many datasets, according to different contexts, over many data centers. The storage system provides high availability and ensures reliability and fault tolerance by using various storage models: full columns, key-value, documents, graph, binary format (Avro, ORC, or Parquet), and tuple storage.

Data Analysis Refers to data processing and exploration by using substantial computational resources, advanced processing techniques, and statistical methods to extract useful insights and to find hidden patterns. The various analyses could be applied, such as descriptive, predictive, and prescriptive ones that could be achieved by relying on data mining, machine learning, and deep learning approaches.

Data Visualization Refers to the graphical representation using visual elements and tools such as graphs, maps, alerts, monitoring, reports, and dashboards. Data visualization makes it possible to see and understand significant trends and hidden values in a very accessible way. In the world of Big Data, data visualization tools and technologies are essential for analyzing enormous volumes for better decision-making.

Data Sharing Refers to sharing data either in its raw format or the preprocessed one between different entities operating within the same perimeter. Data sharing can be used for several purposes and is sometimes carried out in the form of a partnership between two companies that offer complementary but non-competitive products or services. Data sharing requires rigorous supervision to avoid any risk of data leakage.

Insight Sharing Refers to the sharing of processed data leading to chunks of information that are discreet and easily understandable. In this level, insights are crafted as small homogeneous information that is quickly interpreted. Organizations could then pollinate those chunks across widely dispersed entities and collaborators so that more collective actions can be taken on the data.

Data Monetization This phase consists of sharing data under different formats: raw, preprocessed, or thoroughly analyzed. Data monetization should rely on specific platforms with dedicated secure interfaces to ensure an efficient and controlled partnership.

It is essential to mention that after having invested significantly in collecting, storing, and processing Big Data by setting up sophisticated BDVCs, organizations would prefer to ensure a return on investment. Data monetization is one of several ways to increase the return on the Big Data investments and ensure the profitability of Big Data management strategies. However, setting up dedicated BDVC with such monetization platforms requires extensive computational and analytical capabilities. Also, the continuous growth of monetization platforms and its principle of controlled openness need scalable technology. Furthermore, advanced concepts of monetization consist of connecting multiple BDVCs to form vast collaborative ecosystems.

Considering the complexity of all these challenges, cloud computing remains a suitable solution to provide an extensive, robust, scalable, and elastic ecosystem to host Big Data monetization with all its constraints.

In the next section, we propose a novel approach to setup BDVC and Big Data monetization based on the cloud topology. Then, we show how this model quickly leads to Big Data monetization as a service concept.

4 Big Data Monetization and Cloud Computing

The BDVC model enables monetization data at different levels according to their maturity. It can offer several sharing stages to expose that data in either tangible or intangible ways, depending on the organizations' data strategies and partners' needs. This section describes the potential of relying on cloud computing technologies, exposition layers, and network capabilities to provide secure and scalable data monetization services.

4.1 Big Data Monetization as a Service

Monetizing data can be performed directly by selling or sharing data or indirectly by creating new ideas and business lines [12]. Monetization mainly relies on different steps, along with a dedicated BDVC. This latter requires a set of tools and techniques for processing and analytics capabilities that are distributed and virtualized. So, to realize Big Data monetization, organizations have to consider some challenges such as easy deployment, scalability, exposition, networking, and enormous resources, and above all, different levels of interactions such as public, private, and community.

According to the National Institute of Standards and Technology (NIST), "cloud computing is a model for enabling ubiquitous, convenient, and on-demand network access to a shared pool of configurable computing resources (e.g., networks, servers, storage, applications, and services)" [19]. Cloud computing infrastructure is rapidly provisioned with minimal management effort or service provider interaction.

Cloud computing is based on three service layers: Software as a service, platforms as a service, and infrastructure as a service to deploy scalable solutions. It can be implemented as public, private, hybrid, or community.

The main reason for which the cloud is the ideal solution for monetization is that BDVC can be adapted to monetize data through several nodes. These nodes correspond to several maturity levels that providing multi-level monetization services.

In the context of BDVC and Big Data monetization, the cloud computing topology could be used in the following way. The layer IaaS could be used to set up the BDVC infrastructure, such as nodes provisioning as well as network and communication topologies with third parties, which could be entities, partners, or other BDVCs.

The PaaS could be used to host all Big Data tools and analytics platforms required by BDVC while taking advantage of the distributivity and the colossal processing capabilities of the cloud. The IaaS could be used to create interfaces and expose data to other parties. On the other hand, the BDVC phases are deployed as follows: the data generated could be hosted by public, private, or hybrid clouds. The private cloud is more suitable for data acquisition, data preprocessing, data analysis, data visualization, data sharing, and insight sharing to protect the core of the BDVC. Finally, the data monetization layer is used to set up sharing policies for either internal or external communities, for so the cloud is the more appropriate. Figure 2 shows the distribution of BDVCs phases by the cloud platform.

It is worth noting that the cloud performance and elasticity are with significant added value for each phase in the BDVC, and especially for the Big Data monetization in terms of exposition and access management. In this regard, it can provide distributed, virtualized and secure storage, processing, and kind of services that enable a full collaboration platform making Big Data management more efficient and the monetization chain more resilient. Moreover, the cloud represents a suitable platform to go beyond traditional hosting and classic sharing and propose (XaaS) for different phases in the BDVC as follows:

- Raw data as a service refers to unprocessed data that is delivered on-demand through APIs to consumers.
- Preprocessing data as a service refers to virtual instances of data curation that can be called as cloud services.

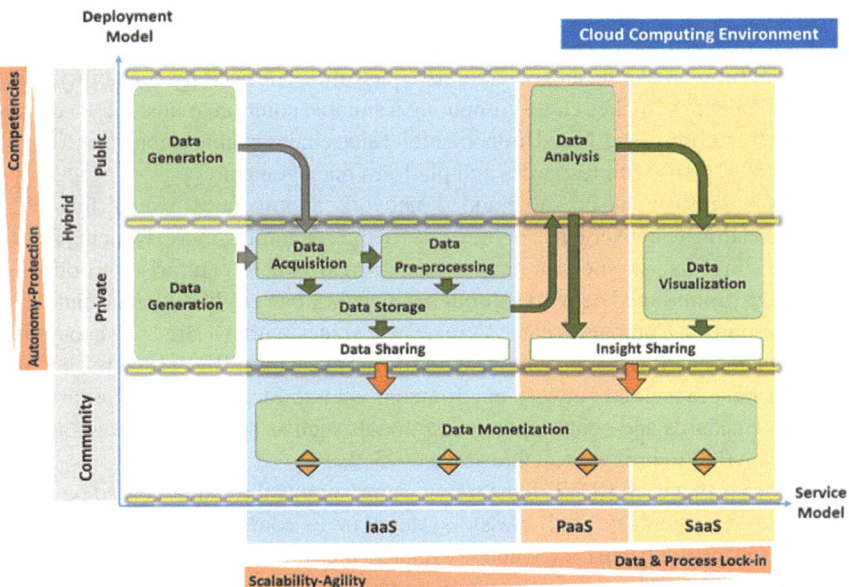

Fig. 2 Big Data monetization as a service under a cloud computing topology

- Storage as a service enables data storage by using dedicated storage architectures and platforms. They allow optimization computing cost, manage, and consume both raw and preprocessed data according to access rules and data sensitivity.
- Analytics as a service refers to Big Data analytics platforms in cloud, which enable to provide data analytics services through customized user interfaces.
- Visualization as a service enables cloud service providers to assert visualization demands at different levels such as platform and application.
- Insight as a service can harvest Big Data by analyzing massive raw and preprocessed data. Insight as a service then proceeds to the analysis of this data through business expertise, which could be customized depending on the context. It interprets the data to transform it into levers that can be explicitly exploited.

All that we have just presented as an added value of the cloud in the context of the BDVC and data monetization makes setting up a BDVC easy and practical choice to track data and to collaborate effectively with partners. But what makes the cloud an efficient choice is its ability to go beyond the borders of an organization and allow several entities to form a single ecosystem by connecting their BDVC. This connection is possible due to cloud networking capabilities and exposition elasticity.

4.2 Big Data Value Networks

The Big Data value network is an ecosystem that allows organizations to work and collaborate. It will enable them to exchange resources, technologies, data, and information to increase their benefits. This ecosystem could use a set of interconnected BDVCs, forming a secure, and collaborative platform. The need for scalability, elasticity, and openness makes cloud computing a suitable solution to host these ecosystems. Furthermore, cloud-based monetization value chains enable enterprises to align their business goals with resources and platform management.

BDVCs deployment in value-added network architectures would link VCs together, leading to convergence of their knowledge, becoming more resilient [7, 20].

Monetizing data in such an environment would enable enterprises to develop appropriate professional networks through which they can easily exchange and share critical resources. Companies can co-create new values within a Big Data ecosystem that no single enterprise can achieve alone [21]. Therefore, BDVCs can share and exchange data in a cloud network environment via interfaces of monetization using metadata standards and communication protocols such as application programming interfaces(APIs) to provide suitable services [5, 7, 10].

The idea is that each BDVC must provide entry points throughout based on APIs, whether to communicate with classic systems or to connect with other BDVCs. These interfaces allow BDVCs and its related platforms to consume and expose services at different levels, in an automatic way, according to data value maturity (e.g., raw, preprocessed, dataset, analysis result, and visualization). The BDVC can thus exchange data in the same ecosystem or interact with several heterogeneous environments.

On the other hand, interconnected BDVCs might be formed around a specific business or by converging close or independent trades. Sectors such as finance, health, pharmacy, logistics, and media have established their data ecosystems. In Fig. 3, we show an example of multiple interconnected internal and external BDVCs where the ecosystem "A" presents a banking platform in which different BDVCs interact with each other. The ecosystem "B" represents the insurance platform with its BDVCs that share all kinds of exposition and analysis capabilities. The important thing is these two ecosystems can exchange and expose their data depending on some collaboration policies. This ability to transfer data across multiple branches and interfaces allows monetizing this data in either raw format or as insight between different platforms with less complexity since the cloud offers necessary resources. This multitude of flows leads to what we could call a monetization grid.

The exposition and data sharing over this grid play an excellent basis for cross-disciplinary data monetization. Partners of the same ecosystem, with similar or complementary activities, have to combine their analytical capabilities to produce a global monetization scheme. However, building a reliable and cost-effective ecosystem requires the right tools for data governance as well as the establishment of regulatory agencies to address legal and data privacy issues.

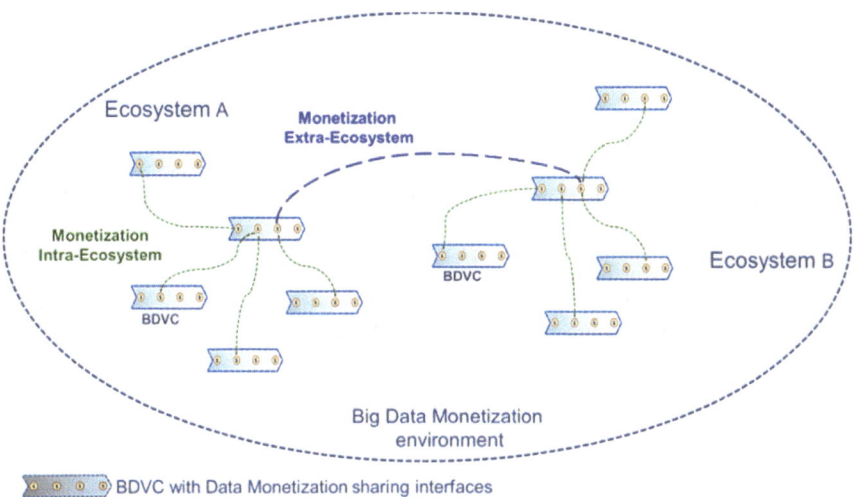

Fig. 3 Big Data monetization for interconnected and expandable ecosystems

5   Conclusion

The Big Data monetization is an attractive research area, especially when it is linked to BDVC. This research field remains promising as there are many challenges to be considered, such as Big Data management, global and efficient BDVC, as well as controlled monetization strategies. But the most critical problem is to manage the BDVC ecosystems in terms of scalability, sharing performances, and expandable collaborations. In this paper, we tackled monetization throughout BDVC and its related challenges. We firstly presented an overview regarding existing research on value chains, monetization, and cloud computing combined. Then, we proposed a generic model of Big Data monetization-driven value chain that aims to share both data and insight at different levels of data maturity.

Moreover, we presented how it is possible to rely on cloud capabilities in terms of scalability, elasticity, and flexible exposition to set up a global system of interconnected BDVCs. This ecosystem natively enables efficient collaboration by sharing both data and insights. Finally, we show how such systems can easily lead to Big Data monetization as a service, which is an advanced sharing concept. The future directions consist of challenging and detailing more aspects of combining cloud computing and Big Data monetization throughout BDVC. One of the most attractive elements that we are working on is data privacy and exchanges' security policies.

References

1. Schmidt, R., Mohring, M.: Strategic alignment of cloud-based architectures for big data. In: 2013 17th IEEE International Enterprise Distributed Object Computing Conference Workshops, pp. 136–143. IEEE, Vancouver, BC, Canada (2013). https://doi.org/10.1109/EDOCW.2013.22
2. Kacfah Emani, C., Cullot, N., Nicolle, C.: Understandable big data: a survey. Comput. Sci. Rev. **17**, 70–81 (2015). https://doi.org/10.1016/j.cosrev.2015.05.002
3. Mikalef, P., Pappas, I.O., Krogstie, J., Giannakos, M.: Big data analytics capabilities: a systematic literature review and research agenda. Inf. Syst. E-Bus. Manag. **16**, 547–578 (2018). https://doi.org/10.1007/s10257-017-0362-y
4. Demchenko, Y., de Laat, C., Membrey, P.: Defining architecture components of the big data ecosystem. In: 2014 International Conference on Collaboration Technologies and Systems (CTS), pp. 104–112. IEEE, Minneapolis, MN, USA (2014). https://doi.org/10.1109/CTS.2014.6867550
5. Curry, E.: The big data value chain: definitions, concepts, and theoretical approaches. In: Cavanillas, J.M., Curry, E., Wahlster, W. (eds.) New Horizons for a Data-Driven Economy, pp. 29–37. Springer International Publishing, Cham (2016). https://doi.org/10.1007/978-3-319-21569-3_3
6. Miller, H.G., Mork, P.: From data to decisions: a value chain for big data. IT Prof. **15**, 57–59 (2013). https://doi.org/10.1109/MITP.2013.11
7. Visconti, R.M., Larocca, A., Marconi, M.: Big data-driven value chains and digital platforms: from value co-creation to monetization. SSRN Electron. J. (2017). https://doi.org/10.2139/ssrn.2903799

8. Lamba, H.S., Dubey, S.K.: Analysis of requirements for big data adoption to maximize IT business value. In: 2015 4th International Conference on Reliability, Infocom Technologies and Optimization (ICRITO) (Trends and Future Directions), pp. 1–6. IEEE, Noida, India (2015). https://doi.org/10.1109/ICRITO.2015.7359268
9. Serhani, M.A., El Kassabi, H.T., Taleb, I., Nujum, A.: An hybrid approach to quality evaluation across big data value chain. In: 2016 IEEE International Congress on Big Data (BigData Congress), pp. 418–425. IEEE, San Francisco, CA (2016). https://doi.org/10.1109/BigDataCo ngress.2016.65
10. Kasim, H., Hung, T., Li, X.: Data value chain as a service framework: for enabling data handling, data security and data analysis in the cloud. In: 2012 IEEE 18th International Conference on Parallel and Distributed Systems, pp. 804–809. IEEE, Singapore (2012). https://doi.org/10. 1109/ICPADS.2012.131
11. Sharma, S.: Expanded cloud plumes hiding big data ecosystem. Future Gener. Comput. Syst. **59**, 63–92 (2016). https://doi.org/10.1016/j.future.2016.01.003
12. Najjar, M., Kettinger, W.: Data monetization: lessons from a retailer's journey. MIS Q. Exec. **12** (2013)
13. Cao, H., Dong, W.S., Liu, L.S., Ma, C.Y., Qian, W.H., Shi, J.W., Tian, C.H., Wang, Y., Konopnicki, D., Shmueli-Scheuer, M., Cohen, D., Modani, N., Lamba, H., Dwivedi, A., Nanavati, A.A., Kumar, M.: SoLoMo analytics for telco big data monetization. IBM J. Res. Dev. **58**, 9:1–9:13 (2014). https://doi.org/10.1147/JRD.2014.2336177
14. Bataineh, A.S., Mizouni, R., Barachi, M.E., Bentahar, J.: Monetizing personal data: a two-sided market approach. Procedia Comput. Sci. **83**, 472–479 (2016). https://doi.org/10.1016/j.procs. 2016.04.211
15. Liu, C.-H., Chen, C.-L.: A Review of Data Monetization: Strategic Use of Big Data, p. 7. Hong Kong (2015)
16. Wixom, B.H., Ross, J.W.: How to monetize your data. https://sloanreview.mit.edu/article/how-to-monetize-your-data/. Last accessed 2019/06/07
17. Wells, A.R., Chiang, K.W.: Monetizing Your Data: A Guide to Turning Data into Profit Driving Strategies and Solutions. Wiley, Hoboken, NJ (2017)
18. Tanner, J.F.: Analytics and Dynamic Customer Strategy: Big Profits from Big Data. Wiley, Hoboken, NJ (2014)
19. Mell, P., Grance, T.: The NIST Definition of Cloud Computing, p. 7
20. Moro Visconti, R.: Public private partnerships, big data networks and mitigation of information asymmetries. Corp. Ownersh. Control. **14**, 205–215 (2017). https://doi.org/10.22495/cocv14 i4c1art3
21. Adner, R.: Match your innovation strategy to your innovation ecosystem. Harv. Bus. Rev. **84**, 98–107; 148 (2006)

Ensemble Feature Selection Method Based on Bio-inspired Algorithms for Multi-objective Classification Problem

Mohammad Aizat Basir, Mohamed Saifullah Hussin, and Yuhanis Yusof

Abstract Feature selection is a challenging task, specifically in achieving an optimal solution. This is due to the difficulties in choosing the most suitable feature selection method as they tend to work individually and cause incorrect feature selection, which in turn affect the classification accuracy. The objective of this research was to utilise the potential of ensemble methods (boosting) with bio-inspired techniques in improving the performance of multi-objective algorithms (ENORA and NSGA-II) in term of optimum features set. The vital stage in this research was the optimisation of both algorithms using suitable bio-inspired search algorithms with an ensemble method. The next step was the confirmation of the optimal selected feature set by performing a classification task. The evaluation metrics were determined based on the number of selected features with good classification accuracy. Eight benchmark datasets with various sizes were carefully chosen to experiment. Experimental results revealed that both algorithms that used the selected bio-inspired search algorithms with an ensemble method have successfully achieved a better solution with an optimum set of features, that is, less number of features with higher classification accuracy on the selected datasets. This discovery implies that the combination of bio-inspired algorithms with an ensemble method (boosting) improves the performance of both algorithms in attribute selection and classification task.

Keywords Ensemble feature selection · Bio-inspired · Classification · ENORA · NSGA-II

M. A. Basir (✉) · M. S. Hussin
Faculty of Ocean Engineering Technology and Informatics, Universiti Malaysia Terengganu, Kuala Terengganu, Malaysia
e-mail: aizat@umt.edu.my

M. S. Hussin
e-mail: saifullah@umt.edu.my

Y. Yusof
School of Computing, Universiti Utara Malaysia, Sintok, Malaysia
e-mail: yuhanis@uum.edu.my

F. Saeed et al. (eds.), *Advances on Smart and Soft Computing*, Advances in Intelligent Systems and Computing 1188, https://doi.org/10.1007/978-981-15-6048-4_15

167

1 Introduction

In recent years, the dimensionality of datasets used in data mining applications increased exponentially as discussed in [1, 2]. This scenario leads to several challenges in algorithms learning due to the large search space, which makes it difficult to capture valuable features of the interest domain. Through the understanding of this problem, it reduces the number of insignificant features in the machine learning model, which is a crucial task to achieve. In practical circumstances, the elimination of unimportant and unnecessary dimensions is strongly advisable to reduce the cost of labour and processing time [3]. It is important to highlight that a large dataset that consists of a huge number of attributes is known as a high dimensionality dataset. This contributes to a phenomenon that is known as the curse of dimensionality, which means the time of computation is an exponential function of the number of dimensions. In addition, the high dimension of searching space contributes to the redundancy of features in a model. The problems of attribute reduction can be categorised into two parts, (1) the optimality degree of the attributes, which refer to the dependency degree and subset size, (2) time needed to attain the optimality of the attributes. Many researchers have investigated the strength and weaknesses of the available feature selection approaches [4–7]. However, finding a suitable approach for a given problem remains difficult. With the aim of pursuing better trade-off stability and performance accuracy, more advanced feature selection techniques are being experimented [2, 8].

One of the convincing ideas is utilising the ensemble approaches as a framework to improve the robustness of the selection process, especially in the high-dimensional and less sample size settings where the extraction of stable feature subsets is fundamentally more difficult. The ensemble selection methods were presented in multiple recent studies [9]. It can be categorised into two main groups, namely functionally heterogeneous and functionally homogenous approaches. The first group involves the utilisation of different selection algorithms to the same dataset while the same selection algorithm is used to different perturbed versions of the original data, which is similar to the bagging and boosting techniques in the context of multi-classifier systems [10]. In both approaches, different outcomes can be acquired, which are subsequently combined to generate a single feature subset. This contributes to a better approximating, which expectantly achieves an optimal solution for the problem stated.

2 Related Works

A new ensemble construction method was suggested by [11]. They applied ACO on the stacking ensemble construction process to generate domain-specific configurations and the results indicated that the new approach generates better stacking ensembles. ACO was also studied with genetic algorithm for ensemble feature selection in [12]. The new artificial bee colony stacking (ABC) was applied for the selection of the optimal base classifiers configuration and the meta-classifier [13]. For the handling of the multi-objective problem, [14] proposed a novel multi-strategy ensemble ABC (MEABC) algorithm. In MEABC, a pool of distinct solution search strategies coexists throughout the search process and competes to produce offspring.

A novel multi-objective algorithm, which combines hybrid flower pollination with the bat search algorithm, was proposed by [15] to forecast the wind energy in power grids. This hybrid algorithm was used to search for the optimal weight coefficients based on the previous step, while Pareto optimality theory provides the necessary conditions to identify an optimal solution. In the field of engineering, a hybrid binary bat algorithm (HBBA) that hybridised with machine learning algorithm was presented by [16] to reduce the dimensionality as well as to select the predominant features of gear faults, which contains the necessary discriminative information. A multi-objective algorithm based on the cuckoo search algorithm was applied in problem optimisation in the medical field [17]. Similarly, a new classification algorithm entitled "CS-Boost" that employs cuckoo search algorithm to optimise the performance of the AdaBoost algorithm was proposed by [18]. Experimental results showed that the CS-Boost algorithm enhances the accuracy of AdaBoost in predicting rheumatoid arthritis. In the field of engineering, an intelligent fault identification method of rolling bearing was proposed by [19].

Yang [20] invented the firefly algorithm that was later applied in many areas, especially in feature selection application. Recently, in the medical field, a new firefly algorithm ensemble-based was presented by [21] to differentiate healthy and diseased brain conditions by analysing the brain MRI images. This system uses a weighted classifier combination-based ensemble model to classify the brain images. In the area of network intrusion, a new ensemble data pre-processing approach for intrusion detection system using a variant of the firefly algorithm and bagging k-NN techniques was proposed by [22]. The two popular multi-objective feature selection methods, which are frequently used, are ENORA and NSGA-II. ENORA is one of the multi-objective evolutionary algorithms based on the selection strategy for the random search method [23, 24]. ENORA has two main goals, that are, minimises the root mean squared error (RMSE) and the number of selected features of the model, which learned through the random forest (RF), and lowers the influence to over-fitting and increases accuracy [25]. Besides, an NSGA-II wrapper approach was presented in [26] in recognition of the named entities. The modified dominance relation was introduced in [27] as a classification algorithm. The diagnosis problem in medicine was studied using multi-objective feature selection in [28].

Motivated by the competency of various bio-search techniques, this study aimed to produce an optimised ENORA and NSGA-II algorithms by deploying several types of bio-search algorithms with an ensemble boosting method, which is the AdaBoost algorithm. The optimised multi-objective algorithms were then used to obtain the optimal number of attributes for selected datasets. The detailed steps taken are described in the next section.

3   Methodology

The methodology of this paper is represented in Fig. 1. WEKA software was used for the experimentation task. The overall process was translated into four main steps: (1) Selection of data [different sizes and domain], (2) pre-processing of data [handling missing values], (3) optimal dimensionality reduction [optimal reduction], (4) improvement of ENORA and NSGA-II.

Step 1—Selection of data: Datasets were selected from the UCI Machine Learning Repository [29] (refer to Table 1 for profile details). These datasets contain different sizes and mix domains to test and confirm the fitness of algorithms while performing feature selection task.

Step 2—Pre-processing of data: Missing values in the dataset were pre-processed for experimentation. The original dataset with no value, which was symbolised with the symbol '?', was converted to 0 value. Next, all the clean datasets were discretised

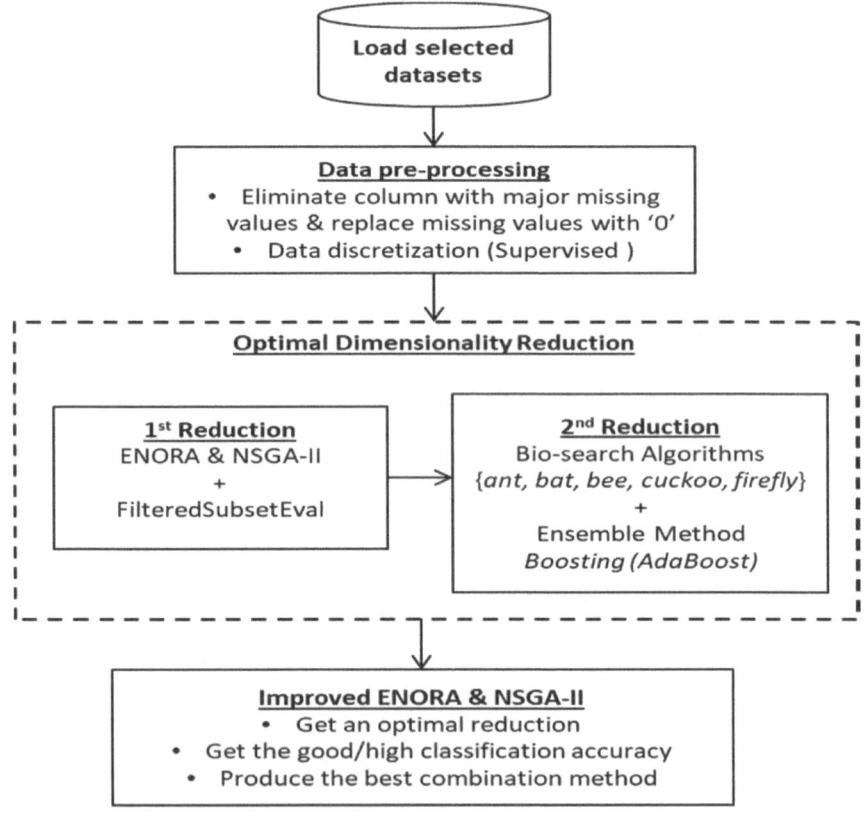

Fig. 1 Methodology

Table 1 Profile of the selected datasets

Dataset	#Attribute	#Classes	#Instances
Arcene	10,000	2	900
Breastcancer	9	2	367
Clean1	166	2	476
Emails	4702	2	64
Gisette	5000	2	13,500

Table 2 Detailed setting

Search algorithm	Detailed setting
Ant	Pheromone rate: 2.0	Heuristic rate: 0.7	Evaporation rate: 0.9
Bat	Frequency: 0.5	Loudness: 0.5
Bee	Radius mutation: 0.80	Radius damp: 0.98
Cuckoo	Sigma rate: 0.70	Pa rate: 0.25
Firefly	Beta zero: 0.33	Absorption Coefficient: 0.001

Fixed setting for all bio-search algorithms: iteration: 20, mutation probability: 0.01, population size: 20 except for bee = 30

into nominal attributes. This discretisation was done based on Fayyad and Irani's MDL method [30].

Step 3—Optimal dimensionality reduction: In this step, two reduction processes were applied. The first reduction applied was the ENORA and NSGA-II algorithms with a filtered method. The output of the first reduction was then further reduced. Five bio-search methods with an ensemble method, namely boosting (AdaBoost algorithm) were applied for optimal search. This process balanced the exploitation and exploration mechanisms for the solution space. Three popular learning algorithms, namely decision tree (DT), K-nearest neighbour (K-NN) and Naïve Bayes (NB) were deployed with wrapper methods.

Step 4—Improvement of ENORA and NSGA-II: To optimise the ENORA and NSGA-II algorithms, different matching of the bio-search techniques and wrapper with an ensemble method (boosting) were explored. To indicate the best combination list of the ENORA and NSGA-II model, a smaller number of attributes (optimal) with good classification accuracy were considered as the important criteria options (Table 2).

4 Results and Discussion

In Table 3, the initial result showed that ENORA + Filtered technique reduced the attributes of the seven datasets, which include Arcene, Clean1, Emails, Gisette,

Table 3 Comparison of attribute reduction: (ENORA + filtered) versus (ENORA + AdaBoost + bio-search)

Dataset	#Attr Original	First selection #Attr ENORA (filtered)		Second selection # Attr [ENORA + AdaBoost (bio-search + wrapper)]				
				Ant	Bat	Bee	Cuckoo	Firefly
Arcene	10,000	391	(96.1%)[a]	184	127	98	173	206
Breastcancer	9	9	(0.0%)[a]	3	4	4	4	4
Clean1	166	22	(86.7%)[a]	13	14	15	13	19
Emails	4702	79	(98.3%)[a]	19	27	13	17	32
Gisette	5000	66	(98.7%)[a]	21	28	16	21	24
Ozone	72	12	(85.7%)[a]	6	7	6	6	6
Parkinson	22	9	(59.1%)[a]	3	3	3	5	3
Semeion	265	5	(98.1%)[a]	4	4	4	4	4

[a] % of reduction from original attributes

Ozone, Parkinson and Semeion, but not the Breastcancer dataset where the original attributes remained the same. Semeion, Emails and Gisette datasets achieved almost 99% in reduction. Nevertheless, these enormous attributes reduction cannot confirm the best solution still. An extended experiment was conducted to optimise the ENORA algorithms with five bio-search algorithms and wrapper method with AdaBoost algorithm. Results showed that further reduction was observed in all datasets. Overall, the combination of the bio-search algorithm with boosting method has outperformed the filtered approach in reducing attributes. The result confirmed the competency of bio-search algorithm-based boosting techniques in selecting optimal features for the ENORA algorithms. It is important to highlight that the existing random search function in bio-search algorithm has contributed significant strengths to the algorithm structure to possibly selects the best features. On the other hand, boosting method enhanced the performance of bio-search algorithms in term of the best features selection and classification accuracy.

In Table 4, a similar pattern as the ENORA was observed in the NSGA-II + filtered method where it reduced the attributes of the seven datasets, which include Arcene, Clean1, Emails, Gisette, Ozone, Parkinson and Semeion, but not the Breastcancer dataset where the original attributes remained the same. NSGA-II performed better and reduced more than 97% attributes in half of the selected datasets, which are Semeion, Emails, Gisette and Arcene. Even though the performance of NSGA-II was better than ENORA in term of reducing the selected attributes in the first reduction, it is not sufficient to obtain the optimal set of attributes still. A further experiment was conducted to optimise the NSGA-II algorithms with five bio-search algorithms and wrapper with an ensemble method (boosting). Results showed a superior reduction in all datasets compared to ENORA. Ozone dataset maintained and achieved the same outstanding reduction result as ENORA where the number of attributes was reduced from twelve to one in the second reduction. Generally, bio-search algorithms

Table 4 Comparison of attribute reduction: (NSGA-II + filtered) versus (NSGA-II + AdaBoost + bio-search)

Dataset	#Attr original	First reduction		Second reduction				
		#Attr NSGA-II (filtered)		# Attr [NSGA-II + AdaBoost (bio-search + wrapper)]				
				Ant	Bat	Bee	Cuckoo	Firefly
Arcene	10,000	216	(97.8%)[a]	90	85	52	72	95
Breastcancer	9	9	(0%)	3	4	4	4	4
Clean1	166	19	(88.6%)[a]	16	15	15	11	13
Emails	4702	40	(99.1%)[a]	4	7	2	3	13
Gisette	5000	49	(99.0%)[a]	15	16	18	12	15
Ozone	72	6	(91.7%)[a]	3	3	3	3	3
Parkinson	22	7	(68.2%)[a]	5	5	5	5	5
Semeion	265	7	(97.4%)[a]	7	7	7	7	7

[a]% of reduction from original attributes

performed better than filtered approach with ENORA. This scenario confirmed the adaptive behaviour of bio-search algorithms and the additional lift from the AdaBoost algorithm in optimal features selection for NSGA-II algorithms.

Referring to Table 5, surprisingly, the attributes selected from all datasets by ENORA in the first reduction did not improve the classification accuracy, but maintained the same accuracy as of the original datasets. However, attributes selected in the second reduction by ENORA and bio-search algorithms with boosting method (AdaBoost algorithm) have successfully increased the accuracy of the classifier. The five bio-search algorithms achieved a better accuracy in all datasets except for the

Table 5 Comparison of classification: (ENORA + filtered) versus (ENORA + AdaBoost + bio-search)

Dataset	Acc (%) [before reduction]	First reduction	Second reduction				
		Acc (%) ENORA	Acc (%) [ENORA + ADABOOST (bio-search + wrapper)]				
			Ant	Bat	Bee	Cuckoo	Firefly
Arcene	70.6	85.3	82.4	82.4	88.2	88.2	88.2
Breastcancer	96.2	96.2	95.8	96.6	96.6	96.2	96.2
Clean1	85.8	75.9	80.9	75.9	81.5	81.5	80.9
Emails	72.7	77.3	72.7	77.3	72.7	72.7	77.3
Gisette	91.5	88.2	87.1	86.7	87.1	87.1	87.1
Ozone	93.3	93.7	93.9	93.9	93.9	93.9	93.9
Parkinson	84.8	89.4	90.9	90.9	90.9	90.9	90.9
Semeion	94.5	92.4	92.4	92.4	92.4	92.4	92.4

Gisette dataset. Even so, the classification accuracy of the Gisette dataset was still considered acceptable and good as the percentage of reduction achieved was more than 70% (Table 4). Reduction using the firefly algorithm achieved good classification results for datasets of all sizes. This reflects the capability of the search features in the firefly algorithm, which have an absorption coefficient parameter to evaluate the light intensity for new solution optimisation. In addition, the AdaBoost algorithm improved the classification accuracy when it is combined with all the bio-search algorithms, especially the firefly algorithm.

Interestingly, in the first reduction, Table 6 highlights that NSGA-II was inconsistent in attributes selection in all the datasets. However, it improved the accuracy of most of the datasets except for the Clean1 and Gisette datasets, which showed a decrease in classification accuracy. The first reduction results by NSGA-II algorithms need to be optimised in order to get a better classification accuracy. While in the second reduction, NSGA-II and bio-search algorithms with boosting method (AdaBoost algorithm) showed a significant increment in all the datasets. Similar to ENORA, firefly algorithm showed superior dominant where it boosted the classification accuracy in all the datasets. The important parameter, which named as brightness intensity, in firefly algorithm shows that the brighter fireflies will attract the less bright ones towards them. While in the case where the fireflies are similar in brightness, they will then move randomly (a feature to be selected).

Table 7 shows the ideal bio-search algorithms and boosting method (AdaBoost algorithm) for ENORA and NSGA-II with datasets of different sizes. This alternative guideline can be used to perform feature selection task, especially when dealing with data and domain of different sizes.

Table 6 Comparison of classification accuracy of (NSGA-II + filtered) versus (NSGA-II + AdaBoost + bio-search)

Dataset	Acc (%) [before reduction]	First reduction	Second reduction				
		Acc (%) NSGA-II	Acc (%) [NSGA-II + AdaBoost (bio-search + wrapper)]				
			Ant	Bat	Bee	Cuckoo	Firefly
Arcene	70.6	85.3	79.4	79.4	82.4	76.5	88.2
Breastcancer	96.2	96.2	95.8	96.6	96.6	96.2	96.2
Clean1	85.8	82.1	83.3	82.7	81.5	84.0	84.6
Emails	72.7	77.3	77.3	77.3	77.3	77.3	77.3
Gisette	91.5	86.8	88.2	88.2	88.2	88.8	88.8
Ozone	93.3	93.9	93.9	93.9	93.9	93.9	93.9
Parkinson	84.8	87.9	87.9	87.9	87.9	87.9	87.9
Semeion	94.5	93.4	93.4	93.4	93.4	93.4	93.4

Table 7 Ideal bio-search algorithms for ENORA and NSGA-II for feature selection	Multi-objective algorithm	Ensemble	Bio-search algorithm
	ENORA	Boosting (AdaBoost)	Ant, Bat, Bee, Cuckoo, Firefly
	NSGA-II	Boosting (AdaBoost)	Bat

5 Conclusion and Future Work

In summary, this alternative technique provides a better understanding of the implementation of various bio-inspired techniques in manipulating exploration and exploitation of the search space, especially the optimisation of multi-objective algorithms. This study discovered a new optimised technique for ENORA and NSGA-II, which were compared and tested on eight datasets. The suitable bio-search algorithms and boosting method, which is the AdaBoost algorithm, for ENORA and NSGA-II were determined based on the high classification accuracy generated with relevant features. Future studies may consider to study various bio-search algorithms and formulate the right parameters' setting for new optimisation techniques.

Acknowledgements The authors would like to recognise the Universiti Malaysia Terengganu (UMT), Universiti Utara Malaysia (UUM) and Ministry of Education Malaysia (MOE) for their supports in term of services and facilities. This research was supported by the UMT and MOE under TAPE-RG research grant (Vot No. 55133).

References

1. Iguyon, I., Elisseeff, A.: An introduction to variable and feature selection. J. Mach. Learn. Res. (2003)
2. Bolón-Canedo, V., Sánchez-Maroño, N., Alonso-Betanzos, A.: Recent advances and emerging challenges of feature selection in the context of big data. Knowl.-Based Syst. (2015)
3. Jensen, R., Shen, Q.: Computational Intelligence and Feature Selection: Rough and Fuzzy Approaches (2008)
4. Bolón-Canedo, V., Sánchez-Maroño, N., Alonso-Betanzos, A.: A review of feature selection methods on synthetic data. Knowl. Inf. Syst. (2013)
5. Tang, J., Alelyani, S., Liu, H.: Feature selection for classification: a review. In: Data Classification: Algorithms and Applications (2014)
6. Khalid, S., Khalil, T., Nasreen, S.: A survey of feature selection and feature extraction techniques in machine learning. In: Proceedings of 2014 Science and Information Conference, SAI 2014 (2014)
7. Vergara, J.R., Estévez, P.A.: A review of feature selection methods based on mutual information. Neural Comput. Appl. (2014)
8. He, Z., Yu, W.: Stable feature selection for biomarker discovery. Comput. Biol. Chem. (2010)
9. Guan, D., Yuan, W., Lee, Y.K., Najeebullah, K., Rasel, M.K.: A review of ensemble learning based feature selection. IETE Tech. Rev. (2014)

10. Bühlmann, P.: Bagging, boosting and ensemble methods. In: Handbook of Computational Statistics: Concepts and Methods, 2nd edn. (2012)
11. Chen, Y., Wong, M.L., Li, H.: Applying ant colony optimization to configuring stacking ensembles for data mining. Expert Syst. Appl. (2014)
12. Santana, L.E.A., Silva, L., Canuto, A.M.P., Pintro, F., Vale, K.O.: A comparative analysis of genetic algorithm and ant colony optimization to select attributes for an heterogeneous ensemble of classifiers. In: 2010 IEEE World Congress on Computational Intelligence, WCCI 2010—2010 IEEE Congress on Evolutionary Computation, CEC 2010 (2010)
13. Shunmugapriya, P., Kanmani, S.: Optimization of stacking ensemble configurations through artificial bee colony algorithm. Swarm Evol. Comput. (2013)
14. Wang, H., Wu, Z., Rahnamayan, S., Sun, H., Liu, Y., Pan, J.S.: Multi-strategy ensemble artificial bee colony algorithm. Inf. Sci. (Ny) (2014)
15. Qu, Z., Zhang, K., Mao, W., Wang, J., Liu, C., Zhang, W.: Research and application of ensemble forecasting based on a novel multi-objective optimization algorithm for wind-speed forecasting. Energy Convers. Manag. 154, 440–454 (2017)
16. Rajeswari, C., Sathiyabhama, B., Devendiran, S., Manivannan, K.: Diagnostics of gear faults using ensemble empirical mode decomposition, hybrid binary bat algorithm and machine learning algorithms. J. Vibroeng. (2015)
17. Wang, Y., Liu, B., Ma, Z., Wong, K.C., Li, X.: Nature-inspired multiobjective cancer subtype diagnosis. IEEE J. Transl. Eng. Health Med. (2019)
18. Shiezadeh, Z., Sajedi, H., Aflakie, E.: Diagnosis of rheumatoid arthritis using an ensemble learning approach (2015)
19. Huang, H., Fan, Q., Wei, J., Huang, D.: An intelligent fault identification method of rolling bearings based on SVM optimized by improved GWO. Syst. Sci. Control Eng. (2019)
20. Yang, X.-S.: Firefly algorithms. In: Nature-Inspired Optimization Algorithms (2014)
21. Pati, S.P., Mishra, D.: Classification of brain MRIs using improved firefly algorithm based ensemble model. Int. J. Innov. Technol. Explor. Eng. (2019)
22. Shona, D., Senthilkumar, M.: An ensemble data preprocessing approach for intrusion detection system using variant firefly and Bk-NN techniques. Int. J. Appl. Eng. Res. (2016)
23. Nandi, G.: An enhanced approach to Las Vegas filter (LVF) feature selection algorithm. In: Proceedings—2011 2nd National Conference on Emerging Trends and Applications in Computer Science, NCETACS-2011 (2011)
24. Vafaie, H., De Jong, K.: Genetic algorithms as a tool for feature selection in machine learning. In: Proceedings— International Conference on Tools with Artificial Intelligence, ICTAI (1992)
25. Breiman, L.: Random forests. Mach. Learn. (2001)
26. Ekbal, A., Saha, S., Garbe, C.S.: Feature selection using multiobjective optimization for named entity recognition. In: Proceedings—International Conference on Pattern Recognition (2010)
27. Reynolds, A.P., Corne, D.W., Chantler, M.J.: Feature selection for multi-purpose predictive models: a many-objective task. In: Lecture Notes in Computer Science (including subseries Lecture Notes in Artificial Intelligence and Lecture Notes in Bioinformatics) (2010)
28. Gaspar-Cunha, A.: Feature Selection Using Multi-Objective Evolutionary Algorithms: Application to Cardiac SPECT Diagnosis (2010)
29. Asuncion, A., Newman, D.J.: UCI Machine Learning Repository. University of California, School of Information and Computer Science, Irvine, CA (2017). [Online]. Available: https://archive.ics.uci.edu/ml
30. Fayyad, U.M., Irani, K.B.: Multi-interval discretization of continuos-valued attributes for classification learning. In: Proceedings of the International Joint Conference on Uncertainty in AI, pp. 1022–1027 (1993)

Automated Binary Classification of Diabetic Retinopathy by Convolutional Neural Networks

Ayoub Skouta, Abdelali Elmoufidi, Said Jai-Andaloussi, and Ouail Ochetto

Abstract This paper proposes a binary classification method for high-resolution retinal images by using convolutional neural networks. The convolutional neural network is first, formed to recognize and classify the fundus images of the eye as normal retina or proliferative diabetic retina. We are training proposed network by using a graphics processor and a public dataset available on the Kaggle Web site. An increase in data makes possible the identification of complex characteristics involved in the classification task. Our proposed convolutional neural networks achieve an accuracy of 95.5% on 1000 images of the normal retina and 1000 images of proliferative diabetic retina for training and 200 images of the normal retina and 200 images of proliferative diabetic retina for testing.

Keywords Deep learning · Convolutional neural networks · Diabetic retinopathy · Binary classification of images

1 Introduction

The eye is an important organ for the human body. The majority of external information obtained by humans comes from the eye [1]. The global prevalence of diabetes

A. Skouta (✉) · S. Jai-Andaloussi
Computer Science, Modeling Systems and Decision Support, Department of Mathematics and Computer Science, Faculty of Science Aïn Chock, Hassan II University, Casablanca, Morocco
e-mail: ay.skouta@gmail.com

S. Jai-Andaloussi
e-mail: andaloussi.said@gmail.com

A. Elmoufidi
Laboratory of Information Processing and Decision Support, Department of Computer Science, Faculty of Sciences and Technics, Sultan Moulay Slimane University, Beni Mellal, Morocco
e-mail: elmoufidi10@gmail.com

O. Ochetto
Laboratory of Computer Science, Modeling Systems and Decision Support, FSJES Ain-Chock, Hassan II University, Casablanca, Morocco
e-mail: ouail.ouchetto@gmail.com

F. Saeed et al. (eds.), *Advances on Smart and Soft Computing*, Advances in Intelligent Systems and Computing 1188, https://doi.org/10.1007/978-981-15-6048-4_16

177

mellitus has reached epidemic proportions. According to the International Diabetes Federation (IFD), there would be 642 million people with diabetes worldwide in 2040. It is extremely important de rank the gravity of diabetic in order to recommend the appropriate treatment. In fact, ophthalmologists use fundus photography techniques to show the eyeball background in a very high-resolution image (up to 3500–3500 pixels). They make a diagnosis after the analysis of these images based on the knowledge acquired during previous experiments [2]. The problem is that the lesions are small and difficult to detect. Carrying out such a diagnosis is useless because it takes time and depends on the physician's expertise. Therefore, the automation of this procedure is very desirable. Currently, artificial intelligence, machine learning, convolutional neural networks [3] have produced better results for image classification and its ability to analyze the most discriminating image features. Such an approach allows automatic retinal screenings [4, 5]. Indeed, a sequence of convolutional and grouping layers has used to train an encoder, to convert the input image to feature cards. The lure comes from the ability to learn from examples, or learning datasets, rather than hard coded rules. The structure of our article is as follows: The second section tackles the classification of diabetic retinopathy, formerly ALFEDIAM, and the third section is a description of deep learning and the convolutional neural network. The fourth section displays our methodology to analyze the dataset; the fifth one presents the results of our experience with the discussion. The final section is conclusion, which summarizes our findings. The aim of this study is to classify diabetic retinopathy images by using convolutional neural network to recognize images according of the following categories (0: normal retina, 1: retina proliferative diabetic).

2 Related Previous Work

Many computer-aided diagnoses have been developed to help physician improving the diagnosis results in several medical fields, such as breast cancer [6], brain cancer [7] and retinopathies diabetic [8]. In this section, we are focusing of retinopathies diabetic previous work. For example, in [9], the authors proposed an interpretable computer-aided diagnosis (CAD) pipeline for diagnosing glaucoma using fundus images and convolutional neural networks (CNNs), Li et al. [10] have developed a deep learning (DL) system for retinal hemorrhage (RH) screening using ultra-widefield fundus (UWF) images. Ortiz-Feregrino et al. [11] have published a method for classification of proliferative diabetic retinopathy using deep learning.

3 Classification of Diabetic Retinopathy

The classification of DR aims to establish different stages of severity and prognosis in the evolution of this pathology. Many classifications have followed one another. The

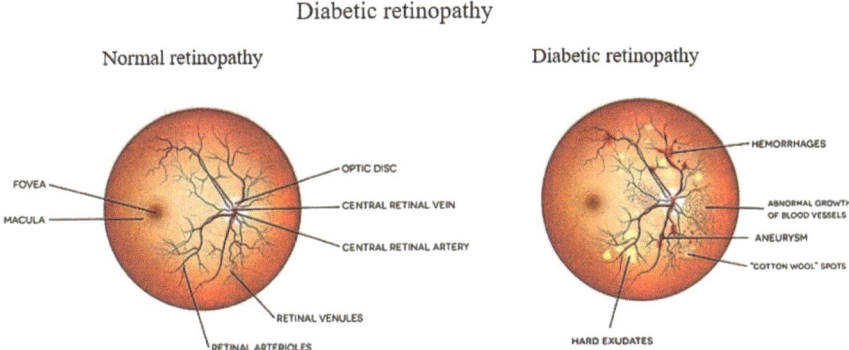

Fig. 1 Normal retina and retina with diabetic retinopathy [13]

first ones based on the natural history of DR were qualitative, sufficient for current practice. The Airlie House classification is the basis of contemporary classifications, the DR has differentiated into two groups: non-proliferating and proliferating [12]. Figure 1 shows one image from each group, and Table 1 contains more details about non-proliferating and proliferating classes.

4 Convolutional Neural Networks

A convolutional neural network (CNN) or deep neural network (DNN) developed in 2012 is an ANN artificial neural network with several intermediate layers. CNNs have an operating principle similar to that of traditional supervised learning algorithms: They receive input images, automatically detect the characteristics of each of them and then train a classifier on them. However, the features are learning automatically. The CNNs themselves carry out the extraction and description of characteristics: During the training phase, the classification error is minimized in order to optimize the classifier parameters as well as the characteristics. One of the strengths of convolutional neural networks are the automatic extraction of features, which adapt to the given problem. Among the major advantages of DNNs is that their performance increases when the size of the learning dataset continues to increase (Fig. 2) [14]. Unlike traditional supervised learning techniques, which tend to become saturated with large amounts of data [15].

The main idea is that a deep neural network performs both tasks at the same time, learns to extract the features from the given classification problem and to classify them (Fig. 3).

The architecture is suitable for image processing with deep learning this one based on convolutional neural networks. CNNs use a mathematical filtering technique called convolution, which produces data that proceed as detectors of special and more descriptive features. The detectors have adjusted to detect the precise and

Table 1 No-proliferative diabetic retinopathy (NPDR) and proliferative diabetic retinopathy (PDR)

No-proliferative diabetic retinopathy (NPDR)	Minimal	There are some microaneurysms at the back of the eye, and some areas of capillary micro occlusions and intraretinal diffusions localized in angiography. There are no neovessels or preproliferative signs
	Moderate	There is at the back of the eye microaneurysms, hemorrhages in sparks, punctate or intraretinal moderately numerous and extensive, cottony nodules, not very marked venous anomalies and few AMIR Angiography of small or small enough localized areas of retinal ischemia in the periphery and/or at the posterior pole
	Severe	There are preproliferative signs without new vessels. The risk of progression to proliferative diabetic retinopathy is high RDNP moderate: at least one of the following three signs: • Microaneurysms and severe deep hemorrhages in four quadrants • Venous anomalies in two or more quadrants • AMIRs in a quadrant or more Very severe RDNP: at least two of the 3 signs described in moderate RDNP
Proliferative diabetic retinopathy (PDR)	Uncomplicated	Presence of preretinal and/or peripapillary new vessels
	complicated	• Vitreous hemorrhage • Neovascular glaucoma • Traction retinal detachment

Fig. 2 Advantage of deep learning performance compared to classic machine learning [14]

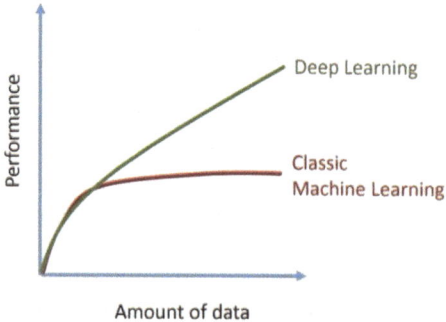

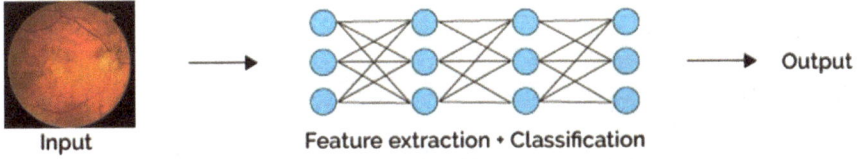

Input   Feature extraction + Classification   Output

Fig. 3 Deep learning models learn to extract features directly from the inputs data and classify them

necessary image features for resolving the classification task. Deep learning quickly demonstrated its capabilities in medical applications. For some time, CNNs have trained to screen for diabetic retinopathy, and the results obtained by these programs are comparing with the diagnosis given by ophthalmologists experts in this field [8]. The obtained performance is equivalent to that obtained by six ophthalmologists who used a similar dataset.

5 Proposed Method

In this work, we develop a classified model for images of the normal retina or prolif-erative diabetic retina. We implemented the visual geometry group (VGG) of Oxford architecture, which contains four blocks in order to build the proposed classification model. Each block is composed of a convolution layer with a 3×3 filter, followed by a maximum grouping layers. Each conventional layer uses the ReLU activation function. The depths of our architecture for the four blocks are, respectively, 32, 64, 128 and 256. Next, we have a dense layer fully connected. Since this is a binary classification task, a value of zero (0) or one (1) must be predicted. The final output layer to a sigmoid type activation derived regularization has used to regularize the deep neural network by canceling or possibly deleting the inputs of a layer. The first role of regularization is to simulate as many networks as possible by changing the structure of the network; the second role is to increase the strength of the nodes of the network. The next step is to compile. Since this is a binary classification, the loss function, we use is binary cross entropy. We use the optimizer Adam and finally form a classifier on the training data.

The execution of our model has been adapted to 250 epochs. The result obtained is 95.5% accuracy with four blocks (Fig. 4).

6 Experience and Results

In this section, we present the simulation platform used to train and test the proposed method. We start by the hardware used in this word: Operating system: Windows 10 Professional 64-bit; Processor: Inter (R) Core (TM) i7-4710MQ CPU @ 2.50 GHz,

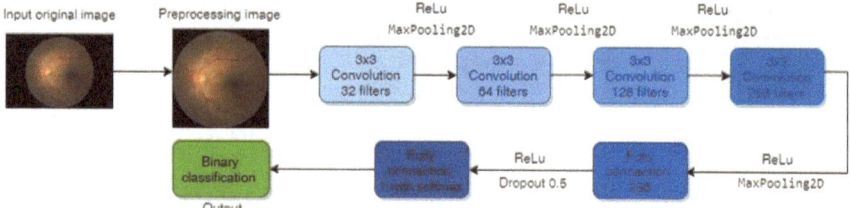

Fig. 4 Network architecture of our proposed model

×64 processor; installed memory (RAM): 16.0 GB Graphics card: NVIDIA GeForce GTX 765 M. Second, we have used the Jupyter Notebook development environment with Keras and TensorFlow.

6.1 Dataset

The data for this experiment has extracted from a "diabetic-retinopathy-classified" dataset provided via Kaggle on diabetic retinopathy for training and testing. The dataset contains 35,108 images taken under various conditions and from different models and cameras affecting the visual appearance of the left eye versus the right one. The images are divided into five categories (normal: level_0, light: level_1, moderate: level_2, severe: level_3 and proliferating: level_4). The images based on these scales have distributed in five separate folders (level_0, level_1, level_2, level_3, level_4) according to the breakdown of a clinician who assessed the presence of diabetic retinopathy. A left and right field has provided for each subject. Images have libeled with a subject identifier and the image orientation (right or left). For example, the image named 1_left.jpeg is the left eye of the patient identified by the number 1. In empirical dataset, there is noise in images and labels. Images may contain artifacts, be blurred, underexposed or overexposed (Fig. 5).

In this work, we have used a sub dataset contains 2400 images, in our case, we have based on the total number of proliferative retina images existing in the "diabetic-retinopathy-classified" database. There are 1200 proliferative retina images exactly, so we have taken the same amount of normal retina images 1200. Our selected dataset has devised on two groups. The first one contains 2000 images (1000 normal retina images and 1000 proliferative retina images) for training used CNN classifier and 400 images (200 normal retina images and 200 proliferative retina images) for testing the building classifier model. Then, the images have cropped to eliminate the black borders, reduce images size, and normalize the local contrast, for unified the processing. The dataset has supplemented by artificially increasing the size of our dataset [16] (Fig. 6).

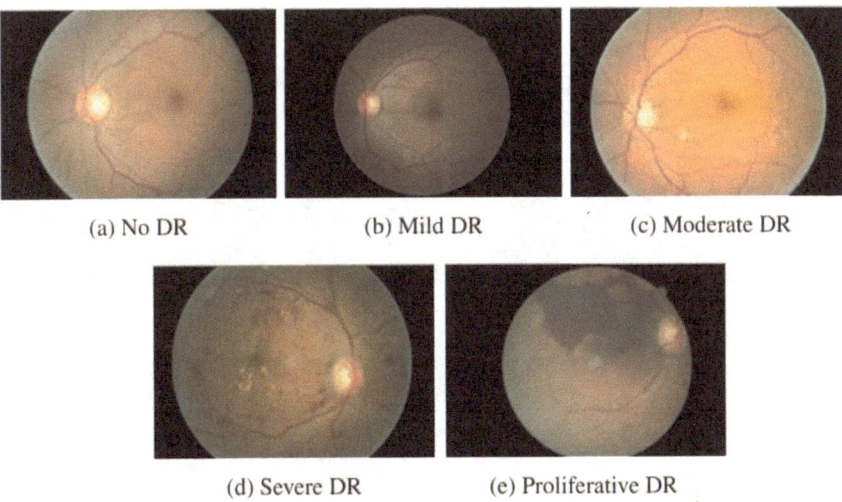

(a) No DR   (b) Mild DR   (c) Moderate DR

(d) Severe DR   (e) Proliferative DR

Fig. 5 Diabetic retinopathy (DR) levels with increasing severity

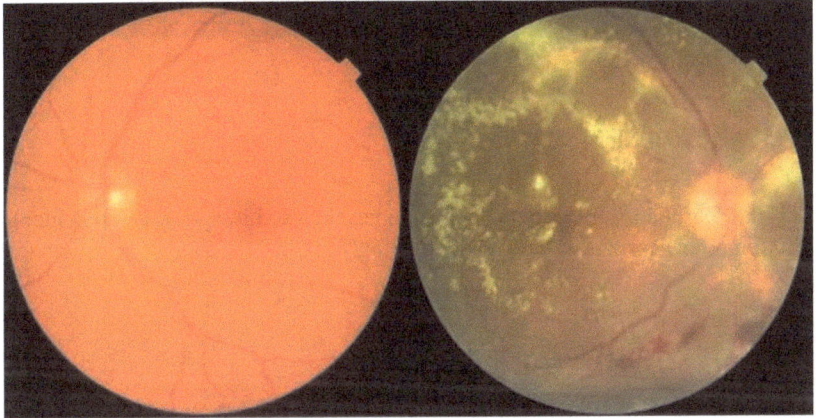

Fig. 6 Fondus image: normal and proliferative Diabetic retinopathy (DR)

6.2 Image Augmentation

Generally, a good classify model based on deep learning techniques requires much data for training and testing it. For this reason, we are using a data augmentation technique to enlarge the quantity of the used dataset, by creating modified versions of the original images. Therefore, the new obtained images do not need to be stored on hard disk. Increasing data can be also used as a regularization technique and helping the model to learn the same features, regardless of their position in the inputs [17]. The goal of this approach is to improve the ability of the classify model to generalize

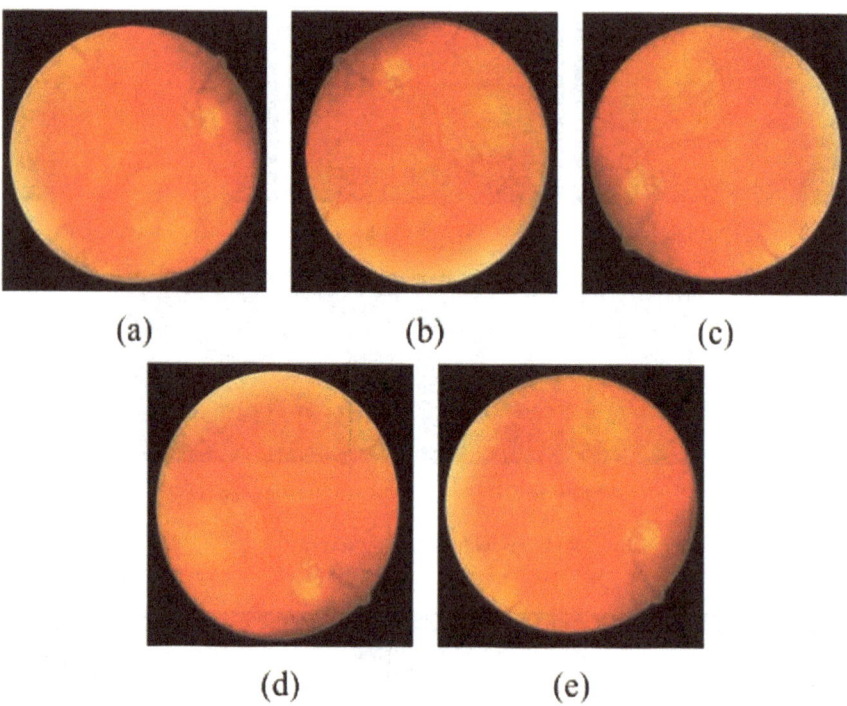

Fig. 7 Example of data augmentation

what it has learned to new inputs. In this paper, the used images are supplementing by small random horizontal and vertical shifts (Fig. 7).

6.3 Results

In this paper, we have proposed a classified model by implementing the deep learning approach; exactly, we have built a model based on VGG architecture and a "diabetic-retinopathy-classified" dataset for training and testing it. In Fig. 8, the screenshot shows a part of obtained experimental results, and Fig. 9 represents details of obtained accurate and the loss error. The accuracy of the proposed model achieved a maximum of 98.08%, and a mean accuracy is 95.5%. Therefore, experimental results demonstrate the robustness and efficiency of the used model.

For the validation step, we recorded 400 images for the testing step, 200 images of the normal retina and 200 images of the proliferative retina. The measures that we can calculate from the confusion matrix, which is a table that describes the performance of a classification model, are:

```
Epoch 244/250
62/62 [==============================] - 32s 514ms/step - loss: 0.0701 - acc: 0.9768
Epoch 245/250
62/62 [==============================] - 32s 518ms/step - loss: 0.0601 - acc: 0.9803
Epoch 246/250
62/62 [==============================] - 32s 517ms/step - loss: 0.0632 - acc: 0.9808
Epoch 247/250
62/62 [==============================] - 32s 519ms/step - loss: 0.1000 - acc: 0.9637
Epoch 248/250
62/62 [==============================] - 33s 528ms/step - loss: 0.1199 - acc: 0.9536
Epoch 249/250
62/62 [==============================] - 37s 589ms/step - loss: 0.0619 - acc: 0.9763
Epoch 250/250
62/62 [==============================] - 32s 513ms/step - loss: 0.0567 - acc: 0.9788
```

Fig. 8 This screenshot represents a part of the training process of our proposed model

Fig. 9 Linear plots of the loss and accuracy learning curves for the basic model with abandonment on the dataset

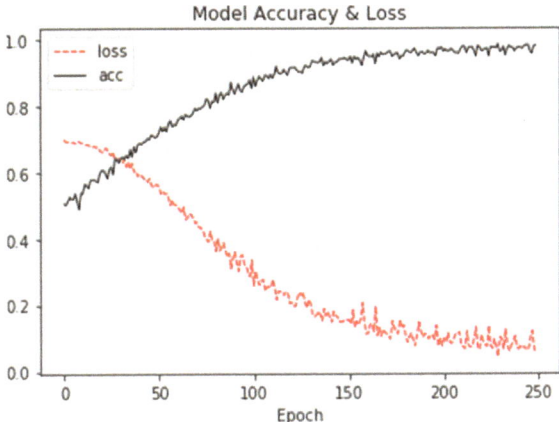

$$Sensitivity = \frac{True\ Positive}{True\ Positive\ +\ False\ Negative}$$

$$Specificity = \frac{True\ Negative}{True\ Negative\ +\ False\ Positive}$$

$$Accuracy = \frac{True\ Positive\ +\ True\ Negative}{True\ Positive\ +\ False\ Positive\ +\ True\ Negative\ +\ False\ Negative}$$

For our binary classification problem, the specificity is the number of images, which have correctly classified as proliferative diabetic retina divided by the total number of images of the proliferative diabetic retina. Sensitivity is the number of images correctly classified as normal retina divided by the total number of images of normal retina. The accuracy is the number of images correctly classified divided by the number of total used images. The results obtained by our CNN network are

94.5% specificity, 96.5% sensitivity and 95.5% accuracy. The proposed method has compared with the method proposed by N. H. Harun et al. [18] for binary classification of fundus images for diabetic retinopathy using artificial neural network. Their method has learned that MLP trained with BR provides a better classification performance with 72.11% (training) and 67.47% (testing).

7 Conclusion and Future Work

The model was able to differentiate normal retina images from proliferative diabetic retina images by looking for the presence of the main differentiating or similar factors. The neural network was able to recognize the class of images with an accuracy of 95.5%. Given the very large number of patients with diabetic retinopathy, as well as the insufficient number of specialists worldwide, generally, images have sent to clinicians to classify them and have not classified when the patient presents for screening. Therefore, an automatic screening aid system allows rapid diagnosis and instant response for a patient [1]. Therefore, these systems great practical importance in helping doctors diagnose diseases of the funds. In the future work, we plan to perform a reliable pretreatment method in order to make the lesions clearer so that the proposed classifying model could better learn and extract the precise features that this network has trouble detecting and extracting them. We will also look to use other CNN architectures as AlexNet, VGG16, GoogLeNet, ResNet and Inception-v3 to make a comparison with the results obtained. In conclusion, computer-aided screening based on CNNs algorithms will have the potential to be incredibly useful for helping DR clinicians. These tools and datasets continue improving and offering real-time classifications. It is notable that the performance of our classify model degrades if one of conventional layers has moved. Depth is therefore important to achieve meaningful results. In the end, we would like to use convolutional networks with a very large and depth to make a multi-class classification (normal, light, moderate, severe and proliferating).

References

1. Liang, Z., Bin, X.W., Yong, K.W.: Information identification technology. Mechanical Industry Press, New York, NY, USA (2006)
2. Srivastava, S., et al.: Visualizing the indicators of diabetic retinopathy learnt by convolutional neural networks. In: 2017 IEEE International Conference on Computational Intelligence and Computing Research (ICCIC). IEEE (2017)
3. Alam, M., Le, D., Lim, J.I., Chan, R.V., Yao, X.: Supervised machine learning based multi-task artificial intelligence classification of retinopathies. J. Clin. Med. 8(6), 872 (2019)
4. Petrov, D., Marshall, N., Cockmartin, L., Bosmans, H.: First results with a deep learning (feed-forward CNN) approach for daily quality control in digital breast tomo-synthesis. In 14th International Workshop on Breast Imaging (IWBI 2018), vol. 10718, p. 1071819. International Society for Optics and Photonics, July, 2018

5. Krizhevsky, A., Sutskever, I., & Hinton, G.E.: Imagenet classification with deep convolutional neural networks. In: Advances in Neural Information Processing Systems, pp. 1097–1105 (2012)

6. Elmoufidi, A., El Fahssi, K., Jai-Andaloussi, S., et al.: Anomaly classification in digital mammography based on multiple-instance learning. IET Image Process. 12(3), 320–328

7. Elmoufidi, A., El Fahssi, K., Jai-Andaloussi, S., et al.: Detection of regions of interest's in mammograms by using local binary pattern, dynamic k-means algorithm and gray level co-occurrence matrix. In: 2014 International Conference on Next Generation Networks and Services (NGNS). IEEE, pp. 118–123 (2014)

8. Gulshan, V., Peng, L., Coram, M., Stumpe, M.C., Wu, D., Narayanaswamy, A., Kim, R.: Development and validation of a deep learning algorithm for detection of diabetic retinopathy in retinal fundus photographs. Jama 316(22), 2402–2410 (2016)

9. Martins, J., Cardoso, J.S., Soares, F.: Offline computer-aided diagnosis for Glaucoma detection using fundus images targeted at mobile devices. Comput. Methods Programs Biomed. 105341 (2016)

10. Li, Z., et al.: Development and evaluation of a deep learning system for screening retinal hemorrhage based on ultra-widefield fundus images. Transl. Vis. Sci. Technol. 9.2, 3 (2020)

11. Ortiz-Feregrino, R., et al.: Classification of proliferative diabetic retinopathy using deep learning. In: 2019 IEEE Colombian Conference on Applications in Computational Intelligence (ColCACI). IEEE (2019)

12. Wu, L., Fernandez-Loaiza, P., Sauma, J., Hernandez-Bogantes, E., Masis, M.: Classification of diabetic retinopathy and diabetic macular edema. World J Diabetes. 4(6), 290–294 (2013). https://doi.org/10.4239/wjd.v4.i6.290

13. https://www.houstoneye.com/retinal-disorders-houston/diabetic-retinopathy/

14. Schmidt-Erfurth, U., Sadeghipour, A., Gerendas, B.S., Waldstein, S. M., & Bogunović, H.: Artificial intelligence in retina. Prog. Retinal Eye Res. (2018)

15. LeCun, Y., Bengio, Y., Hinton, G.: Deep learning. Nature 521(7553), 436 (2015)

16. Kaggle Diabetic Retinopathy Classified 2019. Kaggle website. Available at: https://www.kaggle.com/dola1507108/diabetic-retinopathy-classified. Accessed 06 Nov 2019

17. Xiuqin, P., Zhang, Q., Zhang, H., Li, S.: A fundus retinal vessels segmentation scheme based on the improved deep learning U-Net model. IEEE Access 7, 122634–122643 (2019)

18. Harun, N., Yusof, Y., Hassan, F., et al.: Classification of fundus images for diabetic retinopathy using artificial neural network. In: 2019 IEEE Jordan International Joint Conference on Electrical Engineering and Information Technology (JEEIT), pp. 498–501. IEEE (2019)

Feature Selection and Classification Using CatBoost Method for Improving the Performance of Predicting Parkinson's Disease

Mohammed Al-Sarem, Faisal Saeed, Wadii Boulila, Abdel Hamid Emara, Muhannad Al-Mohaimeed, and Mohammed Errais

Abstract Several studies investigated the diagnosis of Parkinson's disease (PD), which utilized machine learning methods such as support vector machine, neural network, Naïve Bayes and K-nearest neighbor. In addition, different ensemble methods were used such as bagging, random forest and boosting. On the other hand, different feature ranking methods have been used to reduce the data dimensionality by selecting the most important features. In this paper, the ensemble methods, random forest, XGBoost and CatBoost were used to find the most important features for predicting PD. The effect of these features with different thresholds was investigated in order to obtain the best performance for predicting PD. The results showed that CatBoost method obtained the best results.

M. Al-Sarem (✉) · F. Saeed · W. Boulila · A. H. Emara · M. Al-Mohaimeed
College of Computer Science and Engineering, Taibah University, Medina, Saudi Arabia
e-mail: msarem@taibahu.edu.sa

F. Saeed
e-mail: fsaeed@taibahu.edu.sa

W. Boulila
e-mail: wadii.boulila@riadi.rnu.tn

A. H. Emara
e-mail: abdemara@gmail.com

M. Al-Mohaimeed
e-mail: mmohimeed@taibahu.edu.sa

W. Boulila
RIADI Laboratory, National School of Computer Sciences, University of Manouba, Manouba, Tunisia

A. H. Emara
Computers and Systems Engineering Department, Al-Azhar University, Cairo, Egypt

M. Errais
Research and Computer Innovation Laboratory, Hassan II University of Casablanca, Casablanca, Morocco
e-mail: mahammed.errais@gmail.com

189
F. Saeed et al. (eds.), *Advances on Smart and Soft Computing*, Advances in Intelligent Systems and Computing 1188, https://doi.org/10.1007/978-981-15-6048-4_17

Keywords CatBoost method · Ensemble methods · Features importance · Feature selection · Parkinson's disease

1 Introduction

Parkinson's disease (PD) is a progressive neurodegenerative disorder characterized by the degeneration of dopaminergic nerve cells in the substantia nigra [1, 2], leading to deficiency of dopamine in neural cells, which causes four cardinal motor symptoms including tremor, rigidity, bradykinesia and postural instability [3, 4]. Other non-motor manifestations such as depression, sleep problems, muscle and joints pain as well as dementia can also occur [5]. PD is one of the global age-related progressive neurological disorders, correlated to different risk factors including environmental and genetic factors [6]. More than 6 million persons suffer from PD all over the world. Despite the availability of many anti PD medicines which give rise to some relief from symptoms of the disease but not a complete recovery of PD, and in some advanced cases, surgical treatment may be required [7].

Recently, clinical datasets are continuously increasing in size and become more complex, leading to high-dimensional big datasets. Dimensionality reduction is aiming at decreasing the number of variables under consideration. Both feature extraction and feature selection techniques are used for the dimensionality reduction process. Dimensionality reduction plays an important role in reducing the time and storage space required for experiments and in improving medical data interpretation of PD [8].

In machine learning, several methods were applied to improve the prediction of PD, including ensemble methods [9]. In addition, feature selection techniques play an important role in reducing the training time and decreasing the number of features from the original ones, by choosing only informative and relevant features and eliminating the redundant ones to increase the performance of the classifiers. Feature selection algorithms can be divided into three techniques; wrappers (use search techniques to search and estimate which subset of features are important), filters (choose a subset of features without any evaluation) and hybrid methods (a combination of the previous methods).

In this paper, the most important features for the PD dataset were identified and used by several ensemble learning methods including random forest, extreme gradient boosting (XGBoost) and categorical boosting method (CatBoost) to enhance the prediction of PD.

This paper is organized as follows. Section 2 presents related works. Section 3 discusses the methods. Section 4 presents the experimental design, and Sect. 5 showed the experimental results and discussion. Section 6 presents the conclusion and future works.

2 Related Works

In the literature, feature importance (FI) is gaining a continuous interest in the research community. Different feature ranking methods have been used to reduce the dimensionality by selecting the optimal features. Saeys et al. [10] suggested using feature selection methods to propose an early diagnosis of PD. The proposed method combined several methods namely support vector machine (SVM), Naïve Bayes, logistic regression, boosted trees and random forests. The authors evaluated the effect of combining these methods on classification performance. They concluded that using an ensemble technique provides better results especially for high-dimensional domains with small data size, which can help to determine early diagnosis of PD.

Khan et al. [11] used random forest to estimate FI that influences PD prediction. The authors started by extracting features from speech samples based on the unified PD rating scale motor-speech examination. Then, they performed a statistical analysis to compare groups. Finally, random forest was applied to identify the severity of speech impairment using cepstral separation difference features. They proposed to use permutation importance score for FI in classification. The results depicted that the proposed method based on random forest and FI succeeded to characterize the severity of speech impairment of non-native language speakers with high accuracy.

In [12], the authors proposed a novel measurement criterion called mean decrease impurity (MDI) for estimating features importance of random forest using simulated data and a genomic ChIP dataset. The results showed that the proposed MDI approach achieved a good performance in feature selection for both deep and shallow trees.

In addition, Lai et al. [13] studied the impact of different FI methods in text classification. They compared, in a systematic manner, the features produced from built-in and post-hoc methods. The results showed that no matter what method is used to compute the important features since traditional models such as SVM and XGBoost are more similar to each other. Post-hoc methods also tend to generate more similar important features. Additionally, they noted that even if two models agreed on the predicted label, the important features do not always resemble each other. In [14], the authors presented two methods based on the measurement of the sensitivity for incremental ranking of features in classification tasks. The proposed methods were designed also to handle the concept of drift problem.

In [15], Hoyle et al. showed the importance of feature selection in the field of photometric redshift. They proposed a machine learning approach based on a decision tree combined with an ensemble learning technique (AdaBoost). The goal was to identify FI using the Gini criteria to determine features that have the most predictive power for redshift estimation. After that, in order to enhance the estimation of redshift, the authors applied artificial neural networks to combine the selected features with the standard magnitudes and colors.

In addition, Elghazel and Aussem [16] illustrated a method called random cluster ensemble (RCE) to evaluate FI in unsupervised learning. They suggested finding a solution for supervised and unsupervised feature selection by using ensemble learning. The proposed method was evaluated on several datasets coming from the UCI repository and compared to several state-of-art methods. The results depicted that the RCE method can significantly improve the clustering accuracy, especially in very large domains.

Moreover, Prakash and Kankar in [17] proposed to use XGBoost and ReliefF as a feature ranking technique. They compared the results of both techniques and their implication on two different activation functions. The results showed that XGBoost gives higher performance.

3 Methods

In this section, we present the details of the proposed model for improving the performance of predicting PD. The proposed method is summarized into two main steps, as shown in Fig. 1:

- **Finding the Features Importance**: Here, we used several tree-based classifiers such as random forest, XGBoost and CatBoost to compute the feature importance for PD dataset.
- **Examining the Effect of Number of Features and the Threshold**: At this step, each ensemble classifier uses different sets of features (obtained in the previous step) and different thresholds in order to obtain the best accuracy.

4 Experimental Design

The dataset used in this study is available online at UCI Machine Learning Repository,[1] which includes acoustic features of 80 patients. Half of these patients are suffering from PD. The dataset has 240 recordings with 46 acoustic features extracted from three voice recording replications per patient.

The experimental protocol was designed for finding those features that enhance the performance of predicting PD (see Fig. 1). The experiments were carried out on Jupyter Notebook (Anaconda3) with Python version 3.8 and 64-bit Windows 10 Pro operating system with 8 RAM and Intel® Core™ i7-4600U CPU @ 2.10 GHz 2.70 GHz. Besides the standard Python libraries, we installed XGBoost version 0.90 and CatBoost 0.14.2. The size of the training set and the testing set was 67% and 33%, respectively. To evaluate the performance of each classifier, the results have been reported in terms of accuracy. Finally, we analyzed the results achieved from the experimentations.

---

[1] https://archive.ics.uci.edu/ml/datasets/Parkinson+Dataset+with+replicated+acoustic+features+

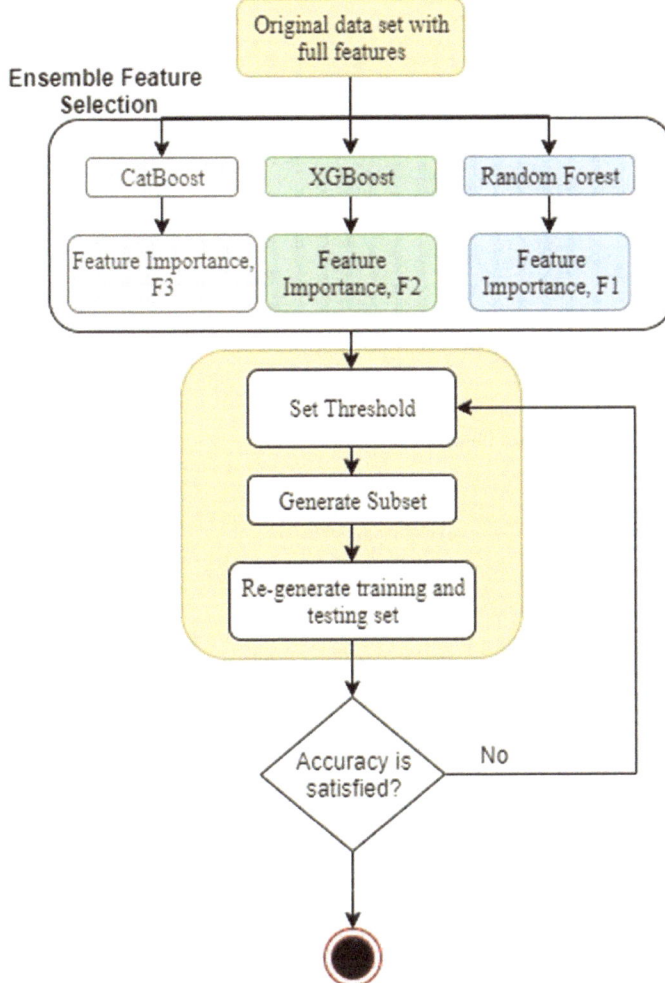

Fig. 1 Proposed ensemble feature selection

5 Experimental Results and Discussion

5.1 Finding the Important Features Using Random Forest

Random forest (RF) classifier is a type of ensemble method that generates multiple decision trees randomly. Besides the robustness of the model, RF classifier has a helpful feature called *information importance*. Information importance helps to find which of variables have the most effect on the classification [18]. Figure 2 shows the obtained feature importance for all features engaged in the RF model. To find

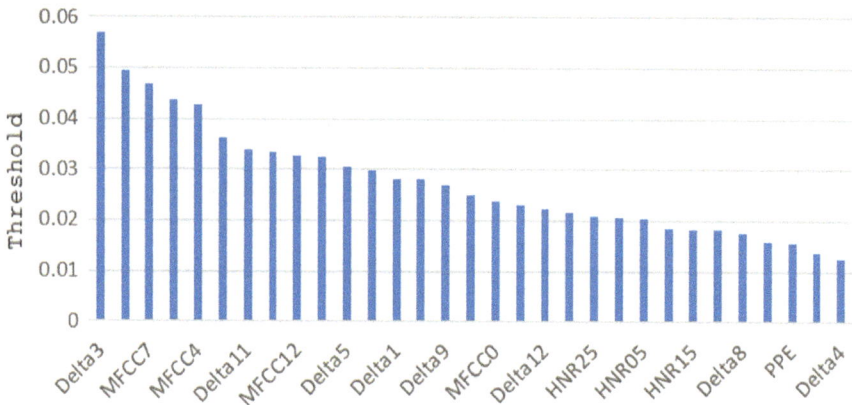

Fig. 2 Random forest features importance

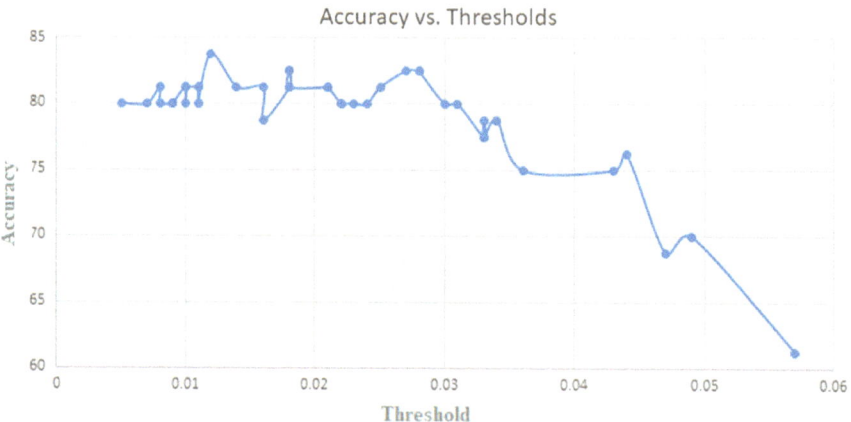

Fig. 3 Accuracy of RF model respect to the thresholds value

the most important features that extremely influence the model, the *threshold* values
that yield the highest model accuracy were investigated. Figures 3 and Fig. 4 showed
that with a threshold of 0.012 and using 31 features, the accuracy of the RF model
was 83.75% which enhances the model +3.75%.

5.1.1 Finding the Important Features Using XGBoost

XGBoost is an improvement of the traditional gradient boosting decision tree [19].
XGBoost can perform the three major gradient boosting techniques, which are
regularized boosting, stochastic boosting and gradient boosting [20]. Similarly, the
XGBoost provides the list of the important features that speed up the construction of

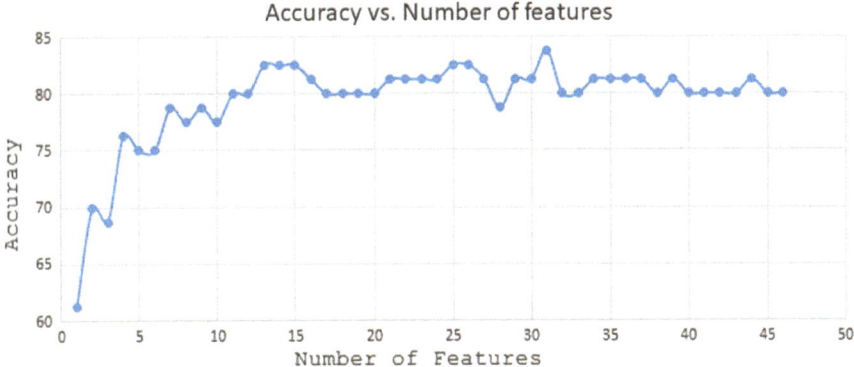

Fig. 4 Accuracy of RF model respect to the number of selected features

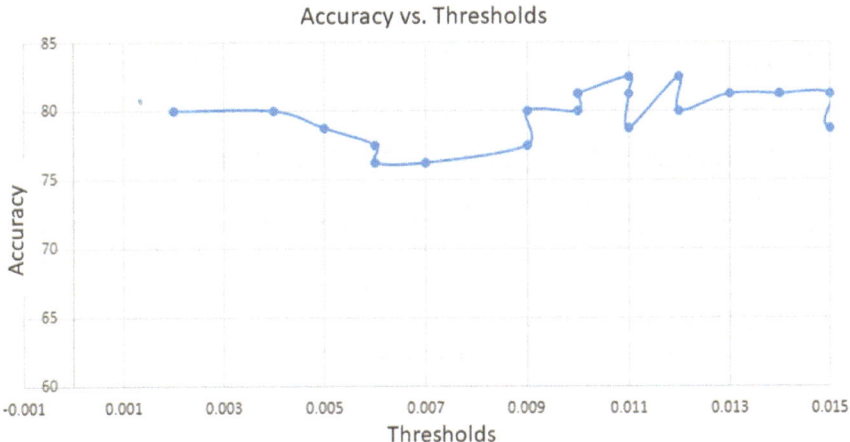

Fig. 5 Accuracy of XGBoost model respect to the thresholds value

the boosted trees. Figures 5 and 6 showed that the best threshold that yields the highest accuracy is 0.017. Accordingly, the number of important features was reduced from 46 to 23, which is less than the number of features that were produced by the random forest.

5.1.2 Finding the Important Features Using CatBoost

CatBoost is a powerful gradient boosting machine learning technique that achieves state-of-the-art results in a variety of practical tasks [21, 22]. Although CatBoost is designed mainly to deal with categorical features, however, it is possible to run CatBoost over a dataset with continuous features. The importance of features is

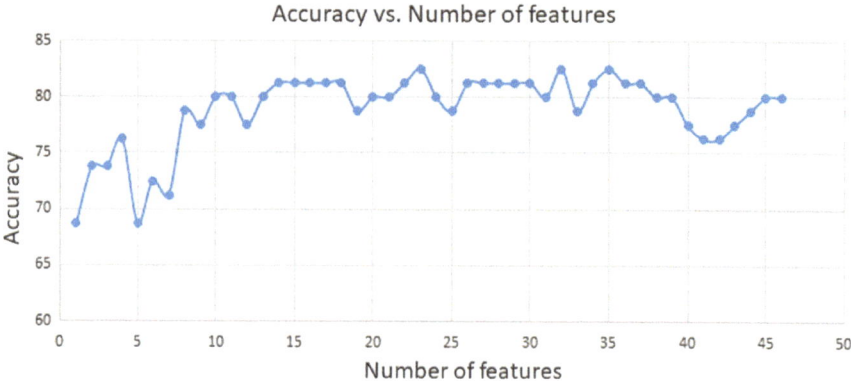

Fig. 6 Accuracy of XGBoost model respect to the number of selected features

depicted in Fig. 7. According to Fig. 8, the best accuracy (86.25%) was achieved after 56 iterations.

5.1.3   Comparison of the Performances of the Three Methods

As stated earlier, the proposed approach computes the important features using three different ensemble classifiers namely random forest, XGBoost and CatBoost. Table 1 shows that the CatBoost reaches its highest accuracy (86.25%) when the number of features was 23 and the threshold was adjusted to be 2. In addition, the results showed that different models yield different number of features.

6   Conclusion

This paper examined the performance of three ensemble methods which are random forest, XGBoost and CatBoost. The methods were used to obtain the most important features for Parkinson's disease dataset; the effect of different sets of important features on the predicting PD was investigated. The results showed that random forest used the highest number of features (31 features) in order to achieve its best accuracy, while the CatBoost obtained the best accuracy with 23 features. Overall, CatBoost showed superior results comparing to the other methods. In future work, other ensemble methods could be applied to obtain the most important features. In addition, the performance of predicting PD could be investigated using the obtained important features and deep learning methods [23]. A challenging perspective to this work would be investigating the performance of ensemble methods for big data in case of PD [24].

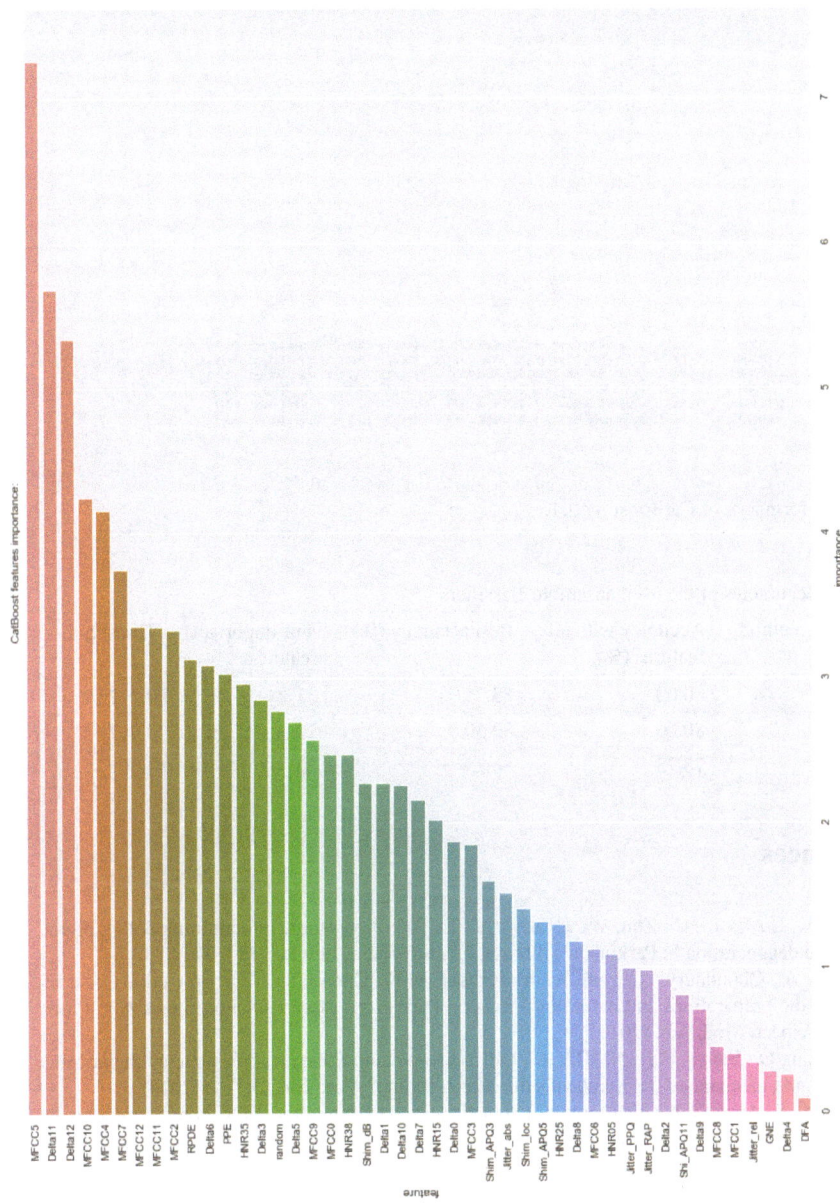

Fig. 7 Importance of features obtained by CatBoost

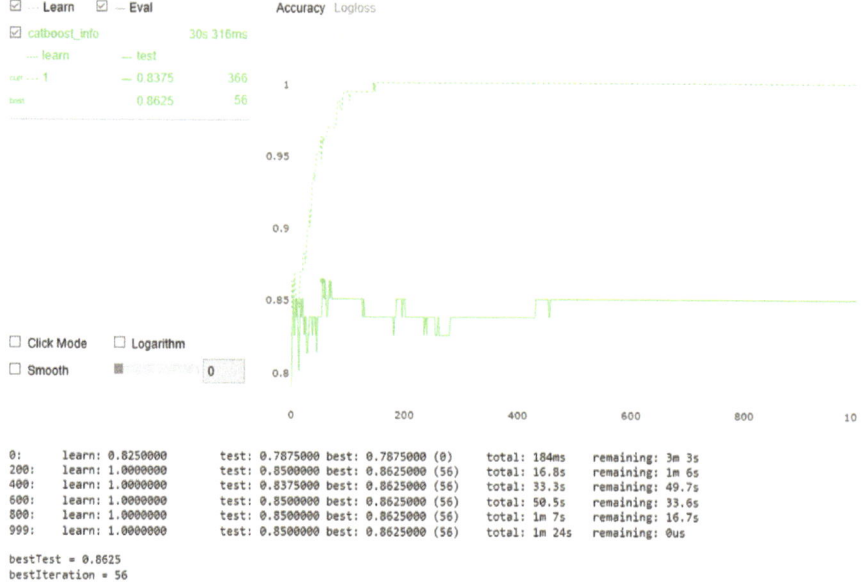

0:        learn: 0.8250000   test: 0.7875000 best: 0.7875000 (0)   total: 184ms   remaining: 3m 3s
200:      learn: 1.0000000   test: 0.8500000 best: 0.8625000 (56)   total: 16.8s   remaining: 1m 6s
400:      learn: 1.0000000   test: 0.8375000 best: 0.8625000 (56)   total: 33.3s   remaining: 49.7s
600:      learn: 1.0000000   test: 0.8500000 best: 0.8625000 (56)   total: 50.5s   remaining: 33.6s
800:      learn: 1.0000000   test: 0.8500000 best: 0.8625000 (56)   total: 1m 7s   remaining: 16.7s
999:      learn: 1.0000000   test: 0.8500000 best: 0.8625000 (56)   total: 1m 24s remaining: 0us

bestTest = 0.8625
bestIteration = 56

Fig. 8 Performance of CatBoost model

Table 1 Accuracies of the used ensemble classifiers

Ensemble method	Accuracy with all features (%)	Best accuracy (%)	# of important features	Threshold
RF	80.00	83.75	31	0.012
XGBoost	80.00	82.50	23	0.017
CatBoost	78.75	86.25	23	2.00

References

1. Xie, W., Li, X., Li, C., Zhu, W., Jankovic, J., Le, W.: Proteasome inhibition modeling nigral neuron degeneration in Parkinson's disease. J. Neurochem. **115**(1), 188–199 (2010)
2. Borah, A., Choudhury, A., Paul, R., Mazumder, M.K., Chetia, S.: Neuroprotective effect of ayurvedic preparations and natural products on Parkinsons disease. Neuroprotect. Nat. Prod. Clin. Aspect. Mod. Act. (2017)
3. Olalekan, O., Sanya, O.J.: NMDA R/VDR in fish melanocytes; receptor targeted therapeutic model and mechanism in Parkinson's disease: J. Biomol. Res. Ther. **3**(114) (2014)
4. Pallanti, S., Marras, A.: Transcranial magnetic stimulation treatment for motor symptoms in Parkinson's disease: a review of two decades of studies. Alzheimers Dis. Parkinsonism **5**(191), 2161–2460 (2015)
5. Ford, B.: Pain in Parkinson's disease. Clinical Neurosci. **5**(2), 63–72 (1998)
6. Benitez, A., Edens, H., Fishman, J., Moran, K., Asgharnejad, M.: Rotigotine transdermal system: developing continuous dopaminergic delivery to treat Parkinson's disease and restless legs syndrome. Ann. N. Y. Acad. Sci. **1329**(1), 45–66 (2014)

7. Olanow, C.W., Rascol, O., Hauser, R., Feigin, P.D., Jankovic, J., Lang, A., Tolosa, E., et al.: A double-blind, delayed-start trial of Rasagiline in Parkinson's disease. New Engl. J. Med. **361**(13), 1268–1278 (2009)

8. El Moudden, I., Ouzir, M., ElBernoussi, S.: Feature selection and extraction for class prediction in dysphonia measures analysis: a case study on Parkinson's disease speech rehabilitation. Technol. Health Care **25**(4), 693–708 (2017)

9. Patra, A. K., Ray, R., Abdullah, A. A., & Dash, S. R.: Prediction of Parkinson's disease using Ensemble Machine Learning classification from acoustic analysis. J. Phys Conf. Ser. **1372**(1) (2019)

10. Saeys, Y., Abeel, T., Van de Peer, Y.: Robust feature selection using ensemble feature selection techniques. In: Joint European Conference on Machine Learning and Knowledge Discovery in Databases, pp. 313–325. Springer, Berlin (2008)

11. Khan, T., Lundgren, L. E., Anderson, D. G., Nowak, I., Dougherty, M., Verikas, A., Aharonson, V.: Assessing Parkinson's disease severity using speech analysis in non-native speakers. Comput. Speech Lang. **61**(101047) (2020)

12. Li, X., Wang, Y., Basu, S., Kumbier, K., & Yu, B.: A Debiased MDI feature importance measure for random forests. arXiv preprint arXiv 1906 (10845) (2019)

13. Lai, V., Cai, J.Z., Tan, C.: Many faces of feature importance: comparing built-in and post-hoc feature importance in text classification. arXiv preprint arXiv 1910 (08534) (2019)

14. Razmjoo, A., Xanthopoulos, P., Zheng, Q.P.: Feature importance ranking for classification in mixed online environments. Ann. Oper. Res. **276**(1–2), 315–330 (2019)

15. Hoyle, B., Rau, M.M., Zitlau, R., Seitz, S., Weller, J.: Feature importance for machine learning redshifts applied to SDSS galaxies. Mon. Not. R. Astron. Soc. **449**(2), 1275–1283 (2015)

16. Elghazel, H., Aussem, A.: Unsupervised feature selection with ensemble learning. Mach. Learn. **98**(1–2), 157–180 (2015)

17. Prakash, J., Kankar, P. K.: Health prediction of hydraulic cooling circuit using deep neural network with ensemble feature ranking technique. Measurement **151**(107225) (2020)

18. Al-Sarem, M., Saeed, F., Alsaeedi, A., Boulila, W., Al-Hadhrami, T.: Ensemble methods for instance-based arabic language authorship attribution. IEEE Access **8**, 17331–17345 (2020)

19. Chen, Z., Jiang, F., Cheng, Y., Gu, X., Liu, W., Peng, J.: XGBoost classifier for DDoS attack detection and analysis in SDN-Based cloud. In: 2018 IEEE International Conference on Big Data and Smart Computing (BigComp), pp. 251–256. Shanghai, China (2018)

20. Dhaliwal, S., Nahid, A.A., Abbas, R.: Effective intrusion detection system using XGBoost. Information **9**(7), 149 (2018)

21. Prokhorenkova, L., Gusev, G., Vorobev, A., Dorogush, A. V., & Gulin, A.: CatBoost: unbiased boosting with categorical features. In: 32nd Conference on Neural Information Processing Systems, pp. 6638–6648. Montréal, Canada (2018)

22. Dorogush, A.V., Ershov, V., Gulin, A.: CatBoost: gradient boosting with categorical features support. arXiv preprint arXiv 1810(11363) (2018)

23. Al-Sarem, M., Boulila, W., Al-Harby, M., Qadir, J., Alsaeedi, A.: Deep learning-based rumor detection on microblogging platforms: a systematic review. IEEE Access **7**, 152788–152812 (2019)

24. Chebbi, I., Boulila, W., Farah, I.R.: Big data: concepts, challenges and applications. In: Computational collective intelligence, pp. 638–647. Springer, Cham (2015)

Lean 4.0, Six Sigma-Big Data Toward Future Industrial Opportunities and Challenges: A Literature Review

Hanane Rifqi, Abdellah Zamma, and Souad Ben Souda

Abstract Industry 4.0 technologies are generating a huge amount of data more than ever before. Heterogeneous and voluminous data puts a project's decision makers at risk of making poor decisions. For a Lean Six Sigma methodology that uses traditional statistical tools such as Analysis of Variance (ANOVA), design of experiments, etc., we found a risk to carry out both the right data collection and data processing. In our article, we discuss the advantage of Big Data analytics for good Lean Six Sigma project processing and show its benefits for the methodology and also on all elements of the supply chain.

Keywords Lean · Six Sigma · LSS · Big Data · Industry 4.0 · Industrial Big Data · Big Data analytics BDA

1 Introduction

In the era of the Fourth Industrial Revolution based on cyber-physical systems, the industrial world will face an unprecedentedly huge production of data. Many companies have experienced an exponential growth of their data and traditional information systems and techniques are no longer sufficient to process it efficiently [1]. These data require the use of powerful computational techniques to reveal trends and patterns within and between these extremely large socio-economic data sets [2].

With the advent of Big Data technology, greater volumes of data can be collected and analyzed faster than ever before [3]. Big Data contains a set of analytical techniques that help companies to obtain and suggest the best measurement practice [4].

H. Rifqi (✉) · A. Zamma · S. Ben Souda
SSDIA Laboratory, ENSETM, Hassan II University, Casablanca, Morocco
e-mail: hanane.rifqi.14@gmail.com

A. Zamma
e-mail: ph_cnd@yahoo.fr

S. Ben Souda
e-mail: souadbensou@hotmail.com

F. Saeed et al. (eds.), *Advances on Smart and Soft Computing*, Advances in Intelligent Systems and Computing 1188, https://doi.org/10.1007/978-981-15-6048-4_18

201

Among many performance measures, Lean Six Sigma (LSS) is the most widely practiced [5]. The philosophy is a fusion of two famous approaches that appeared in two different cultures; the Lean Manufacturing of the Toyoda family and Six Sigma invented by Bill Smith of Motorola. LSS is a very powerful tool that helps in the financial, strategic, and operational improvement of a company through the use of a huge selection of management tools and statistics.

These traditional statistical methods, generally used in LSS projects, such as linear regression and t-testing, are less effective with larger datasets [6] However, these analyses can be consolidated by more advanced analyses.

This is a review paper; its purpose is to discuss a hybrid model of LSS and Big Data together in order to gain more outcomes, and achieve and maintain business performance [7]. For the article methodology, we followed and adopted the process and methodology of Gupta, Modgil, and Gunasekaran in their systematic literature review [8]. The methodology followed to conduct our study includes three main phases, namely (i) the review preparation in which we searched using all keywords related to our study on the Science Direct platform, (ii) literature classification of articles related to the Lean Six Sigma approach and Big Data, and (iii) comment on the review and synthesize valorize the use of Big Data analytics with Lean Six Sigma.

To treat this subject, we followed this structure; we present a literature review of LSS, its tools, and its link with Industry 4.0, and in the third part, we will present industrial Big Data. The synergy between Big Data and Lean Six Sigma will be discussed in the fourth part and the final part will be devoted to a general conclusion of the article and a research perspective.

2 Lean Six Sigma (LSS)

2.1 Lean Six Sigma (LSS) Concept and Tools

Lean Manufacturing and Six Sigma are two very famous methods in the industrial and service world. They are also often used in an integrated way, known as Lean Six Sigma [9].

These two approaches complement each other to provide a powerful strategy that allows a process without waste and variation (see Fig. 1).

In the literature, we find implementations and real case studies of Lean Six Sigma in various sectors: transactional services [11], manufacturing [12], health [13], construction [14], airlines [15], and aeronautics companies [16, 17]. LSS has been used in education [18], project management [19], and in working toward a more sustainable environment [20]; these applications have demonstrated the effectiveness of this approach through tangible organizational and financial results, as LSS requires a well-structured belt system that requires extensive training and support on a large scale.

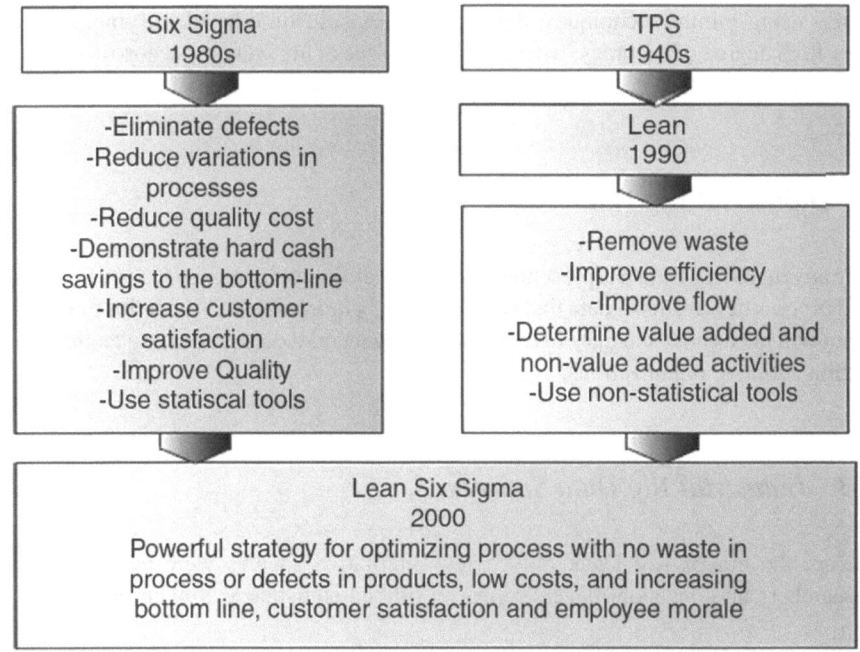

Fig. 1 Integration of lean manufacturing and six sigma [10]

In the above-mentioned LSS applications, the authors do not follow the same Lean Six Sigma implementation framework. However, they use almost the same tools [21] to achieve the objectives, namely customer satisfaction, process optimization by minimizing waste, and quality improvement by eliminating variation.

2.2 Lean Six Sigma Industry 4.0—LSS 4.0

Today, during the Industry 4.0 revolution, there are many continuous improvement methodologies such as LSS where the majority of its tools rely on data to extract the root causes of problems [22].

Although LSS is based on many traditional data analysis techniques that help improve product and process quality, there is a lot of data collected by Industry 4.0 technologies that needs to be processed by robust data analysis methods to generate relevant results [23].

LSS 4.0 dramatically optimizes processes through the structured Define, Measure, Analyze, Improve and Control (DMAIC) approach, but the huge amount of data collected requires the use of different mining techniques, such as Big Data Analytics (BDA), data mining, and process mining.

By using mining techniques, decision makers save time by identifying, with a very high degree of accuracy, what is difficult to see at first glance because decisions are based on data [23].

3   Industrial Big Data

The advent of Big Data has presented opportunities for companies to record, retrieve, and process new forms of data that were previously unavailable [24] but this development will necessitate changes in hardware, software, and data processing techniques within adoptive organizations.

3.1   *Industrial Big Data Sources*

Before the era of Big Data, only a few alternative sources were available and researchers and decision-makers relied on official statistics as a main data source [25].

Currently, industrial Big Data are produced via different sources in different manufacturing areas and can be described under five elements [26]:

Large-scale device data: New types of data produced by new Mobility and Cyber-Physical System (CPS) related devices (sensors, actuators, video cameras, and radio frequency infrared (RFID) readers) that will serve both humans and machines in making the most relevant and valuable decisions.

Production lifecycle data: Includes production requirements, design, manufacturing, testing, sales, and maintenance.

Manufacturing value chain: To understand the drivers of competitiveness, companies use data from customers, suppliers, partners, and other data.

Business operating data: These types of data include organizational structure, business management, production, devices, marketing, quality control, production, procurement, inventory, objectives, plans, e-commerce, and other data.

External collaboration data: in order to respond to external environmental changes induced by the risk, it is necessary to take into account data related to economy, industry, market, competitors, and other data.

4 Big Data Analytics (BDA)

Today, there are a variety of data analysis techniques available. BDA was considered one of the most important technologies because of its ability to explore large and heterogeneous datasets to uncover hidden patterns and knowledge and other useful information that could help industry leaders make more informed business decisions, optimize the entire product life cycle and ensure more sustainable production [27].

BDA also helps to identify transformers at risk and detect abnormal behaviors of connected devices [28].

Today's BDA includes several techniques to gain in-depth knowledge where many of them test the combination of various algorithms and technologies [23].

Among these, we found data mining techniques in which integrated data processing is carried out using statistical and mathematical techniques. Among the many data mining techniques, the most frequently used by business people are: Statistics, clustering, visualization, decision tree, association rules, neural networks, and classification.

Despite the advantages of BDA, Big Data remains difficult to interpret and analyze [28] for many reasons (e.g., its complex nature, the need for scalability, and performance to analyze heterogeneous and real-time data sets). To face these challenges, it is necessary to know how to ensure a rapid response when the volume of data is very large [29].

5 Synergy Between LSS and Big Data

Most LSS research focuses either on frameworks and models for applying the methodology in different types of industries, or on the specific tools and analytical techniques used in the LSS approach.

There are several types of study and a variety of case studies on LSS, but there are too few that directly highlight the relevance of the link between LSS and Big Data primarily through comprehensive theoretical or empirical research [7].

5.1 Traditional LSS Versus LSS with Big Data Analytics

In LSS projects with a low data volume, the traditional implementation structure of DMAIC projects has to be followed in conjunction with Six Sigma statistical tools. However, in the case of the availability of large amounts of data to be exploited, the LSS project must be executed according to the iterative phases of DMAIC and the presence of a data scientist on the project team is essential [6]. Figure 2 shows the need for a data scientist on a project with an unknown solution, a high level

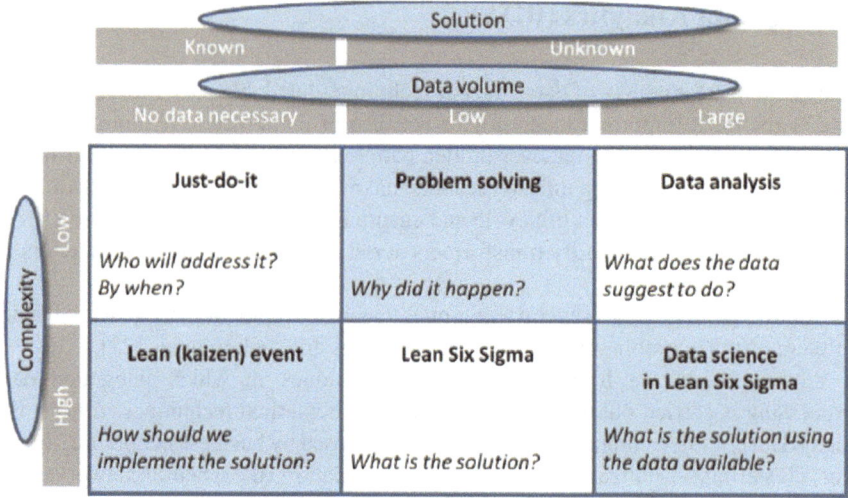

Fig. 2 Improvement methodology selection matrix [6]

of complexity, and a very high data volume. In this project, LSS cannot solve the problem with relevance through its traditional toolbox.

Industry 4.0 technologies will produce voluminous and heterogeneous data which forces decision makers to store as much data as possible in order to avoid making bad decisions, however, modeling or optimization is not easy with quality problems containing voluminous data.

Big Data has the ability to process and analyze heterogeneous data and extract relevant information at an acceptable cost. While facilitating companies that are looking to gain competitive advantage through Big Data [3] and supporting the agility and sustainability of their supply chain [30].

During 2018 and 2019, articles that have dealt with Big Data and LSS are not numerous. An analytical study of the results presented a research framework for the application of BDA in the DMAIC LSS phases [8] (see Table 1). For each phase of

Table 1 Lean six sigma framework with Big Data analytics [8]

Phases of LSS				
Define	Measure	Analyze	Improve	Control
• Text mining • Video mining • Process discovery	• Conformance checking • Confidence interval • Process sigma	• Decision trees • Association rules • Clustering • Classification • Machine learning	• Artificial intelligence • Machine learning • Predictive analytics • Flow diagrams	• Graphing • Visualization • Causality
BDA techniques and technologies for all dimensions				

the DMAIC cycle, a number of Big Data techniques are dedicated to facilitate the processing and the classification of the data.

Through the exploitation of BDA in LSS projects, there are huge opportunities to benefit from Big Data and data science tools and techniques advantages within the framework of LSS implementation [31].

BDA offers advantages and suggestions to achieve the best performance measures for all types of firms [4, 32] Inopportunely the procedure of integration of the two approaches still remains ambiguous and abstruse [6].

5.2 Benefits of Big Data in LSS

LSS enables the successful integration of Big Data processing, because it is a strategy addressing the entire process system, aimed to reduce non-value added activities by processing a huge amount of data [33].

The LSS framework provides a structured and measured DMAIC approach, while advanced methods and BDA have the ability to predict and analyze business problems more appropriately [34]. This advantage of Big Data can improve the response of LSS projects.

In the previous section, we cited some of the reasons that led researchers to incorporate Big Data into LSS projects; below is presented a non-exhaustive list of the Big Data benefits for a successful execution of an LSS project:

- Extracting information from event logs that can help minimize variation and waste in the processes [35].
- The collection of manufacturing data and its integration with customer data can lead to the improvement of products and processes [31].
- Value creation in a variety of operations from manufacturing production and value chain optimization to make more efficient use of the workforce and improve customer relations through the effective use of data [36].
- The adoption of Big Data will facilitate real-time information sharing, optimize the supply chain and human resources, better understand customers, improve financial metrics, and develop the knowledge essential for decision making in LLS projects [32].
- Big Data can provide unique information on market trends and customer buying habits [37].
- Big Data offers the opportunity for leaders of LSS projects to solve problems that were previously thought to be unsolvable [31].
- The use of different mining techniques helps to reduce lead times, to produce better quality products, and to make effective decisions [23].
- Big Data provides maintenance cycle data, as well as ways to reduce costs and enables more targeted business decisions [37].
- The use of surveys that help to better understand customers' needs and experiences [31], especially as LSS focuses on customer satisfaction.

The literature has contributed to the improvement of the LSS approach by integrating it with other tools or techniques to make it more sophisticated. However, currently, little is known about the integration of the two domains, LSS and Big Data analytics [6], while there is a huge opportunity for LSS to take advantage of Big Data and data science.

6 Conclusion and Perspective

Industry 4.0 has a positive impact on several operational performance measures [38]. However, industry has been involved in the generation of the extremely voluminous data collected by its technologies, which must be processed by robust data analysis methods in order to generate relevant results.

During this Fourth Industrial Revolution, many continuous improvement methodologies such as LLS are used; a powerful method based on statistical analysis and management tools for multiple objectives:

- Improving product and process quality by minimizing the variation.
- Reduction of waste that prevents the maximum profit of the organization.
- Customer satisfaction.
- Focus on value-generating activities.

Over time, the descriptive and the exploratory data analysis tools and techniques taught in green and black belt training will be unable to model or optimize quality problems, especially as project managers are forced to store as much data as possible to avoid making bad decisions.

In this article, we have started to address the issues related to the voluminous data produced by Industry 4.0 technologies by showing the advantages and benefits of BDA in LSS projects, which contribute to the improvement of quality, processes, supply chain, profit, customer satisfaction, and more.

Finally, Big Data and its techniques facilitate the tasks of data storage, analysis, and processing in favor of LSS projects, but LSS can also assist to solve the problems related to Big Data during the extraction, sorting and analysis phase, adding automation and optimization of Big Data functions (categorization, classification, grouping, digitization, …).

The connection between LSS and Big Data has been increasingly established through its contribution to accelerate the process of extracting key information from Big Data, while highlighting how Big Data can bring new light and innovation to projects requiring the use of Lean Six Sigma [39].

From a research perspective, we will try to work on a model that combines Big Data and its techniques with LSS methodology in a way that is more relevant in both theoretical and empirical frameworks. In this model, the links of the different techniques of Big Data with those of LSS in the DMAIC approach phases should be studied. This work will be based on a survey dedicated to industrial companies that have implemented LSS and that have started to encounter problems related to large data volumes.

Another perspective is to carry out a case study to find out the effect of this new concept on production in three dimensions: plant structure, plant digitization, and plant processes, also to conduct a survey to illustrate how leading industry suppliers are testing new concepts and Industry 4.0 technologies.

References

1. Montgomery, D.C.: Big data and the quality profession. Qual. Reliab. Eng. Int. **30**(4), 447 (2014)
2. George, G., Haas, M.R., Pentland, A.: BIG data and management: from the editors. Acad. Manag. J. **57**(2), 321–326 (2014)
3. Sun, S., Cegielski, C., Jia, L., Hall, D.J.: Understanding the factors affecting the organizational adoption of big data. J. Comput. Inf. Syst. **58**(3), 193–203 (2016)
4. Gupta, M., George, J.F.: Toward the development of a big data analytics capability. Inf. Manag. **53**(8), 1049–1064(2016)
5. Antony, J.: Six sigma vs lean: some perspectives from leading academics and practi-tioners. Int. J. Prod. Perform. Manag. **60**(2), 185–190 (2011)
6. Zwetsloot, I.M., Kiuper, A., Akkerhuis, T.S., Koning, H.D.: Lean six sigma meets data science: integrating two approaches based on three case-studies. Qual. Eng. **30**, 419–431 (2018)
7. Antony, J., Gupta, S., Sunder, M.V., Gijo, E.V. : Ten commandments of lean six sigma: a practitioners. Int. J. Prod. Perform. Manag. **67**(6), 1033–1044 (2018)
8. Gupta, S., Modgil, S., Gunasekaran, A.: Big data in lean six sigma: a review and further research directions. Int. J. Prod. Res. 22 p (2019)
9. Walter, O.M.F.C., Paladini, E.P., Kalbusch, A., Henning, E.: Lean six sigma and sustainability: literature review analysis. In: 7th Workshop Advances in Cleaner Production, pp. 1–10 (2018)
10. Albliwi, S., Antony, J., Lim, S.: A systematic review of lean six sigma for the manufacturing industry. Bus. Process Manag. J. **21**(3), 665–691 (2015)
11. Laureani, A.: Lean six sigma in the service industry. In: Advanced Topics in Applied Operations Management. InTech, Rijeka (2012)
12. Indrawati, S., Ridwansyah, M.: Manufacturing continuous improvement using lean six sigma: an iron ores industry case application. Ind. Eng. Serv. Sci. IESS 2015. Procedia Manufacturing, pp. 528–534 (2015)
13. Cima, R., Brown, M., Hebl, J., Moore, R., Rogers, J., Kollengode, A., Amstutz, G., Weisbrod, C., Narr, B., Deschamps, C.: Use of lean and six sigma methodology to improve operating room efficiency in a high-volume tertiary-care academic medical center. J. Am. Coll. Surg. **213**, 83–92 (2011)
14. Bos, A.V.D., Kemper, B., Waal, V.D., : A study on how to improve the throughput time of Lean Six Sigma projects in a construction company. Int. J. Lean Six Sigma **5**(2), 212–226 (2014)
15. Psychogios, A., Tsironis, L.: Towards an integrated framework for Lean Six Sigma application: lessons from the airline industry. Total Qual. Manag. Bus. Excell. **23**, 397–415 (2012)
16. Thomas, J., Francis, M., Fisher, R., Byard, P.: Implementing lean six sigma to over-come the production challenges in an aerospace company. In: Prod. Plann. Control **27**(7), 591–603 (2016)
17. Panagopoulos, I., Atkin, C.J., Sikora, I.: Developing a performance indicators Lean-Sigma framework for measuring aviation system's safety performance. In: 19th EURO Working Group on Transportation Meeting, EWGT2016, Istanbul, Turkey. Transportation Research Procedia, vol. 22, pp. 35–44 (2016)

18. Antony, J., Ghadge, A., Ashby, S., Cudney, E.: Lean six sigma journey in a UK higher education institute: a case study "lean six sigma journey in a UK higher education institute: a case study. Int. J. Qual. Reliab. Manag. **35**(2), 510–526 (2018)
19. Teneraa, A., Pinto, L.C.: A lean six sigma (LSS) project management improvement model. In: 27th IPMA World Congress. Procedia—Social and Behavioral Sciences, vol. 119, pp. 912–920 (2014)
20. Powell, D., Lundeby, S., Chabada, L., Dreyer, H.,: Lean six sigma and environmental sustainability: the case of a Norwegian dairy producer. Int. J. Lean Six Sigma **8**(1), 53–64 (2017)
21. Antony, J.: Implementing the lean sigma framework in an Indian SME: a case study. Prod. Plann. Control. **17**(4), pp. 407–423 (2006)
22. Arcidiacono, G., Alessandra, P.: The revolution Lean Six Sigma 4.0. Int. J. Adv. Sci. Eng. Inf. Technol. **8** (2018)
23. Dogan, O., Gürcan, O.F.,: Data perspective of Lean Six Sigma in Industry 4.0 Era: a guide to improve quality. In: Proceedings of the International Conference on Industrial Engineering and Operations Management Paris, France, pp. 943–953 (2018)
24. Purcell, B.M.: Emergence of "Big Data" technology and analytics Article. J. Technol. Res. 6 p (2012)
25. Upadhyaya, S., Kynclova, P.: Big data—Its relevance and impact on industrial statistics. United Nations Industrial Development organization, Vienna. Inclusive and Sustainable Industrial Development Working Paper Series WP 11 (2017)
26. Wang, J., Zhang, W., Shi, Y., Duan, S.: Industrial big data analytics: challenges, methodologies, and applications. IEEE Trans. Autom. Sci. Eng. abs/1807.01016.13 p (2016)
27. Ren, S., Zhang, Y., Liu, Y., Sakao, T., Huisingh, D., Almeida, C.M.V.B.: A comprehensive review of Big Data analytics throughout product lifecycle to support sustainable smart Manufacturing: a framework, challenges and future research directions. J. Cleaner Prod. **210**, 1343–1365 (2019)
28. Oussous, F.Z., Benjelloun, A., Ait Lahcen, A., Belfkih, S.: Big Data technologies: A survey. J. King Saud Univ. Comput. Inf. Sci. 1–8 (2017)
29. Fang, H., Zhang, Z., Wang, C.J., Daneshmand, M., Wang, C., Wang, H.: A survey of Big Data research. IEEE Netw. **29**, 6–9 (2015)
30. Hazen, B.T., Skipper, J.B., Ezell, J.D., Boon, C.A.: Big data and predictive analytics for supply chain sustainability: a theory-driven research agenda. Comput. Ind. Eng. **101**, 592–598 (2016)
31. Antony, J., Snee, R., Hoerl, R. : Lean Six Sigma: yesterday, today and tomorrow. Int. J. Qual. Reliab. Manag. **34**(7), 1073–1093 (2017).
32. Wamba, S.F., Akter, S., Edwards, A.J., Chopin, G., Gnanzou, D.: How „Big Data" can make Big impact: Findings from a systematic review and a longitudinal. International Journal of Production Economics, Forthcoming. **165**, 234–246 (2015)
33. Arcidiacono, G., De Luca, E.W., Fallucchi, F., Pieroni, A.: The use of lean six sigma methodology in digital curation. In: 1st Workshop on Digital Humanities and Digital Curation (2016)
34. Fogarty, D.J.: Lean six sigma and advanced analytics: integrating complementary activities. Global J. Adv. Res. **2**(2), 472–480 (2015)
35. Rovani, M., Maggi, F.M., de Leoni, M., van der Aalst, W.M.: Declarative process mining in healthcare. Exp. Syst. Appl. **42**(23), 9236–9251 (2015)
36. Matthias, O., Fouweather, I., Gregory, I., VERNON, A.: Making sense of Big Data—can it transform operations management? Int. J. Oper. Prod. Manag. **37**(1), 37–55 (2017)
37. Wang, G., Gunasekaran, A., Ngai, E., Papadopoulos, T.: Big Data analytics in logistics and supply chain management: certain investigations for research and applications. Int. J. Prod. Econ. **176**, 98–110 (2016)
38. yMoeuf, A., Pellerin, R., Lamouri, S., Tamayo, S., Barbaray, R.: The industrial management of SMEs in the era of Industry 4.0. Int. J. Prod. Res. **56**(3), 1118–1136 (2017)
39. Fogarty, D.: Lean six sigma and big data: continuing to innovate and optimize business processes. J. Manag. Innov. **1**(2), 2–20 (2015)

Fog Computing in the Age of Big Healthcare Data: Powering the Medical Internet of Things

Hayat Khaloufi, Karim Abouelmehdi, and Abederrahim Beni-Hssane

Abstract Today's global marketplace for Internet of Things (IoT) remains fiercely competitive within the healthcare sector. This can be attributed to many exciting achievements. It is expected to have a positive influence on all aspects of healthcare such as reducing prevalence and incidence of chronic conditions, improving the patient treatment and lowering in per capita healthcare costs, resulting from the forecasted increases in the adoption of remote monitoring devices. These smart devices along with wearables and Internet-connected sensors are expected to generate a great amount of data every second. However, the present model used to handle these unprecedented data named as cloud computing becomes inadequate. The solution must be seen in fog computing. The strong need of this model was mainly driven by a growing demand for real-time analysis and low latency. Motivated from these facts, this paper defines fog computing for IoT-driven smart healthcare while comparing it with cloud computing. It also presents an architecture of fog-based healthcare system containing three layers: edge, fog, and cloud. Finally, this paper highlights opportunities and top challenges facing smart healthcare organizations in the coming years.

Keywords Fog computing · Big data · Internet of things (IoT) · Smart healthcare · Cloud computing

H. Khaloufi (✉) · A. Beni-Hssane
Department of Computer Science, LAROSERI Laboratory, Faculty of Science El Jadida, Chouaib Doukkali University, El Jadida, Morocco
e-mail: hayat.khaloufi@gmail.com

A. Beni-Hssane
e-mail: abenihssane@yahoo.fr

K. Abouelmehdi
Laboratory of Applied Mathematics and Information System, Faculty Polydisciplinary of Nador, Mohamed First University, Nador, Morocco
e-mail: karim.abouelmehdi1@gmail.com

F. Saeed et al. (eds.), *Advances on Smart and Soft Computing*, Advances in Intelligent
Systems and Computing 1188, https://doi.org/10.1007/978-981-15-6048-4_19

1 Introduction

Over the last few years, the IoT has been undoubtedly the most exciting technology that gains popularity for the next generation of healthcare systems.

However, the increasing integration of this emerging technology in health facilities has highlighted several issues and challenges facing smart healthcare organizations, firstly, storage and analysis of big medical data. Using traditional architectures and algorithms is, therefore, inadequate and inefficient for storing and processing these data. The second challenge is the exchange and interoperability of data between devices. Achieving meaningful health information interoperability becomes a constraint due to differences in operating systems, hardware, and programming language for those devices. The third challenge is security and privacy. Considering the health environment, securing big healthcare data collected from millions of things, and protecting the confidentiality and illegal disclosure during transfer of these data between different systems and organizations are a point of major concern. Moreover, comprehensive, unified, and ubiquitous access to medical information to authorized users at any point of care or decision making and any time presents another serious challenge.

It is believed that traditional cloud computing architectures cannot address all of these challenges [1]. Indeed, sending large amount of heterogeneous variety and high velocity data to the cloud for further analysis and storage is impractical [2]. Furthermore, mainly due to its inherent latency, cloud could not achieve real-time interaction and location awareness along with supporting mobility and large-scale networks [3, 4], that remain essentially the requirements for a successful decision-making relating health. Fog computing technology is seen as a possible solution that can perfectly play a crucial role in processing big health data generated from billions of distributed sensors.

Actually, fog computing is as an intermediate layer between the cloud (backend) and the edge devices [5]. It is designed to augment cloud efficiency and not to replace it. This is reinforced by the fact that fog extends cloud services to the edge of the network and ultimately reducing demand for big data analysis in the cloud.

This paper has attempted to review and discuss fog computing for IoT-based healthcare applications while comparing it with cloud computing in several aspects. The remainder of the paper is organized as follows. Section 2 discusses related work including latest industry trends, concepts, applications, and services regarding medical IoT and fog-based smart healthcare. Characteristics and architecture of fog-based healthcare system—where IoT, fog, and cloud are interacting with each other—are carefully explained in Sect. 3. The last section contains the concluding remarks and highly recommends the integration of IoT, fog, and cloud into healthcare for open research directions.

2 Motivation and Related Work

2.1 Wearable Big Data in Healthcare

The ceaseless augmentation of health information coming from intelligent edge devices such as wearables results in an incredible volume of high velocity and high variety healthcare data. These wearables come in the form of wristbands, smart-watches, smart textiles, skin patches, smart shoes and socks, and even contact lenses. The information offered by these wearables are absolutely necessary for the patients to increase their control over their own health status over time and make informed decisions. Actually, the greatest benefit of these wearables in healthcare is that they allow patients improving, monitoring, and tracking their health in non-clinical settings such as private homes, nursing homes, and assisted living.

This section discusses a wide variety of latest medical sensors in different stages of technology readiness, from heart monitors to temperature, emotion, and even menstrual cycle trackers and many more.

Table 1 presents some existing wearables that are usually used to manage specific condition with a focus on describing in detail how they transmit data [6].

Today, an ever-increasing number of wearable sensors are constantly gathering and transmitting too much data to the cloud where they can be analyzed and stored. Hence, gaining knowledge from these complex data becomes tedious and cumber-some. At the moment, big data can take considerable time (days, weeks, or even months) to travel from the edge to the cloud for analysis, which is unacceptable for the management of a patient health data that demands a quicker approach. Ideally, real-time big data analytics could be accomplished in only minutes, but in the life or death situation, it may result in catastrophic consequences [16]. Any delay in noti-fication of parameters such as high pressure or temperature may result in irrecover-able losses. The processing latency is then the most crucial issue for assessing the cloud efficiency. Compounding the challenge, the generated big data from medical devices contains valuable information, along with, non-deterministic errors such as motion artifacts, data corruption issues, and unwanted signals that are also uploaded increasing storage requirements and power consumption significantly [17]. Another concern is the ability to act in real time to incoming transactions and working within set limits of bandwidth. Actually, traffic that is being generated from thousands of medical sensors and sent to the cloud outstrips bandwidth capacity [18].

The model of sending raw medical data to the cloud for big data analysis is becoming inefficient, time consuming, and expensive [17]. These magnitude chal-lenges can be successfully dealt with by acting on at the edge. That is to say, data should be operated closer to the devices where it is produced instead of transmitting everything to the cloud, thus permitting faster data processing, improved response time, and scaling down the need of bandwidth. This ultimately led to lowering costs and enhancing efficiency [18]. This approach is called fog computing. The evolu-tionary shift from cloud to fog makes a lot of sense and could potentially revolutionize IoT-driven healthcare solutions.

Table 1 Examples of existing sensors used for specific conditions and detailed descriptions of data transmission

Wearable sensors	Condition/IoT roles	Data transmission
Glucose sensor Contextual sensor Dexcom CGM System Guardian connect system	**Diabetes** [7] Monitoring blood glucose and delivering appropriate insulin doses	The sensor data are sent to an Android gateway to be locally stored and pre-processed. The results are then sent to the cloud for further analysis and notifying the responsible person concerned (healthcare providers, physicians) if abnormalities occur
Optical heart rate sensor Owlet baby monitor KardiaMobile BP sensor; ECG sensor	**Heart conditions** [8] Measuring heart rates and oxygen saturation	The sensor's output is wirelessly transmitted to a microcontroller and automatically sent to the servers via an appropriate smartphone gateway
Pulse sensor Gobe activity monitor Temperature sensor MonBaby SpO2 sensor	**Asthma** [9] **and respiratory conditions** [10] Pulse oximetry and pulmonary ventilation monitoring Activity tracking and air quality assessment	The sensors signals are fed to a microcontroller for processing. Data are then sent wirelessly to the cloud for storage and further analysis via hypertext transfer protocol (HTTP)
Thermopile infrared (IR) sensor; Wearable thermometry; TempTraq sensor	**Hyperthermia and hypothermia** [11] Tracking temperatures during an illness	Sensors of wireless body area network (WBAN) are connected using an appropriate gateway to send their data to the server through WiFi, where they can be analyzed and stored
Microelectromechanical sensors Robot arms	**Teleconsultation/telesurgery** [12] Telementoring, teleproctoring, teleconferencing, and teleconsulting	Microcontroller processes sensors values. Then, data are rapidly fed to connected Zigbee system in real-time without data loss

(continued)

Table 1 (continued)

Wearable sensors	Condition/IoT roles	Data transmission
Miotherapy Force sensor Distance sensor Radio frequency identification reader (RFID)	**Rehabilitation/Medication non-compliance** [13] Improve outpatient rehabilitation, monitor and personalize treatment, increase patient compliance, and reduce recovery time	The raw sensor data are sent to the controller integrated into the node for processing and then transmitted to the remote system via a smartphone gateway or through WiFi connection
EMG sensors GymWatch Fitbit; Apple Watch Garmin	**Fitness/Neuromuscular diseases** [14] Performing exercises correctly, avoiding injuries, and maximizing workouts. Counting steps, distance, intensity, and calories	The sensors signals are transmitted to the controller for classifying and processing to reveal neuromuscular disorders. The results are sent to the cloud for storage
Connected MOM Owlet band Ava bracelet Oura ring Fitbit Versa	**Menstrual cycle** [15] Monitor fertility, pregnancy and hormonal cycles, reveal menses-driven changes in physiology, and detect phase-based shifts in pulse rate and wrist skin temperature	The sensor data are recorded by WBAN to detect menses disorders. The results are then sent to the cloud via an appropriate gateway for further analysis and storage

The following subsection presents a summary of the main contributions from recent fog-based healthcare studies.

2.2 Fog-Based Smart Healthcare

The aim of an IoT-driven healthcare is to facilitate instantaneous connection between patients and their providers and, ultimately, delivering value-based care required by populations. Many research contributions in different fields of healthcare show that fog computing is the foundational infrastructure for transforming the healthcare IoT from novelty to reality. Following some examples of healthcare fields in which fog computing is successfully operated.

- **Smart health**

Smart health area, that covers intelligent, networked technologies for improved health provision, has the potential to deeply transform the healthcare sector.

Authors in [19] have implemented a fog-enabled smart health to improve data sharing services. While in [20], authors introduced a novel fog-cloud framework for healthcare services in smart office. They have also implemented an application scenario for predicting healthcare and generating alert. Further, in [21], authors have introduced a self-adaptive filter to recollect missing or incorrect data automatically to improve network reliability and speed.

- **Smart homes and hospitals**

With the development of information technology, the concept of smart homes and hospitals has gradually come to the fore.

Smart homes technologies are designed to assist the homes' individual inhabitants accomplishing their daily-living activities and thus having personalized services based on their unique routines and needs while preserving their privacy. Meanwhile, smart hospital is a digitally organized hospital that intelligently connects all departments and entities based on optimized and automated processes in order to provide extended patient care such as increasing remote health services, reducing costs, and maintaining the required level of data privacy. Authors in [22] have investigated the noteworthy features of fog computing. Moreover, they have introduced a cloud to fog U-healthcare monitoring system in smart homes and hospitals.

- **Wearable healthcare monitoring system**

Wearable healthcare monitoring system is very important and necessary especially in remote areas. Authors in [23] have proposed a cloud-based user authentication scheme for secure authentication of medical data, while creating a secret session key for future secure communications.

- **Healthcare system privacy**

A patient privacy is the fundamental component of a well-functioning of healthcare system.

Accessing and storing private medical data securely in the cloud through a fog computing facility were the primary objectives of authors in [24]. They have proposed a tri-party one round authenticated key agreement protocol based on the bilinear pairing cryptography while implementing a decoy technique.

- **Diabetes**

It is widely known that diabetes is one of the most serious epidemics and costly conditions to manage in healthcare. It is a chronic disease associated with abnormally high levels of blood glucose (BG).

In [25], for instance, a fog-based health framework to diagnose and remotely monitor diabetic patients in real-time has been proposed. Risk assessment of diabetes patients has also been performed at regular intervals.

- **Hypertension attack**

A hypertension attack is a severe increase in blood pressure that can lead to a stroke. Paper [26] has proposed an IoT-fog-based healthcare system for ceaselessly monitoring patients and remotely predicting the risk of high level hypertension attack. Data collected from patients were sent to the cloud to be stored and shared with domain experts.

- **Acute illnesses**

Acute illnesses tend to have very quick onsets and typically last for only a brief period. For such diseases, authors in [27] have created a smart e-Health gateway prototype called UT-GATE with several higher-level features. It uses fog computing to generate a geo-distributed intermediary layer between sensor nodes and cloud.

Through this subsection, different healthcare fields in which fog computing could play a significant role are discussed while summarizing recently published papers that apply fog computing in healthcare applications.

3 Key Technologies of Fog Computing in Healthcare System

Fog computing is considered as a bridge linking IoT devices and far off data centers. It allows edge devices to conduct substantial amount of computation, analytics, storage, and communication on their own. Requirements of reliable low latency responses are therefore capable of being met using a fog computing facility. Such tech open a myriad of possibilities to improve the value, efficiency, and purpose of IoT-driven healthcare system.

3.1 How Does Fog Differ from Cloud Computing?

Both cloud computing and fog computing are created with virtual systems and offer similar functionalities in terms of flexibility and scalability of on demand supplies of computation, storage, and networking resources [28].

The mean difference between cloud computing and fog computing is the location where data processing occurs [29]. Accordingly, the communication with a cloud server can take up to several minutes, while it may only take up to a few milliseconds when interacting with "nodes" placed near the device.

Smart applications and IoT-based healthcare devices require instant decision-making tools. Since healthcare organizations are adding new, enhanced, and much better features that help in quick decisions, there is always some degree of latency, which requires subsequent implementation on the basis of fog computing.

Cloud computing is most appropriate for long term in-depth analysis of data, while fog computing is more suitable for the quick analysis required for real-time response.

While fog computing has limited processing power, comparing with cloud computing that still remains the first preference for providing superior and advanced data storing and processing technological capabilities, companies are gradually moving toward fog computing to reduce costs. It extends the cloud services at the

Table 2 Comparison between fog and cloud computing

Main features	Fog computing	Cloud computing
Server nodes location	At the local network edge	Within the Internet
Distance between devices and server	Single step	Multiple steps
Response time	Milliseconds to sub-second	Minutes, days, weeks
Latency	Low	High
Distribution	Distributed	Centralized
Geographic coverage	Very local (e.g., one city block)	Global
Length of time IoT data can be stored	Transient	Month or years
Scalability	More (micro-fog centers—easy to deploy)	Less (big cloud centers)
Mobility support	Supported	Limited
Real-time interactions	Supported	Supported
On the fly analysis	Data aggregation partially and remaining to cloud	Data aggregation at cloud
Security	More secure	Less secure
Attack on moving data	Very low probability	High probability

local network edge to address applications that require nodes in the vicinity to meet their delay requirements [30], including:

- Applications that require very low and predictable latency
- Geographically distributed applications
- Fast mobile applications
- Large-scale distributed control systems

Table 2 summaries the differences between fog and could computing [31]:

3.2 Fog-Based Healthcare System Architecture

Lately, numerous researches have introduced several fog architectures. Among them, the three tiers architecture is considered as the most predominant structure nowadays. Fog can therefore be seen as a third-party between cloud and devices (Fig. 1).

Actually, what happens in fog and cloud computing, below points helps in understanding [2]:

Fog Nodes:

- Receive feeds from IoT devices using any protocol, in real time.
- Run IoT-enabled applications for real-time control and analytics, with millisecond response time.

Fig. 1 Fog extends the cloud closer to the devices producing data [32]

- Provide transient storage often.
- Transmit periodic data summaries to the cloud.

Cloud Platform:

- Receives and aggregates data summaries from many fog nodes.
- Performs analysis on the IoT data and big data from other sources to gain business insight.
- Send new application rules to the fog nodes based on these insights.

Medical data are acquired from intelligent edge devices such as wearables and sent to an Android gateway (e.g., smartphone or tablet) to be locally stored and pre-processed. The extraction of clinical features was done locally on fog device that was kept in patient's home (or near the patient in care-homes). Finally, the results are sent to the cloud for further analysis and notifying the responsible person concerned (healthcare providers, physicians) if abnormalities occur or just for long-term comparative study.

4 Conclusions and Future Work

Healthcare system can be operative and proficient due to the growing use of various IoT medical sensors. However, the promises of the healthcare IoT are generally very exciting, and they should be used carefully, as there are still legitimate concerns associated to patient privacy, cost-effectiveness, consistency, and so on. As cloud models are not designed for the escalating volume, high-variety, and high-velocity of health-related data created by IoT devices, implementing a fog computing facility has been seen as a better approach. Fog may solve problems with the bandwidth, a considerable amount of time, and latency in a case of cloud computing that are currently stunting the growth utilization of IoT in healthcare. As a matter of fact, fog computing comes up with multiple benefits including: low latency, low bandwidth, heterogeneity, inter-operability, scalability, security and privacy, real-time processing, and actions. Even if with such model holding the potential to make IoT-based healthcare systems reliable, simpler, scalable, and exceptionally high performance, its ability to store and process big medical data, comparing to cloud, is limited. Combining fog and cloud computing is often promoted as a recipe for success in various aspects of the healthcare domain.

In further work, it would be interesting to use a fog computing facility to propose a security model for preserving the privacy of medical big data in a IoT-based healthcare cloud.

References

1. Nandyala, C.S., Kim, H-K.: Cloud to fog and IoT-based real-time U-healthcare monitoring for smart homes and hospitals. Int. J. Smart Home **10**(2), pp. 187–196 (2016)
2. Cisco: Fog computing and the internet of things: Extend the cloud to where the things are. White Paper (2015)
3. Milunovich, S., Passi, A., Roy, J.: After the cloud comes the fog: fog computing to enable the internet of things. IT hardware & data networking to enable the Internet of Things. IT Hardware & Data Networking, 4 Dec, 2014
4. Pramanik, P., PalBengal, S., Brahmachari, A., Choudhury, P.: Processing IoT data: from cloud to fog-it's time to be down to earth. In: Applications of Security, Mobile, Analytic and Cloud (SMAC) Technologies for Effective Information Processing and Management (Chap. 7). IGI Global, pp. 124–148, May, 2018
5. Westbase Technology: Fog computing vs cloud computing: what's the difference? https://www.westbaseuk.com/news/fog-computing-vs-cloud-computing-whats-thedifference/#sth ash.5j0QkbxG.5HRRJSNc.dpuf. Last accessed 01 Dec 2019
6. Minh Dang, L., Jalil Piran, Md., Han, D., Min, K., Moon, H.: A survey on Internet of Things and cloud computing for healthcare. Electronics **8**(7), 768. https://doi.org/10.3390/electroni cs8070768
7. Gia, T.N., Ali, M., Dhaou, I.B., Rahmani, A.M., Westerlund, T., Liljeberg, P., Tenhunen, H.: IoT-based continuous glucose monitoring system: a feasibility study. Procedia Comput. Sci. **109**, 327–334 (2017)
8. Xin, Q., Wu, J.: A novel wearable device for continuous, non-invasion blood pressure measurement. Comput. Biol. Chem. **69**, 134–137 (2017)
9. AL-Jaf, T.G., Al-Hemiary, E.H.: Internet of Things based cloud smart monitoring for asthma patient. In: Proceedings of the 1 st international conference on information technology (ICoIT'17), Erbil, Iraq, p. 380, 10 Apr 2017
10. Fu, Y., Liu, J.: System design for wearable blood oxygen saturation and pulse measurement device. Procedia Manuf. **3**, 1187–1194 (2015)
11. Huang, M., Tamura, T., Tang, Z., Chen, W., Kanaya, S.: A wearable thermometry for core body temperature measurement and its experimental verification. IEEE J. Biomed. Health Inform. **21**, 708–714 (2017)
12. Shabana, N.; Velmathi, G.: Advanced tele-surgery with IoT approach. In: Intelligent Embedded Systems. Springer, Berlin, pp. 17–24 (2018)
13. Yang, G., Deng, J., Pang, G., Zhang, H., Li, J., Deng, B., Pang, Z., Xu, J., Jiang, M., Liljeberg, P.: An IoT-enabled stroke rehabilitation system based on smart wearable armband and machine learning. IEEE J. Transl. Eng. Health Med. **6**, 1–10 (2018)
14. Subasi, A., Yaman, E., Somaily, Y., Alynabawi, H.A., Alobaidi, F., Altheibani, S.: Automated EMG signal classification for diagnosis of neuromuscular disorders using DWT and bagging. Procedia Comput. Sci. **140**, 230–237 (2018)
15. Goodale, B.M., Lisa Falco, M.S., Dammeier, F., Hamvas, G., Leeners, B: Wearable sensors reveal menses-driven changes in physiology and enable prediction of the fertile window: observational study. J. Med. Internet. Res. 21(4): e13404 (2019)

16. Bresnick, J.: How fog computing may power the healthcare internet of things. Health IT Analytics, August 23, 2016. https://healthitanalytics.com/features/how-fog-computing-may-power-the-healthcare-internet-of-things. Last accessed 12 Dec 2019
17. Dubey, H., et al.: Fog data: enhancing telehealth big data through fog computing. In: Proceedings of ASE BigData Social Informatics, pp. 1–6 (2015)
18. Joshi, N.: Why is fog computing beneficial for IoT? [Online]: https://www.linkedin.com/pulse/why-fog-computing-beneficial-iot-naveen-joshi?articleId=8166335329880272527#commen ts8166335329880272527&trk=sushi_topic_posts_guest. Last accessed 12 Dec 2019 (2016)
19. Tang, W., Zhang, K., Zhang, D., Ren, J., Zhang, Y., Shen, X.S.: Fog-enabled smart health: toward cooperative and secure healthcare service provision. IEEE Commun. Mag. **57**, 42–48 (2019)
20. Bhatia, M., Sood, S.K.: Exploring temporal analytics in fog-cloud architecture for smart office healthcare. Mob. Netw. Appl. 1–19 (2018)
21. Wang, K.; Shao, Y.; Xie, L.; Wu, J.; Guo, S.: Adaptive and fault-tolerant data processing in healthcare IoT based on fog computing. IEEE Trans. Netw. Sci. Eng. (2018)
22. Nandyala, C.S., Kim, H.K.: From cloud to fog and IoT-based real-time U-healthcare monitoring for smart homes and hospitals. Int. J. Smart Home **10**, 187–196 (2016)
23. Srinivas, J., Das, A.K., Kumar, N., Rodrigues, J.: Cloud centric authentication for wearable healthcare monitoring system. IEEE Trans. Depend. Secur, Comput (2018)
24. Hadeal Abdulaziz, A., Sk Md Mizanur, R., Shamim, H. M., Almogren, A., Alamri, A: A security model for preserving the privacy of medical big data in a healthcare cloud using a fog computing facility with pairing-based cryptography. IEEE Access, 1–1. 10.1109/ACCESS.2017.2757844
25. Devarajan, M.; Subramaniyaswamy, V.; Vijayakumar, V.; Ravi, L. Fog-assisted personalized healthcare-support system for remote patients with diabetes. J. Ambient Intell. Humaniz. Comput. 1–14 (2019)
26. Sood, S.K., Mahajan, I.: IoT-fog based healthcare framework to identify and control hypertension attack. IEEE Internet Things J **6**, 1920–1927 (2018)
27. Rahmani, A.M., Gia, T.N., Negash, B., Anzanpour, A., Azimi, I., Jiang, M., Liljeberg, P.: Exploiting smart e-Health gateways at the edge of healthcare Internet-of-Things: a fog computing approach. Future Gener. Comput. Syst. **78**, 641–658 (2018)
28. Sukanya, C.N., Kim, H.: From cloud to fog and IoT-based real-time U-healthcare monitoring for smart homes and hospitals. Int. J. Smart Home **10**(2), 187–196 (2016). https://dx.doi.org/10.14257/ijsh.2016.10.2.18
29. Miller, V.: Cloud, fog, and edge computing: 3 differences that matter, Nov. 05, 19 Cloud-Zone-Analysis. https://dzone.com/articles/cloud-vs-fog-vs-edge-computing-3-differences-that. Last accessed 22 Dec 2019
30. Bonomi, F., Milito, R., Zhu, J., Addepalli, S.: Fog computing and its role in the internet of things. In: Proceedings of the First Edition of the MCC Workshop on Mobile Cloud Computing, ser. MCC' 12, pp.13–16. ACM (2012)
31. Vaquero, L.M., Rodero-Merino, L.: Finding your way in the fog: towards a comprehensive definition of fog computing. ACM SIGCOMM Comput. Commun. Rev. **44**(5), 27–32 (2014)
32. Kumar, Y., Mahajan, M.: Intelligent behavior of fog computing with IOT for healthcare system. Int. J. Sci. Technol. Res. **8**(07) (2019)

Prediction of Diabetes Using Hidden Naïve Bayes: Comparative Study

Bassam Abdo Al-Hameli, AbdulRahman A. Alsewari, and Mohammed Alsarem

Abstract Classification techniques performance varies widely with the techniques and the datasets employed. A process performance classifier lies in how accurately it categorizes the item. The technique of classification finds the relationships between the value of the predictor and the values of the goal. This paper is an in-depth analysis study of the classification of algorithms in data mining field for the hidden Naïve Bayes (HNB) classifier compared to state-of-the-art medical classifiers which have demonstrated HNB performance and the ability to increase prediction accuracy. This study examines the overall performance of the four machine learning techniques strategies on the diabetes dataset, including HNB, decision tree (DT) C4.5, Naive Bayes (NB), and support vector machine (SVM), to identify the possibility of creating predictive models with real impact. The classification techniques are studied and analyzed; thus, their effectiveness is tested for the Pima Indian Diabetes dataset in terms of accuracy, precision, F-measure, and recall, besides other performance measures.

Keywords Classification · Hidden Naïve Bayes · Naïve Bayes · Decision tree · Support vector machine · Data mining · Pima Indian Diabetes dataset

B. A. Al-Hameli (✉) · A. A. Alsewari
Faculty of Computing, IBM Centre of Excellence, Universiti Malaysia Pahang, 26300 Kuantan, Malaysia
e-mail: bassam566606@gmail.com

A. A. Alsewari
e-mail: alsewari@gmail.com

B. A. Al-Hameli
Computer and Information Technology College, University of Sheba, Sana'a, Yemen

M. Alsarem
College of the Computer Science and Engineering, Taibah University, Medina, Saudi Arabia
e-mail: mohsarem@gmail.com

223
F. Saeed et al. (eds.), *Advances on Smart and Soft Computing*, Advances in Intelligent Systems and Computing 1188, https://doi.org/10.1007/978-981-15-6048-4_20

1   Introduction

Data mining (DM) has emerged as a more popular method of synthetic intelligence techniques in healthcare and business communities. It is also known as knowledge extraction and discovery, the collection of facts, data/pattern analysis, etc. [1]. Due to its several and varied uses, it is developed in the field of health care through the employment of smart devices as well as the development of intelligent systems. All of those serve as a human expert, summarize the experience, and provide consultations to specialists in this area.

Nowadays, hospitals are collecting a large amount of information in the form of patient records, called electronic health records (EHR). Data mining is used to discover knowledge for predictive purposes, which is a technique of study that helps to make inferences [1, 2].

Decision support systems (DSSs), as well as the expert systems based on machine learning, are now a persistent need in medical and health applications [3], which depend on early detection of diseases [4]. The presence of the decision tree classification, because it helps in setting up decision support systems and distinguished by its simplicity, requires less effort for data preparation during preprocessing.

Numerous researches in the medical field have depended on the use of many types of classification algorithms. The largest share of these algorithms tends to cope with the binary classification based on the type of data being handled such as the Naive Bayes (NB) [5], and it is an extension of the hidden Naive Bayes (HNB) and the support vector machine (SVM) [6]. In addition to that, the previous classifiers fall within the structural and statistical classifier, in order to make a qualitative comparison about performance and effectiveness.

In this paper, a set of algorithms for the evaluation of early detecting diabetes disease classifiers on a PID dataset of University of California, Irvine (UCI) repository has been studied. The classification algorithms of HNB, decision tree (DT) called C4.5, NB, and SVM are examining on how to use a pre-existing dataset to predict whether a patient is suspected of having diabetes. Results are compared with performance metrics of the classifiers and evaluation techniques, then sent from both methods. Over the years, several other models have been developed that they are used to predict diabetes.

2   Materials and Methods

Researchers have studied cases of diabetics and other caused or affected diseases, and studies have been followed by the use of techniques with high performance and accuracy in the field of health care and diabetes and were verified dataset for patients with diabetes mellitus in the Type-1 and Type-2 [7].

In this study, we examine the quality of performance of the HNB technique for sweating on the appropriate accuracy, as well as the revision of the previous attempts with NB, DT C4.5, and SVM. Besides, the Pima dataset as diabetic data has been used [8].

2.1 Classification Techniques

Classification is one of the important applications of DM and the most common technique in data mining filed for machine learning, in addition to one of the necessary tools for evaluation and prediction. The objective of classification is to find a class that can detect a set of attributes and features belongs to a class or represent a class or lead to identify and discover new classes that assist to determine the disease or not [9]. Classification techniques of medical domain use to classify whether a patient has a disease or not, so they are called binary classifiers. Here, we review four common classification techniques as shown in Table 1.

Researchers developed many approaches for the classification of diabetes disease by applying several methods. Table 1 shows four data mining techniques used in PID dataset prediction with accuracy the performance comparison between different classification techniques on the PID dataset.

By comparing, the results are recorded in Table 1 based on the accuracy percentage column. The maximum accuracy rate of the NB classifier reached 79.6%, while the C4.5 classifier reached 76.7% and the SVM classifier reached 77.1%. Finally, the HNB classifier reached 76%.

Existing studies have shown that data mining techniques that used PID dataset prediction are insufficient, and a proper examination is required for early diagnosis and detection of disease and data mining techniques that will improve the performance. Therefore, researchers and data analysts still seek toward improving the performance and effectiveness of classifiers in the variant domains to reach high accuracy for developing predictive models. Some of them have also customized a particular age category of cases belonging to the dataset and subject them to training and testing sets.

2.2 Experimental Design

In the first phase, preprocessing of data is necessary to prepare data to be used in the training and testing stages including data cleaning, data reduction, data transformation, and data integration. There are two important operations for the given dataset, one used to build model, besides, analyze classification process, called the training set, and others used to validate it and to determine and estimate the accuracy of the model, called the test set.

Datasets Description. The dataset consists of several medical predictor (independent) variables and one target (dependent) variable; 768 instances and 9 attributes.

Table 1 Values of accuracy of classification techniques on PID dataset by literature review

Classification techniques used	Accuracy %	Method and tools	Resources/Reference
NB	74.9	He used Type-2 diabetes mellitus (T2DM) with cross-validation method in Weka	[10]
	76.302	Apply Weka as API in MATLAB	[11, 12]
	73.57	R program	[13, 14]
	75.51	Apply function based on the principle of minimal description length (MDL) in MLC++ system	[15]
	75.68	Used the 36 UCI datasets selected by Weka	[16–18]
	75.75	Authors have used of AdaBoost with the difference that in each iteration of AdaBoost for increasing the prediction accuracy of the Naive Bayes model in Weka	[19]
	79.565	Split data into 70% for training and 30% for testing in Weka	[20]
DT (C4.5)	76.7	He used Type-2 diabetes mellitus (T2DM) with cross-validation method in Weka	[10]
	73.828	Apply Weka as API in MATLAB	[11, 12]
	72	It used the redundancy cut value and use the Bootstrap validation to minimize error rate in Tanagara	[21, 22]
	68	Weka	[9]
	74.63+	R program	[13, 14]
	73.89	Weka	[17]
	75.13	Apply function based on the principle of minimal description length (MDL) in MLC++ system	[23]
	74	ID3 with OLAP	[4]

(continued)

Table 1 (continued)

Classification techniques used	Accuracy %	Method and tools	Resources/Reference
SVM	74.8	Tanagara	[24]
	70.67	It used the redundancy cut value and use the Bootstrap validation to minimize error rate in Tanagara	[21, 22]
	72.17+	R program	[13, 14]
	77.07	Authors have used of AdaBoost with the difference that in each iteration of AdaBoost for increasing the prediction accuracy of the SVM in Weka	[19]
HNB	75.83	Used the 36 UCI datasets selected by Weka	[18, 25]
	74.57	Authors have used of AdaBoost with the difference that in each iteration of AdaBoost for increasing the prediction accuracy of the hidden Naive Bayes model in Weka	[19]
	76	Weka	[16, 17]

Table 2 shows the attributes description of the Pima Indian Diabetes (PID) dataset and represents its value and data type. To classify diabetic patients, it is necessary to know the behavior and the frequency of characteristics of this type of disease [14]. Independent variables include the number of pregnancies the patient has had, their BMI, insulin level, age, and so on [26].

PIDD set contains missing or incorrect value or containing zero values ranging from (4%-Pres) and (48%-Insu) [26].

Preprocessing of data is a frequently neglected but important step in the mining of data process, so we tend to perform steps in data preprocessing are:

1. Import PID dataset from UCI machine learning repository, we download in the format of "*arff*" which the model takes PID dataset as input in First phase.
2. Checking out of missing data in PID dataset, we used two ways to handle missing values (removing the data, replacing value), taking into consideration no addition of bias.
3. Discretizing numeric attribute values, to convert categorical variable into numerical data and assist to data Interpretation. Data discretization used in HNB

Table 2 Features and parameters of the Pima Indians Diabetes Dataset (PIDD)

Attributes	Abbre	Type	Mean	Stnd dev.	Min	Max	Distinct	Unique
Number of times pregnant	Preg	Numeric-continuous	3.8	3.4	0	17	17	2 (0%)
Plasma glucose concentration, 2 h in an oral glucose tolerance test	Plas	Numeric-continuous	120.9	32.0	0	199	136	19 (2%)
Diastolic blood pressure (mm Hg) (BP)	Pres	Numeric-continuous	69.1	19.4	0	122	47	8 (1%)
Triceps skin fold thickness (mm)	Skin	Numeric-continuous	20.5	16.0	0	99	51	5 (1%)
2-h serum insulin (mu U/mL)	Insu	Numeric-continuous	79.8	115.2	0	846	186	93 (12%)
Body mass index (kg/m^2) (BMI)	Mass	Numeric-continuous	23.0	7.9	0	67.1	248	76 (10%)
Diabetes pedigree function (DPF)	Pedi	Numeric-continuous	0.5	0.3	0.078	2.42	517	346 (45%)
Age (years)	Age	Numeric-continuous	33.2	11.8	21	81	52	5 (1%)
Class variable (diabetes on set within five years)	Class	Numeric-continuous	–	–	0	1	2	0

classifier to transform quantitative data into qualitative data [27]. We used the supervised filter named discretize in Weka. This procedure used only for HNB classifier.

4. Splitting the dataset into training set and test set, we had divided our PID dataset into two sets. The first phase is training set and the next phase is test set, and then have applied to each classifier.

The samples size of the PID dataset represents 0.65 as a negative and 0.35 as a positive case from entire dataset. Therefore, we carried out DT on the PID dataset based on the percentage split method to split 76.2% as a training set and 23.8% as a test set. In the last nodes of the tree, the leaves of the tree represent final decisions, whether the predicted results are negative or positive. The filter named the normalized polynomial kernel "NormalizedPolyKernel" in SVM has been chosen to perform the classification to apply to the training data. The overall accuracy of SVM with using percentage split technique is 78% as a training set and 22% as a test set. In NB based on percentage split technique applied on PID dataset, the dataset is divided into 78.6% for the training set and 21.4% for the testing set. Then, discretization method is carried out in PID dataset based on HNB classifier with training set 62.7% and training set 37.3%.

2.3 Performance Measures

True positive (TP) represents the number of correct predictions that an instance is a sensitivity. False negative (FN) represents several incorrect predictions that were misclassified as negative that an instance is a specificity. True negative (TN) represents the number of correct predictions that an instance is a specificity. False positive (FP) represents the number of incorrect predictions that were misclassified as positive that an instance is a sensitivity. The confusion matrix table for four classification techniques has been calculated, in Table 3.

Receiver operator characteristic (ROC) curve and the area under the curve (AUC) can be used as a detection performance measure. Kappa statistics is a metric that compares the accuracy observed with the accuracy predicted (random chance).

From the outcome of detailed performance measures, we present some significant indicators as follows:

$$\text{Accuracy} = (TP + FN)/(TP + FP + FN + TN) \tag{1}$$

$$\text{Error Rate} = (FP + FN)/(TP + FP + FN + TN) \quad \text{or} \quad 1 - \text{Accuracy} \tag{2}$$

$$\text{Precision} = TP/(TP + FP) \tag{3}$$

$$\text{Recall} = TP/(TP + FN) \tag{4}$$

$$\text{MCC} = (TP * TN) - (FP * FN)/\sqrt{((TP + FP)(TP + FN)(TN + FP)(TN + FN))} \tag{5}$$

Table 3 Confusion matrix table for classifiers on PIDD

Classifier	No. samples	TN	FN	FP	TP
NB	164	97	15	19	33
DT	183	112	13	22	36
SVM	169	106	9	22	32
HNB	286	170	21	31	64

$$\text{Sensitivity} = TP/(TP + FN) \tag{6}$$

$$\text{Specificity} = TN/(FP + TN) \tag{7}$$

$$F - \text{Measure} = (2 * \text{Recall} * \text{Precision})/(\text{Recall} + \text{Precision}) \tag{8}$$

Table 3 shows the confusion matrix which is obtained to calculate the accuracy of classification techniques.

Negative test results and positive test results are instances that are actually non-diabetic (negative) and diabetic (positive), respectively, which have then been classified as any of the two classes, non-diabetic (negative condition), or diabetic (positive condition). Therefore, we describe predicted values as positive and negative and actual values as true and false, and results obtained are in Table 3.

3 Results and Discussion

Based on the confusion matrix Table 3 for the four classifiers, the false-positive value is misclassified as positive (diabetic) for NB, C4.5, SVM, and HNB classifiers and, respectively, equals (19, 22, 22, 31), whereas the false-negative value is misclassified as negative (non-diabetic) to same classifiers and, respectively, equals (15, 13, 9, 21). Since the misclassification tends to be false positive by, respectively, about (%56, %63, %71, %60) of the misclassified instances, this means that much of the accuracy of the resulting classifiers were lost because of the false-positive error, which is of no risk compared with a false-negative error.

Accuracy is the proportion of correct predictions. Precision is the precision factor given a particular class is expected. The recall is the percentage of positive marked cases expected to be predicted as positive [22].

Dataset is computed and determined to evaluate the performance criterion using the above Eqs. (1)–(8) for the classifiers in precision, accuracy, recall, and ROC of disease detection. NB, C4.5, SVM, and HNB techniques gradually increased through the performance measures.

As observed in Table 4, the results indicate that HNB outperforms the SVM, DT, and NB classifiers, as well as the ROC curve of SVM, is a lower rate than other techniques.

Results obtained in Table 4 with the results of previous attempts for improving the performance of classifiers in Table 1 show that NB has a nuance is 0.26969, and this is illustrated using features selection technique, which is not covered in this paper. In other words, this paper does not conduct a feature selection test on the PID dataset for each classification technique in the article and used the full features in training and testing.

All classification techniques have been evaluated in this study based on the accuracy, ROC region, recall, precision, and kappa statistics discussed above-mentioned.

Table 4 Results performance measures for PID dataset Acc, Err, and other measurements for classification techniques

Classifiers	Recall	Precision	MCC	Kappa	Accuracy (%)	Specificity	ROC	F-measure
NB	0.793	0.789	0.512	0.5112	79.2683	0.836	0.843	0.790
DT (C4.5)	0.809	0.804	0.543	0.5391	80.8743	0.836	0.862	0.804
SVM	0.817	0.813	0.559	0.5494	81.6568	0.828	0.757	0.809
HNB	0.818	0.815	0.581	0.5791	81.8181	0.846	0.876	0.815

In comparison, the most significant factor was predictive accuracy as performance evaluation methods. These methods are commonly accepted to summarize and compare classifiers' performance.

Figure 1 shows the prediction accuracy obtained from studies mentioned in Table 1, referred to as ACC_d, beside to the results obtained in this study as in Table 4, and referred to as ACC_n, as well as error rate from past studies referred to as ERR-d and results obtained in this study ERR-n.

Based on the results, the accuracy and precision rate are improved as well as the error rate is decreased compared between previous and obtained results. Figure 1 shows the proportion of improvement of error and accuracy based on the using of the percentage split method.

Table 5 shows the performance of each of the classification technique in terms of time processing during building, testing the model, as well as the percentage of training, and testing.

While C4.5 and HNB are more than 80% precision, they have major variations (0.04–0.24s) in the results of their timely response, because their models of classification depending on the amount of data in each dataset. In contrast, Naïve Bayes achieves more than 84.3% of precision, with a time response varying from 0.01 to 0.02 s. Thus, the difference between HNB as the most accurate technique and C4.5 as the most stable in time is about 5.5% in precision; a difference that NB recovers, without raising the precision, decreases the time response around 5 s.

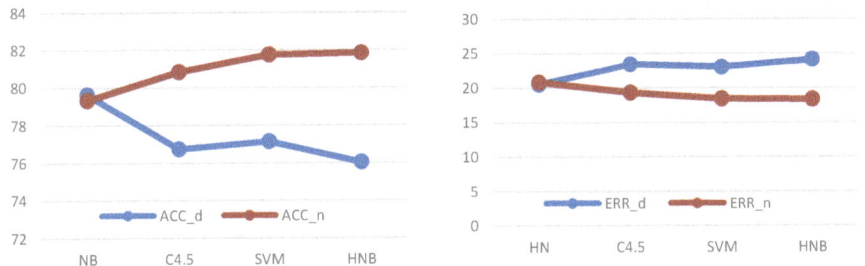

Fig. 1 Comparing accuracy rate on PID dataset between previous attempts results and the results obtained

Table 5 Performance evaluation for classification on PID dataset

Classifiers	Time of build model (ms)	Time of test model (ms)	No. of training examples (%)	No. of testing examples (%)	Technique	Filter
NB	10	20	78.6	21.4	Percentage split	–
DT(C4.5)	240	50	76.2	23.8		–
SVM	30	10	78	22		–
HNB	30	20	62.7	37.3		Discretization

From the analysis of accuracy measures of NB, DT, SVM, and HNB classifiers, HNB performs well when compared to all accuracy measures, namely TP rate, F-measure, ROC area. As a result, HNB outperforms well when compared to other classifiers.

4 Conclusion

Modern studies seek the automatic diagnosis of diabetes for the detection of diabetes in its early stages; this issue is an important real-world medical problem. In comparison to previous attempts for improving the performance of classifiers (results/examination) and by checking the quality standards and matching the previous techniques, we found that the three techniques achieve the closest results with the superiority and preference of the hidden Naive Bayes over NB and DT.

Though the results have been reached, the study seeks to find ways for the detection and early diagnosis of diabetes diseases using data mining technology by using the HNB technique. The HNB technique demonstrated better efficiency then NB, DT, and SVM.

Acknowledgments The research supported by KPT (FRGS/1/2018/TK03/UMP/02/25) (RDU190102): A Novel Hybrid Kidney-inspired algorithm for Global Optimization, and Universiti Malaysia Pahang (UMP) (PRGS190301): Discretization Method based on Hidden Naive Bayes Algorithm for Diabetes Datasets Classification.

References

1. Jantawan, B., Tsai, C.-F.: The application of data mining to build classification model for predicting graduate employment. arXiv preprint arXiv:1312.7123 (2013)
2. Foroughi, F., Luksch, P.: Data science methodology for cybersecurity projects. arXiv preprint arXiv:1803.04219 (2018)
3. Salih, S.Q., Alsewari, A.A.: A new algorithm for normal and large-scale optimization problems: nomadic people optimizer. Neural Comput. Appl. 1–28 (2019)

4. Bagdi, R., Patil, P.: Diagnosis of diabetes using OLAP and data mining integration. Int. J. Comput. Sci. Commun. Netw. **2**(3), 314–322 (2012)
5. Kaviani, P., Dhotre, S.: Short survey on Naive Bayes algorithm. Int. J. Adv. Eng. Res. Dev. **04**(11) (2017)
6. Shankaracharya, D.O., Samanta, S., Vidyarthi, A.S.: Computational intelligence in early diabetes diagnosis: a review. Rev. Diabet. Stud. RDS **7**(4), 252 (2010)
7. Hemant, P., Pushpavathi, T.: A novel approach to predict diabetes by cascading clustering and classification. In: 2012 Third International Conference on Computing, Communication and Networking Technologies (ICCCNT'12), pp. 1–7. IEEE (2012)
8. Novaković, J.: Toward optimal feature selection using ranking methods and classification algorithms. Yugoslav J. Oper. Res. 21(1) (2016)
9. Ashwinkumar, U., Anandakumar, K.: Predicting early detection of cardiac and diabetes symptoms using data mining techniques. Proc. Comput. Des. Eng. **49**, 106–115 (2012)
10. Wu, H., Yang, S., Huang, Z., He, J., Wang, X.: Type 2 diabetes mellitus prediction model based on data mining. Inf. Med. Unlocked **10**, 100–107 (2018)
11. Kaur, G., Chhabra, A.: Improved J48 classification algorithm for the prediction of diabetes. Int. J. Comput. Appl. **98**(22) (2014)
12. Koklu, M., Unal, Y.: Analysis of a population of diabetic patients databases with classifiers. Hum. Res. **1**(2) (2013)
13. Mareeswari, V., Saranya, R., Mahalakshmi, R., Preethi, E.: Prediction of diabetes using data mining techniques. Res. J. Pharm. Technol. **10**, 1098 (2017). https://doi.org/10.5958/0974-360X.2017.00199.8
14. Steffi, J.A., Balasubramanian, D.R., Kumar, M.K.A.: Predicting diabetes mellitus using data mining techniques comparative analysis of data mining classification algorithms. Int. J. Eng. Dev. Res. (IJEDR) **6**(2), 460–467 (2018)
15. Lv, L., Zhang, Q., Zeng, S., Wu, H.: Traffic classification of power communication network based on improved hidden Naive Bayes algorithm. In: 2016 4th International Conference on Electrical & Electronics Engineering and Computer Science (ICEEECS 2016) 2016. Atlantis Press
16. Yu, L., Jiang, L., Wang, D., Zhang, L.: Attribute value weighted average of one-dependence estimators. Entropy **19**(9), 501 (2017)
17. Zhang, H., Jiang, L., Su, J.: Hidden Naive Bayes. In: Proceedings of the Twentieth National Conference on Artificial Intelligence, pp. 919–924 (2005)
18. Liangxiao, J., Zhang, H., Zhihua, C.: A Novel Bayes Model: hidden Naive Bayes. IEEE Trans. Knowl. Data Eng. **21**(10), 1361–1371 (2009)
19. Kotsiantis, S., Tampakas, V.: Increasing the accuracy of Hidden Naive Bayes model. In: 6th International Conference on Advanced Information Management and Service (IMS), pp. 247–252. IEEE (2010)
20. Iyer, A., Jeyalatha, S., Sumbaly, R.: Diagnosis of diabetes using classification mining techniques. arXiv preprint arXiv:1502.03774 (2015)
21. Lakshmi, K., Kumar, S.P.: Utilization of data mining techniques for prediction of diabetes disease survivability. Int. J. Sci. Eng. Res. **4**(6), 933–940 (2013)
22. Lingaraj, H., Devadass, R., Gopi, V., Palanisamy, K.: Prediction of diabetes mellitus using data mining techniques: a review. J. Bioinf. Cheminf. **1**(1), 1–3 (2015)
23. Friedman, N., Geiger, D., Goldszmidt, M.: Bayesian network classifiers. Mach. Learn. **29**(2–3), 131–163 (1997)
24. Radha, P., Srinivasan, B.: Predicting diabetes by cosequencing the various data mining classification techniques. Int. J. Innov. Sci. Eng. Technol. **1**(6), 334–339 (2014)
25. Li, K.-H., Li, C.T.: Locally weighted learning for Naive Bayes Classifier. arXiv preprint arXiv:1412.6741 (2014)
26. Larabi-Marie-Sainte, S., Aburahmah, L., Almohaini, R., Saba, T.: Current techniques for diabetes prediction: review and case study. Appl. Sci. **9**(21), 4604 (2019)
27. Jin, R., Breitbart, Y., Muoh, C.: Data discretization unification. Knowl. Inf. Syst. **19**(1), 1 (2009)

Toward Data Warehouse Modeling in the Context of Big Data

Fatimaez-Zahra Dahaoui, Lamiae Demraoui, Mohammed Reda Chbihi Louhdi, and Hicham Behja

Abstract Data modeling was and still one of most organizations' big interest, which adds significant value to their decision-making process. In the last decades, we are observing an astonishing growth of data volume and availability, caused by the multitude of sources that are continuously producing structured, semi-structured, or unstructured data. The thing is traditional data modeling techniques are no longer efficient to handle big data, due to its complex structures. Hence, modeling should not be limited to relational databases; we can use it to design data structures at various levels from conceptual to physical in a big data warehouse cycle. Previous works have been limited to model few parts of the data warehouse cycle in big data context. The aim of our work is modeling the whole data warehouse cycle in big data era. With the major focus on the missing concept in the existing approaches, namely the reusability plays a vital role in big data warehouse decision-making process.

Keywords Big data modeling · Data warehouse · Reusability · Decision making

1 Introduction

Big data covers 85% of unstructured data [1], which gives data modeling a crucial role to play in big data analytics, and this data should be modeled as required to the

F.-Z. Dahaoui (✉) · L. Demraoui · H. Behja
Engineering Research Laboratory, National School of Electricity and Mechanics, Hassan II University of Casablanca, Casablanca, Morocco
e-mail: Fatimaez.zahra.dahaoui@gmail.com

L. Demraoui
e-mail: demraoui.lam@gmail.com

H. Behja
e-mail: h_behja@yahoo.com

M. R. Chbihi Louhdi
Faculty of Science Aîn Chock, Hassan II University of Casablanca, Casablanca, Morocco
e-mail: chbihi@gmail.com

© The Editor(s) (if applicable) and The Author(s), under exclusive license to Springer Nature Singapore Pte Ltd. 2021
F. Saeed et al. (eds.), *Advances on Smart and Soft Computing*, Advances in Intelligent Systems and Computing 1188, https://doi.org/10.1007/978-981-15-6048-4_21

organization's needs. Besides, the data model is an important element of a big data application; it is essential, where big data resides [1].

Over the last few years, the term big data appeared to express the huge data sets that are continuously been produced from different sources in various structures. Handling this kind of data represents new challenges, since the traditional relational database management systems (RDBMSs) and data warehouses (DWs) expose serious limitations in terms of performance and scalability when dealing with such a volume and variety of data. Consequently, it is needed to reinvent the ways in which data is represented and analyzed, in order to be able to extract value from it.

Big data modeling is the new way to model data, which takes into consideration the different types and sources of data in the aim to provide what traditional data modeling could not achieve.

This paper takes a look at modeling the data warehouse cycle in the big data context and sheds new light on the reusability of realized models in a decision-making process.

The paper is structured as follows: Sect. 2 discusses the limitations of traditional data warehouse modeling when facing the big data era. Section 3 gives a vision of the existing approaches to model data warehouse in big data age. Section 4 presents our comparative study of the existing approaches of big data modeling, and Sect. 5 concludes the paper and gives our perspectives.

2 Limitation of Traditional Data Warehouse Cycle

Over the years, in order to identify relationships/patterns in the data, organizations have been using data warehouses to integrate data coming from various sources and use the unified repository for data mining. However, with the complexity and volume of data that is increasing over time, data warehouses today are unable to provide the flexibility and agility required for quick decision making [2].

Table 1 shows the difference between traditional data and big data.

Modeling in big data is very different; in traditional data, you model first and then you put data, but in big data modeling, you can bring out data even before modeling it. Then back to models, since traditional systems cannot handle unstructured data, its models also will not have the ability to surround all types of data.

3 Modeling in Data Warehouse Cycle: NoSQL and UML

Nowadays, big data is present everywhere and the focus on information taken from data is increasing; the existing software engineering models are insufficient to fulfill the requirement process. Requirement engineering involves analyses to identify what users want from a software system and to understand what their needs mean in terms

Table 1 Difference between traditional data and big data [3]

Category	Big data	Traditional small data
Volume of data generated Data variety	Terabytes Petabytes Exabytes Zettabytes	Gigabytes Terabytes
Data variety	It may be structured, semi-structured, or unstructured	It may be structured, semi-structured. Does not support unstructured data
Data velocity	Often real-time data generated on hourly or daily basis. It requires immediate response	Data generated in batch mode or near real time, more rapidly than big data. It does not require immediate response
Data storage	NoSQL, HDFS	RDBMS
Data integration	Difficult	Easy
Data access	Interactive	Batch or near real time

HDFS, Hadoop Distributed File System, https://hadoop.apache.org/

of software design and architecture [4]. However, data scientists are the ones who should know what customers want, not customers themselves. In coming lines, we will see the gathered works about modeling the data warehouse cycle in big data context.

3.1 Modeling in Big Data with ETL

The approach of conceptual modeling in the big data extract process intends to promote the use of UML[1] for modeling big data scenarios. The approach has selected the deployment diagramthis—diagram depicts the system's architecture, the devices, their connections, and the software used to model the big data extract process.

The idea of the approach is to extend UML process on the extract process and then propose a set of stereotypes—an extensibility mechanism in UML that allows designers to create new model elements, derived from existing ones—[5] to create the icons for conceptual modeling from the existing UML elements, that should be familiar with software developers.

Table 2 shows an example of a big data type stereotypes and icons; it presents a set of icons for each type of data.

As we know, the extract process is a part of the ETL process, and Table 3 represents stereotypes for the ETL process in the data warehouse.

The example below shows the result of modeling one of big data solutions: Sqoop, which is a big data tool that imports different types of data (structured,

---

[1]UML, Unified Modeling Language, https://www.uml.org/.

Table 2 Big data type stereotypes and icons [5]

Stereotype	Description	Icon
Structured data	Data ordered by rows and columns	
Semi-structured data	Data with structure but not organized in relational databases	
Unstructured data	Data without any organization	
Text files	Text data format	
Binary files	Binary data format	

Table 3 Big data type stereotypes and icons in ETL process [5]

Stereotype	Description	Icon
Conversion	Changes data type and format or derives new Data from existing data	A → B
Loader	Loads data into a target	

semi-structured, or unstructured) from a different sources to Hadoop HDFS with UML (see Fig. 1).

The defined stereotypes conversion and loader are been used to depict the functionality of Scoop to transform the data in the HDFS format and to load it into the file system.

3.2 Modeling in Big Data with OLAP

OLAP for the big data world must be a powerful concept that consists of pre-aggregation of immense volumes of data into multidimensional cubes and then query them to get faster results. Nevertheless, it is not a fresh concept; the thing is traditional OLAP solutions are unable to deal with big data.

In order to handle this issue, OLAP should has been built on the big data platform; by doing so, we ensure that the cubes can handle plenty of dimensions considering the variety of data, high volume, and data velocity [7].

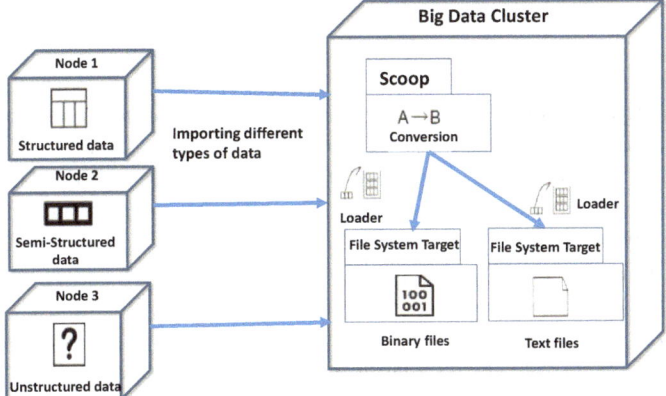

Fig. 1 Modeling Sqoop with UML [6]

3.3 Types of NoSQL Data Bases

The limitation of a relational model is having some scalability issues: the performance degrades while data volumes increase. This led to the development of a new data model called "NoSQL" (Not only SQL) and not (NoSQL); this kind of database is an emerging alternative to the most widely used type of databases (i.e., relational databases). NoSQL—as the name proposes—does not completely replace SQL, though it compliments it in a manner that they can both coexist [8].

- Key-Value Store data bases
  Key-value store databases store data in a schemaless way, with all of the data consisting of an indexed key and value, such as Oracle NoSQL and Redis [7] (see Fig. 2).
- Column store
  This type of NoSQL is been also known as wide-column store; Google's Big Table has motivated it and HBase is a popular tool for this type of databases. Data

Fig. 2 Key-value data model [6]

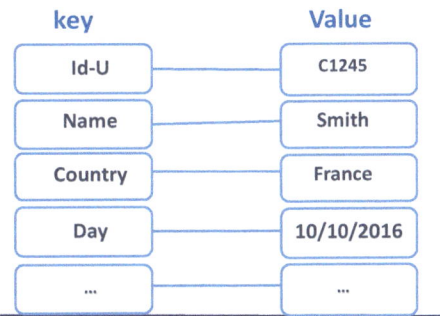

tables are stored as sections of columns of data, instead of storing data in row [7] (see Fig. 3).

- Document store

  Concerning document type, "it is a collection of key-value stores; where the value is a document, each document has a unique key, which is assigned to retrieve the document" [6] (see Fig. 4). Any collection of data can be stored such as nested structures, maps, collections, and scalar values. There can be secondary indexes to access a component of a document. Document stores are been found in MongoDB, CouchDB, and others [7].

- Graph database

  The last type is based upon graph theory (set of nodes, edges, and properties) and useful for interconnected relationship data such as communication patterns, social networks, and biographical interactions (see Fig. 5). Graph databases allow asking deeper, more complex, questions and express queries as traversals; one of the most useful tools for graph databases is Neo4j [9].

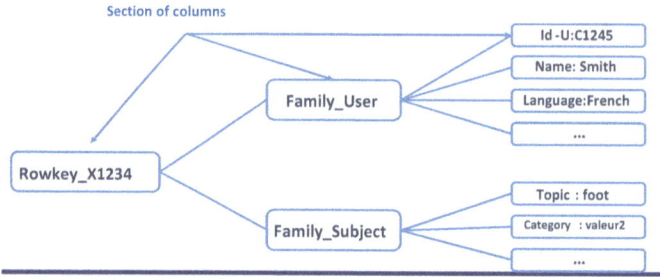

Fig. 3 Column store data model [6]

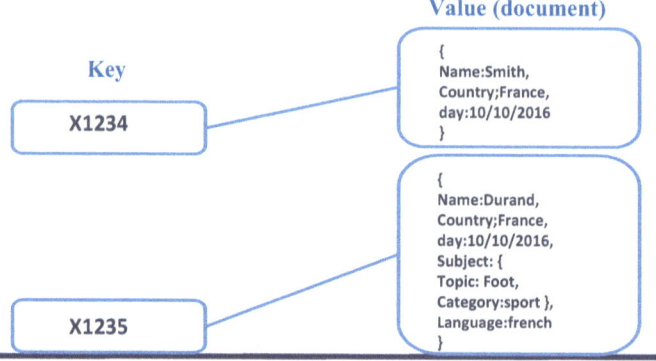

Fig. 4 Document store data model [6]

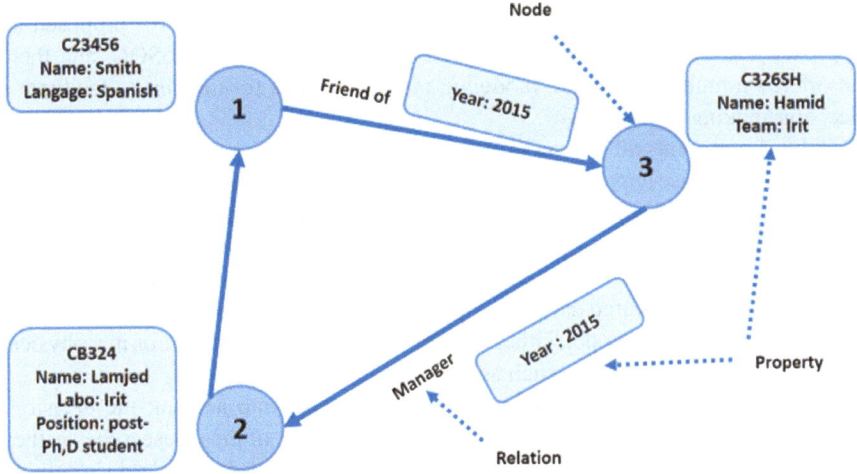

Fig. 5 Graph store data model [6]

4 Existing Approaches

Numerous suggestions exist that propose modeling in big data since this latter is less predictable than traditional data and therefore requires special consideration when building models during Data Warehouse cycle.

According to [10], data collection is the process of gathering quantitative and qualitative information on specific variables with the aim of evaluating outcomes or gleaning actionable insights. Good data collection requires a clear process to ensure the data you collect is clean, consistent, and reliable. For this aim, [4] has proposed a new model for requirement engineering for big data. This model allows scientists and software engineers to work together using an adapted use case.

Besides, several works have treated the extract, transform, and load (ETL), which is an essential technique for integrating data from multiple sources into a data warehouse. ETL applies data warehousing, big data, and business intelligence. In order to benefit from the use of big data technologies, [11] proposed an approach that use big data technologies in the ETL process, from the other hand [12] thought about the other approach to extend the ETL workflow designed for big data.

In data warehouse cycle, data mining is an essential phase for extracting hidden, unknown, but potentially useful information from massive data with the aim of reusing the previous analysis thanks to the generated metadata [13, 14], and big data has great impacts on scientific discoveries and value creation. In this context, [15] proposed a schema on read modeling approach as a basis of big data analytics integration in enterprise information system.

Furthermore, in a recent paper by [16], NoSQL databases are becoming increasingly popular and have some interesting strengths such as scalability and flexibility. With the advent of big data, new challenges have emerged regarding the evaluation of decision support systems (DSS). In order to handle this issue, [6] proposed an approach to implement multidimensional data warehouse into NoSQL, and then, he offered a multidimensional modeling of NoSQL data for the aim to facilitate a decision-making process.

Table 4 represents all the works mentioned previously, from one hand, specifying data types that these approaches work on tools and concepts employed by each work, from the other hand, answering the two questions if these works support decisions making and the guarantee of the reusability of successful models.

Previous studies have widely shown that they can handle big data structures (structured data, semi-structured data, and unstructured data). Each one of these works has its way to present data model, either in a conceptual way using UML or in a physical way using big data platforms such as Hadoop.

However, most of the previous studies do not take into account the decision-making support process to help users. The key limitation of these researches is they not address the reusability of successful models, which has an essential role in helping novice users to integrate easily data analysis world and decision-making process.

5 Conclusion and Future Work

Big data is all around us; its technologies need to be understood, including software and hardware approaches, as well as approaches to big data analytics using sophisticated and automated tools. In order to achieve this, the beginning should be a good modeling of the data warehouse cycle in the age of big data. In this paper, we have compared the existing approaches of big data warehouse modeling. As known, a decisional process was usually being hold by one or more expert users who manipulate several knowledge and expertise, and our immediate future work consists of proposing an approach to model the data warehouse cycle in the context of big data, with the focus on the reusability of the existing succeeded models in order to ensure a better decision making.

Table 4 A comparative table of big data modeling approaches

	Approach	Big data			Big data modeling tools and concepts	Data model structure	Decision-making support	Reusability
		Structured E.g., relational databases….	Semi-structured E.g., NoSQL…	Unstructured E.g., real-time data				
Data collection	A requirement engineering model for big data [4]	X	X	X		UML use case	Absent	Absent
ETL	Conceptual modeling of big data extract processes with UML [5]	X	X	X	Flume Scoop DataClick	UML conceptual model	Absent	Absent
	Next-generation ETL framework to address the challenges posed by big data [12]	X	X	X	MapReduce Spark Hadoop	–	Absent	Their perspective in coming work
	ETL in the big data era [11]	X	X	X	CloudETL PDI Hadoop MapReduce	Big data platforms (Hadoop)	Helps users to work with right data	Absent
Data mining	Intelligent data engineering for migration to NoSQL-based secure environments [17]	X	X	X	Oracle NoSQL concepts	–	Absent	Absent

(continued)

Table 4 (continued)

Approach		Big data			Big data modeling tools and concepts	Data model structure	Decision-making support	Reusability
		Structured	Semi-structured	Unstructured				
		E.g., relational databases…	E.g., NoSQL…	E.g., real-time data				
	Modeling and distribution of big data warehouse [18]	X	X	X	Hbase	–	Yes	Absent
	Schema on read modeling approach as a basis of big data analytics integration in EIS [15]	X		X	Hadoop Hive	Big data platforms (Hadoop)	Helps for better data integration	Absent
OLAP	NoSQL modeling of multidimensional massives data warehouses [6]	X	X	X	CouchDB MongoDB Hive	Big data platforms (Hadoop)	Yes	Absent

References

1. Jyothi, B.S., Jyothi, S.: A study on Big Data modelling techniques. Int. J. Comput. Netw. Wirel. Mob. Commun. (IJCNWMC) **5**(6), 19–26 (2015)
2. Pandey, A., Mishra, S.: Moving from traditional Data Warehouse to entreprise data management: a case study. Issues Inf. Syst. **15**(II), 130–140 (2014)
3. Nagaraj, K.S., Sharvani, G.S., Sridhar, A.: Emerging trend of Big Data analytics in bioinformatics: a literature review. Int. J. Bioinform. Res. Appl. **14**(1/2), 63 (2018)
4. Altarturi, H.H., Ng, K.Y., Ninggal, M.I.H., Nazri, A.S.A., Ghani, A.A.A.: A requirement engineering model for Big Data software. In: 2017 IEEE Conference on Big Data and Analytics (ICBDA), pp. 111–117. IEEE, Kuching, Malaysia, Nov 2017
5. Martinez-Mosquera, D., Lurjan-Mora, S., Recalde, H.: Conceptual modeling of Big Data extract processes. In: 2017 International Conference on Information Systems and Computer Science (INCISCOS). Editador (2017)
6. El Malki, M.: Modélisation NoSQL des entrepôts de données multidimensionnelles massives. Doctoral dissertation (2016)
7. Kyvos: https://www.kyvosinsights.com/understanding-olap-on-big-data-why-do-you-need-it/. Last accessed: 14 Jan 2020. Storey, V.C., Song, Y I.: Big Data technologies and management: what conceptual model can do. Data Knowl. Eng. **108**, 50–67 (2017)
8. Nayak, A., Poriya, A., Poojary, D.: Type of NOSQL databases and its comparison with relational databases. Int. J. Appl. Inf. Syst. (IJAIS) **5**, 16–19 (2013)
9. Storey, V.C., Song, Y.I.: Big Data technologies and management: what conceptual model can do. Data Knowl. Eng. **108**, 50–67 (2017)
10. Data Collection Guide: Dimagi, https://www.dimagi.com/data-collection/. Last accessed: 10 Jan 2020
11. Silva, M.M.L.: ETL in the Big Data era, Portugal (2015)
12. Ali, F.M.S.: Next-generation ETL framework to address the challenges, the workshop. In: Proceedings of the EDBT/ICDT 2018 Joint Conference (2018). ISSN 1613-0073
13. Demraoui, L., Behja, H., Zemmouri, E.M., BenAbbou, R.: A viewpoint based extension of the common warehouse metamodel to support the user's viewpoint approach. Int. J. Metadata Semant. Ontol. **11**(3), 137–154 (2016)
14. Demraoui, L., Behja, H., Zemmouri, E.M.: Metadata management in the DW process: a viewpoint-based approach. Project-ID (178290). LAP LAMBERT Academic Publishing, July 2018. ISBN 978-613-9-86714-1
15. Jankovic, S., Mladenovic, S, Mladenovic, M.D., Veskovic, S.: Schema on read modeling approach as a basis of Big Data analytics integration in EIS. In: 7th International Conference on Information Society and Technology ICIST (2017)
16. El Malki, M., Kopliku, A., Sabir, E., Teste, O.: Benchmarking Big Data OLAP NoSQL. In: UNet. LNCS, vol. 11277, pp. 82–94. Springer Nature Switzerland AG (2018)
17. Ramzan, S., Bajwa, S.I., Ramzan, B., Anwar, W.: Intelligent data engineering for migration to NoSQL based secure environments. Adv. Softw. Data Eng. Secure Soc. **7**, 69042–69057 (2019)
18. Ghourbel, M., Tekaya, K., Abdellatif, A.: Modélisation et répartition d'un Big Data Warehouse. In: Big Data and Applications 12th edition of the Conference on Advances of Decisional Systems (ASD 2018). Marrakech, Morocco (2018)
19. Pandya, R., Sawant, V., Mendjoge, N., D'silva, M.: Big Data vs traditional data. Int. J. Res. Appl. Sci. Eng. Technol. (IJRASET), **3**, 05 (2015). ISSN 2321-9653
20. Aggarwal, D.: https://www.projectguru.in/publications/difference-traditional-data-big-data/. Last accessed: 08 Jan 2020

Internet of Things (IoT) and Smart Cities

Internet of Things Industry 4.0 Ecosystem Based on Zigbee Protocol

Hamza Zemrane, Youssef Baddi, and Abderrahim Hasbi

Abstract In the context of the Smart City, the Internet of Things is a major asset for the development of different sectors of activity; in our study, we focused on the new industrial revolution, the Industry 4.0 Ecosystem, which is based on the digitization of all the industrial processes. For this, the deployment of a large number of sensors in the digital factory is an indispensable thing; the communication between the different components of the network becomes something that must be taken seriously. Our simulation consists of studying the performance of the ZigBee protocol in different modes: Mesh, star, and tree topologies.

Keywords Smart City · Internet of Things (IoT) · Industry 4.0 Ecosystem · ZigBee protocol · Sensor network

1 Introduction

One of the most important axes of Smart City is the development of the industry sector; the development of the industry depends on the development of the Internet of Things; the Internet of Things are based on a lot of sensors that capture the information on the environment in which they are deployed; and transmit it to the Internet of further processing. In our work, we applied the IoT to the industry sector to come out with the IoT Industry 4.0 Ecosystem, that digitalize the industrial processes inside a digital factory, and the sensor in the digital factory are deployed with a specific network protocol and a specific network topology. We focus in our performance study on the ZigBee protocol that is used a lot in the Industry 4.0 Ecosystem, and

H. Zemrane (✉) · A. Hasbi
Lab RIME, UM5, Mohammadia School of Engineering, Rabat, Morocco
e-mail: zemranehamza93@gmail.com

A. Hasbi
e-mail: ahasbi@gmail.com

Y. Baddi
Lab STIC, UCD, High School of Technology Sidi Bennour, El Jadida, Morocco
e-mail: baddi.y@ucd.ac.ma

F. Saeed et al. (eds.), *Advances on Smart and Soft Computing*, Advances in Intelligent Systems and Computing 1188, https://doi.org/10.1007/978-981-15-6048-4_22

we compared its performance in case of a mesh, star, and tree network topologies. In Sect. 2, we talks about the Internet of Things and we give a definition, its architecture, and we focus on its network layer, and the possible network typologies. In Sect. 3 we talks about the Industry 4.0 Ecosystem, the industrial revolutions, the digital factory and we focus on its component: the sensors, PLC, and the numerical control. In the Sect. 4, we give a performance analysis based on the MAC Data Traffic sent, the Media Access Delay, the Management Traffic sent, and the Application Traffic received of the ZigBee protocol when it use the mesh, star, and tree typologies.

2 Internet of Things (IoT)

2.1 Definition of IoT

We talk about the Internet of Things when the number of devices exceeds the number of people connected to the Internet, and the goal of the Internet of Things is to facilitate human life by building a smart environment using smart objects that can autonomously generate data from the environment in which they are deployed and transmit this data to the Internet for decision making. These devices are used to collect information from the physical environment, and send it to the network edge for further processing. These devices are deployed with a network architecture and a separate data processing application according to the specific task in a particular area, like Ehealth [1], Smart Homes [2], intelligent transportation system [3], and others.

2.2 Architecture of IoT

The layers of the Internet of Things architecture are:

- Perception layer: represents the sensors, RFID, wearables, smartphones, and others that can capture information from the physical environment like: temperature, pressure, movement, radiation, and others, and convert them on digital magnitudes, process this information, and transfer it to the sink or the network gateway.
- Network layer: represents new network protocols like: Low Energy Bluetooth [4], ZigBee [5], Lorawan [2], WiFi [6], WIMAX [7] and others, that can transmit the information collected from the perception layer and send it to the sink or the network gateway for further processing.
- Middleware layer: this layer can extract information sent from different devices of the perception layer, on the same domain of activity, and convert it into a service information, for the denomination of the requested service, to transmit it on the core network (ATM, MPLS, …) with a guarantee of the quality of service.

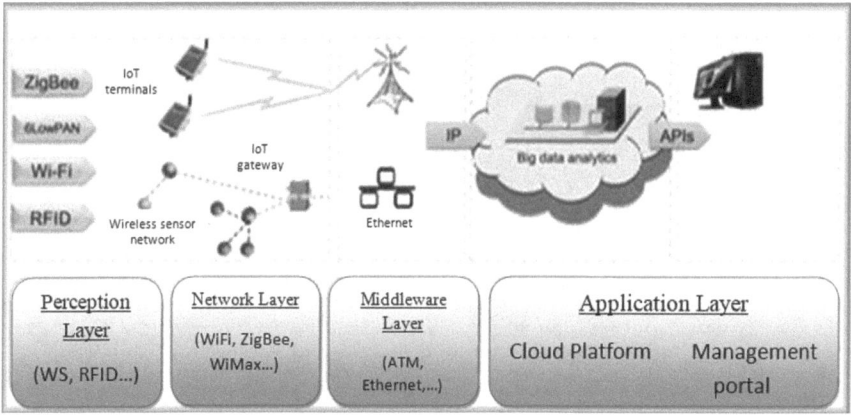

Fig. 1 Architecture of the Internet of Things [8]

- Application layer: contains the processing system that process the collected information of a specific domain of activity, and permits to the user to manage this information using an application interface (Fig. 1).

2.3 The Network Layer of the IoT Architecture

In the field of wireless and wired networks, a communication protocol defines the rules and the communication procedures of the physical and MAC layers of the OSI model on a medium or physical channel. It allows to connect an object to a wired or wireless network. In addition, if this network includes a gateway, that is to say a device connected to both the network and the Internet, then this object can transmit and receive data to the Internet (Fig. 2).

2.4 Network Topology

Mesh Topology [9] In a Mesh topology, a node is connected to one or more other nodes of the same network. This forms a mesh in which an emitted data is relayed potentially by several nodes before reaching its destination, and this is called a route. It is also possible to prioritize a mesh topology to de ne levels in order to manage the network more easily. In this case, the parent node is the master of the network. This is called cluster tree (Fig. 3).

Star topology [10]: The objects are connected to a gateway called hub or router. Its role is to ensure communication between the nodes of the network. This type of topology makes it easy to add or remove nodes without impacting the rest of the network. In addition, all the intelligence of the network is concentrated on a single

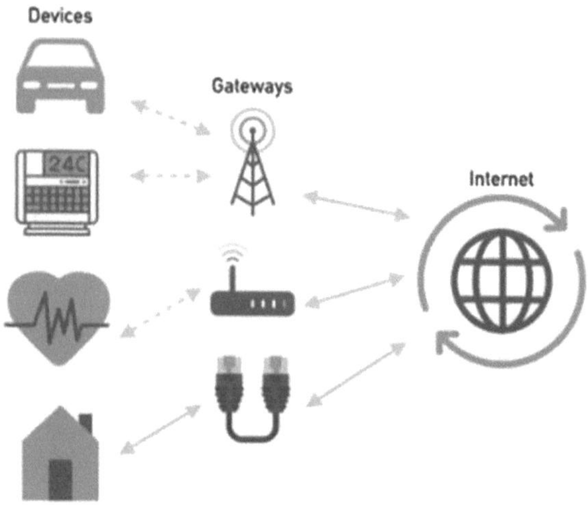

Fig. 2 Principle of a
connected object

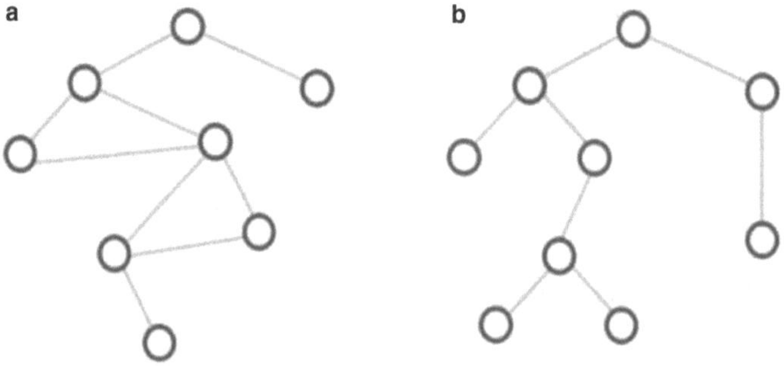

Devices

Gateways

Internet

Fig. 3 Mesh topology classical (**a**) and cluster tree (**b**)

node which makes it possible to manage the network more easily. However, if the
hub is experiencing a technical problem, then the entire network is down (Fig. 4).

Cell topology [11]: A cellular topology is based on a division of a territory into
zones called cells. The radius of a cell can range from a few hundred meters (urban)

Fig. 4 Star topology

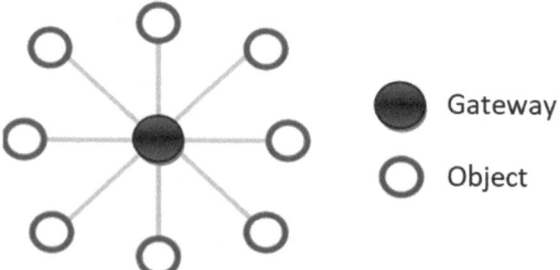

Gateway

Object

Fig. 5 Cell topology

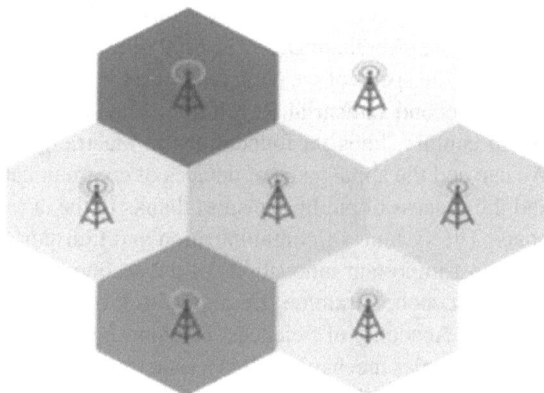

Fig. 6 Broadcast topology

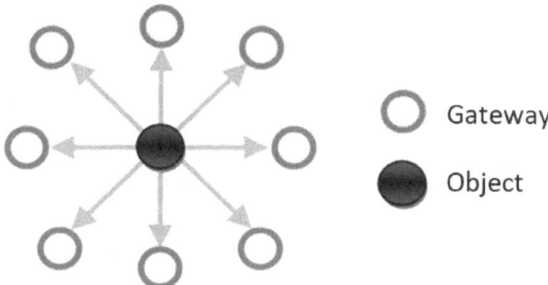

to several kilometers (rural). At the heart of the cell, an antenna provides the radio link between objects and the Internet. This type of topology is the basis of mobile networks (e.g., 2G/GSM, 3G/UMTS, 4G/LTE, 5G and 6G) (Fig. 5).

Broadcast topology [12]: In this type of topology, an object transmits a message without specifying a particular recipient. What makes the message is analyzed by all the objects that have received the message correctly, that is why we talk about broadcast as shown in Fig. 6.

3 The IoT Industry 4.0 Ecosystem

3.1 Industrial Revolutions

The First Industrial Revolution: The First Industrial Revolution is based on coal, metallurgy, textiles, and the steam engine. It began in Britain at the end of the eighteenth century, then spread in France in the early nineteenth century before expanding to Germany, the USA, Japan and Russia. The highlights of this period are: the improvements made in 1705 by Thomas Newcomen to the steam engine and the extension of its use to the industry, the first use of coke instead of wood to melt iron ore (Abraham Darby in 1709), the development of the flying shuttle which increases the speed of weaving (John Kay in 1733), and others.

The Second Industrial Revolution: The second, started at the end of the eighteenth century, finds its foundations in electricity, mechanics, oil, and chemistry. We can add the appearance of devices of communication (telegraph and telephone), and the success of public transport thanks to the development of railways or steamboats. The systems of communication and transport favor international exchanges. The most important innovations of the second revolution are: The invention of the Belgian Zenobe Gramme, the magneto Gramme, presented on July 17, 1871, at the Paris Academy of Sciences, is major, because it makes the production of electricity becomes mechanical. It is a rotary machine driven by a crank. Later improvements made it an industrial dynamo (1873) generating continuous electric current. In 1878, Thomas Edison developed the incandescent lamp. No more electric-arc lamps, kerosene, and gas lamps for public lighting. In 1881, one of the Edison Company's engineers, Lewis Howard Latimer, improved the process by patenting the first incandescent lamp with carbon lament, and other innovations.

The Third Industrial Revolution: The third revolution occurs in the middle of the twentieth century, whose dynamics come from electronics, telecommunications, computer, audiovisual, and nuclear. They make possible the production of miniaturized materials, robots and advanced automation of production, the development of space technologies and biotechnologies. Part of the USA, then Japan and the European Union, the Third Industrial Revolution also saw the birth of the Internet, at the twilight of the twentieth century. The real start of the miniature electronics dates from the arrival of the transistor (and integrated circuits). It came out from the Bell Labs in 1948. The industrial robot, out of the imagination of Georges Devol and Joseph Engelberger, was first intended for handling operations, then for production tasks: welding, assembly, etc. The first industrial robot was installed in 1959 at the General Motors plant in Trenton, New Jersey.

The Fourth Industrial Revolution, the last industrial revolution, is taking shape under our eyes, at the dawn of the twenty-first century. It will be ripe at the earliest around 2020. All the technological bricks on which it is built are there. It can be summed up by the extreme digitization of economic and productive exchanges. Industry 4.0 assumes horizontal integration. We realize everything from A to Z in interaction between the products and the machines, and the machines between them. We are in an interconnected global system. The finished product, which will be personalized, will also be able to communicate with the machines in its production phase. This is called "Smart Product" (Fig. 7).

3.2 The Digital Factory

The digital factory [13] can be defined as the set of software tools and methodologies present at different levels to design, simulate, implement, and optimize the production

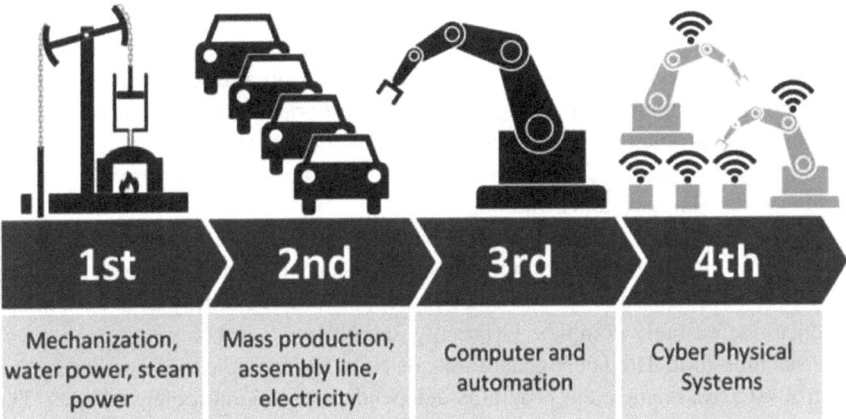

Fig. 7 Industrial revolutions

systems from the early stages of design to piloting the plant. The digital factory [13] is made up of all the digital tools that make it possible to design the manufacturing process of a product. The main objective is to build the process in parallel with the product design, as part of the simultaneous engineering. The construction of the process is done by describing in an exhaustive way the operations allowing to obtain, step by step, the finished product.

Sensors: Sentinels of Industry 4.0. By becoming more and more intelligent, the sensors gain in relevance (the right measurement), in precision (the finesse of the measurement), and in decisional autonomy (the order to act can start from the sensor). Their role as guardian of preventive maintenance will grow. In fact, they are a key element of the digital factory [13], or even Industry 4.0. A sensor is an organ that retrieves a measurement (tank level, oven temperature …) or a state (open-closed, on–off). The sensor reads information, digital or analog. In the latter case, it digitizes it by translating it into bits (0011100011). The sensor is connected with a link to the object to be measured and with another link (wired or wireless) to a device, often PLC (automat) [14] or an input module, to which it transmits the collected information. Thus, the PLC gives instructions to the actuator based on the information received by the sensor. The sensors are more and more sophisticated, even basic, they are made intelligent by giving them the ability to diagnose any information they have just collected. Example of an optical sensor positioned on a production line. It analyzes the light reflected by the product passing in front of him. It can identify the product and detect possible defects, thanks to a repository contained in a database.

PLC: the guarantor of production 4.0. The PLC [14] is able to control a manufacturing, and inform if need, in real time, the technician responsible for monitoring the smooth running of the factory. The PLC position, as close as possible to the productive process, between sensors and supervisor, makes it the effective and guarantor of the digital factory. Thanks to its intelligence, it progresses in decisional autonomy. The PLC (can be a traditional automat, industrial PC, or even a PAC) [14] is programmed to control a manufacturing. It sends commands to the actuators, who execute them, according to the pre-programmed scenario, and production data gived by the sensors. The PLC has the double advantage, a scalable and modular deployment; it ensures the control of diverse applications: machining program of a machine-tool, robot control, entire production islands, process chemical transformation (today, it has the capabilities of a traditional DCS system), the energy consumption of machines, and buildings, the alarm systems, and others.

Numerical control in Industry 4.0 is used exclusively for the programming and operation of machine-tools, and the numerical control (NC) [15] has experienced a steady improvement in its performance (speed, accuracy, exibility of use), to make products increasingly complex. Originally, the NC was used to program "3 axes" of a machine-tool. The latest generations of NCs, for example those of Fanuc, can control 40 axes, manage ten programs independently, and interpolate 24 axes. The numerical control [15] integrates a program equivalent to a PLC [14] that manages various functions of the machine tool, such as watering, opening and closing of doors, and a specific program that controls the axes of the machine-tool. The information collected by the NC on each product of the factory can also be recorded on a chip (RFID or other), or on a server and used for traceability, especially in the aeronautical and automotive. With this information, we can know the place of manufacture, the machining conditions, and others.

4 Performance Analysis of the ZigBee Protocol

4.1 Simulation Scenario

The following scenario simulates the behavior of a ZigBee network in a digital factory [13], our network is composed of: a ZigBee router, seven ZigBee coordinators, and forty-four sensors. The following simulation is to compare network performance when it is in: Mesh topology mode, star topology mode, and tree topology mode, for a period of 30 min starting at 05:18:20 pm and ending at 05:50:00 pm (Fig. 8).

4.2 The MAC Data Traffic Sent

It is information rate sent in the MAC layer by the different sensors, coordinators, and ZigBee router in bits per seconds. The curve that represents the tree topology is in first position, it makes a rapid increase that takes 1,400,000 bit/s at 05:22:00 pm, then it continues its increase to reach a little more than 1,500,000 bit/s toward the end of the simulation. The curve that represents the mesh topology also makes an increase which is 1,200,000 bit/s at about 05:23:00 pm, then continues to increase to 1,275,000 bit/s toward the end of the simulation. The curve that represents the

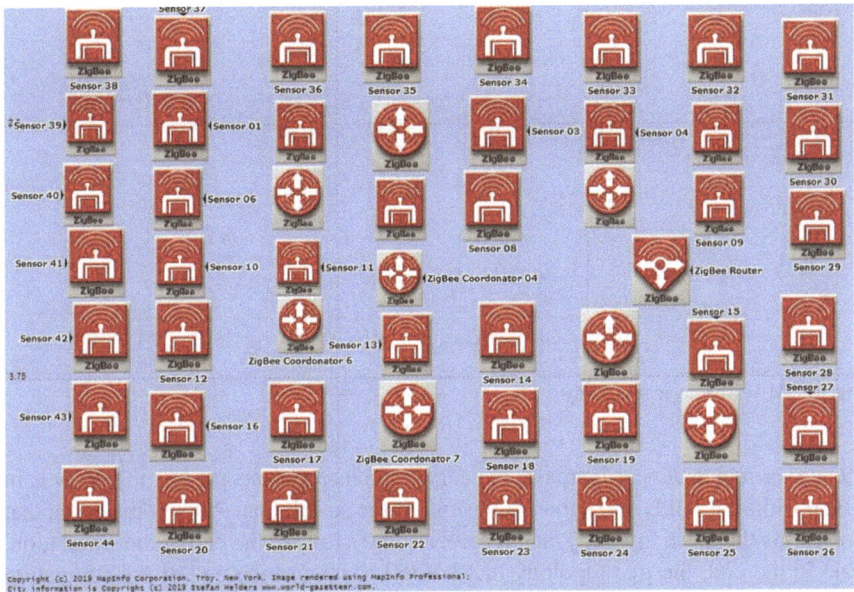

Fig. 8 Simulation scenario of the ZigBee network

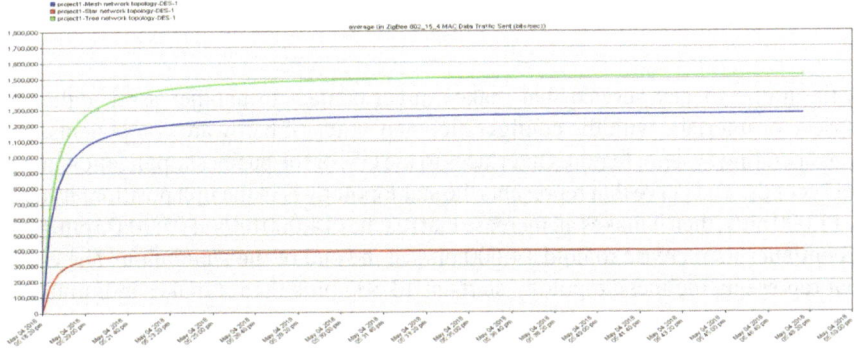

Fig. 9 MAC data traffic sent

star topology makes a slow increase that gets 400,000 bit/s toward the end of the simulation (Fig. 9).

4.3 The MAC Media Access Delay

It is the time from when the data reaches the MAC layer until it is successfully transmitted out on the wireless medium for all the sensors in seconds. This time,

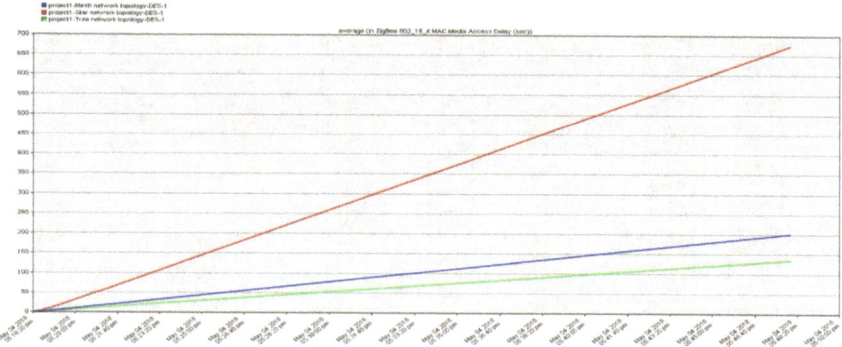

Fig. 10 MAC media access delay

the curves make a linear increase, the curve which represents the star topology is in first position; it holds 650 s toward the end of the simulation. After is the curve that represents the mesh topology, and it receives 200 s toward the end of the simulation. And in the end, the tree topology that holds 140 toward the end of the simulation (Fig. 10).

4.4 The MAC Management Traffic Sent

Is the traffic of any service that makes direct contact to another asset, either to retrieve or interface with the configuration and status of hardware components, the core operating system, features interfaces to the OS, or the business application, in bit per seconds. The curve that represents the star topology is in the first position; it starts at 11,000 bit/s, then it makes an increase that reaches 14,000 bit/s at 05:20:00 pm, then it continues its increase to hold 14,600 bit/s at the end of the simulation. The curve that represents the mesh topology starts at 13,500 bit/s then it makes a diminution to reach 12,900 bit/s toward the end of the simulation. And finally, the curve that represents the tree topology that starts at 13,500 bit/s then it makes a diminution that holds 12,000 bit/s toward the end of the simulation (Fig. 11).

4.5 The ZigBee Application Traffic Received

It is the information rate send at the application level by the different sensors, in bit per second. The curves that represent the star topology make increase that reach 6000 bit/s at 05:22:00 pm, then it continue the increase slowly until the end of the simulation. The curve that represents the tree topology is in the second position, it makes a fast increase that reach 110 bit/s at 05:40:00 pm, and then it makes a slow

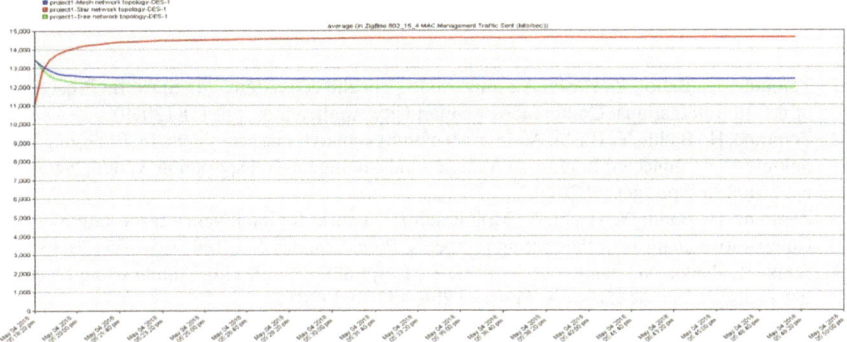

Fig. 11 MAC management traffic sent

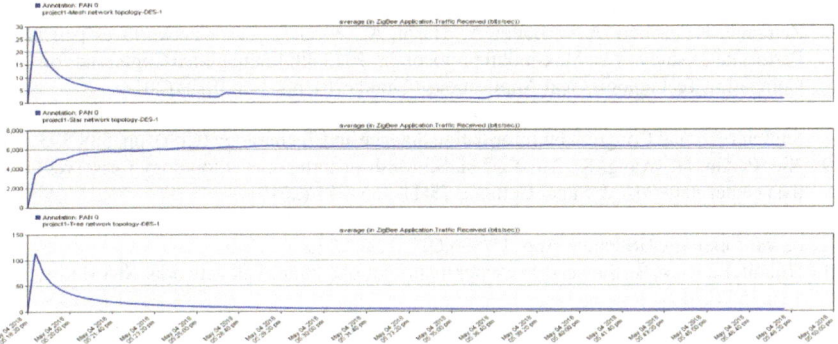

Fig. 12 ZigBee application traffic received

decrease until the end of the simulation. In the last position, the curve that represent the mesh topology makes a fast increase that takes 29 bit/s at 05:40:00 pm then it decease in a variable way until the end of the simulation (Fig. 12).

5 Conclusion

The digital factory is an indispensable component for the Industrial Revolution 4.0; the good functioning of this factory is based on a set of sensor and actuator that allows the digitization of industrial possesses, to ensure high productivity with better quality. For the communication between the different sensor and actuator, we are based on the ZigBee protocol, and we study its performances in the case where it uses the mesh, star, and tree topologies. Our future work will focus on mobile ad hoc networks without infrastructure MANETs and more specifically the routing of information.

References

1. Zemrane, H., Baddi, Y., Hasbi, A.: Improve IoT ehealth ecosystem with SDN. In: Proceedings of the 4th International Conference on Smart City Applications, pp. 1–8 (2019)
2. Zemrane, H., Baddi, Y., Hasbi, A.: Internet of things smart home ecosystem. In: Emerging Technologies for Connected Internet of Vehicles and Intelligent Transportation System Networks, pp. 101–125. Springer (2020)
3. Zemrane, H., Baddi, Y., Hasbi, A.: Mobile adhoc networks for intelligent transportation system: Comparative analysis of the routing protocols. Proc. Comput. Sci. **160**, 758–765 (2019)
4. Janik, P., Pielka, M., Janik, M., Wróbel, Z.: Respiratory monitoring system using bluetooth low energy. Sens. Actuators A: Physical **286**, 152–162 (2019)
5. Zemrane, H., Baddi, Y., Hasbi, A.: Comparison between IoT protocols: Zigbee and WiFi using the OPNET simulator. In: Proceedings of the 12th International Conference on Intelligent Systems: Theories and Applications, pp. 1–6, (2018, October)
6. Zemrane, H., Baddi, Y., Hasbi, A.: IoT smart home ecosystem: Architecture and communication protocols. In: 2019 International Conference of Computer Science and Renewable Energies (ICCSRE), pp. 1–8. IEEE (2019)
7. Zemrane, H., Abbou, A.N., Baddi, Y., Hasbi, A.: Wireless sensor networks as part of IoT: Performance study of WiMax-mobil protocol. In: 2018 4th International Conference on Cloud Computing Technologies and Applications (Cloudtech), pp. 1–8. IEEE (2018)
8. Zemrane, H., Baddi, Y., Hasbi, A.: SDN-based solutions to improve IoT: survey. In: 2018 IEEE 5th International Congress on Information Science and Technology (CiSt), pp. 588-593. IEEE
9. Xie, P., Gu, H., Wang, K., Yu, X., Ma, S.: Mesh-of-torus: a new topology for server-centric data center networks. J. Super Comput. **75**(1), 255–271 (2019)
10. Khanna, R., Pisharody, G., Carlson, C.R.: Coordinator for low power sensor network with tree or star topology. US Patent App. 15/636,699, 3 Jan 2019
11. Nijholt, E., Rink, B., Sanders, J.: Center manifolds of coupled cell networks. SIAM Rev. **61**(1), 121–155 (2019)
12. Hsu, Y.P., Modiano, E., Duan, L.: Scheduling algorithms for minimizing age of information in wireless broadcast networks with random arrivals. IEEE Trans. Mob. Comput. (2019)
13. Leeson, D.: The digital factory in both the modern dental lab and clinic. Dent. Mater. (2019)
14. Pinto, J.E.M.G., de Azevedo Dantas, A.F.O., Maitelli, A.L., da Silva Dantas, A.D.O., Dorea, C.E.T., Campos, J.T.L.S., de Castro Rego, E.J.: PLC implementation of piecewise affine PI controller applied to industrial systems with constraints. J. Control Autom. Electr. Syst. **30**(3), 311–322 (2019)
15. Fu, C.B., Tian, A.H., Li, Y.C., Yau, H.T.: Active controller design for precision computerized numerical control machine tool systems. J. Low Freq. Noise Vib. Active Control **38**(3–4), 1149–1159 (2019)

IOT Search Engines: Study of Data Collection Methods

Fatima Zahra Fagroud, El Habib Ben Lahmar, Hicham Toumi, Khadija Achtaich, and Sanaa El Filali

Abstract Internet of Things (IOT) presents the trend in the world of computing science, which occupies a great interest in research and industry, offers a new field of creativity and innovation and can be applied in different areas such as education, agriculture, and industry. The number of IOT solutions developed attained 25.2 billion according to study of GSMA, their deployment in different areas improve the quality of life and will make the environment intelligent and conducive to all human activity. Searching information about connected devices and information about IOT topic in the world became possible by a search tools dedicated to IOT. In this work, we are interested in data collection methods used by IOT search engines, more precisely network port scanning and data generated by this method.

Keywords Search engine · IOT · IP · Port · Shodan · Censys

1 Introduction

The growth in the number of devices connected to the Internet is one result of advances in software, hardware, and network connectivity. These devices will offer advanced and accessible services remotely such as access to installed control devices.

F. Z. Fagroud (✉) · E. H. Ben Lahmar · K. Achtaich · S. El Filali
Laboratory of Information Technology and Modeling, Faculty of Sciences Ben M'sik, Hassan II University-Casablanca, BP 7955, Sidi Othman, Casablanca, Morocco
e-mail: fagroudfatimazahra0512@gmail.com

E. H. Ben Lahmar
e-mail: h.benlahmer@gmail.com

K. Achtaich
e-mail: kachtaich@gmail.com

S. El Filali
e-mail: elfilali.sanaa@gmail.com

H. Toumi
Higher School of Technology—Sidi Bennour, Chouaib Doukkali University, El Jadida, Morocco
e-mail: toumi.h@ucd.ac.ma

© The Editor(s) (if applicable) and The Author(s), under exclusive license to Springer Nature Singapore Pte Ltd. 2021
F. Saeed et al. (eds.), *Advances on Smart and Soft Computing*, Advances in Intelligent Systems and Computing 1188, https://doi.org/10.1007/978-981-15-6048-4_23

It is an exponential evolution in the number of connected objects that according to forecasts by 2020, almost 50 billion devices will be connected to the Internet.

With the emergence of the Internet of Things, challenges relating to network security, management, device status, access control, and anomaly detection bring managers and administrators of the IOT infrastructure to think of the creation of new support and mechanism. The use of IOT search engines and network port analysis tools can alleviate the IOT challenges mentioned because they have the capacity to identify devices and services directly connected to the Internet as well as vulnerable devices.

Since the appearance of the Internet of Things, various types of IOT devices have been appeared, such as the thermostat, camera, smart bulb, etc. This diversity makes the recognition of the devices important and a considerable necessity. IOT network security has become crucial given the number of infected devices, for example, according to [1]. Mirai (one of the most popular of existing IOT malware) has infected 600,000 devices worldwide in order to use them for DDoS attacks. Indeed, the connected devices type recognition can contribute to the security reinforcement and to block devices considered vulnerable to access networks.

In this work, we will focus in data collection methods used to discover IOT devices by IOT search engine and information generated by network port scanning tools. We believe our study allows to specify which tool has the best to use and the informations which their visibility causes security problems.

The rest of this paper is organized as follows: Sect. 2 presents IOT search engine, in Sect. 3 we introduce network port scanning and present some tools. Section 4 presents the information generated by network port scanning tools and comparison between them. Finally, in Sect. 5, a conclusion.

2   IOT Search Engine

Search engines are one of the main mechanisms by which users obtain information on the Web. For any individual wanting information about something on the Internet, search engines are the first choice to help find what the user wants [2]. IOT search engine is a type of search engines that can be classified as a specialized search engine. IT serve to solve the issues of IOT device discovery, search of information about IOT and reinforcement of security mechanisms. We can distinguish two subcategories of this search engine: The first define it has a search tool that allow the research of the connected devices in the world. It enable to their users to computers (webcam, power plants, refrigerator …) connected list on the Internet that are identified by city, country, latitude/longitude, OS, n IP…. The second think that is dedicated to Internet of Things topic which groups Web pages accessible via the internet that their content is related to IOT and make enable to their users the search of information and learn about this topic.

In this section we present the most popular IOT search engine who belong to the first subcategories.

2.1 Censys

Censys.io is a search engine designed by Zakir Durumeric in 2015. Censys is based on daily Zmap analyzes in order to collect its data, its database has more than 3 billion IPv4 addresses. This tool allows its users to find what makes up the internet and can also interact with Censys data using scripts (this request can be made using an SQL engine).

Censys is a search engine used to query information of hosts and networks stored in daily ZMap scans. Censys enables querying data from the Internet-wide scan repository (scans.io), a data repository hosting the periodic scan results as collected by the ZMap scanner. Censys tags the collected data with security-related properties and device types, allowing easy but powerful search queries through its online search interface and REST API. As an example application for Censys, the prevalence of the unauthenticated Modbus protocol among SCADA systems has been studied. Numerous such systems have been found across the globe. However, non-SCADA devices, specifically, the TLS ecosystem for those devices have not been studied [3].

2.2 Shodan

Shodan.io is a search tool (search engine) dedicated to IoT devices, offers a set of information relative to these devices but full access requires paid subscriptions. The Shodan developers used a loophole in the NTP pool project that allowed them to scan millions of IPv6 addresses. This result represents a plus for the search engine Shodan compared to Censys.

Shodan has been recognized as one of the most popular search engines on the market in recent years for exploring the net and indexing discovered services. Launched in 2009 by John Matterly, this tool allows the visualization of devices and services accessible on the Internet while using a graphic computer program. Shodan scans the net for global devices and services and stores the collected information, including the IP address, port, and Net Protocol repair banner for info accessible via https://www.shodanhq.com or via the Shodan application programming interface (Shodan API), for each service discovered, Shodan analyzes and stores the results several times over time [4]. Using Shodan, we can compile market statistics for the different services and allow security specialists to perform additional processes and analyzes (Fig. 1).

2.3 Thingful

Thingful can be described as a "discoverable search engine" which allows its users to have a geographic index of connected objects around the world (https://thingful.

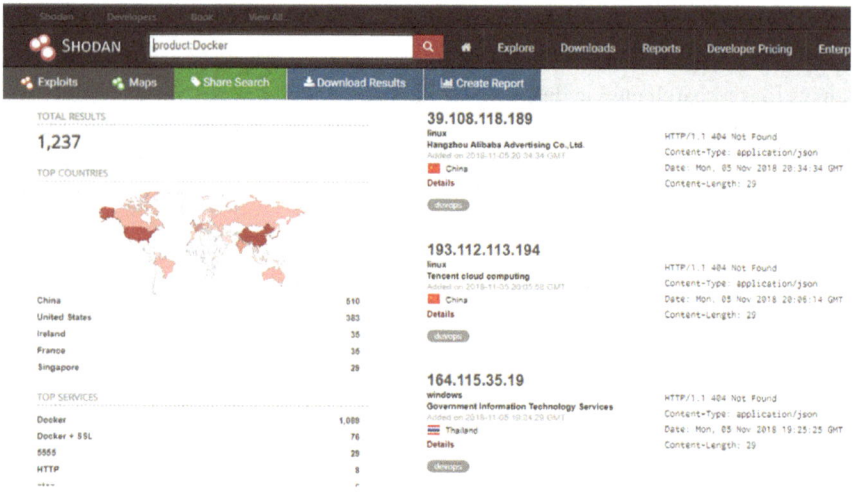

Fig. 1 Shodan: example of scan results

net/). As such, Thingful boasts that it can index across multiple IoT networks and infrastructures, because this search engine can locate the geographical position of objects and devices [5]. Personal information access especially those related to the way how users act and communicate with their connected devices and objects (as in the case of owner's human activity with "smart" objects in their home) is one of the critics of using Thingful.

2.4 Qwant IOT

Technological partnership between the French search engine Qwant and the Montpellier start-up Kuzzle launches at the innovation and tech fair, which opens in Paris on May 24, 2018. This cooperation aims to optimize the collection and retrieval of data from connected objects.

Search engine based on the IOT application development solution of the Kuzzle start-up, based in Montpellier, to collect and retrieve this information.

2.5 Comparison of IOT Search Engines

Several IOT search engines have been developed in the aim to solve some issues of Internet of Things like security and device discovery. In Table 1, we present a comparison of the following IOT search engines: Shodan, Censys, Thingful and Qwant IOT.

Table 1 Comparison of IOT search engines

	Founding date	Website	Devices detected	Data collection	Data stores in database	Access	Filters
Shodan	2009	www.sho dan.io	All devices connected on IPv4 and IPv6	Port scanner: SYN scan	Device summary (IP address, location, latitude/longitude coordinates) and available services	Web interface, API	IP, Port, Location, version, hostname …
Thingful	2014	www.thingf ul.net	IOT devices	Added by owner of IOT devices	All available data	Web interface, API	Application area
Censys	2015	www.Cen sys.io	All devices connected on IPv4	Port scanner: ZMap (+ZGrab)	Very complete data about IOT devices	Web interface, API, or texts to download	Keyword, IP, protocol used, certificate …
Qwant IOT	2018	–	IoT solution of the Kuzzle start-up	Data of IoT solution developed by Kuzzle start-up	Data from various IOT sensors available in open data	–	–

From these comparison we can conclude that:

- The information from different IOT search engines can be a help to hackers
- censys and Shodan detect all devices connected on the Internet (IOT devices and no IOT devices)
- Shodan and censys use the port scanners on the other hand Thingful and Qwant IOT are based on open data from a specific infrastructure
- IOT search engines offer an API furthermore the Web interface to access and use their data
- Need to support real-time update of data
- Need development of other IOT search engine that respond to all needs of users.

3 Network Port Scanning

Network analysis is a technique that is used primarily for security assessment, system maintenance, and in addition for attacks by hackers. The objective of network analysis is to analyze network ports as a vulnerability analysis. This technique is one in all three necessary ways employed by associate degree assailant to assemble data.

Network port scanning refers to the way of dispatch information packets via the network to a computing system's given service port numbers (for example, port twenty three for Telnet, port eighty for protocol so on). This is often to spot on the world network services on it explicit system. Network port scanning moreover as vulnerability scanning is associate degree information-gathering technique, however once applied by anonymous people, these are viewed as a prelude to associate degree attack. Network scanning processes, like port scans and ping sweeps, come details regarding that information science addresses map to active live hosts and therefore the kind of services they supply.

3.1 Nmap

Nmap (Network Mapper) is an open-source port scanner that can be used to detect hosts and services connected to a computer network and designed to detect a range of information related to scanned devices such as open and closed ports, operating system, device status, manufacturer, device type, IP address, and name of host. It is a software that has become a reference for administrators' networks because the audit of Nmap's results provides indications on the security of a network (it represent a famous tool to use in order to find and exploit vulnerabilities in a network).

Nmap is a command line utility, in addition that there are several graphical interfaces that have been developed for this tool like Zenmap. The use of Nmap is not necessarily user-friendly, can even be cumbersome and its results may not be understandable by non-experts.

Nmap can be installed on Windows, Linux, BSD Unix, and Mac OS and has a graphical interface called Zenmap. Zenmap is a tool that does not have the current network monitoring capabilities of the other tools on this list (Fig. 2).

3.2 Advanced Port Scanner

Advanced port scanner is fast, free port scanning software that scans network devices and allows to its users to find open ports quickly and retrieve versions of programs running on these identified ports in easy way.

This program is very easy to use, even for beginners, thanks to its intuitive interface and its assistant. The interface of the program is both rich and intuitive and it displays the list of default ports. Their users can analyze all ports or only those in the ranges they define from all networked devices and not only on their main PC.

Advanced port scanner offer to their users:

- Fast scanning of network devices
- Identification of programs running on detected ports
- Easy access to resources found: HTTP, HTTPS, FTP, and shared folders
- Remote access to computers via RDP and Radmin (Fig. 3).

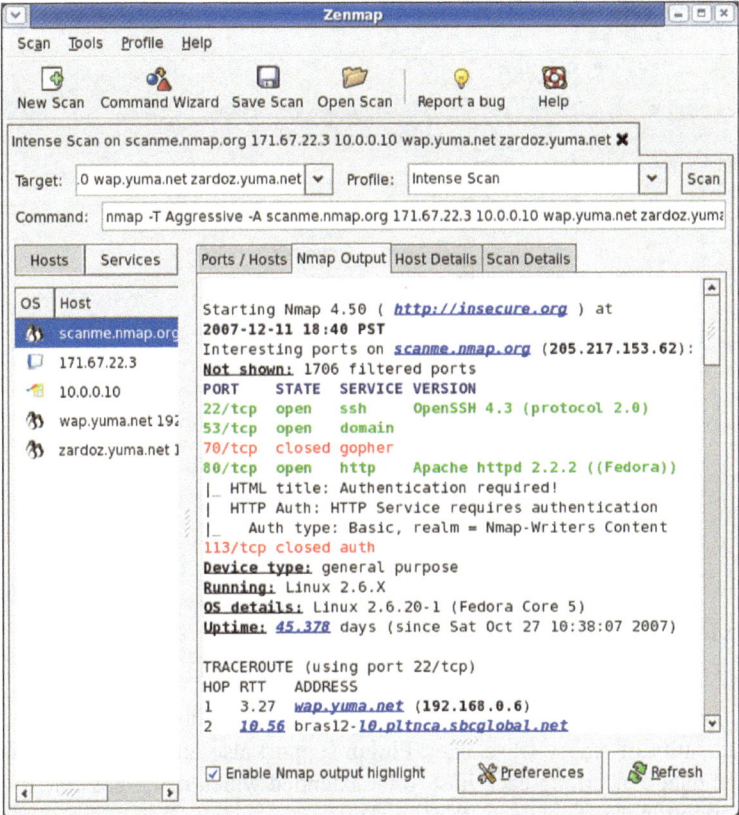

Fig. 2 Zenmap: example of scan results

3.3 Angry IP Scanner

Angry IP scanner is a tool (network scanner) which allows to their users to scan IP addresses in any range as well as any their ports. It is cross-platform, open source, and any installations are needed. This tool use a multithreaded approach in order to increase scanning speed by creating an independent scanning thread for each IP address scanned.

This tool is wide utilized by network administrators and simply curious users round the world, as well as a giant and tiny enterprises, banks, and government agencies.

Angry IP scanner pings every IP address to examine if it is active or not, at that point alternatively it breakdown its hostname, determines the raincoat address, scans ports, etc. It conjointly has extra options, like NetBIOS info (computer name, workgroup name, and windows user which are currently logged), preferred IP address ranges, net server detection, customizable openers, etc.

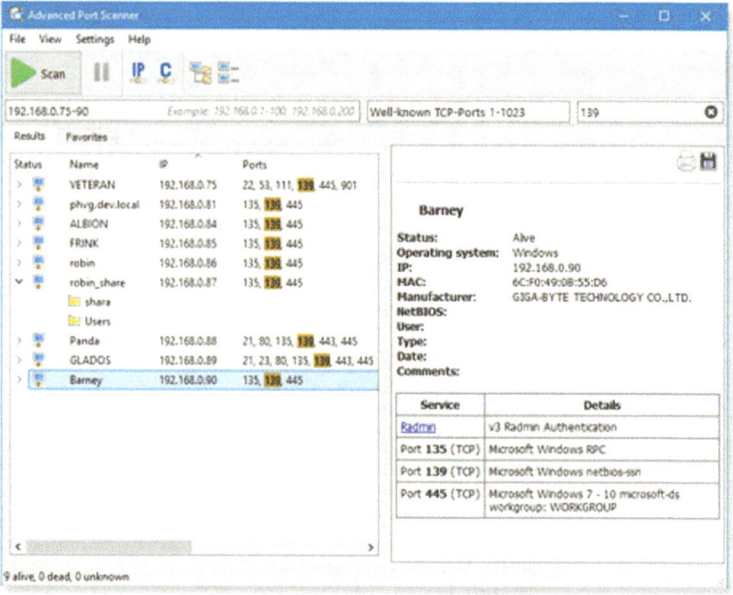

Fig. 3 Advanced port scanner: example of scan results

Anybody that can code with Java language may write down plugins and extend the practicality of Angry IP scanner. Plugin support also allows for the amount of collected data concerning each host to be extended which make possible to gather any information about scanned IPs (Fig. 4).

3.4 PortScan and Stuff

PortScan and Stuff is network scanning tool for Windows operating system that is a free and portable tool. It allow to their users to exploit and recognize network devices as well as perform various speed tests over the Internet with selectable server rundown and Who is capacities.

PortScan and Stuff is a tool designed to find all active devices, shows all open ports and supplementary data, for example, MAC address, HTTP, FTP, SMTP, SNMP, host name, and SMB services.

This tool is multithreaded to increase the performance when scanning wide networks. It will assign to a hundred coinciding threads. Scanning filter represents one of the best features of this tool. It permits one to limit the filtering to explicit criteria, for example, a port number. PortScan and Stuff can also obtain an inherent propelled ping and traceroute utilities. Its ping, for example, can ping a computer with three standard-sized packets, with three estimated ones, ping the computer at

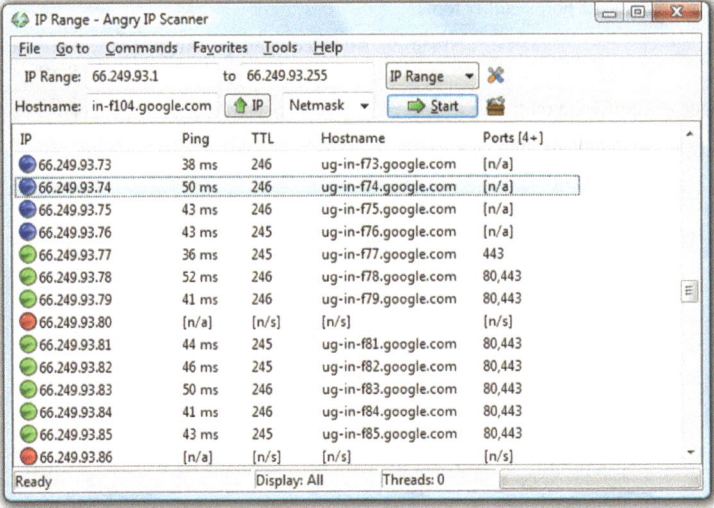

Fig. 4 Angry IP scanner: example of scan results

multiple times, or ping it continuously. Moreover, DNS and Who is queries are also performed by this tool.

4 Results and Discussion

This section has been focused on the study of information related to discovered devices generated by network port scanning tools who we present a comparison of network port scanning tools.

Until now, a set of tools has been designed to network scanning in the aim to facilitate this task and enable to their users to have all informations that can be retrieved. In Table 2, we present a comparison between the following port scanners: Nmap, advanced port scanner, Angry IP scanner, and PortScan and Stuff.

The realized comparison shown that the tools nmap and angry ip scanner are the best network scanner tools to use because:

- Enable scanning of connected devices in IPv4 and IPv6
- Offer a significant amount of information
- Nmap and Angry IP scanner have an API which make possible to integrate them in different programs
- Nmap use NSE (Nmap Scripting Engine) which provide to Nmap a flexible infrastructure to extend its capabilities
- Angry IP scanner allows to extend their functionalities by writing a new plugins (Java code).

Table 2 Comparison of port scanner tools

		Nmap	Advanced port scanner	Angry IP scanner	PortScan and Stuff
Available on	Software tool	✓	✓	✓	✓
	Command-line	✓			
	API (java)			✓	
	Portable		✓	✓	✓
IP address	IPv4	✓	✓	✓	✓
	IPv6	✓		✓	✓
Input	IP/IP range	✓	✓	✓	✓
	Server name				✓
	Host name			✓	
	Ports		✓		✓
	Scan profile	✓			
	Scan speed				✓
	Command	✓			
	Netmask			✓	
Output	IP	✓	✓	✓	✓
	Host name	✓	✓	✓	✓
	Ports	✓	✓	✓	✓
	MAC address	✓	✓	✓	✓
	Traceroute	✓			✓
	OS	✓	✓		
	NetBIOS information		✓	✓	
	Who is				✓
	Manufacturer		✓	✓	
	Device type	✓	✓		
	Services informations	✓	✓		✓
	State	✓	✓		
	User		✓		
	Web server detection			✓	
	DNS				✓
Save as	.txt			✓	
	.csv		✓	✓	
	.xml	✓	✓	✓	✓
	List of IP: port			✓	

(continued)

Table 2 (continued)

	Nmap	Advanced port scanner	Angry IP scanner	PortScan and Stuff
.nmap	✓			
.html		✓		
By extension	✓			✓

5   Conclusion

Today, device discovery issues of the Internet of Things can be solved by using IOT search engines which make possible to search informations related to these devices. The informations obtained by IOT search engines must be search engines, it is very difficult to obtain the relevant, accurate, and updated in real-time. This task is very difficult because the data related to connected devices are dynamic and the appearance of a large number of connected devices per day. Finally, our research will focus on the development of a new way to collect informations related to IOT devices that allow to collect all available informations, detect more IOT devices and enable to have accurate data.

References

1. Shahid, M.R., Blanc, G., Zhang, Z., Debar, H.: IoT devices recognition through network traffic analysis. In: 2018 IEEE International Conference on Big Data (Big Data), pp. 5187–5192. IEEE (2018)
2. Faqeeh, M., Al-Ayyoub, M., Wardat, M., Hmeidi, I., Jararweh, Y.: Topical search engine for Internet of Things. In: 2014 IEEE/ACS 11th International Conference on Computer Systems and Applications (AICCSA), pp. 829–835. IEEE (2014)
3. Samarasinghe, N., Mannan, M.: Short paper: TLS ecosystems in networked devices vs. web servers. In: International Conference on Financial Cryptography and Data Security, pp. 533–541. Springer, Cham (2017)
4. Genge, B., Enăchescu, C.: ShoVAT: Shodan-based vulnerability assessment tool for Internet-facing services. Secur. Commun. Netw. **9**(15), 2696–2714 (2016)
5. Search Engines and Ethics. https://stanford.library.sydney.edu.au/entries/ethics-search/. Last accessed 20 Nov 2019
6. Liang, F., Qian, C., Hatcher, W.G., Yu, W.: Search engine for the Internet of Things: lessons from web search, vision, and opportunities. IEEE Access **7**, 104673–104691 (2019)
7. Arnaert, M., Bertrand, Y., Boudaoud, K.: Modeling vulnerable Internet of Things on SHODAN and CENSYS: an ontology for cyber security. In: Proceedings of the Tenth International Conference on Emerging Security Information, Systems and Technologies (SECUREWARE 2016), pp. 24–28. Nice, France (2016)
8. Rahman, A., Kawshik, K.R., Sourav, A.A., Gaji, A.A.: Advanced network scanning. Am. J. Eng. Res. (AJER) **5**(6), 38–42 (2016)
9. Angry IP scanner. https://angryip.org/about/. Last accessed 02 Dec 2019
10. Nmap. https://nmap.org/. Last accessed 21 Dec 2019
11. Advanced Port Scanner. https://www.advanced-port-scanner.com/fr/. Last accessed 10 Nov 2019

12. Qwant IOT. https://www.lemondeinformatique.fr/actualites/lire-avec-kuzzle-le-moteur-qwant-restitue-les-donnees-iot-en-open-data-71871.html. Last accessed 18 Aug 2019
13. Qwant IOT. https://objectif-languedoc-roussillon.latribune.fr/entreprises/tic/2018-05-22/iot-le-montpellierain-kuzzle-s-allie-a-qwant-779227.html. Last accessed 20 Aug 2019
14. Santos, M.R., Andrade, R.M., Gomes, D.G., Callado, A.C.: An efficient approach for device identification and traffic classification in IoT ecosystems. In: 2018 IEEE Symposium on Computers and Communications (ISCC), pp. 00304–00309. IEEE (2018)
15. Datta, S.K., Bonnet, C.: Search engine based resource discovery framework for Internet of Things. In: 2015 IEEE 4th Global Conference on Consumer Electronics (GCCE), pp. 83–85. IEEE (2015)
16. Feng, X., Li, Q., Wang, H., Sun, L.: Acquisitional rule-based engine for discovering Internet-of-Things devices. In: 27th USENIX Security Symposium (USENIX Security 18), pp. 327–341 (2018)
17. Fagroud, F.Z., Ben Lahmar, E.H., Amine, M., Toumi, H., El Filali, S.: What does mean search engine for IOT or IOT search engine. In: Proceedings of the 4th International Conference on Big Data and Internet of Things, pp. 1–7 (2019)

An Agent-Based Architecture Using Deep Reinforcement Learning for the Intelligent Internet of Things Applications

Dhouha Ben Noureddine, Moez Krichen, Seifeddine Mechti, Tarik Nahhal, and Wilfried Yves Hamilton Adoni

Abstract Internet of Things (IoT) is composed of many IoT devices connected throughout the Internet, that collect and share information to represent the environment. IoT is currently restructuring the actual manufacturing to smart manufacturing. However, inherent characteristics of IoT lead to a number of titanic challenges such as decentralization, weak interoperability, and security. The artificial intelligence provides opportunities to address IoT's challenges, e.g., the agent technology. This paper presents first an overview of ML and discusses some related work. Then, we briefly present the classic IoT architecture. Then, we introduce our proposed intelligent IoT (IIoT) architecture. We next concentrate on introducing the approach using multi-agent DRL in IIoT.

Keywords Internet of Things · Intelligent IoT · Software agent · Deep reinforcement learning · Intelligent IoT architecture

D. B. Noureddine
LISI, INSAT, University of Carthage, Tunis, Tunisia
e-mail: dhouha.bennoureddine@fst.utm.tn

FST, University of El Manar, Tunis, Tunisia

M. Krichen (✉)
FCSIT, Albaha University, Albaha, Saudi Arabia
e-mail: moez.krichen@redcad.org

REDCAD, University of Sfax, Sfax, Tunisia

S. Mechti
Miracl Laboratory, University of Sfax, Sfax, Tunisia
e-mail: mechtiseif@gmail.com

T. Nahhal · W. Y. H. Adoni
Hassan II University of Casablanca, Casablanca, Morocco
e-mail: t.nahhal@fsac.ac.ma

W. Y. H. Adoni
e-mail: adoniwilfried@gmail.com

© The Editor(s) (if applicable) and The Author(s), under exclusive license to Springer Nature Singapore Pte Ltd. 2021
F. Saeed et al. (eds.), *Advances on Smart and Soft Computing*, Advances in Intelligent Systems and Computing 1188, https://doi.org/10.1007/978-981-15-6048-4_24

273

1   Introduction

The Internet of Things is a network of systems where many objects are interconnected to each other through the Internet. It provides new applications to enhance the life quality of humans, including the waste management, urban traffic control, environmental detection, social interaction gadgets, sustainable urban environment, emergency services, mobile shopping, smart metering, smart home automation, healthcare, transportation and logistics, etc. Nonetheless, many study issues in the fields of software architecture, application development, network security, and reliability need to be addressed to attain the complete potential of IoT in the aforementioned fields. As more of the infrastructure, both virtual and physical, "get interconnected through the IoT, the opportunities for more disruptive and widespread cyber-attacks grow concomitantly" [1]. As a further matter, the real circumstance is that these objects remain far off from reaching natural intelligence boundaries and there are still many other problems resulting from the interoperability of things and edge devices in a dynamic environment.

In this context, many researchers [2, 3] have integrated the artificial intelligence (AI) with IoT technologies, for example, when a large number of devices connected to the Internet generate big quantities of sensory data to represent the physical world's status. These data can be interpreted and analyzed using machine learning (ML) algorithm, which makes good decisions to monitor devices' reactions to the physical world. ML is considered to be one of the most suitable computational paradigms to provide embedded intelligence in the IoT devices [4]. The integration process of IoT and AI, therefore, plays a crucial role in the technology and makes the IoT became intelligent and autonomous by seeing system capabilities grow, including increasing operational efficiency.

The intelligent IoT has a more complicated system involving identification, sensing, perception, communication, computation, and services. We use in our approach the software agents that have been already proposed for distributed IoT application execution in such distributed context [5]. Agents are designed and programmed to execute tasks as components of distributed applications. Similarly, agents related to things and edge devices in IoT improve interoperability, autonomy, behavior, reasoning, learning, etc. Accordingly, the development of the IoT system architectures has been facilitated by the communication and cooperation between these agents representing things, thus reducing at the least human interventions through the use of autonomous behavior of agents in the IoT applications control or execution. Agents have so far largely unexplored, specifically in the IoT interoperability context.

This paper aims at providing new approach integrating the learning agent that is able to act and adapt based on new information into the classic architecture IoT. The presented research provides an IIoT architecture based on software agent framework that applied machine learning to tackle many IoT applications. The rest of this paper is structured as following. Section 1 includes some current research related to this area. Section 2 presents the machine learning algorithms. After that, in Sect. 3 an architecture of IIoT and a proposed approach using multi-agent DRL in IIoT will be depicted in detail. Then, a simulated autonomous car driving case study is illustrated as well as its related simulations in Sect. 4. Finally, we discuss and conclude our work in Sect. 5.

2 Related Work

Most applications of IoT in the literature focus on the consolidation of the other approaches such as Internet of agents, Internet of robots, Internet of services, Internet of drones, and Internet of people. Software agents can interact with the other agents in the environment to accomplish their goals [6]. Some research using the agents creates an embodied agents, e.g., virtual robots, wireless devices, ubiquitous computing, etc., to design the interactions with the sensors and actuators of the IoT applications [7]. To create an embodied agent, the IT experts provides this agent with a behavior description that is consistent with its body and the task to be performed. It is very hard, however, to completely define a physical system's behaviors at the time of design and to recognize and promote characteristics that contribute to attractive collective behavior. In order to solve these issues, some approaches [8, 9] have suggested using evolving machine learning such as neural networks to enable an embodied agent to learn how to adapt their behavior in a dynamic environment. In fact, through learning skills, agents will be able to reason conveniently, create a suitable policy, and make good decisions [10]. During the research, many contributions have been proposed in the field of machine learning, in particular, approaches from deep reinforcement learning (DRL) for intelligent IoT.

Others have proposed an architecture based on multi-agent system (MAS) [11] that aims to coordinate IoT devices. This architecture is distinguished by its capabilities to allow dialogues between IoT devices using rational agents. This work focuses on the infrastructure of IoT services, while our work is not dependent on these kind of services. Moreover, the focus of [11] was on traditional reasoning and dialogue capabilities, whereas this paper focuses on advanced and DRL techniques.

Some researchers [12] have suggested an architecture based on autonomous technical assistance of agent-based objects (detection of failures) associated with safety and medical assistance in an ambient environment. However, that work did not consider machine learning, which is the focus of our approach.

3 Introduction to Machine Learning

In this section, we first present the best-known machine learning examples such as reinforcement learning (RL), deep learning (DL), and their combination deep reinforcement learning (DRL). Then, we introduce the MASs using DRL, whereby communicating with the environment, each DRL agent learns to develop his own strategy and the other agents to attain rewards.

3.1 Reinforcement Learning

RL [10] is defined as a type of ML algorithm that can optimally control a Markov decision process (MDP). RL presents ambient intelligence in IIoT systems by offering a class of sequential experience for processing sensory data in order to produce reaction control decisions. More specifically, the RL agents learn to maximize their expected future sum of rewards by interacting with an environment and learn from sequential experience.

3.2 Deep Learning

DL provides a subset of ML approaches that attempt to model with a high level of data abstraction by articulating architectures of various nonlinear transformations. DL is a subset of techniques that exploit artificial neural networks (ANNs) to learn autonomously from a large amounts of data. Significant and rapid progress has been made in the fields of audio and visual signal analysis, including facial recognition, speech recognition, computer vision, and automatic language processing. DL is able to perform tasks such as regression and classification. There are many different DL architectures available in the literature such as convolutional neural networks (CNN), recurrent neural networks (RNN), Boltzmann machine (BM), long short-term memory (LSTM) networks, feedforward deep networks, (FDN) and deep belief networks (DBN).

3.3 Deep Reinforcement Learning

DRL is a subset of RL. Rather, it is the combination of DL and RL. As previously stated in Sect. 3.1, for simple cases in the RL, it is possible to memorize and refine the reward estimator for all combinations of states or actions. But the problem arises when this number of possible states/actions takes huge proportions that a computer is no longer able to process. In this case, it is necessary to find a more compact way than the exhaustive storage to store this function. This compression of information amounts to saying that certain sets of states/actions are actually similar and will produce the same reward.

This is where DRL comes in as a way of compressing information. The idea is to use neural networks as can be used in DL. These networks are extremely efficient for generalizing and approximating any function. The purpose of the game is, therefore, to use such a network which takes as input the current state and an action, and output an estimate of the reward that can be obtained. Once the network is trained, selecting the right action is like comparing the potential rewards of each one and choosing the best one.

The RL and DRL models for real-world intelligent IoT systems is not as simple as it might seem. The RL and DRL models are only interested in the environment and the agent. On the one hand, the RL/DRL environment can only reflect the physical system or be expanded to include wireless networks, edge/fog servers, and cloud servers. And it would be so because of the communication deadline, computing delay, power consumption, and network reliability, etc., that will have significant consequences on the physical system's control efficiency. The control actions in RL/DRL can, therefore, be divided into two levels: control of actuators and control of resources. It is possible to separate or jointly learn and optimize the two levels of control. On the other hand, the agent in RL is an entity that makes decisions about action selection. In IIoT systems, the intelligence agencies can reside in IoT devices, edge servers, fog servers, and/or cloud servers.

3.4 Multi-agent Deep Reinforcement Learning

In Sect. 3.3, we aim attention at the DRL methods for single-agent cases. In reality, situations exist where many agents require to interact together to achieve the general aim of the system, e.g., in the multi-robot systems, in the network vehicles, in the cloud robotics, etc. DRL methods are conceived for multi-agent systems in these cases. A MAS is composed of multiple interacting agents within a common environment. The agents in the system can interact directly with each to cooperatively or competitively accomplish their tasks, and thus the theory of stochastic MDP into the multi-agent setting.

In multi-agent RL, by interacting with the environment to attain rewards, each agent learns to improve their own policy. The environment is generally dynamic, and the system can detect the problem of explosion space action. Because of many agents learn simultaneously, the optimal policy of itself may also change for a certain agent when the policies of the others change. This can affect the learning algorithm's convergence and cause instability. Using, therefore, independent learning agents is the simplest method to learning in multi-agent settings. For instance, independent Q-learning is an algorithm where each agent learns its own policy independently, while the other agents are viewed as part of the environment'.

Independent Q-learning, nevertheless, cannot handle the non-stationary environment problem. Throughout recent years, single-agent DRL techniques have been expanded to cases involving multiple agents.

4 Proposed Approach Using Multi-agent DRL in IIoT

The IIoT presents different concepts of AI using the data needed to apply both timely analytics and informed action predicated based on this continuous data transmission. For IIoT, the combination of IoT with cognitive science is an evolution in advanced science that will have implications in the near future for scientific areas. AI's capabilities are seen in the transformation from stand-alone IoT devices to the interactive collaboration of things. IoT software agents extend the capabilities of devices and software components across all levels of IoT device architectures for autonomous intelligent behavior [5]. We present, in this section, the description of a classic IoT architecture existing in the literature followed by our proposed architecture based on agents using the deep reinforcement learning.

4.1 Classic IoT Architecture

The IoT architecture is layered, simple, able to integrate networked objects and transfer valuable data to the computer system, which can then be processed to give a concrete operational vision. A classic IoT system consists of four main blocks:

- Block 1: It includes sensors and actuators. On the one hand, it consists of sensors capable of detecting physical parameters, gathering data from the environment, and identifying other intelligent objects. On the other hand, it consists of actuators that can influence environmental change.
- Block 2: It contains the data from block 1 (from sensors). It is necessary to aggregate and convert these received data in analog form to digital form.
- Block 3: It contains the incoming pre-treated data for advanced processing analytic in the computational computing systems.
- Block 4: In this block, the data on high-performance computer systems are processed, controlled, managed, and stored.

We can also describe by analogy with the IoT blocks architecture, a four-layered architecture such as:

- The perception layer: presents essentially the physical autonomous layer of sensors and actuators interacting with the environment for data acquisition and control actions.
- The network layer: connects IoT communication network devices including wireless access networks, the Internet, and servers. It essentially transfers perception sensor data to the processing layer through networks. In other words, this layer allows the IoT devices to be discovered and connected for data and control command transmission to the edge/fog servers and cloud servers.
- The processing layer: stores, interprets, and processes enormous data using different technologies, e.g., big data. This refers to the commitment of data processing/storage and analysis actions by IoT edge/fog computing systems.

- The application layer: provides the client with application-specific services. This refers to the commitment of data processing/storage and control actions determination by IoT cloud computing systems.

4.2 Proposed IIoT Architecture

Our goal in this paper is to make an IoT (that is composed of many devices connected with the network) intelligent. But these dropping costs, combined with AI, mean that we are at the forefront of the next wave of the IoT, which is to make these devices detect, understand, adapt, and react to the world. IoT devices became smarter, and they can further manage real-time data processing, analytic, sensing, acting in the environment. The decision quality is improving, but most importantly, the timing of these decisions is also improving as well. Consequently, the challenge for future IoT developers will not be to use the analysis in their system, but rather to know what analytical decisions they need in real time, and what additional localized contextual data can improve the decision making. It, therefore, seems that the long-term ideal will be to move away from centralized control and more localized and real-time decision making.

We propose an IIoT architecture [13] as shown Fig. 1 by integrating smart agents using ML to learn from their past experiences to improve system performance. Each layer is coupled with an intelligent agent(s) allowing devices and machines to be able to make decisions autonomously and intelligently through ML. When this happens, humans can be excluded from the equation, generating value and profitability. In the IIoT architecture, therefore, one or more layers may be included in the environment. IoT devices, IoT edge/fog servers, IoT cloud servers, and IoT Internet gateway can be located by the agent(s).

Perception layer. Assuming the environment only contains the perception layer, so the physical system is designed as a stochastic process where the state, action, and reward are defined as follows:

- Physical observation (external device state) s_{layer1}, for example, the observation o_t in the time-step t consists of the readings of the system RGB images o_t^{RGB}, the relative goal position o_t^g, and the current velocity of the robot o_t^v, the locations of the robot o_t^l, etc.
- Action output is a set of permissible actions in continuous space a_{layer1}, for example, control of robot motion, switching ON/OFF the device, moving a robotic arm, processing of workpieces, etc.
- Reward (physical system performance) r_{layer1}: a reward feature is designed to direct a team of robots to achieve the goal of the system, for example, How easily a robot or a vehicle can move, or is it away from obstacles?

There is much research in the literature on DRL applications in the perception layer. We do not describe in this section in detail the existing work, but we will choose one for more explanations, such as the work of [14]. Where, the authors

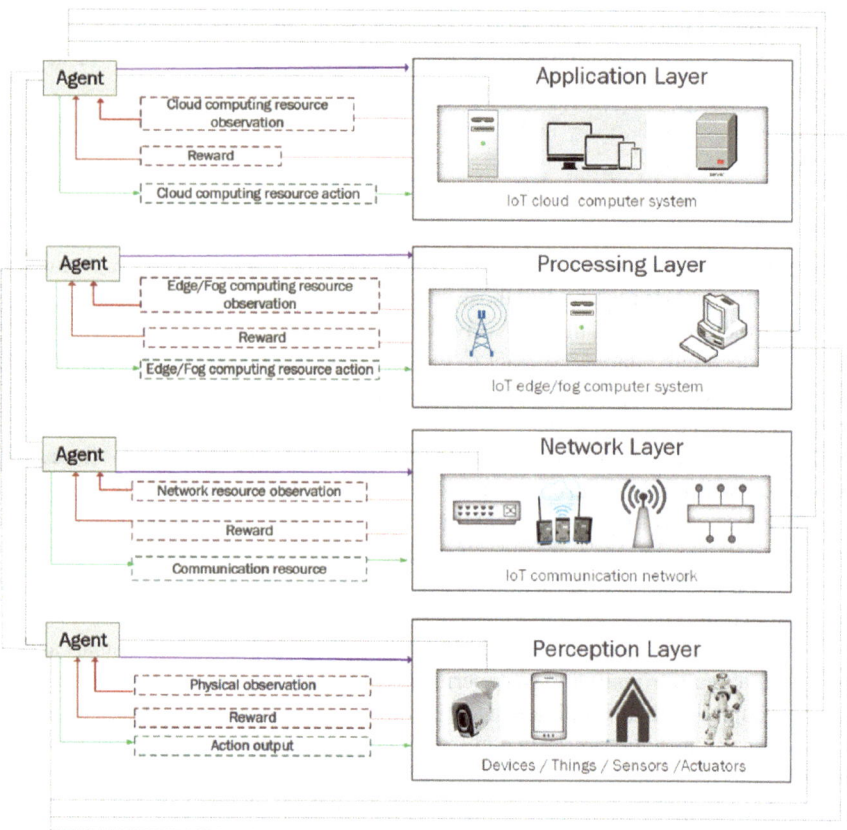

Fig. 1 Intelligent Internet of Things architecture based on agents using DRL

apply Deep Q-network DQN to the robot behavior learning simulation environment to allow mobile robots to learn how to achieve good behaviors, for example, avoid the wall and move along the centerline by using high-dimensional visual information as input data.

Network layer. Assuming the environment only contains the network layer, so the network dynamic system is designed as a stochastic process where the state, action, and reward defined as follows:

- Network resource observation s_{layer2}, for example, the bandwidth allocation, the signal to interference and noise ratio, etc.;
- Communication resource action a_{layer2}, for example, the planning of multiple users, etc.;
- Reward (Network performance) r_{layer2}, for example, the probability of transmission failure, the power consumption of transmission, etc.

Processing layer. Assuming the environment only contains the processing layer, so the edge/fog computing dynamic system is designed as a stochastic process where the state, action, and reward defined as follows:

- Edge/fog computing resource observation s_{layer3}, for example, the number of virtual machines at runtime, the fault detection, and the state analysis of devices in assembly lines, etc.;
- Edge/fog computing resource action a_{layer3}, for example, the computational and storage services, etc.
- Reward (Edge/fog computing performance) r_{layer3}, for example, the utilization rate of the edge/fog computing resources, etc.

Application layer. Assuming the environment only contains the application layer, so the cloud computing system is designed as a stochastic process where the state, action, and reward defined as follows:

- Cloud computing resource observation s_{layer4}, for example, the number of task scheduling, the number of tasks for processing buffered in the queue, etc.
- Cloud computing resource action a_{layer4}, for example, the selection of caches, the task offloading decisions, the virtual machine allocation, the task allocation, the resource allocation, etc.
- Reward (Cloud computing performance) r_{layer4}, for example, the utilization rate of the cloud computing resources, etc.

Combination of all the layers. Assuming the environment contains the four layers, so the DRL models typically contain elements (state, action, reward) described as following:

- IIoT state s_{IIoT} includes the aggregation of physical observation (external device state), network resource observation, edge/fog computing resource observation and cloud computing resource observation, i.e., $s_{IIoT} = \{s_{layer1}, s_{layer2}, s_{layer3}, s_{layer4}\}$;
- IIoT action a_{IIoT} includes the aggregation of action output, communication resource action, edge/fog computing resource action and cloud computing resource action, i.e., $a_{IIoT} = \{a_{layer1}, a_{layer2}, a_{layer3}, a_{layer4}\}$;
- IIoT reward r_{IIoT} is usually set to maximize the system performance that can be defined as a function of the network performance, edge/fog computing performance and cloud computing performance, i.e., $r_{IIoT} = r_{layer1}(r_{layer2}, r_{layer3}, r_{layer4})$.

In our model, since the agent in DRL is a software concept, the agent related to each layer can solve the DRL problem by observing the states and rewards of its environment and by applying learning strategies. These strategies aim at automatic learning of tasks by the gradual improvement of a function making it possible to estimate a future reward according to an action chosen at a given instant to determine the appropriate actions.

5 Conclusion

AI, IoT, ML, and agents are now considered to be highly advanced science that cover a variety of concepts and technologies. The basic computing machines have been evolved to smartphones, the portable devices to wearables, the cloud to IoT, that causes data management, data processing, and data analytic to move to the next tier. Nowadays, multi-agent systems play a crucial role in many fields, including robotics, healthcare, video game, cellular biology models and genomics, etc. The complex problems will be executed in virtually an autonomous and intelligent way without any kind of human intervention, and by combining MAS, IoT, and ML. For IoT, several agents can be combined in a highly collaborative and cooperative way to perform tasks. In this paper, we have proposed a new intelligent IoT architecture. The paper comprehensively reviews machine learning as well as the classic architecture of IoT systems. In the near future, we would like to demonstrate our approach by experiences the application of deep reinforcement learning methods in autonomous robots, especially for autonomous car driving scenarios and our focus will also be to propose some novel and intelligent IoT applications integrating the multi-agent systems with intelligent edge computing using machine learning. In addition, we will adapt existing formal approaches [15–18] for testing and validating our framework.

References

1. Leong, P., Lu, L.: Multiagent web for the Internet of Things. In: Proceedings of the International Conference on Information Science and Applications, pp. 271–350 (2014)
2. Mzahm, A.M., Ahmad, M.S., Tang, A.Y.C.: Towards a design model for things in agents of things. In: Proceedings of the International Conference on Internet of Things and Cloud Computing, pp. 3–7 (2016)
3. Savaglio, C.,Fortino, G., Ganzha, M., Paprzycki, M., Bădică, C., Ivanović, M.: Agent-based computing in the Internet of Things: a survey. In: Intelligent Distributed Computing XI, pp. 302–307 (2017)
4. Hussain, F., Hussain, R., Hassan, S.A., Hossain, E.: Machine learning in IoT security: Current solutions and future challenges. arxiv:1904.0573v1 (2019).
5. Fortino, G.: Agents meet the IoT: toward ecosystems of networked smart objects. IEEE Syst. Man Cybernet. Mag. 2(2), 43–47 (2016)
6. Noureddine, D.B., Gharbi, A., Ahmed, S.B.: A social multi-agent cooperation system based on planning and distributed task allocation: real case study. In: Proceedings of the 13th International Conference on Software Technologies (IC-SOFT) (2018)
7. Nol, S., Bongard, J., Husbands, P., Floreano, D.: Evolutionary Robotics, pp. 2035–2068. Springer International Publishing, Cham (2016)
8. do Nascimento, N.M., de Lucena, C.J.P.: Engineering cooperative smart things based on embodied cognition. In: Proceedings of the International Conference on Adaptive Hardware and Systems (AHS) NASA/ESA IEEE, pp. 109–116 (2017)
9. Marocco, D., Nol, S.: Emergence of communication in embodied agents evolved for the ability to solve a collective navigation problem. Connect. Sci. 19(1), 53–74 (2007)
10. Noureddine, D.B., Gharbi, A., Ahmed, S.B.: Multi-agent deep reinforcement learning for task allocation in dynamic environment. In: Proceedings of the 12th International Conference on Software Technologies (ICSOFT), pp. 17–26 (2017)

11. Nieves, J.C., Andrade, D., Guerrero, E.: MAIoT—an IoT architecture with reasoning and dialogue capability. Appl. Futur. Int. 179 (2017)
12. Zouai, M., Kazar, O., Haba, B., Saouli, H., Benfenati, H.: IoT approach using multi-agent system for ambient intelligence. Int. J. Softw. Eng. Appl. **11**(9), 15–32 (2017)
13. Noureddine, D.B., Gharbi, A., Ahmed, S.B.: An approach for multi-robot system based on agent layered architecture. Int. J. Manag. Appl. Sci. **2**(12), 135–143 (2016). ISSN: 2394-7926
14. Sasaki, H., Horiuchi, T., Kato, S.: A study on vision-based mobile robot learning by deep Q-network. In: Proceedings of the 56th Annual Conference of the Society of Instrument and Control Engineers of Japan (SICE). IEEE, pp. 799–804 (2017)
15. Lahami, M., Fakhfakh, F., Krichen, M., Jmaiel, M.: Towards a TTCN-3 test system for runtime testing of adaptable and distributed systems. In: IFIP International Conference on Testing Software and Systems, pp. 71–86. Springer (2012)
16. Lahami, M., Krichen, M., Jmaiel, M.: Safe and efficient runtime testing framework applied in dynamic and distributed systems. Sci. Comput. Program. **122**, 28 (2016)
17. Krichen, M.: A formal framework for black-box conformance testing of distributed real-time systems. Int. J. Crit Comput. Based Syst. **3**(1–2), 26–43 (2012)
18. Krichen, M.: Improving formal verification and testing techniques for Internet of Things and smart cities. Mob. Netw. Appl., 1–12 (2019)

Toward a Monitoring System Based on IoT Devices for Smart Buildings

Loubna Elhaloui, Sanaa Elfilali, Mohamed Tabaa, and El Habib Benlahmer

Abstract The Internet of Things (IoT) is a technological revolution in the field of IT and telecommunications. It aims to connect intelligent objects to a network that can be remotely accessed anywhere and at any time. This study focuses on the use of the IoT for smart buildings and its benefits. In this paper, the main objective is to simulate and model a connected smart building network implemented on Cisco Packet Tracer that uses a set of intelligent devices through an IP addressing network to allow them to transmit valuable information to homeowners remotely. When an intelligent device is connected to a wireless or wired media, it can send and receive information via a gateway to which users can control it.

Keywords Smart building · Internet of Thing · Smart cities · Wireless communication · IP network · Cisco

1 Introduction

Today, the IoT refers to a variety of detection equipment and information systems, such as sensor networks, reading devices (RFID, barcode, LoRa, Bluetooth and others), location and short-range communication systems based on M2M communication, through the Internet to form a larger and more intelligent network [1, 2]. The IoT seeks to offer many benefits that will improve all aspects of daily lives. The conceptualization of the IoT environment consists of making the Internet more ubiquitous by providing intelligent solutions, allowing heterogeneous devices to connect to a network, as remotely transmit the data generated by these objects and interact with each other [3].

L. Elhaloui (✉) · S. Elfilali · E. H. Benlahmer
Laboratory of Information Technologies and Modelling, Faculty of Sciences Ben M'sik, Hassan II University, Casablanca, Morocco
e-mail: l.elhaloui@gmail.com

M. Tabaa
Pluridisciplinary Laboratory of Research and Innovation (LPRI), EMSI Casablanca, Casablanca, Morocco

F. Saeed et al. (eds.), *Advances on Smart and Soft Computing*, Advances in Intelligent Systems and Computing 1188, https://doi.org/10.1007/978-981-15-6048-4_25

285

In recent years, it has been applied to the creation of smart buildings to improve people's living conditions by optimizing energy consumption, promoting comfort, user safety, control of access to the building and ensuring Internet connectivity. A smart building is defined as a residence equipped with ambient IT technologies whose is to increase the environmental interests in the building sector and have the ability to improve building performance, but also increase the comfort of residents, such as, through a Web interface that manages and controls shared resources or electronic devices. The management of energy resources is another challenge for smart building. It is thus possible to install motion sensors when users are absent or to automatically adapt the use of the electric resources according to the user's needs in order to reduce the waste of energy resources.

Smart building systems are composed of multiple technologies, intelligent objects, namely network, software, protocols and storage services. In terms of securing, it is necessary for a smart building to take into account the vulnerable aspects of each of these elements. Vulnerabilities in smart objects can be linked to poor application security and illegitimate authentication allowing attackers to gain access and control. Protection of information communications passes encryption and authentication of communications. Hardware encryption solutions can be quite limited in the case of certain smart objects due to their low computing power and their reduced storage space. Authentication is one of the critical requirements for the security of connected objects [4]. It consists in identifying a user or a device in order to authorize this entity's access to resources such as applications.

In this paper, the focus is mainly on the application of the IoT to the design and to simulate of a smart building, testing the ecosystem, the connectivity of smart objects and the automation of different types of activities. When testing the connectivity of smart objects that have IP addresses deployed to smartphone application interfaces. Which also allows monitoring and control of different devices that must send in permanent valuable information, can be accessed by owners on the Internet, depending on configuration conditions and predefined scenarios allowing security, reducing energy use costs and increasing energy efficiency.

This paper is organized as follows: the first section presents the smart cities applications. The second presents a deal with the architecture, where as the third section explains a global discussion on the different implementation of the smart building. Toward the end, a conclusion and perspectives are provided.

2 Internet of Things

Nowadays, the Internet is experiencing an unprecedented expansion with the development of connected objects. It no longer only allows people to communicate at any time and in any place; but in addition to the notion of connected objects, the physical world can now communicate, be it from person to person, from person to object or from object to object.

The IoT emerged as a result of the standardization applied to the automation of document and information processing, first on hardware and then on digital media. Its extension to other fields of application was very rapid. The objects were modified to become "connected objects" linked to centralized servers and/or capable of communicating with each other. The important capabilities offered by the IoT are still, for the most part, to be discovered and explored [5].

Data and its exploitation are indeed at the heart of the IoT. They come from a variety of terminals and sensors and allow users to be informed in real time of the evolution of their environment. More than that, the management of the multiplicity of these data from various sources makes it possible to quantify the connected environment in order to take advantage of it and enrich the uses or envisage new uses.

In the literature, there is a set of applications using the IoT which we cite for example:

- **Smart Home**: The main advantage of home automation is the improvement of everyday life in the home, in terms of comfort, safety and energy management. Home automation allows energy savings to be achieved through automatic management of heating, air conditioning and lighting and the programming of household appliances in off-peak hours. It also has the advantage of improving security through alarms; automatic door opening systems (voice recognition, magnetic card...). In case of attempted intrusion into the house, an automatic phone call can contact the owner or a security company. Finally, these different technologies are a valuable aid for dependent and disabled people [6].
- **Smart Health**: An important application of IoT that uses special devices such as cochlear implants, insulin pumps and pacemakers with sensors to observe, treat, monitor and evaluate the health status of patients and actuators to respond to life-threatening situations. A typical use case is diabetes monitoring when a patient is equipped with a wireless blood glucose meter and injection device to check their blood sugar and receives notifications on their mobile phone for updates of hypoglycemia and insulin doses of his or her treating physician [7].
- **Smart Traffic**: The main advantage of the smart traffic application is the significant reduction in daily traffic congestion by smoothing traffic flows and prioritizing traffic according to real-time demand [8]. Sensors connected to traffic lights continue to send information to a central server on the number of vehicles stacking and to coordinate the ambulance driver in order to determine the signal status and choose the path in which the traffic flow can be dynamically controlled where traffic violations are identified by traffic officers on site, through monitoring or centralized control [9].

3   Our Architecture

The smart building is a particularly important place for everyone; etymologically it is the place where we stay, where we return, the place of sedentarization. The majority of people, especially seniors, spend much of their time at home, resulting in a significant influence of habitat on quality of live. Improving the feeling of security and comfort in the home, therefore, appears to be a task of great social importance.

This research work has used smart devices in an IoT network connected via a wireless and wired link to interconnect them through the HomeGateway gateway. The smart devices of the design will be randomly dispatched around the smart building, they will have dynamic IP addresses to connect to the HomeGateway gateway, and the smart objects will then pass the authentication data to an IoT server for users can identify and control them. The HomeGateway adapts to binding objects such as sensor nodes in an IoT environment, and also allows Internet users to interact directly with each thing connected, it will also perform the translation from IPv4 protocol to IPv6 protocol for all smart objects connected to it in the smart building. An architecture diagram of an IoT network for the smart building is illustrated in Fig. 1.

In an IoT environment, intelligent devices can interconnect to each other for different topologies, such as point-to-point, tree and mesh. The topology used for the IoT network model is a point-to-point topology where each intelligent object is managed as a separate device, therefore, the management of rest on a small environment. This is a direct connection between the objects and the network HomeGateway gateway. Most smart building networks are also point-to-point; each object of the model is configured with its own security, its own identify and all can access the Internet, or other simply by connecting. So smart building is probably the most

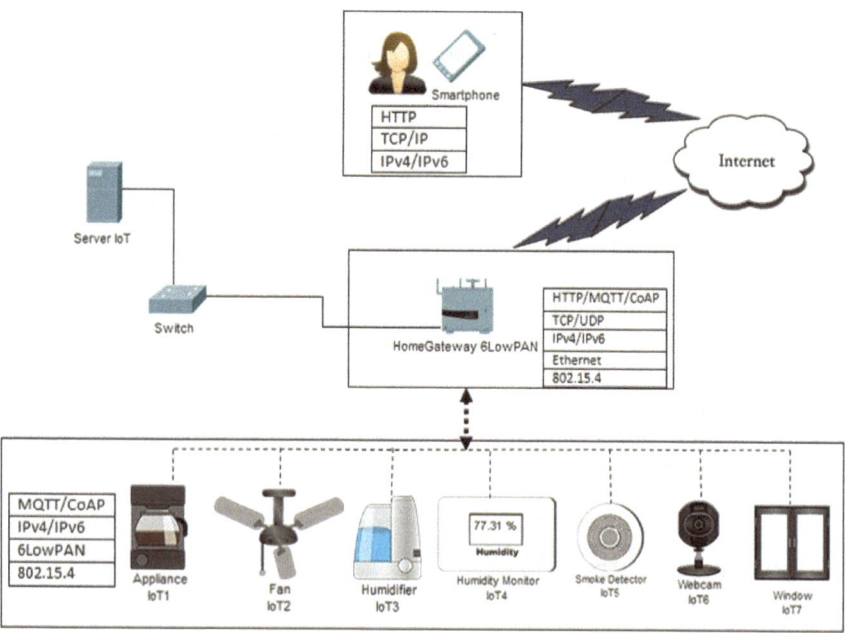

Fig. 1 IoT architecture design for smart building

commonplace for a point-to-point management model. With the mesh topology, the objects are connected with each other, each intelligent object being able to connect and interact with its neighbor. There is no concept of coordinator that serves as a central point of communication for the transmission of information. The mesh topology is fairly dominant for IoT networks. One of the main advantages of point-to-point topology is that the data does not have to make its way through different sensor nodes which ensures the data transfer is fast. Another advantage of this topology is that it is quite flexible and easy on adding new objects or replacing old ones compared to other topologies, as the disturbance of the entire network is not required to facilitate same.

4 Discussion

In this passage, we will first present the smart building network modeling approach using the Cisco Packet Tracer simulator and then will discuss the results of the simulation. We used Cisco Packet Tracer version 7.2, which allows visualizing and analyzing the flow of data packets transmitted over the IoT network has been used.

This section describes the proposed smart building simulation and analysis of IoT network performance, such as data packets within the IoT network. Cisco Packet Tracer [10] allows us to visualize the data traffic between objects connected in the IoT network, to test the behavior of network topologies, to design IoT network models and to put hypotheses into practice. Network performance evaluation consists of visualizing the data transmitted and received by IoT devices.

The main objective of this modeling experience is to allow users to control, identify and access the smart devices in a smart building through generated data and exchanges between them. The devices used for this simulation study are smart windows, smoke detector, smart camera, motion detector, temperature sensor, 6LowPAN gateway, switch, smartphone and IoT server.

In the model configuration as shown in Fig. 2, the smart devices of the design will be randomly dispatched in the IoT environment; they will have dynamic IP addresses to connect to the HomeGateway via a wireless link. The smartphone is configured to connect to the HomeGateway, it contains a desktop interface that has a Web browser interface to manage connected objects to detect and control a set of intelligent devices in a smart building network. Users from anywhere in the smart building can get access to smart objects via the HomeGateway on the Internet using a smartphone.

Smart devices use dynamic IP address range that can be addressed in the smart building. These intelligent devices can filter messages by transferring the data flow to the HomeGetway. Below you will find details of the addresses used for this simulation, as presented in Tables 1 and 2.

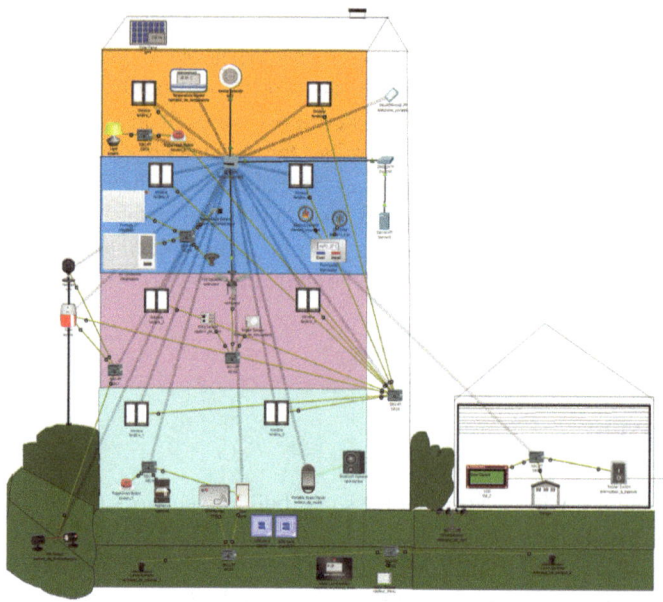

Fig. 2 IoT model for smart building

In the smart building simulation, when the motion sensor captures a threat, the alarm and camera are activated and the windows close and transmit valuable information to the owners. For other sensors either motion or wind are activated, so the fan turns on, otherwise, it is off.

During the simulation process, the modeled IoT network has been tested by sending a ping request to the IoT devices using their IP address of the smartphone to display the sent and received datapackets. Figure 3 shows that the ping request is established and that the number of datapackets sent and received is 4; there is no loss of information in the transmission. The approximate average round trip time is 83 ms, which illustrates the rapid transmission and reception of data packets to the user smartphone. At some point, the HomeGetway gateway sends ping requests to test the connectivity of IoT devices as shown in Fig. 4.

5   Conclusion

This paper has presented IoT architecture for future smart city applications. The operation of a connected smart building in a Packet Tracer environment has been simulated in order to provide reliable information for good decision-making. Throughout this work, some challenges arise and are described as follows. The first challenge was the detection of interconnected physical objects in a heterogeneous network, to allow users to discover and identify the type and functionality of each intelligent object

Table 1 Dynamic IP address for objects intelligent in the smart building

Smart objects	IP addresses
Window 1	192.168.10.101
Window 2	192.168.10.102
Window 3	192.168.10.103
Window 4	192.168.10.104
Window 5	192.168.10.105
Window 6	192.168.10.106
Fan	192.168.10.110
Lampe	192.168.10.117
Temperature monitor	192.168.10.119
Furnace	192.168.10.120
Air conditioner	192.168.10.121
Fir sprinkler	192.168.10.122
Thermostat	192.168.10.123
Motion sensor	192.168.10.125
Portable music player	192.168.10.126
Door	192.168.10.128
Trip sensor	192.168.10.129
Water level monitor	192.168.10.130
Garage door	192.168.10.131
LCD	192.168.10.132
Camera	192.168.10.133
Siren	192.168.10.135

Table 2 Custom smart object model the smartphone

Port	Wireless 3G/4G cell 1
IP address (wireless)	192.168.10.10
IP address (3G/4G cell 1)	169.254.233.176
DNS server	0.0.0.0
MAC address	00D0.BA3A.E9B0
SSID	Home gateway
Bandwidth	300 Mbps

capable of providing information and perceiving in real time. The second challenge was the deployment of many vulnerable devices that were not authenticated and not secured in the IoT system. As for future work, we want to further expand our architecture and deploy a security strategy.

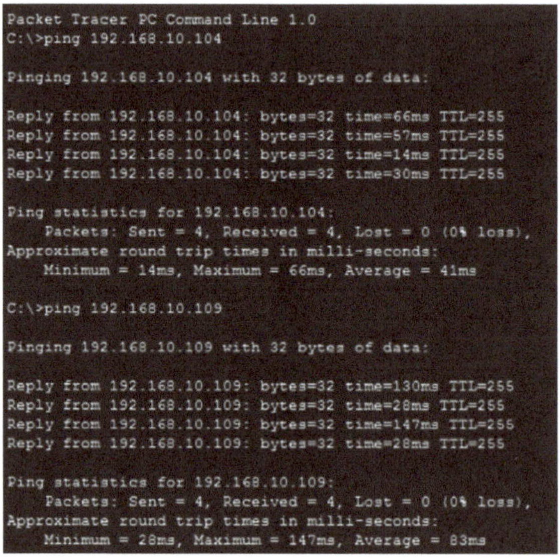
Packet Tracer PC Command Line 1.0
C:\>ping 192.168.10.104
Pinging 192.168.10.104 with 32 bytes of data:
Reply from 192.168.10.104: bytes=32 time=66ms TTL=255
Reply from 192.168.10.104: bytes=32 time=57ms TTL=255
Reply from 192.168.10.104: bytes=32 time=14ms TTL=255
Reply from 192.168.10.104: bytes=32 time=30ms TTL=255
Ping statistics for 192.168.10.104:
Packets: Sent = 4, Received = 4, Lost = 0 (0% loss),
Approximate round trip times in milli-seconds:
Minimum = 14ms, Maximum = 66ms, Average = 41ms
C:\>ping 192.168.10.109
Pinging 192.168.10.109 with 32 bytes of data:
Reply from 192.168.10.109: bytes=32 time=130ms TTL=255
Reply from 192.168.10.109: bytes=32 time=28ms TTL=255
Reply from 192.168.10.109: bytes=32 time=147ms TTL=255
Reply from 192.168.10.109: bytes=32 time=28ms TTL=255
Ping statistics for 192.168.10.109:
Packets: Sent = 4, Received = 4, Lost = 0 (0% loss),
Approximate round trip times in milli-seconds:
Minimum = 28ms, Maximum = 147ms, Average = 83ms

Fig. 3 Pinging the smart devices from a user's smartphone

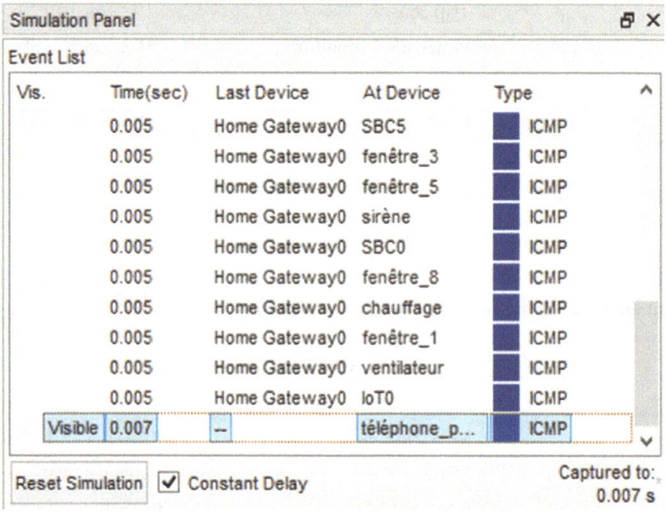
Simulation Panel
Event List
Vis. Time(sec) Last Device At Device Type
0.005 Home Gateway0 SBC5 ICMP
0.005 Home Gateway0 fenêtre_3 ICMP
0.005 Home Gateway0 fenêtre_5 ICMP
0.005 Home Gateway0 sirène ICMP
0.005 Home Gateway0 SBC0 ICMP
0.005 Home Gateway0 fenêtre_8 ICMP
0.005 Home Gateway0 chauffage ICMP
0.005 Home Gateway0 fenêtre_1 ICMP
0.005 Home Gateway0 ventilateur ICMP
0.005 Home Gateway0 IoT0 ICMP
Visible 0.007 -- téléphone_p... ICMP
Reset Simulation
Constant Delay
Captured to:
0.007 s

Fig. 4 Data packets successfully arrive at the smart objects

References

1. Ratasuk, R., Vejlgaard, B., Mangalvedhe, N., Ghosh, A.: NB-IoT system for M2M communication. In: 2016 IEEE Wireless Communications and Networking Conference (WCNC), pp. 1–5. IEEE, Apr 2016
2. Severi, S., Abreu, G., Sottile, F., Pastrone, C., Spirito, M., Berens, F.: M2M technologies: enablers for a pervasive Internet of Things. In: The European Conference on Networks and Communications (EUCNC2014) (2014). https://doi.org/10.1109/EuCNC.2014.6882661
3. Bawa, A., Selby, M.L.: Design and simulation of the Internet of Things for Accra SmartCity, p. 14(2018)
4. Jan, M.A., Khan, F., Alam, M., Usman, M.: A payload-based mutual authentication scheme for Internet of Things. Future Gen. Comput. Syst. **92**, 1028–1039 (2019)
5. Minerva, R., Biru, A., Rotondi, D.: Towards a Definition of the Internet of Things (IoT). IEEE Internet Initiative, Torino, Italy (2015)
6. Khan, T.: A Wi-Fi based architecture of a smart home controlled by smartphone and wall display IoT device. Adv. Sci. Technol. Eng. Syst. J. **3**, 180–184 (2018)
7. Ravidas, S., Lekidis, A., Paci, F., Zannone, N.: Access control in Internet-of-Things: a survey. J. Netw. Comput. Appl. **144**, 79–101 (2019)
8. Saikar, A., Parulekar, M., Badve, A., Thakkar, S., Deshmukh, A.: TrafficIntel: smart traffic management for smart cities. In: 2017 International Conference on Emerging Trends and Innovation in ICT (ICEI), pp. 46–50 (2017)
9. Ghazal, B.: Smart traffic light control system. In: Third International Conference on Electrical, Electronics, Computer Engineering and Their Applications (EECEA). IEEE (2016)
10. CISCO Networking Academy, CISCO Packet Tracer (2018). Available: https://www.netacad.com/courses/packet-tracer-download/. Accessed: 17 July 2019
11. IEEE Green Computing and Communications (GreenCom) and IEEE Cyber, Physical and Social Computing (CPSCom) and IEEE Smart Data (SmartData). IEEE (2017)
12. Tendeng, R., Lee, YoungDoo, Koo, I.: Implementation and measurement of spectrum sensing for cognitive radio networks based on LoRa and GNU radio. Int. J. Adv. Smart Conver. **7**(3), 23–36 (2018)

10 Gbps MMW UFMC-Based Backhaul with FiWi Access Network

Abdulaziz Mohammed Al-Hetar and Abdulnasser Abdulgaleel Abdulgabar

Abstract Millimeter Wave Universal Filtered Multicarrier (MMW-UFMC) is proposed to achieved high flexible bandwidth transmission over fiber-wireless (FiWi) system for 5G wireless communication. The proposed system transmitted 10 Gbps UFMC over passive optical network with 60 GHz MMW link. MATLAB and OPTISYSTEM platforms were used to simulate the proposed system, and the performance has investigated by calculation bit error rate (BER) and error vector magnitude (EVM) for B2B, 15, 25, 40, 60 and 75 km optical length transmission. With B2B and 15 km optical distance transmission constant, EVM and BER performance has measured for several received optical power as well. The results show that hybrid oMMW-UFMC-PON are a potential solution toward 5G networks.

Keywords Millimeter wave · Fiber-wireless · Universal filter multicarrier · PON · EVM · BER

1 Introduction

As the industrial is going toward 5G wireless technology trends including a potential application such as machine type communications (MTC) and Internet of things (IoT), whose increasing by traffic loads and optical is becoming more necessary as backhaul to provide huge bandwidth for wireless communication [1]. In fact, the integration of wireless technologies with optical access networks makes us take advantage of fiber wideband with low attenuation and also can transmit high-frequency wireless signal over long distances, especially millimeter waves (MMW) signal that cover small area due to attenuation and absorption which is affected by it high frequency. This fusion "also called fiber-wireless (FiWi) access network" is achieved

A. M. Al-Hetar (✉) · A. A. Abdulgabar
Communication and Computer Department, Faculty of Engineering and IT, Taiz University, Taiz, Yemen
e-mail: alhetaraziz@taiz.edu.ye

A. A. Abdulgabar
e-mail: Abdulnasser2019abdu@gmail.com

© The Editor(s) (if applicable) and The Author(s), under exclusive license to Springer Nature Singapore Pte Ltd. 2021
F. Saeed et al. (eds.), *Advances on Smart and Soft Computing*, Advances in Intelligent Systems and Computing 1188, https://doi.org/10.1007/978-981-15-6048-4_26

by integrating PONs and wireless access technologies in form radio over fiber (RoF) [2, 3]. FiWi access network systems in MMW bands based on advance modulation waveform would also be very suitable for 5G wireless services [4]. Cyclic prefix (CP)-orthogonal frequency division multiplexing (OFDM) transmission over FiWi system has been study in many papers [5, 6]. OFDM filters the complete signal in single shot which have large out-of-band emission as well as reduce the spectrum efficiency due to CP [7]. Universal filtered multicarrier (UFMC) without CP is proposed candidate for 5G waveforms as a better alternative to OFDM to minimize overhead system and suppress out of band power with good spectral efficiency transmission over optical access network which the performance of system is improved [8].

Furthermore, the UFMC modulation format shows the best spectral localization, which the required frequency difference is reduced by 18.2% less than OFDM [9]. A FiWi with UFMC modulation system has been proposed, capable of asynchronous transmission with diverse applications in 5G communication [10]. Here, 10 Gbps UFMC downstream data is proposed to transmit over FiWi, which the UFMC is modulated optically with MMW link at optical line terminal (OLT) and transmitted over a long-distance single mode fiber (SMF) via remote node (RN) for de-multiplexing to several small coverage BS areas called wireless access point (WAP). In one BS receiver part, optical modulated signal is detected by photodiode (PD) to 60 GHz UFMC wireless signal, then amplified and propagate over wireless channel to end user as shown in Fig. 1. The performance of proposed system evaluates both BER and EVM for 10 Gbps UFMC received signal over multiple optical transmission. With B2B and 15 km optical transmission constant, EVM and BER versus several received optical power is measured as well.

The organization of this paper shows as following: Sect. 2 gives the structure of UFMC transceiver. Simulation setup of the system given in Sect. 3. Section 4 hows the result and analysis of simulation setup. Finally, conclusion of this paper is given in Sect. 5.

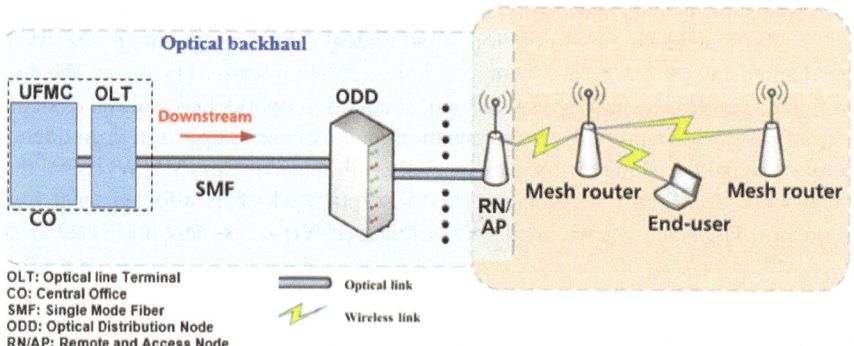

Fig. 1 The proposed FiWi access network architecture

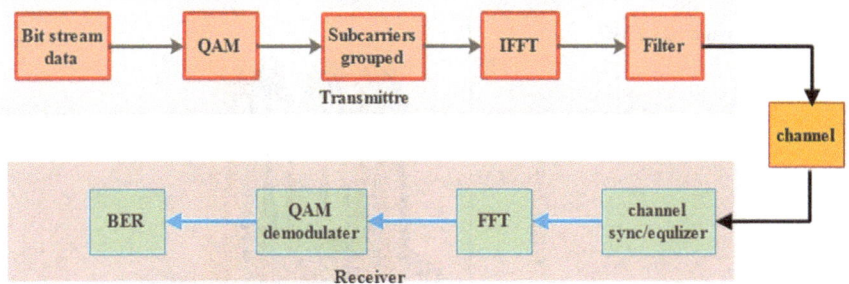

Fig. 2 The simplified block diagram of UFMC

Table 1 Parameters that used in UFMC generation

Parameter	Value	Parameter	Value	Parameter	Value
FFT size	512	Number of sub-bands	10	Leng_UFMC	800
Modulation format	4 QAM	Sub-band size	20	Time window	6553.6 ns
Filter length	43	Sub-band offset	156	Data rate per BS	10 Gbps

2   5G UFMC Waves Structure

The simplified block diagram of UFMC transceiver is shown in Fig. 2 [11, 12]. In transmitter, the input data stream is divided into sub-streams, and then these sub-carriers are grouped to form sub-bands. The sub-carrier of each sub-band is mapping with QAM symbol format and end each sub-band is modulated digitally by N point IFFT, whose output is converted to serial form and applied to respective a pulse shaping filter smooth edges for sub-band that leads to substantial reduction in out-of-band transmission. Following, a digital to analog converter is applied then real and imaginary components of complex UFMC symbol are separate and up-converter outside MATLAB.

The receiver offline processing is opposite to transmitter (see Fig. 2). Recently, the MATLAB platform is developed to design UFMC signal [11–13]. The parameters that used to generate UFMC signals are given in Table 1.

The normalize spectrum of UFMC signal that generated in MATLAB (see Fig. 3).

3   Simulation Setup

Figure 4 illustrates the detail the tools that used in the co-simulation transmitter and receiver circuit, which the real and imaginary component signal from MATLAB is filtered and up-converter to 2.6 GHz as shown in Fig. 5a. A tunnel laser operates

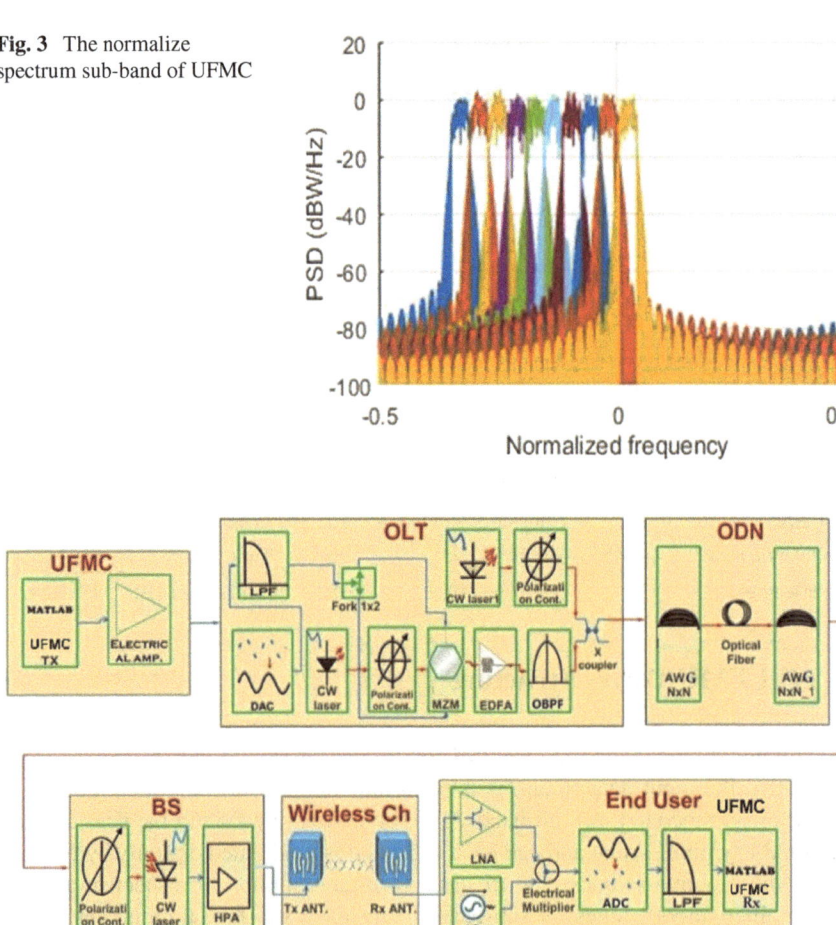

Fig. 3 The normalize spectrum sub-band of UFMC

Fig. 4 The simplified block diagram of the transmitter and receiver system

at 1552.51 nm wavelength with a linewidth of 10 kHz is used to feed the intensity modulator MZM.

The modulator is biasing by 4 V and driven by two electrical UFMC signals at the quadrature point. Then, ODSB is produced at the MZM output as shown in Fig. 5b, which amplified by erbium doped fiber optical amplifier (EDFA) and filtered by 0.9 nm optical band pass filter to reject the amplified spontaneous emission (ASE) noise [14]. Then, the output signal of optical BPF is coupled by optical coupler with another wavelength carrier emits from another tunnel laser operate at 1553 nm wavelength with 900 kHz linewidth. Both tunnel lasers emit 8 dBm optical power and following by polarization controller used to ensure maximum coupling of the

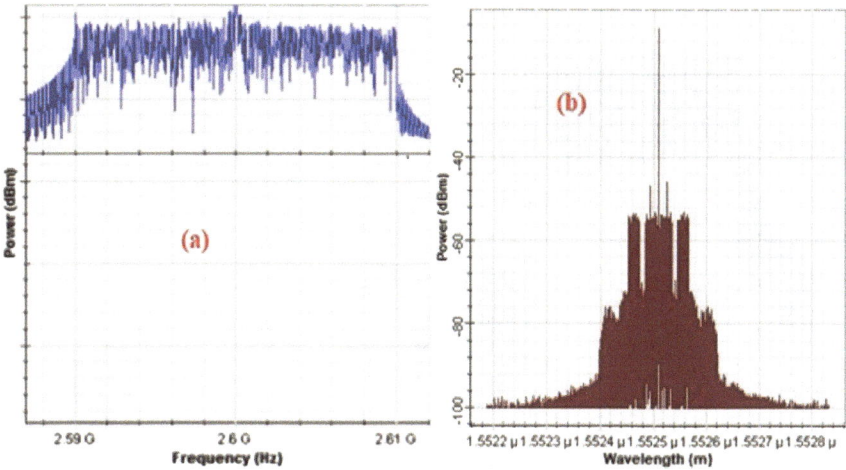

Fig. 5 The spectrum of **a** UFMC signal after quadrature output, **b** optical modulated signal after MZM

two light signals is achieved [9–15]. The optical modulated signal is transmitted over SMF which has 0.2 dB/km, 16 Ps/nm/km attenuation and dispersion, respectively.

The system receiver is shown in Fig. 4, which there is 70 GHz photodetector PIN used for regenerated the electrical downlink signal at 60 GHz frequency. Optical attenuator is used to control the input power to the 70-GHz PID [16]. The spectrum output of PID is shown in Fig. 6a, which it is amplified by a 40 dB gain. After that, electrical attenuator used the same as the wireless channel transmission, which 2 m distance was be computed using free space path loss equation according to Friis transmission formula [16].

The MMW received signal is passed through transimpedance amplifier and downconverter to 2.6 GHz frequency, whose spectrum (see Fig. 6b). Then, the UFMC complex signal is separated to real and imaginary components signal by using quadrature demodulator and then the two components are filtered individual by low pass filter and applied directly to MATLAB receiver for recovered the bit sequence and measured EVM and BER.

4   Results and Discussing

Figure 7a shows the EVM of the performance of the received UFMC signal for back-to-B2B, 15, 25, 40, 60 and 75 km optical length transmission. As we notice in Fig. 7a, EVM is increasing with optical length transmission increasing, Which reach to 9.71% at 25 km optical transmission and, which are in accordance to the Third-Generation Partnership Project (3GPP) requirement [17]. The corresponding BER for those EVM value at the same optical transmission shown in Fig. 7b, which

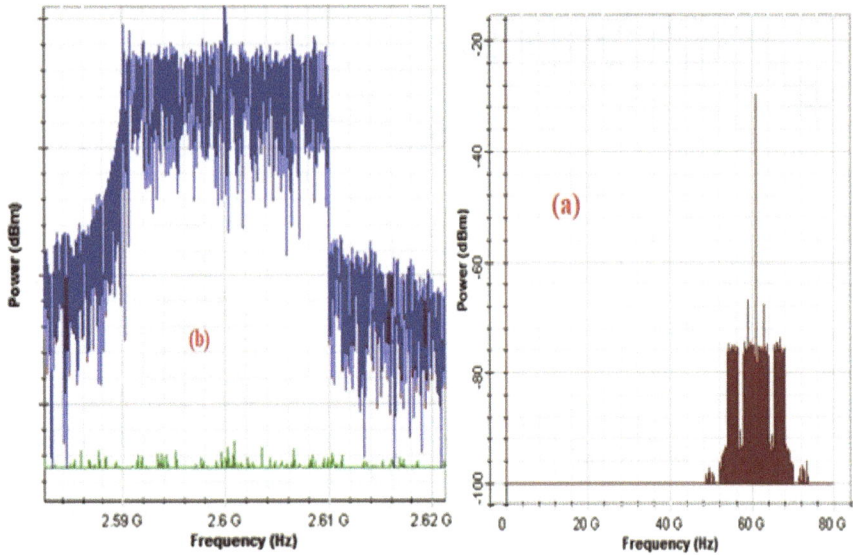

Fig. 6 The spectrum of **a** UFMC MMW signal after PIN, **b** the complex UFMC signal before quadrature demodulator

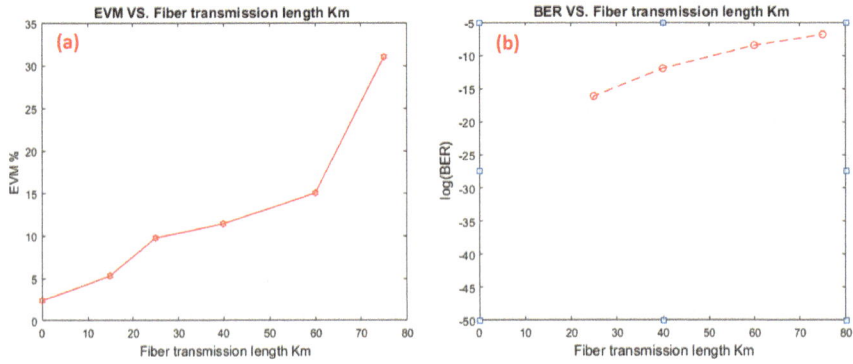

Fig. 7 **a** EVM and **b** BER performances versus various fiber length transmission

have 0.36×10^{-7} value for 25 km optical transmission. Also, with 75 km optical length transmission, EVM and BER have 31% and 2×10^{-3}, respectively. This value can improve if the received optical power has increased.

In order to investigate the effect of input power to photodetector, EVM and BER performance was measured in term of the input power to PD controlled by optical attenuator and set to less than or equal -5 dBm, in case B2B and 15 km configuration, see Fig. 8. Overall, we notice that both EVM and BER at the same received optical power for B2B and 15 km have very close values to be equal. For example, when the received optical power set to -5 dBm for B2B and 15 km fiber transmission, EVM

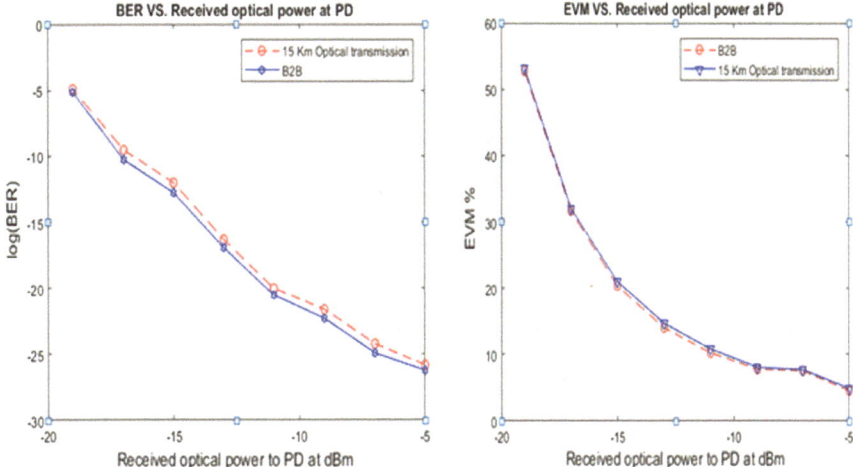

Fig. 8 The **a** EVM and **b** BER performances versus various received optical power in case B2B and 15 km optical transmission constant

has 4.560% and 4.98% value, respectively. This shows us that the very near value for EVM or BER can be obtained in case the optical modulated signal over short or long transmission if and only if the received optical powers are equal. To control the received optical power for different fiber transmission, the transmitted optical power must be increased and attenuated by optical attenuator before photodetector to a require value.

5 Conclusions

In this paper, UFMC signal is proposed to achieved high flexible bandwidth transmission over fiber-wireless system for 5G wireless communication. The proposed system transmits 10 Gbps UFMC modulation with 60 GHz MMW band over PON. MATLAB and OPTISYSTEM platforms were used to simulate the proposed system, and the performance has investigated by calculation bit error rate (BER) and error vector magnitude (EVM) for B2B, 15, 25, 40, 60 and 75 km optical length transmission. With B2B and 15 km optical distance transmission constant, EVM and BER performance has measured for several received optical power as well. The results show that hybrid MMW-UFMC-PON are a potential solution toward 5G networks.

References

1. Rebhi, S., Barrak, R., Menif, M.: Flexible and scalable radio over fiber architecture. Radio Eng. **28**(2) (2019)
2. Tzanakaki, A., Anastasopoulos, M., Berberana, I., Syrivelis, D., Flegkas, P., Korakis, T., Mur, D.C., Demirkol, I., Gutiérrez, J., Grass, E., Wei, Q., Pateromichelakis, E., Vucic, N., Fehske, A., Grieger, M., Eiselt, M., Bartelt, J., Fettweis, G., Lyberopoulos, G., Theodoropoulou, E., Simeonidou, D.: Wireless-optical network convergence: enabling the 5G architecture to support operational and end-user services. IEEE Commun. Mag. **55**(10), 184–192 (2017)
3. Kim, H.: RoF-based optical fronthaul technology for 5G and beyond. In: IEEE Optical Fiber Communications Conference and Exposition (OFC), San Diego (USA), pp. 1–3 (2018)
4. Chang, G.-K., Cheng, L.: Fiber-Wireless Integration for Future Mobile Communications. IEEE (2017)
5. Alavi, S.E., Amiri, I.S., Khalily, M., Fisal, N., Supa'at, A.S.M., Ahmad, H., Idrus, S.M.: W-band OFDM for radio over-fiber direct-detection link enabled by frequency nonupling optical up-conversion. IEEE Photonics J. **6**(6), 1–7 (2014)
6. Lin, C.-Y., Chi, Y.-C., Tsai, C.-T., Wang, H.-Y., Lin, G.-R.: 39-GHz millimeter-wave carrier generation in dual-mode colorless laser diode for OFDM-MMWoF transmission. IEEE J. Sel. Top. Quantum Electron. **21**(6), 1801810 (2015)
7. Zhang, J., Xu, M., Wang, J., Lu, F., Cheng, L., Cho, H., Ying, K., Yu, J., Chang, G.K.: Full-duplex quasigapless carrier aggregation using FBMC in centralized radio-over-fiber heterogeneous networks. J. Lightwave Technol. **35**(4), 989–996 (2017)
8. Zhang, L., Xiao, S., Bi, M., Liu, L., Chen, X.: FFT-based universal filtered multicarrier technology for low overhead and agile datacenter interconnect. In: ICTON 2016. State Key Laboratory of Advanced Optical Communication System and Networks, Shanghai Jiao Tong University, Shanghai, P.R. China (2016)
9. Sarmiento, S., Spadaro, S., Lazaro, J.A., Altabas, J.A.: Experimental assessment of 10 Gbps 5G multicarrier waveforms for high-layer split U-DWDM-PON-based fronthaul. J. Lightwave Technol. **3**(10) (2019)
10. Cho, H.J., Cho, H., Mu, X., Lu, F., Shen, S., Ma, X., Kung, G.: Asynchronous Transmission Using Universal Filtered Multicarrier for Multiservice Applications in 5G Fiber Wireless Integrated Mobile Fronthaul. Optical Society of America (2018). 978-1-943580-38-5
11. Pooja, R., Silki, B., Himanshu, M.: Hybrid PAPR reduction scheme for universal filter multi-carrier modulation in next generation wireless systems. Res. Dev. Mater. Sci. **2**(5), RDMS.000549 (2018). https://doi.org/10.31031/RDMS.2018.02.000549
12. Wang, X., Wild, T., Schaich, F.: Filter optimization for carrier frequency- and timing offset in universal filtered multi-carrier systems. In: IEEE Vehicular Technology Conference Spring (VTC'15 Spring), May 2015
13. Schaich, F., Wild, T.: Waveform contenders for 5G—OFDM vs. FBMC vs. UFMC. In: Proceedings of 6th International Symposium on Communications, Control, and Signal Processing (ISCCSP 2014), Athens, Greece, May 2014 (in press)
14. Wang, H.Y., Chi, Y.C., Lin, G.R.: Remote beating of parallel or orthogonally polarized dual-wavelength optical carriers for 5G millimeter-wave radio-over-fiber link. Opt. Express **24**(16), 17654–17669 (2016)
15. Su, Y.C., Chi, Y.C., Chen, H.Y., Lin, G.R.: All colorless FPLD based bidirectional full-duplex DWDM-PON. J. Lightwave Technol. **33**(4), 832–842 (2015)
16. Beas, J., Castanon, G., Aldaya, I., Aragon-Zavala, A., Campuzano, G.: Millimeter-wave frequency radio over fiber systems: a survey. IEEE Commun. Surv. Tutor. (2013)
17. ETSI: LTE; Evolved Universal Terrestrial Radio Access (E-UTRA); Base Station (BS) Radio Transmission and Reception (2015)

Toward a Real-Time Picking Errors Prevention System Based on RFID Technology

El Mehdi Mandar, Wafaa Dachry, and Bahloul Bensassi

Abstract In order to improve supply chain performance, it is crucial to improve warehouses' performance. To this end, warehousing operations need to be improved. In this paper, we concentrated our improvement efforts on picking operations which are the most costly in terms of resource utilization. In order to do that, we propose a system, based on RFID technology, which can prevent picking errors which can be very expensive, considering penalties, additional transportation, handling costs, and other additional expenses. The proposed system's objective is the prevention of picking errors so as to minimize additional charges and improve the overall warehouse performance. The proposed system is in the prototyping phase and shows promising results in the errors' cost simulation presented in this paper.

Keywords Warehouse · Picking · Picking operator · Picking error · Prevention · RFID

1 Introduction

About 55% of warehousing expenses concern picking operations which makes them the most expensive operations in a warehouse [1]. These operations utilize the workforce heavily since, in general, they are conducted manually by human operators. The fact that picking operations are, in most cases, done manually renders their performance heavily linked to human operators executing them. As a consequence, the picking process is vulnerable to human errors which results in lots of additional

E. M. Mandar (✉) · B. Bensassi
Laboratory of Industrial Engineering, Information Processing, and Logistics, Faculty of Science Ain Chock, Casablanca, Morocco
e-mail: mandar.elmehdi@gmail.com

B. Bensassi
e-mail: Bahloul_bensassi@yahoo.fr

W. Dachry
Laboratory of Engineering, Industrial Management, and Innovation, FST, Hassan I University, Settat, Morocco
e-mail: Wafaa.dachry@gmail.com

© The Editor(s) (if applicable) and The Author(s), under exclusive license to Springer Nature Singapore Pte Ltd. 2021
F. Saeed et al. (eds.), *Advances on Smart and Soft Computing*, Advances in Intelligent Systems and Computing 1188, https://doi.org/10.1007/978-981-15-6048-4_27

expenses. In our quest to find the average picking error's cost, it was found that it differs from one source to another. In [2], a picking error is said to be ranging from 30\$ to 70\$, in [3], it is estimated at 218.14\$, in [4], it is estimated at 30\$ and in [5], it is estimated to be ranging from 50\$ to 300\$. Picking errors' cost naturally varies from one organization to another due to how they use and organize their resources. Furthermore, they may even vary within the same organization. A picking incurs many preventable costs such as penalties costs, re-shipping costs, additional warehouse labour cost, returns costs, repackaging costs, customer service costs and lost sales. In this perspective, it is imperative to find a viable solution to help human operators in performing their picking tasks in order to achieve higher performances. In this paper, a real-time picking errors prevention system based on RFID technology (RTPEPS) is proposed. The system aims to prevent picking errors at each stage of the picking process to avoid additional charges. The rest of this paper is sectioned as follows: In Sect. 2, a presentation and analysis of previous works in the same area of study is given. In Sect. 3, the system's architecture, components and modules are presented. In Sect. 4, a comparison, of the system against other ones in an error cost simulation, is presented, and finally, in Sect. 5, a conclusion of the work and future research perspectives is given.

2 Literature Review

Picking is defined as the process in which product units are picked from specific storage placements in reaction to customers' orders [5]. Over the years, paper-based picking has been replaced by many solutions in order to minimize picking errors. The most known of these solutions are: (a) bar-codes systems, (b) pick to light, (c) pick to voice, (d) pick by HUD.

The optical nature of bar-codes systems requires that the reading laser sees the bar-code. This line of sight need is almost always hard, not practical and sometimes impossible to achieve in industrial environments [6]. The bar-code reader must have a clear view of a clear bar-code in order to obtain a viable reading. Otherwise, this technology can induce errors that can cost enterprises additional costs. Pick to light is a method of guided picking where the picker receives all the necessary picking information from small display boards mounted above or below the shelving bay. The lighting of a lamp indicates the correct removal shelf, a display shows the number of parts to pick and a button is to confirm a removal [7]. A pick by light system can either wired or wireless just like the one presented in [2]. Pick to voice is a procedure in which storage placement codes are read by a terminal that transmits read data wirelessly thanks to radio frequencies to a station generally placed on the waist of the operator. This station translates received data to vocal commands which the operator follows in the execution of the picking tasks. With simple communication, the operator confirms taken actions, and the system responds with the next instructions [3].

Pick by HUD or picking using a head-up display uses glasses, generally Google glasses for the display of picking instructions. It is notably faster than pick to light. However, it tends to induce more errors. These errors can be the result of the operator having to memorize instructions and switch his attention from the HUD to the storage placements [6]. In [4], another approach using the HUD and wearable RFID readers to minimize errors was presented. The approach consisted of Google glasses and two RFID readers worn on the wrists of the operator. The results show that it sustains the HUD speed benefits while also reducing errors making it more reliable. However, this approach misses the count of picked units which represents a significant part of the picking process. In [8], in order to reduce errors, a weight-based verification method was introduced. The results showed reduced errors in picking quantities since all picked quantities are weighted for verification against values stored in the system proposed. However, this approach lacks precision since many items can have identical or close weights. Furthermore, different orders can have the same weights even though their composition is different.

These technologies focus only on the product picking phase. They ignore the steps that precede and succeed it. In [9], a new approach called pick and verify was introduced with the help of a robot that identifies the picked product visually and then verifies its shape thanks to a 3D recognizing technology. This approach made it possible to increase picking reliability by 50%. In the same perspective, in [5], a system capable of validating picked product attributes with the help of sensors (weight, size, shape, colour and other attributes) was proposed. The picker places products on a platform for characteristics comparison against characteristics of the wanted products. However, the use of the technologies of 3D recognition and sensors can induce errors since multiple products can share the same physical attributes, especially externally. Furthermore, the implementation of these technologies requires a more significant investment than the technology presented in this work. In [7], a new approach called pick by projection was presented. It consists of two projectors, a touchscreen and a scale based on a picking trolley. The order selection is made through the touchscreen. The projectors illuminate the storage placements from which to pick the goods as well as the destination bin on the picking trolley. The scale then verifies the weight of the picked product for correctness. This system, however, can be costly to implement as it needs projectors, cameras and touchscreens that can slide along racks as well as a special picking trolley compatible with the projector. Furthermore, the scale verification mechanism can induce errors since different orders with different composition can have the same weight. In [10], a new system coupling the capabilities of a pick to light system and RFID technology was introduced. The picker wears gloves embedded with RFID readers to check that the picker's hands entered the right stock location. This approach sure minimizes the error rate of a traditional pick to light system; however, it still lacks the quantity and identity verification inside the picking bins. Besides, the integration of such a system can be expensive due to the multitude of readers needed (two readers for every picker) and the light displays. The system presented in this paper includes the picking equipment choice, the picked product identity and quantity checking as well as the choice of the packing zone. Additionally, it enables the execution of the picking task in a much more confident way avoiding errors before they happen since it checks at all times the picking task progress and does not rely on the memory of operators to memorize visual instructions (pick by HUD, pick to light).

3   A Real-Time Picking Errors Prevention System Based on RFID Technology

The best way to deal with mistakes is to prevent them. By doing so, the costs of making the mistakes plus the costs of correcting them are optimized. With this in mind, a system capable of preventing picking errors with the help of the capabilities offered by the RFID technology was developed. The architecture of the proposed system is shown in Fig. 1. In this work, a real-time picking errors prevention system based on RFID technology capable of preventing all probable types of errors in the picking process is presented. The proposed system objective is to avoid additional charges incurred when picking errors occur as well as improve the picking process by saving the time spent on making the errors and correcting them. The system in question considers that all picking assignments are executed in one tour, and their volume and weights are equal or less than the capacity of the material handling equipment. Additionally, it considers multiple packing zones in order to reduce the confusion of the packing process and make it easier.

3.1   Data Collection Layer

This layer comprises RFID tags, WiFi-connected RFID readers mounted on operators' personal digital assistants (PDAs) and material handling equipment in order to transmit read data wirelessly to the router. RFID tags are placed on all product units, all storage placements, and all operator badges and all packing zones entrances. RFID readers are mounted on PDAs as well as material handling equipment. They are different in terms of reading range. The readers mounted on PDAs have a narrow reading range while readers mounted on material handling equipment have a more extended range of reading ability in order to read any tag within the perimeter of the equipment comfortably. These readers enable the collection of data such as operators' identity, stock placements' identity, product units' identity and quantity. Operators' PDAs can be attached to their waist. Operators verify storage placements' identity by getting the PDAs closer to the storage placement RFID tag. Once an operator completes this verification successfully, he or she can then proceed to pick the number of units noted on the picking assignment visible on the PDA screen.

Fig. 1 System architecture

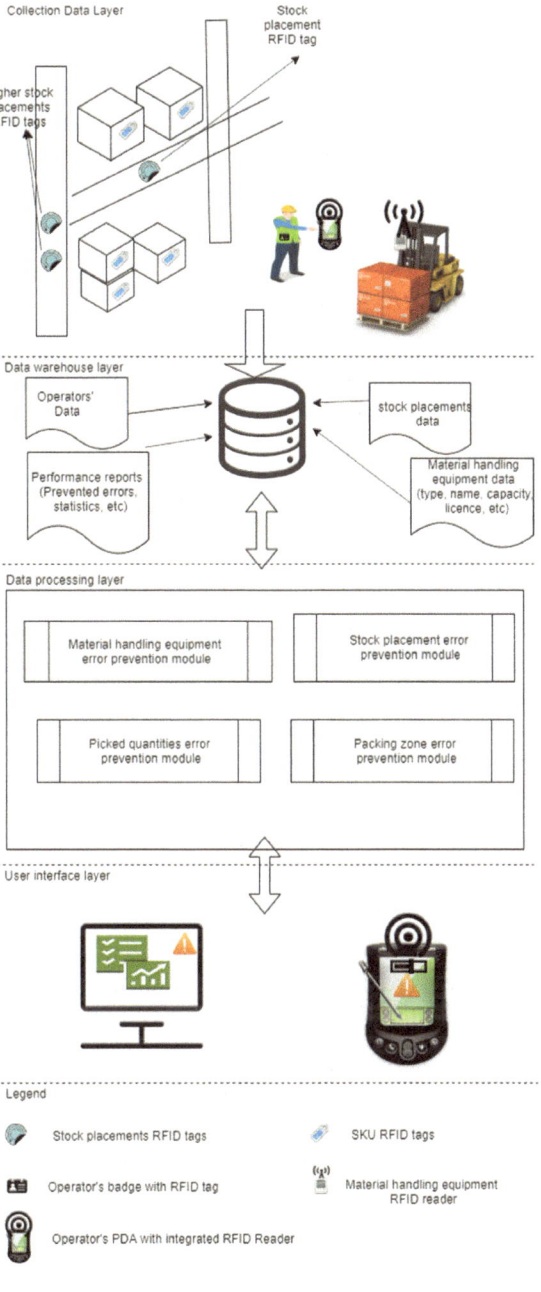

3.2 Data Warehouse Layer

This layer manages all data processed in the system. It consists of a central database that stocks static data like operators' lists, products' lists, PDAs lists and storage placements' lists. As well as dynamic data like error alerts history, error types and statistics.

3.3 Data Processing Layer

This layer is responsible for the treatment of the data present in the data warehouse layer. This module ensures that:

- The material handling equipment chosen by the operator for the picking assignment is the same as the one noted on the picking task.
- Each visited storage placement is indeed present in the picking task.
- The quantities of the units picked of each product are correct.
- The chosen packing zone is correct.

If the material handling equipment chosen by the operator is not suitable for the picking task, the system displays an alert on the PDA of the operator showing the list of correct handling equipment to use. If the RFID reader mounted on the PDA of the operator reads a stock placement RFID tag that is not included in the ongoing picking task, the system displays an alert on the PDA screen showing the correct stock placements from.

3.4 Data Processing Layer

This layer is responsible for the treatment of the data present in the data warehouse layer. This module ensures that:

- The material handling equipment chosen by the operator for the picking assignment is the same as the one noted on the picking task.
- Each visited storage placement is indeed present in the picking task.
- The quantities of the units picked of each product are correct.
- The chosen packing zone is correct.

If the material handling equipment chosen by the operator is not suitable for the picking task, the system displays an alert on the PDA of the operator showing the list of correct handling equipment to use. If the RFID reader mounted on the PDA of the operator reads a stock placement RFID tag that is not included in the ongoing picking task, the system displays an alert on the PDA screen showing the correct stock placements from which to pick. If the operator forgets to pick from one or more stock placements and enters a packing area, an alert is displayed on his PDA screen, showing the stock placements to visit along with the quantities to pick. If the operator enters into the wrong packing zone, an alert is displayed showing the correct packing zone. If an operator ignores an alert, the system elevates it to a higher rank employee. If the higher rank officer also ignores it, the system concludes that the problem is solved. Unique sound signals accompany displayed alerts in order to grab the operators' attention. Each type of alert comes with a different sound signal so that with time operators can learn them and spend less time verifying which error they have made.

1. Material handling equipment error prevention module

The system chooses material handling equipment following their weight and volume capacities. The system firstly checks the picking task total weight and total volume and then assigns it to a matching material handling equipment.

Selecting the right material handling equipment is done following the procedure shown in Fig. 2.

For each picking task class, the system generates a list of corresponding material handling equipment and then explores the list for availability. If the first equipment is not available, the system checks the next one and so on. In Fig. 3, the selection procedure of material handling solutions is presented. The procedure in Fig. 3 also reduces waste in terms of unused weight and volume capacities. If the picking task characteristics surpass the biggest capacities available, the system splits it in two or more picking tours. Figure 4 presents the error prevention procedure in the choice of stock placements.

2. Stock placement error prevention module

This module is responsible for choosing the right stock placement from which the operator picks product units as well as preventing incomplete picking tasks. The operator firstly reads the stock location identity; if it is the right one, the system continues to display the picking task status. In case of an error, the system shows an alert precising the right stock location from which to pick.

This procedure is crucial as it prevents the errors that are the most costly to repair. Figure 5 shows the error prevention procedure concerning forgetting parts of the assigned picking task. This procedure ensures that an operator visit all stock placements included in his or her picking task. When the operator checks in the packing zone, if all stock placements included in the picking task were visited, the system continues to display the task progress. If not, the system displays an alert showing stock placements yet to be visited.

3. Picked quantities error prevention module

With the help of RFID readers mounted on material handling equipment, this procedure ensures that the number of picked items corresponds to the picking task. In the opposite case, the system displays an alert to the operator in question precising the products' IDs and the quantities left to pick or to put back to stock (Fig. 6).

Fig. 2 Material handling
selection procedure

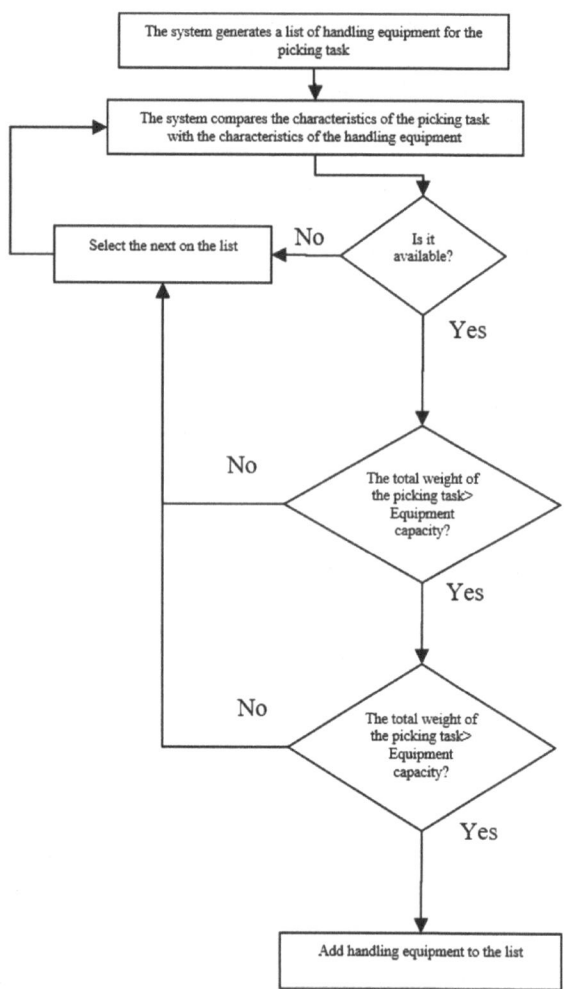

4. Packing zone error prevention procedure

The system assigns each picking task to a packing zone. Before penetrating in a packing zone, the operator reads the packing zone RFID tag. If it is not the right one, the system displays an alert showing the correct packing zone to penetrate. Figure 7 describes the procedure of error prevention concerning the choice of the packing zones.

Fig. 3 Material handling
equipment choice error
prevention procedure

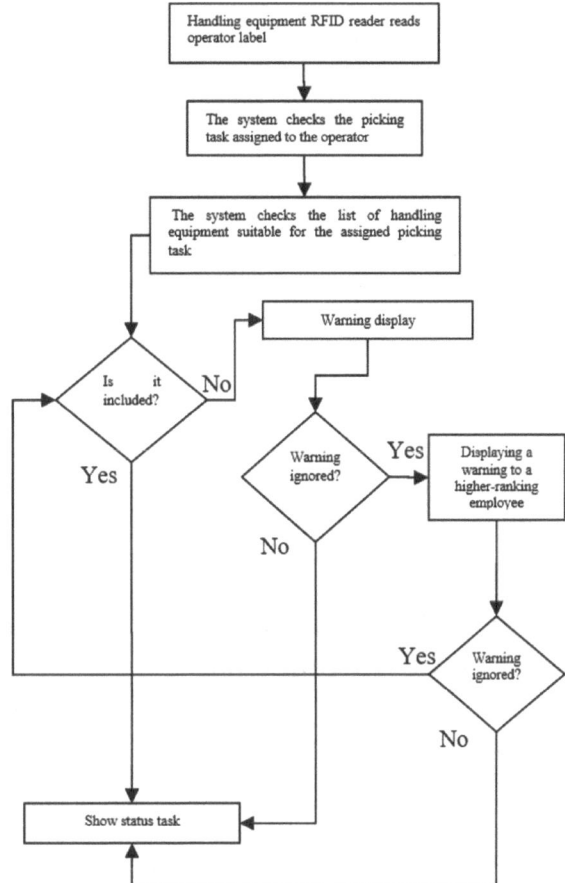

3.5 User Interface

The proposed system is developed in a web environment. Users can be either system administrators, managers, team leaders or operators and can access the system on web navigators on computers or PDAs. After authentication, each user gets an interface matching his or her role on the system. Administrators have read and write privileges. They can modify, add or remove data in the data warehouse. In the managers' user interface, they can access individual and collective performance data. The interface accessible to operators consists of the assigned task and its details and progress in real time. Packing operators access the system on a computer placed in the packing zones. Their user interface includes ongoing picking tasks and their progress in order to prepare the packing in advance.

Fig. 4 Stock placement error prevention procedure

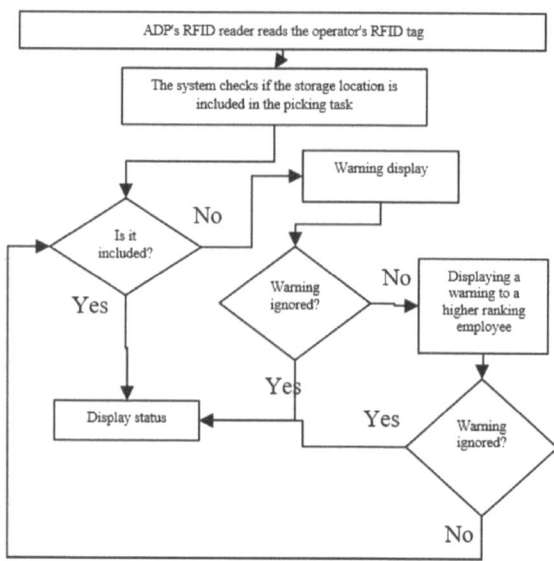

4   Error Cost Simulation and Comparison with Other Picking Technologies

As the proposed system is in the prototyping phase, it was decided to present in this section an error cost simulation and comparison with other picking technologies. This simulation's objective is to give an idea of how costly picking errors are, and how much an upgrade would, from the other picking systems to the one proposed in this paper, benefit companies in financial terms. The calculations are based on 8 h of work per day, 22 working days per month and 25 system users picking at a rate of 100 units per hour. We consider, as in [2], that a picking error costs 50$. In Fig. 8, the graph shows the number of occurring errors per 1000 picked unit, as shown in [9].

Errors cost using other systems are estimated at 50$; the cost of picking errors using the presented system is estimated at 0.115$ since it has two verification steps: the first one is when the items are picked from stock and the second one when the operator places them on a material handling equipment. Thus if somehow an error is past the first firewall, it cannot go past the second one and the estimated error cost is based on the cost of the picker's time considering that the correction of a mispick that made it through to the material handling will cost a maximum of 30 s to correct. The error cost calculation is based on a picker's hourly wage of 13.85$ as in [11].

The price of different picking systems, based on estimates in [9], was included in the error cost simulation. Figure 9 presents the average prices of the most popular picking systems based on a 25 user scenario.

Figure 10 shows how much picking errors cost per 1000 picked units, per hour (at a rate of 100/h) and per day.

Fig. 5 Incomplete picking task prevention procedure

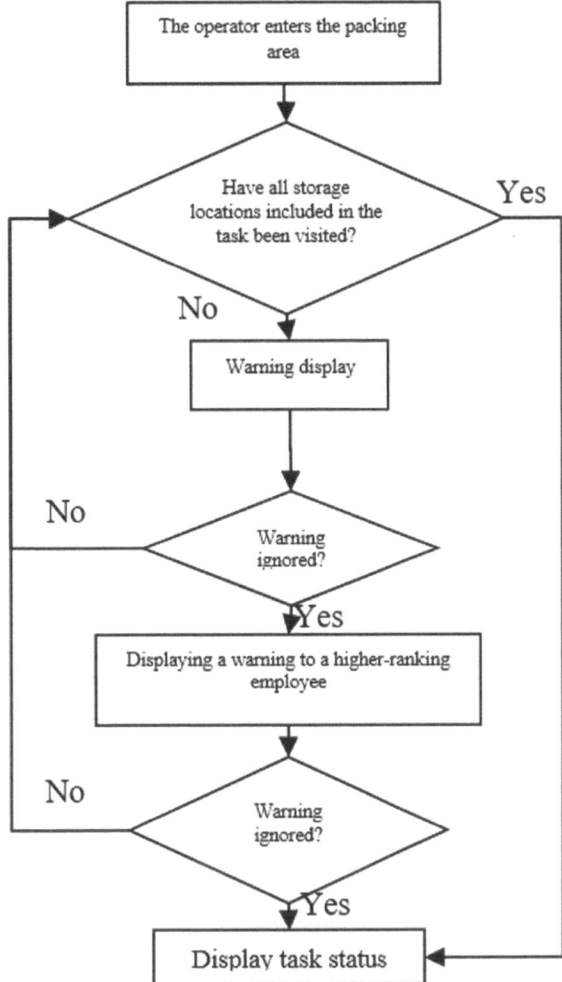

To get a better idea of how much picking errors cost, the monthly and yearly picking errors cost are presented in Fig. 11.

The numbers in the graphs show how much picking is crucial to companies' profitability and how much they lose in picking errors while using the mentioned picking systems.

The proposed system (RTPEPS) is, by all means, the best performing system in this simulation as it leaves no space for picking errors, therefore, cutting the resulting expenses.

The considered scenario is one where companies adopting the other systems decide to upgrade to the RTPEPS system. Figure 12 shows how much time it would take to reach the break-even point and Fig. 13 shows the return on investment after one year of the upgrade to the proposed system (RTPEPS).

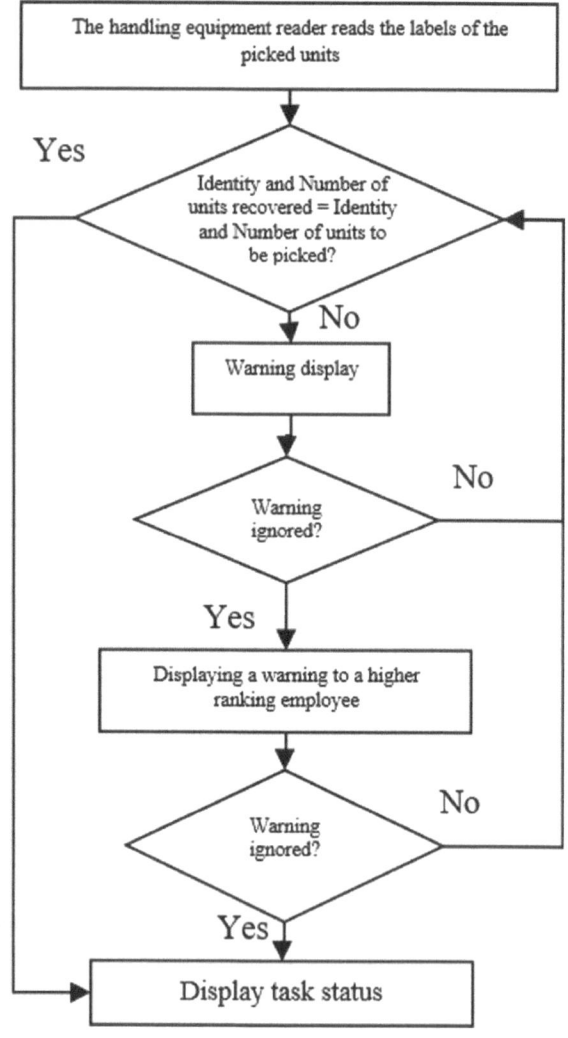

Fig. 6 Products quantity error prevention procedure

If a company upgrades its picking system from a paper-based system to the RTPEPS, it will take the company 13.87 days to reach the break-even point, from the RF-scanning system, it would take 23.12 days, from a pick to light system, it would take 34.68 days, from a pick by voice system, it would take 102.77 days, and from a pick by HUD system, it would take 17.34 days. The return on investment shown in Fig. 13 highlights the value of the proposed system. Regardless of the system that a company is using, the numbers show that upgrading to the RTPEPS system is financially very beneficial.

Fig. 7 Packing area choice
error prevention procedure

The operator reads the label of the
packing area

Packing area
included in the
picking task?

Yes

No

Display an alert
specifying the correct
packing area

Display
status task

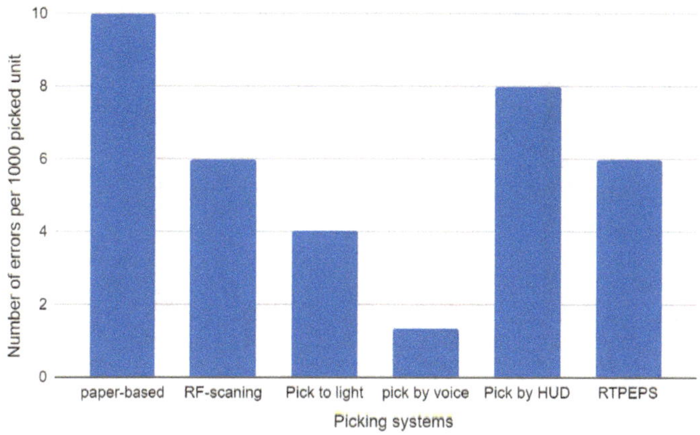

Fig. 7 Packing area choice error prevention procedure

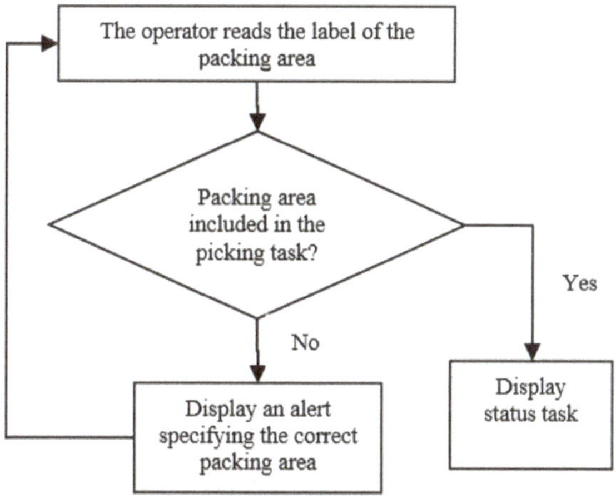

Fig. 8 The number of errors per 1000 picked units for different picking systems

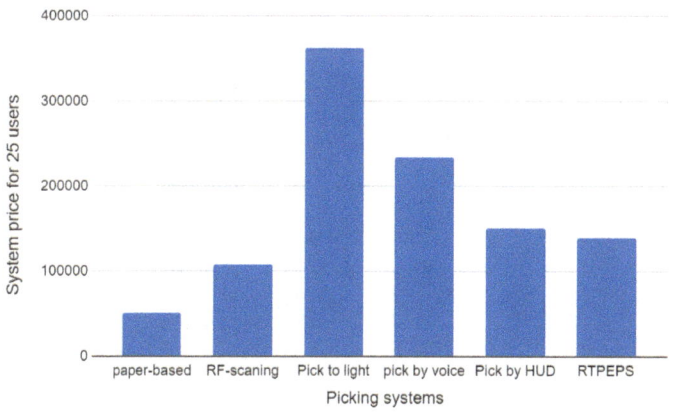

Fig. 9 Picking systems price comparison based on 25 users scenario

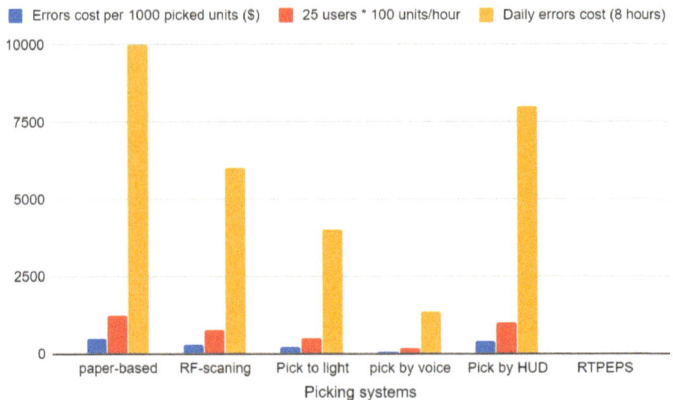

Fig. 10 Picking errors cost comparison, per 1000 units, per hour and per day

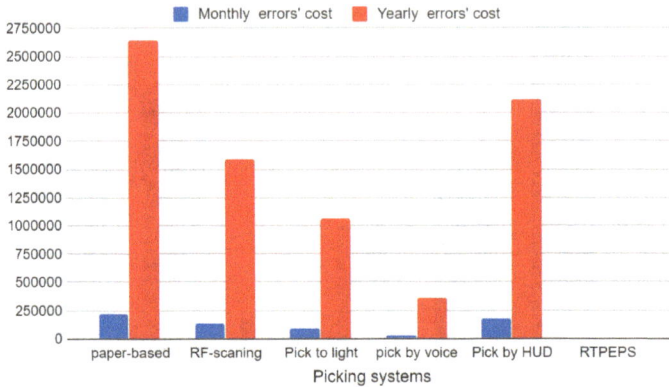

Fig. 11 Monthly and yearly picking errors cost comparison by system used

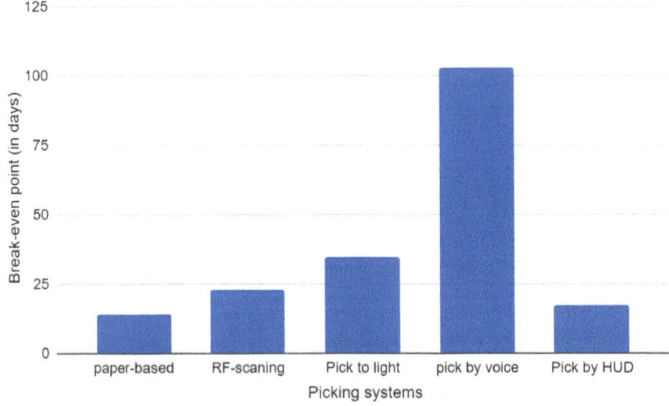

Fig. 12 Break-even point after upgrade from the other picking systems to the proposed system

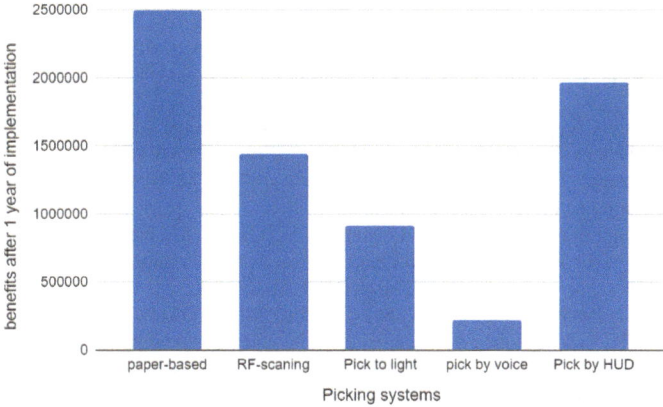

Fig. 13 Returns on investment after 1 year of upgrading to the proposed system

5 Conclusion

This paper presents a real-time picking errors prevention system based on RFID technology which consists of multiple modules that give the possibility to prevent errors at every step of the picking process from the assignment of the picking task to the delivery to packing zones. The proposed system prevents errors firstly, in the choice of the material handling equipment, secondly, in the choice of products as well as their quantities and finally, the choice of the packing zones. This system gives companies the possibility to enhance their warehousing efficiency as well as the efficiency of their supply chain as a whole because it prevents picking mistakes which are costly to make and to repair as they include penalties, returns and may hurt the companies' image.

The proposed system is in the prototyping phase. In the next works, we will focus on its implementation and test in a warehouse environment.

References

1. de Koster, R., Le-Duc, T., Roodbergen, K.J.: Design and control of warehouse order picking: a literature review. Eur. J. Oper. Res. **182**, 481–501 (2007)
2. Asghar, U., Lütjen, M., Rohde, A.-K., Lembke, J., Freitag, M.: Wireless pick-by-light: usability of LPWAN to achieve a flexible warehouse logistics infrastructure. In: Lecture Notes in Logistics, pp. 273–283 (2018). https://doi.org/10.1007/978-3-319-74225-0_37
3. The cost of picking mistakes for small businesses. https://warehouseos.com/the-cost-of-pic king-mistakes-for-small-businesses. 01/06/2020
4. Thomas, C., Panagiotopoulos, T., Kotipalli, P., Haynes, M., Starner, T.: RF-pick. In: Proceedings of the 2018 ACM International Symposium on Wearable Computers—ISWC'18 (2018)
5. Hand Held Products Inc.: System and method for picking validation. US 2016/0101936A1 (2016)

6. Goodall, B.: Measuring the real cost of warehouse picking errors. https://www.linkedin.com/pulse/measuring-real-cost-warehouse-picking-errors-bryan-goodall/. 01/06/2020

7. Picking errors steal money. https://www.consafelogistics.com/picking-errors-steal-money/. 01/06/2020

8. Wu, X., Haynes, M., Guo, A., Starner, T.: A comparison of order picking methods augmented with weight checking error detection. In: Proceedings of the 2016 ACM International Symposium on Wearable Computers (ISWC'16), pp. 144–147. ACM, New York, NY, USA (2016). https://doi.org/10.1145/2971763.2971773

9. Wada, K., Sugiura, M., Yanokura, I., Inagaki, Y., Okada, K., Inaba, M.: Pick-and-verify: verification-based highly reliable picking system for various target objects in clutter. Adv. Robot. (2017). https://doi.org/10.1080/01691864.2016.1269672

10. Andriolo, A., Battini, D., Calzavara, M., Gamberi, M., Peretti, U., Persona, A., Sgarbossa, F.: New RFID pick-to-light system: operating characteristics and future potential. Int. J. RF Technol. 7(1), 43–63 (2016). https://doi.org/10.3233/rft-150071

11. Order picker salary. Retrieved from https://www.indeed.com/career/order-picker/salaries. Last accessed 01/07/2020

12. Baechler, A., Baechler, L., Autenrieth, S., Kurtz, P., Hoerz, T., Heidenreich, T., Kruell, G.: A comparative study of an assistance system for manual order picking—called pick-by-projection—with the guiding systems pick-by-paper, pick-by-light and pick-by-display. In: 2016 49th Hawaii International Conference on System Sciences (HICSS) (2016). https://doi.org/10.1109/hicss.2016.72

13. Flanders, S.: Voice Directed Picking—A Technology That is Ready for Prime Time. Warehouse Management Consultants. A RFID case-based logistics resource management system for managing order-picking operations in warehouses (2002)

UAV Modular Control Architecture

Asmaa Idalene, Khalid Boukhdir, and Hicham Medromi

Abstract This paper presents a novel control prototype for autonomous unmanned aerial vehicle (UAVs) that can operate in dynamic environments. This prototype is called "UAV Modular Control Architecture" or UMCA. It consists of specific components that are organized into two layers: mission planning and mission Supervision. The mission planning layer generates an appropriate flight plan to meet the given mission, and the mission supervision layer interprets the plan and evaluates its execution. The UMCA leads to large advantages such as adaptability, flexibility, robustness, and extensibility.

Keywords UAV · Control architecture · Decision-making · Task planning · Path planning · Mission planning

1 Introduction

Since unmanned aerial vehicles (UAVs) are complex robots operating in a dynamic environment, they share constraints with embedded systems such as limited power resources, limited space, limited processing time, and limited payload [3]. The key challenge problem that faces the implementation of an autonomous UAV is the design of powerful control architecture. Powerful control architecture offers the UAV the capacity to achieve high-level goals taking into account its limitations [9]. The control architecture determines the aircraft's capabilities.

In this paper, we focuses on proposing a new control prototype for autonomous mini-UAV operating in the civil domain. This prototype achieves high-level goals, performs complex tasks, finds an appropriate path, avoids obstacles, and generates

A. Idalene (✉) · K. Boukhdir · H. Medromi
Research and Engineering Laboratory (LRI), National School of Electricity and Mechanics, Casablanca, Morocco
e-mail: Asmaaidalene7@gmail.com

A. Idalene
Pluridisciplinary Laboratory of Research and Innovation (LPRI), EMSI, Casablanca, Morocco

© The Editor(s) (if applicable) and The Author(s), under exclusive license to Springer Nature Singapore Pte Ltd. 2021
F. Saeed et al. (eds.), *Advances on Smart and Soft Computing*, Advances in Intelligent Systems and Computing 1188, https://doi.org/10.1007/978-981-15-6048-4_28

319

a feasible flight plan. The control prototype is called UAV modular control architecture (or UMCA). It consists of two layers: mission planning and mission supervision. The mission planning layer performs decision-making, task generation, and motion planning. The mission supervision layer estimates UAV's state, perceives the environment, supervises task execution, and performs situation prediction.

This article is structured as follows: Sect. 2 provides a comparative study of the state of the art of the control architecture subject. Section 3 describes the proposed control prototype. Finally, we present a conclusion of our work in Sect. 4.

2 Control Architecture: Comparative Study

Various control strategies have been proposed on the robotic control subject, each of which attempts to develop a new robot control system. These strategies can be classified into six approaches [1]: deliberative approach, reactive approach, hybrid approach, behavior approach, hybrid behavior approach, and subsumption strategy. This section discusses and compares these approaches.

The deliberative approach is the more promising strategy for the executing of complex tasks in a static environment where the world model is well known [16]. This approach includes three functionalities: sense, plan, and act [8]. The sense functionality gathers sensor measurement and updates the environment's world model. The plan functionality produces an action plan to meet its mission goal. The last functionality transforms the action plan into feasible low-level commands.

The deliberative approach presents a serious source of weakness in certain cases [10], especially for the autonomous UAV. This approach needs high-performance computational skills. This approach is ineffective in an uncertain environment.

Furthers, if one of the functions fails, the architecture breaks. Moreover, it fails if the world model's representation is either incomplete or inexact.

The reactive control strategy is the best approach choice for applications that needs fast navigation in an uncertain environment [12]. This architecture defines the control strategy as a set of condition-action couples, each of which joins sensor information with a specific robot's action. It needs neither to build the world model nor to perform the planning functionality. It only generates commands based on sensory data. On the other hand, the reactive approach deals neither with high-level objectives nor complicated constraints. So, it just resolves the problems in which the objectives are well known, the environment is well defined, and the robot carries out enough sensors [4].

Applications such as object identification (OI) and intelligent surveillance (IS) require both deliberative and reactive characteristics. That gave birth to the idea of the hybrid architecture. This approach meets high-level objectives in an uncertain environment. Typically, hybrid control architecture contains three hierarchical levels [13].

- The decisive layer (the high level) generates an appropriate mission plan.

- The intermediate layer (the middle level) monitors the communication between the decisive and the reactive layers.
- The reactive layer (the low level) perceives the environment, executes the needed actions, and takes care of the robot's safety.

The hybrid approach presents a drawback for autonomous UAVs. The high level does not have direct access to the low level. So, during the planning process, the higher level may have an old state of the world model. By this means, the planned may produce an invalid plan. Consequently, this approach may fail its mission.

The behavior control strategy is inspired by biological studies, and it performs a fast mapping between sensing and acting functionalities. It divides the control system into a collection of independent behaviors, each of which performs a particular task [6]. The behavior strategy is more efficient than the reactive strategy. Each behavior gives both reactive and deliberative capabilities. The behavior approach meets goals in unpredictable situations. It offers a set of concurrent behaviors which act independently to meet a special goal. Further, it is a good strategy for the robot operating in unknown environments.

The hybrid behavior strategy is a good design for control architecture that needs high flexibility and adaptability to lot of mission kind [11]. This approach solves some drawbacks of hybrid and behavior approaches. It generates the appropriate plan and coordinates the robot's behaviors to reach the mission goals. Essentially for autonomous UAVs, this approach presents many weaknesses:

- First, the hybrid behavior suffers from the dependence on the sensor system.
- Second, the same hybrid behavior architecture cannot be used on different aircraft's types.
- Third, it suffers from the interdependency of layers.

The subsumption approach provides a simple way to build a robust robot. It divides the control system into sorted layers based on the task achieving behaviors [5]. This approach constructs a control system with the ability to improve its autonomy level. It creates a robot dealing with multiple sensors and goals. Further, it implements an extensible layered architecture. Moreover, it offers a unified representation for various robot's type [14].

The subsumption approach presents some limits: While this hierarchical approach depends on the physical architecture, it is neither reusable nor modular. Due to the layer's interdependency, this approach could not integrate additional sensors. During execution, the priority-based mechanism limits the ways the system can be adapted. While higher layers interfere with lower ones, they cannot be developed independently. Vital behaviors cannot be always prioritized.

As a summary, we provide Table 1 that resumes the characteristics of each studied control strategies [1]. Table 1 uses the features below to evaluate these architectures.

- **Global reasoning**: the ability to preserve the robot safety and to choose the suitable decisions.
- **Reactivity**: the capability to recognize changes and to avoid obstacles.

Table 1 Analysis of the studied control approaches

	Deliberative	Reactive	Hybrid	Behavior	Hybrid behavior	Subsumption
Global reasoning			✔		✔	✔
Reactivity		✔	✔	✔	✔	✔
Adaptability				✔	✔	
Flexibility		✔		✔	✔	✔
Adaptability				✔	✔	
Modularity				✔		
Robustness						✔
Sensor integration				✔	✔	
Extensibility						✔
Reusability						

- **Adaptability**: the capacity to meet various mission types with little reconfiguration requirement.
- **Flexibility:** the ability to add (delete) new functionalities or change the existing ones.
- **Modularity**: the capacity to build up the architecture by using a modular approach.
- **Programmability**: the capability to implement the architecture by constructing independently each module.
- **Robustness**: the ability to adapt and repair the mission plan when some sensors fail.
- **Sensor integration:** the capacity to adapt the same approach to new sensors.
- **Extensibility:** the ability to improve the current autonomy level by adding additional modules.
- **Reusability:** the capability to use the same architecture for various robots type.

3 The Proposed Control Architecture

We propose a novel control prototype named UAV modular control architecture (or UMCA). The architecture consists of specific modules organized according to their predefined functions (see Fig. 1). To implement this prototype, we can develop each module independently and then tie them all together to form the UAV control system. UMCA achieves complex mission, performs complicated tasks, finds optimal path, avoids static and dynamic obstacles, and generates a feasible mission plan.

The UAV modular control architecture includes eight modules:

- The decision-making module for making suitable decisions;
- The task planning module for producing a valid task plan;

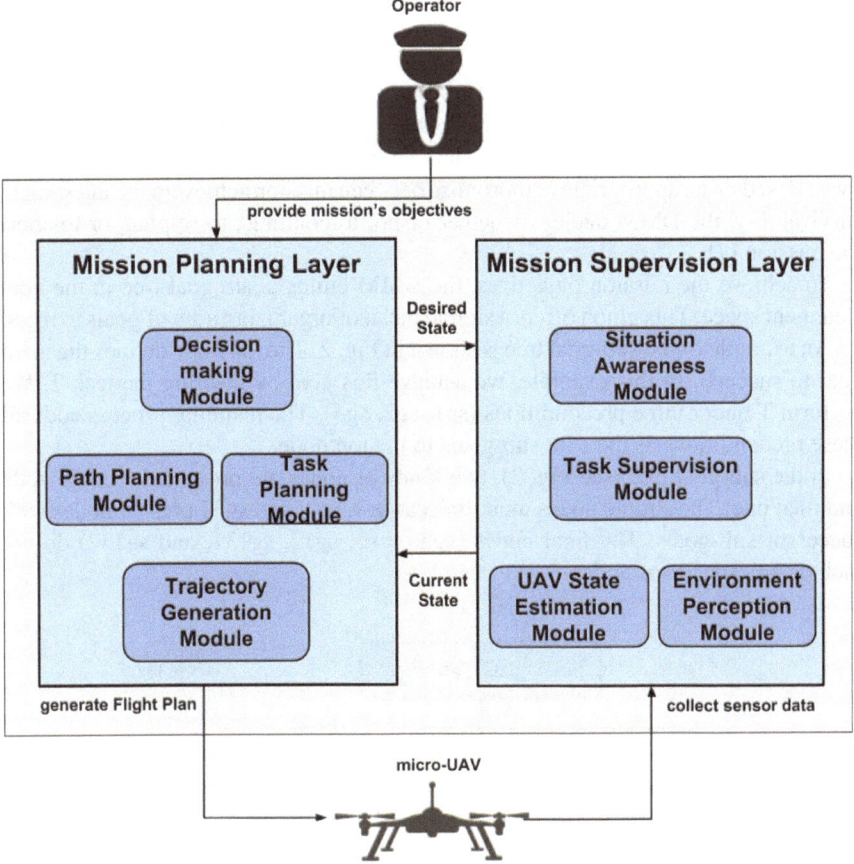

Fig. 1 Proposed prototype "UAV modular control architecture"

- The path planning module for generating an appropriate path to reach the desired locations;
- The trajectory generation module to calculate a feasible and optimal trajectory reference;
- The situation awareness module to provide all the awareness needed to perform the given mission;
- The task supervision module to calculate the current state of the UAV's world model. This module aims to supervise the task execution;
- The environment perception module to provide all perception capabilities;
- And, the UAV state estimation module to calculate the UAV's state. The UMCA structures these modules into two main layers (see Fig. 1);
- Mission planning layer for generating the appropriate flight plan;
- Mission supervision layer for interpreting the flight plan and supervising its execution.

3.1 Decision-Making Module

The decision-making module (or DMM) is the most intelligent component that performs all the needed decisional capabilities [15]. It reasons about available resources and constraints and makes suitable decisions to meet the mission objectives. Based on an appropriate compromise between mission achievement and vehicle survivability, the DMM decides whether or not to continue, to re-plan, or to abort the mission [7].

To achieve the mission objectives, the DMM builds a sub-goal tree in the goal statement space. This graph offers a simple mean of organizing a set of goals to meet.

An example of the sub-goal tree is shown in Fig. 2. The top node defines the main goal to succeed. In this example, we achieve this goal by meeting the task T. We perform T under three preconditions (sg1, sg2, sg3). The planning process adds all these preconditions as the next sub-goals to the top node.

In the sub-goal tree (see Fig. 2), two kinds of nodes are presented: simple node and final one. The simple nodes mentioned as sg1, sg3, and sg31 present nodes with successor sub-goals. The final nodes (sg2, sg11, sg12, sg131, and sg132) do not include any successor nodes.

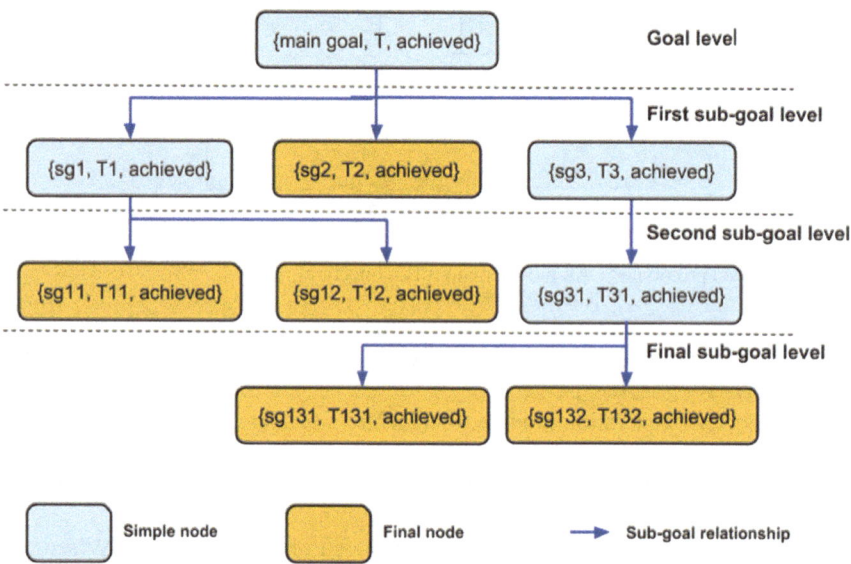

Fig. 2 Sub-goal tree

3.2 Task Planning Module

The main role of the task planning module is to produce a valid task plan. This module uses the sub-goals tree generated by the DMM as its input. A task plan P defines a sorted action sequence $(a_0, a_1 \ldots a_n)$ to execute within a specific time. In the initial world model, the UAV will execute the first action (a_0), then the second (a_1), and so on. A valid task plan must achieve the main goal after executing the last action (a_n).

3.3 Path Planning Module

Path planning Module (or PPM) produces an optimal free-collision path. This path starts from the initial position, visits the desired locations, and reaches the final position [2]. Given the components below:

- w_i: a particular position in a configuration space.
- w_{init}: the initial position of the UAV.
- w_{final}: the final position of the UAV.
- L_G: a set of location.
- C_{space}: the configuration space.
- $C_{obstacle}$: the obstacle space.
- C_{free}: the free space.

Finding an optimal path planning consists of looking for a sorted waypoint sequence (w_i) which starts on w_{init}, reaches all the locations defined in L_G, and goes to w_{final}. This waypoint sequence must define a free-collision path (i.e., in C_{free}).

Figure 3a shows an example of a civil application. The UAV must start its mission from the red point (the initial point), visit all the cyan points (the desired locations), and go to the blue point (the goal point). The operation environment contains a set of obstacles.

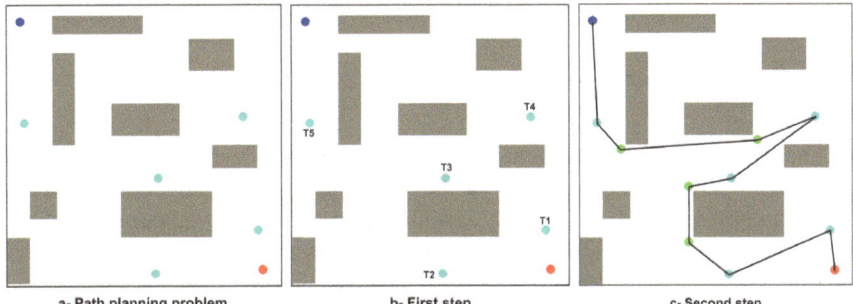

| a- Path planning problem | b- First step | c- Second step |

Fig. 3 Path planning

To find the free-collision path of minimal cost, the PMM computes two processes:

1. The first process sorts the desired locations according to their distance from the initial position (see Fig. 3b).
2. The second process finds the optimal path based on the obstacles' corner (see Fig. 3c).

3.4 Trajectory Generation Module (TGM)

Based on a given waypoints sequence along the desired path, the role of this module is to compute a feasible and optimal trajectory reference, while satisfying the environmental constraints.

3.5 Situation Awareness Module (SAM)

Situation awareness (SA) plays an important role in the success of decision-making. It includes three steps: the perception of UAV and its operating environment, the comprehension of their meaning, and the projection of their future status [17].

3.6 Task Supervision Module (TSM)

This module describes, using a set of predicates, the current state of the real UAV world model to supervise and evaluate the task execution. Each executed task corresponds to a deterministic state transition that transforms the current state of the world model to some new one. The main role of this module is to provide the current state of the UAV's world model to the T.P.M to know actions that have been successfully executed.

3.7 Environment Perception Module (EPM)

To successfully execute the generated task plan in the real world, this module provides all perception capabilities such as obstacle detection, target identification, and environment mapping. Basically, it senses the payload sensor data and estimates the current state of the operating environment. This module sends the computed environment state to the PPM to supervise the mission path.

3.8 UAV State Estimation Module (USEM)

The main purpose of this module is to calculate the state of the UAV based on sensor data. Typically, this module estimates the position, the orientation, the velocity, as well as the acceleration of the UAV. This module continually sends all these measurements to the TGM.

4 Conclusion

In this paper, we have proposed a new control architecture for unmanned aerial systems operating in the civil application. This approach is based on two layers, each of which contains generic components. Each component is an independent module that provides a specific capability. The proposed architecture deals with high goals. It performs complicated tasks. Furthermore, it computes a feasible trajectory. The same architecture can be implemented on various UAV types and can perform a large variety of civil applications. The future work will focus on the development of this prototype. We will develop each module independently and then tie them all together to form the UAV control.

References

1. Asmaa, I., Khalid, B., Hicham, M.: UAV control architecture: review. Int. J. Adv. Comput. Sci. Appl. **10**(11) (2019). https://doi.org/10.14569/IJACSA.2019.0101186
2. Asmaa, I., Khalid, B., Hicham, M.: UAV path planning for civil applications. Int. J. Adv. Comput. Sci. Appl. **10**(12) (2019). https://doi.org/10.14569/IJACSA.2019.0101281
3. Austin, R.: Unmanned Aircraft Systems: UAVS Design, Development and Deployment, vol. 54. Wiley (2011)
4. Belkhouche, F.: Reactive optimal UAV motion planning in a dynamic world. Robot. Auton. Syst. **96**, 114–123 (2017)
5. Brooks, R.: A robust layered control system for a mobile robot. IEEE J. Robot. Autom. **2**(1), 14–23 (1986)
6. Dinh, H.T., Torres, M.H.C., Holvoet, T.: Sound and complete reactive UAV behavior using constraint programming. In: ICAPS Workshop on Planning and Robotics (2018)
7. Estlin, T.A., Volpe, R.D., Nesnas, I., Mutz, D., Fisher, F., Engelhardt, B., Jet, S.C.: Decision-making in a robotic architecture for autonomy (2001)
8. Ghallab, M., Nau, D., Traverso, P.: Automated Planning and Acting. Cambridge University Press (2016)
9. Huang, H.M.: Autonomy levels for unmanned systems (ALFUS) framework: safety and application issues. In: Proceedings of the 2007 Workshop on Performance Metrics for Intelligent Systems, pp. 48–53 (2007)
10. Ingrand, F., Ghallab, M.: Deliberation for autonomous robots: a survey. Artif. Intell. **247**, 10–44 (2017)
11. Liu, J.D., Hu, H.: Biologically inspired behaviour design for autonomous robotic fish. Int. J. Autom. Comput. **3**(4), 336–347 (2006)

12. Nakhaeinia, D., Tang, S.H., Noor, S.M., Motlagh, O.: A review of control architectures for autonomous navigation of mobile robots. Int. J. Phys. Sci. 6(2), 169–174 (2011)
13. Nonami, K., Kendoul, F., Suzuki, S., Wang, W., Nakazawa, D.: Autonomous Flying Robots: Unmanned Aerial Vehicles and Micro Aerial Vehicles. Springer Science & Business Media (2010)
14. Oland, E., Andersen, T.S., Kristiansen, R.: Subsumption architecture applied to flight control using composite rotations. Automatica 69, 195–200 (2016)
15. Rainer, J.J., Cobos-Guzman, S., Galán, R.: Decision making algorithm for an autonomous guide-robot using fuzzy logic. J. Ambient Intell. Humaniz. Comput. 9(4), 1177–1189 (2018). https://doi.org/10.1007/s12652-017-0651-9
16. Simmons, R.G.: Structured control for autonomous robots. IEEE Trans. Robot. Autom. 10(1), 34–43 (1994)
17. Wickens, C.D.: Situation awareness: review of Mica Endsley's 1995 articles on situation awareness theory and measurement. Hum. Factors 50(3), 397–403 (2008)

Trust in IoT Systems: A Vision on the Current Issues, Challenges, and Recommended Solutions

Hanan Aldowah, Shafiq Ul Rehman, and Irfan Umar

Abstract Over the recent overs, the phenomenon of Internet of Things (IoT) has grown rapidly and thus has become the focus of research. IoT involves an important set of technologies which are expected to be the next information technology revolution and the evolution of big data analytics. IoT technologies and applications are intimately associated with people; hence, trust is the major issue. To highlight this key element of IoT, this study reviews the progress of the research work related to trust in IoT and found that several issues need to be addressed. This paper also provides an insight on the challenges of trust in IoT, and some of these challenges from various aspects are precisely discussed. Moreover, existing solutions from academic, technical, and industry aspects are presented. However, many other issues and challenges remain open that needs to be considered while designing and deploying the trusted solutions in IoT environment.

Keywords IoT systems · Trust management · Industry 4.0 · IoT vulnerability · Cybersecurity

1 Introduction

A massive and transformational change will take place that will change the use of the Internet forever. An explosion in connection under the broad descriptor of IoT is presently rolling out around the world, taking advantage of the massive expansion of IP addresses through IPv6 deployment [1]. This new IP protocol transfers Internet

H. Aldowah (✉) · I. Umar
Centre for Instructional Technology and Multimedia (CITM), Universiti Sains Malaysia (USM), Penang, Malaysia
e-mail: hanan_aldwoah@yahoo.com

I. Umar
e-mail: irfan@usm.my

S. Ul Rehman
Singapore University of Technology and Design (SUTD), Singapore, Singapore
e-mail: shafiq_rehman@sutd.edu.sg

© The Editor(s) (if applicable) and The Author(s), under exclusive license to Springer Nature Singapore Pte Ltd. 2021
F. Saeed et al. (eds.), *Advances on Smart and Soft Computing*, Advances in Intelligent Systems and Computing 1188, https://doi.org/10.1007/978-981-15-6048-4_29

addresses from a limited resource that is carefully managed to a new platform without any restriction. This radical change has brought a new creative life, not like the early years of the Internet [2]. Consequently, the number of smart connected devices is expected to increase to 50 billion or more according to some estimates, with more than 200 billion intermittent connections [3, 4]. Considering the fact, new solutions are being designed that can interconnect healthcare, insurance, mass transportation, home energy consumption, and other sectors. Thus, there will be more opportunities for growth, ease of use, and increased efficiency [5].

IoT is a fast-rising model featuring an important set of technologies that are expected to be the next information revolution. However, a research study conducted by Packard [6], reported that more than 70% of the existing IoT systems have severe vulnerabilities due to insecure Web interfaces, lack of encryption for transport, insufficient authorization, and inadequate software protection [7]. All these new possible threats related to data protection and information security must be well thought out. Furthermore, due to insufficient security measures found in IoT technology, make consumers reluctant to its adoption.

IoT is closely related to people, and with the rapid growth in the use of IoT application, many trust issues have risen. When almost everything is connected to the global Internet and things, and these issues become clearer, continuous exposure will reveal more problems with respect to security and privacy such as the integrity of data sensed and exchanged by things, confidentiality, and authenticity. Furthermore, the trust issue plays critical role in IoT to improve user privacy, reliable data integrity, and information security. Thus, it can assist users to embrace this technology. Nevertheless, the existing literature lacks a comprehensive study on trust in IoT systems.

Therefore, this study highlights the more prominent aspect related to IoT in terms of trust. This is an extended work done on our previous paper [8] which specifically addressed the security aspects in IoT. Our motivation for conducting this study is that existing literature primarily discussed IoT solutions proposed from the academic perspective only while ignoring the industrial and technical aspects. In order to come up with robust IoT systems, all IoT stakeholders must work in a cooperative and parallel manner. Hence, this study discusses the main issues and challenges related to trust management in IoT environment and presents some recommendations that can be considered while designing and deploying resilient IoT systems based on the existing solutions developed by academic researchers, technician, and industry experts to ensure trust in IoT environment.

The contribution of this study is twofold. First, we highlight the various key issues and challenges related to trust in IoT systems. Second, we give substantial insight into how specific mechanisms, approaches, and algorithms can be applied to help in solving certain problem of such technologies or minimizing the effect. This is intended to provide future research directions in IoT domain. Rest of the paper is structured as follows: Research scope and methodology are described in Sect. 2. Various issues and challenges associated with IoT technologies and their considerations are stated in Sect. 3. Existing solutions developed for IoT systems are presented in Sect. 4 while the findings of the review are discussed in Sect. 5. Conclusion and recommendations are outlined in Sect. 6.

2 Research Scope and Methodology

This study is intended to critically review the IoT trust issues, challenges, and solutions. Based on the proposed approaches by researchers, specific solutions will be suggested to solve certain problem that will help guide future research. In addition, given the importance of the contribution of industrial sector, special attention will be paid to technical and industrial solutions as well. For this nature of review, the preliminary step was to search databases for recent studies on IoT and its different research topics and issues. Popular databases such as Web of Science, Science Direct, IEEE, Scopus, and Google scholar were utilized for the search. Typical keywords used include Internet of Things, IoT technologies, IoT issues, IoT challenges, IoT trust challenge, trust in IoT, IoT solutions, trust solution in IoT, IoT trust solution. Subsequently, eligibility criteria based on inclusion and exclusion strategies were used.

The scope of this study covered a wider range of topics within IoT and related issues up to date works, for improving trust issues and evaluating current materials. Accordingly, necessary insights will be provided on how to facilitate the use of certain technologies through specific mechanisms, algorithms, and platforms. Articles considered relevant for the study which were based on full-length papers published in peer-review journals and proceedings listed in ISI and Scopus, and present empirical data and provide practical solutions such as algorithms, framework, and mechanisms that are applicable were included.

3 Trust Issues and Challenges in IoT

Trust is prerequisite in the IoT environment because it is widely distributed and dependable on qualitative data [9]. The concept of trust is used with different meanings and different contexts. Despite the fact that trust is an important issue and has been widely recognized, it is a complex conception, and there is no conclusive consensus in the scientific literature [10].

In information and communication technology (ICT), trust is described in a variety of potential meanings and is considered as a crucial dimension of digital interactions by integrating of trust in machines and humans [11]. The IoT is no different in this regard, and security is profoundly associated with the ability of users to trust their environment. Consequently, trust in the IoT can be defined as the expectation that a thing will be done without harming the user [12]. This includes the concept of being safe and resilient to attacks, and the user can understand the distributed services involved [13].

Trust is also known as the level of confidence that an entity can guarantee to other entities for specific services in a particular context [14]. Although trust is frequently used with reference to people, it can also be related to a device or any system, which emphasize the importance of measuring the trust level in a digital community. Accordingly, trust in the IoT is embedded in three layers: user to device, from one device to another, and from device to user [11]. Thus, trust can be fragmented into entity trust, machine trust, and data trust. Machine trust indicates the need to interact with reliable devices such as actuators and sensors [15]. This is a challenge in the IoT environment because it is not always possible to establish trust in devices.

Furthermore, each entity can evaluate trust in any device in a different way, and therefore, IoT architectures have to cope with non-singular views of trust. The IoT entity trust indicates the expected behavior of users or services. Although device trust can be created through reliable computing, mapping such approaches to device trust is more challenging and experimental. Therefore, practical methods such as reliable computing for standardized devices along with computational trust are necessary to build device trust [9].

IoT has added new challenges to the existing ones known from the Internet. Reputation is associated to trust and can also be specified as a measure of trust where each entity maintains trusted information on other entities, hence creating a "web," known as web of trust [11]. Thus, this subsection describes the major issues related to trust and reputation in IoT systems.

One of the most important issues of trust in IoT that needs to be considered is related to heterogeneity. Heterogeneity creates from the concept of the IoT where IoT devices interact with the physical world with many different objects which only have an interface in order to communicate. The differences between those things can be the operating system, I/O channel, connectivity, and performance. The reason for these differences is the used hardware of the things that leads to different storage capacity, computational power, and energy consumption [16]. Dealing with these differences will be a big challenge in the IoT environment, and new approaches should take this issue into account.

The increasing number of devices connected to IoT leads to an increasing number of communications, transactions, and data [17]. For this reason, trust systems must scale with the growing number of devices in order to stay fully functional. Therefore, the development of enforcement techniques to support scalability are key factors, and new approaches must be analyzed based on their capability of dealing with an increasing number of things in the network [11, 16].

Moreover, infrastructure is another challenge in terms of availability and finding other entities to interact with them. Trust and reputation systems should take this challenge into consideration because entities need others to collect information and interact with them [16, 18]. Therefore, the new systems and approaches must be analyzed based on their ability to deal with this challenge.

Identity management is also a vital part of IoT that trust and reputation systems that need to be considered. Important aspects of this challenge are that the identity of things is not the same as the underlying mechanism, and things can have a basic identity and many other identities, and these things can also hide their true identity [11, 16].

Furthermore, meeting trust requirements is closely related to the issues of access control and identity management [19]. This is a key issue since the IoT environment is characterized by different devices that have to process and manipulate the data according to user needs and rights. There should be supported to control the state of the virtual world. Users must have the ability to control their services, and they also should have the appropriate tools that accurately describe all their interactions, so that they can create an accurate mental map of their virtual environment.

In this perspective, trust is more than just mechanisms that reduce the uncertainty of objects as they interact, although these mechanisms are necessary for objects in choosing the right partner according to their needs. In addition, such IoT mechanisms must be able to determine trust in a dynamic and collaborative environment, and understanding the meaning of trust that must be provided during the interaction [20].

Integrity also is not only a challenge for the IoT but for every system which must deal with hardware, software, or data. Integrity concept guarantees the prevention of unauthorized modification of hardware and software, and the unauthorized modification of data is not performed by authorized and unauthorized individuals or processes and that data is consistent internally and externally [21, 22]. This definition can help to identify several challenges which arise from integrity and trust systems have to take into consideration. Under this light, newly distributed and lightweight authentication, and integrity frameworks and approaches must be explored to deal with the particularities of IoT devices.

With the expansion of IoT applications and services in different administrative areas including multiple ownership systems, a trust framework is a requirement to enable the users of the system to completely trust the information and services being exchanged. Such a framework should be able to handle individuals and machines [3, 23]. Developing trust frameworks that meet this need requires progress in areas such as lightweight public key infrastructure (PKI) as a base for trust management as well as self-configuring and decentralized systems as alternatives to PKI to establish lightweight key management systems that enable trust relationships, and new means for evaluating trust in data, individuals, and devices. Based on a series of trust policies, the minimum level of trust is required to grant access to a small part of the information or service, access control, assurance ways and means for trusted platforms such as protocols, hardware, software [3, 24].

Trust also indicates that users receive information that they believe to be true and of a certain quality and timeliness. The received information can be trustworthy (usable immediately), trustworthy with alteration (usable after alteration), or untrustworthy (worthless) [25]. In the absence of trust, the user needs to consider whether it might be beneficial to abstain from using certain services of the IoT. Thus, trust is one of the fundamental considerations for developing IoT laws and regulations that enable user rights. Figure 1 presents the common IoT trust issues and challenges.

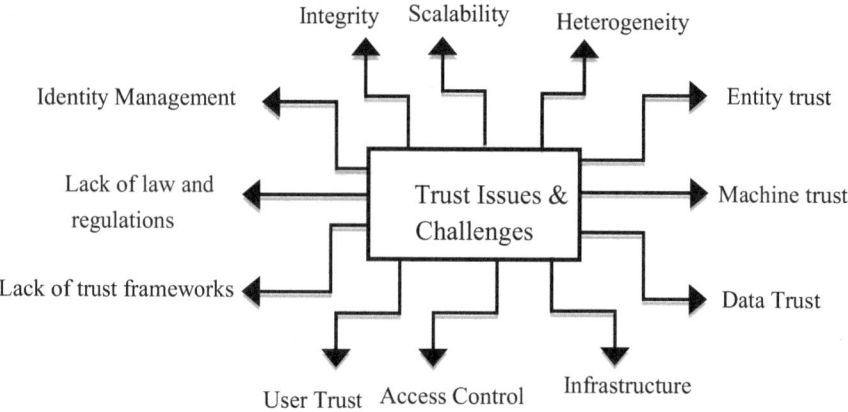

Fig. 1 Common IoT trust issues and challenges

4 Existing Trust Solutions in IoT

To address the trust issues in IoT environment, some of the solutions have been proposed. Such solutions comprise reputation mechanisms, hierarchical model, approaches derived from fuzzy techniques, social networking, and mechanisms based on routing strategies or on nodes past behavior are as follows: Tormo et al. [26] has designed a prototype of a flexible mechanism for selecting the most appropriate model for trust and reputation in heterogeneous environments. This mechanism can be applied on-the-fly, among a set of pre-defined ones, taking into account the current system conditions such as number of users and allocated resources.

Moreover, a flexible trust-aware access control system for IoT (TACIoT) was proposed by Bernabe et al. [27]. This research was conducted to provide a comprehensive and reliable security mechanism for IoT devices. This mechanism was based on a lightweight authorization mechanism and a novel trust model that developed specifically for IoT environments. Such mechanism was successfully implemented and evaluated on a real testbed for constrained and non-constrained IoT devices. On the other hand, another security architecture was presented by Chifor et al. [28] where social networks provide an adaptive sensing system as a service. This system provides a mechanism that allows for autonomic computing where the reputation of smart objects is built on the feedback of human and other nodes. A certain level of trust is transferred to all the nodes that have contributed to the dissemination of reliable information.

Furthermore, a general trust management framework was provided by Ruan et al. [29] that aims to help agents to assess their partners' trustworthiness. In this study, two possible types of attacks were also illustrated and explained how different trust factors or environments together can be used to mitigate the damage. Gu et al. [30] proposed a layered IoT architecture for trust management control mechanism. The IoT infrastructure is composed from three layers that include: sensor layer, core layer, and application layer. Each layer is controlled by a specific trust management under the following purposes: self-organization, routing, and multiservice, respectively. The final decision is implemented by the service requester according to the trust information collected along with requester policy.

Another new IoT intrusion detection model was proposed by Maddar et al. [31], precisely for WSNs. To make sure that users communicate with the appropriate node for each transaction, this model was based on examining the geographic location of the nodes. In addition, a mathematical model for trust calculation was suggested to update trust nodes values and eliminate malicious nodes.

Fernandez-Gago et al. [32] introduced a framework to assist IoT developers embed trust in IoT scenarios. Such proposed framework considers trust and privacy issues as well as identity and other functional requirements that should support trust in IoT to provide the various services that allow for the inclusion of trust in IoT scenarios.

Besides, a smart trust management method was proposed by Caminha et al. [33]. This method was based on machine learning and an elastic slide window technique that automatically evaluates the IoT resource trust and service provider attributes. In the real world and simulated data, this method was able to recognize on-off attackers and fault nodes with an accuracy of up to 96% and low time consumption. Further, based on node behavior detection, another trust system was proposed by Liu et al. [34]. The metrics that evaluated periodically are recommended trust and history statistical trust. They are calculated by evidence combination and Bayes algorithm, respectively.

Moreover, IPv6 a new Internet communication protocol was primarily designed for next-generation smart computing. However, due to its design, it is susceptible to DoS attacks during the device configuration process [7], thus can cause communication disturbance in IoT networks. To address this issue, Rehman and Manickam have proposed two solutions (a) rule-based mechanism [35] to detect any such adversarial attempts at network level and (b) lightweight authentication mechanism known as secure-DAD [36] to prevent such attacks at node level by preconfiguring this scheme on IoT devices during deployment time to ensure that only legitimate devices can communicate in an IoT environment, thus to enable trusted communication in an IoT heterogeneous environment. Table 1 presents some of the existing solutions that can be implemented to build the users trust in IoT deployment.

5 Findings

From the literature review, it is observed that IoT technologies should be designed in a user-friendly manner considering the trust measures. We found that apart from security and privacy concerns, trust also represents fundamental issues in deploying IoT in heterogeneous environment. The concept of IoT is to connect everything to the global network and allow objects remotely communicate with each other, causing new problems related to the authenticity, confidentiality, integrity, and reliability of data exchanged between IoT devices.

Table 1 Summary of proposed trust solutions for IoT systems

Research	Proposed solutions	Contributions
[26]	Dynamic reputation mechanism	Discover and commence the most desirable model based on existing system conditions
[27]	Trust-aware access control system (TACIoT)	Deals with the prevailing nature of the IoT environmental system, while ensuring reliable communication between trusted devices
[28]	System implemented on Eclipse Kura platform	Establish certain level of trust among nodes in IoT environment
[29]	Trust management framework	Support agents to evaluate their partners' trustworthiness
[30]	Trust management control mechanism	Offers a generic framwork for the design of trust models for IoT systems
[31]	A mathematical model for trust calculating (trust model)	Provides a model that can update the trust values of nodes and remove the malicious ones
[32]	Framework for modeling trust dynamics in IoT	Ascertain trust is maintain in all three layers in IoT environment
[33]	A smart trust management method	Potential to detect the on-off attacks and faulty nodes with higher accuracy rates
[34]	A trust system based on behavior detection	Defends against malicious attacks and improves the performance of network
[37]	Google cloud platform	Handles large amount of data processing that is essential for IoT systems
[38]	Privacy by design, data minimization. Notice and choice mechanism	Builds user trust in IoT devices
[35, 36]	Rule-based technique and lightweight authentication mechanism (secure-DAD)	Proactive in DoS attacks detection and prevention, and establishes trusted communication among IoT nodes in distributed networks

To allow authentic users to collect and share sensitive information while preventing the adversaries from obtaining such information, further research is needed that could focus on establishing trust measures in IoT platforms. In maintaining data authenticity, confidentiality, and integrity, appropriate cryptographic algorithms should be applied that meet these measures and consumes less data processing time. In concise, trust in the IoT technologies are very important issues and full of challenges. Therefore, intuitively usable solutions are needed as these solutions must be easily integrated into the real use case scenarios. For that purpose, scholars, academicians, and technicians around the globe are working toward the common goal to design and develop the possible solutions that can support to mitigate such issues and challenges, or the risks associated with IoT technologies.

The findings of the study revealed that the most common issues and challenges that are raised in IoT in terms of trust are related to user trust, identity trust and management, data trust, integrity, scalability, heterogeneity, and infrastructures, and therefore, most of the solutions provided by researchers and IoT developers have been focused to provide a certain degree of trust that ensures reliable communication between trusted devices and build user trust in IoT environment as summarized in Table 1. Both stakeholders and IoT applications can mutually benefit from the establishment of a trusted IoT. Trust means establishing appropriate provisions for security and privacy. In addition, establishing and maintaining trust means meeting today's needs while providing appropriate future provisions to meet the evolving requirements and expectations of stakeholder.

6 Conclusion and Recommendations

In this paper, the most critical aspects of IoT were reviewed with certain focus on trust involved with this technology. Several challenges and issues related to the IoT trust are yet to be encountered. This paper briefly examines several issues such as confidentiality, authenticity, integrity, and other common issues that are associated to the user rights and privacy policies. The study also presents existing solutions from technical, academic, and industry aspects. The recommended solutions came in the form of new approaches, mechanisms, and architecture aimed to improving the quality of trust in IoT environment. Nevertheless, many other issues and challenges remain open that needs to be addressed and requires further thinking and harmonization. The IoT environment encompasses a complex set of technological, social, and policy considerations across a wide range of stakeholders. The technological developments that make it possible to use the IoT are real, growing, and innovative and are here to stay. Thus, academic institutions, governments, and industries should change their perspective toward the trust issues to mitigate the growing threats in IoT ecosystem.

References

1. Rehman, S.U., Manickam, S.: Significance of duplicate address detection mechanism in IPv6 and its security issues: a survey. Indian J. Sci. Technol. 8(30) (2015)
2. Jan, S., et al.: Applications and challenges faced by internet of things—a survey. Int. J. Eng. Trends Appl. (2016). ISSN: 2393-9516
3. Friess, P.: Internet of Things: Converging Technologies for Smart Environments and Integrated Ecosystems. River Publishers (2013)
4. Ul Rehman, S., Manickam, S.: A study of smart home environment and its security threats. Int. J. Reliab. Qual. Saf. Eng. 23(03), 1640005 (2016)
5. Folk, C., et al.: The Security Implications of the Internet of Things, p. 25 (2015)
6. Packard, H.: Internet of Things Research Study. HP Enterprise (2015)

7. Rehman, S.U., Manickam, S.: Denial of service attack in IPv6 duplicate address detection process. Int. J. Adv. Comput. Sci. Appl. **7**, 232–238 (2016)
8. Aldowah, H., Rehman, S.U., Umar, I.: Security in internet of things: issues, challenges and solutions. In: International Conference of Reliable Information and Communication Technology. Springer (2018)
9. Vasilomanolakis, E., et al.: On the security and privacy of internet of things architectures and systems. In: 2015 International Workshop on Secure Internet of Things (SIoT). IEEE (2015)
10. Rose, K., Eldridge, S., Chapin, L.: The Internet of Things: An Overview, pp. 1–50. The Internet Society (ISOC) (2015)
11. Levitt, T.: Internet of Things: IoT Governance, Privacy and Security Issues (2015)
12. Leister, W., Schulz, T.: Ideas for a trust indicator in the internet of things. In: SMART (2012)
13. Fritsch, L., Groven, A.-K., Schulz, T.: On the internet of things, trust is relative. In: International Joint Conference on Ambient Intelligence. Springer (2011)
14. Ion, M., et al.: A peer-to-peer multidimensional trust model for digital ecosystems. In: 2008 2nd IEEE International Conference on Digital Ecosystems and Technologies. IEEE (2008)
15. Daubert, J., Wiesmaier, A., Kikiras, P.: A view on privacy & trust in IoT. In: 2015 IEEE International Conference on Communication Workshop (ICCW). IEEE (2015)
16. Eder, T., Nachtmann, D., Schreckling, D.: Trust and Reputation in the Internet of Things (2013)
17. Evans, D.: The internet of things. In: How the Next Evolution of the Internet is Changing Everything, Whitepaper, vol. 1, pp. 1–12. Cisco Internet Business Solutions Group (IBSG) (2011)
18. Cho, J.H., Swami, A., Chen, R.: A survey on trust management for mobile ad hoc networks. IEEE Commun. Surv. Tutor. **13**(4), 562–583 (2011)
19. Sicari, S., et al.: Security, privacy and trust in internet of things: the road ahead. Comput. Netw. **76**, 146–164 (2015)
20. Roman, R., Najera, P., Lopez, J.: Securing the internet of things. Computer **44**(9), 51–58 (2011)
21. Baars, H., et al.: Foundations of Information Security Based on ISO27001 and ISO27002. Van Haren (2010)
22. Nahari, H.. Krutz, R.L.: Web Commerce Security: Design and Development. Wiley (2011)
23. Saichaitanya, P., Karthik, N., Surender, D.: Recent trends in IoT. Int. J. Electr. Electron. Eng. **8**(02), 9 (2016)
24. Odulaja, G.O., Awodele, O., Shade, K., Omilabu, A.A.: Security issues in the internet of things. Comput. Inf. Syst. Dev. Inform. Allied Res. J. **6**(1), 8 (2015)
25. Leister, W., Schulz, T.: Ideas for a trust indicator in the internet of things. In: SMART 2012: The First International Conference on Smart Systems, Devices and Technologies, p. 31 (2012)
26. Tormo, G.D., Mármol, F.G., Pérez, G.M.: Dynamic and flexible selection of a reputation mechanism for heterogeneous environments. Future Gener. Comput. Syst. **49**, 113–124 (2015)
27. Bernabe, J.B., Ramos, J.L.H., Gomez, A.F.S.: TACIoT: multidimensional trust-aware access control system for the internet of things. Soft. Comput. **20**(5), 1763–1779 (2016)
28. Chifor, B.-C., Bica, I., Patriciu, V.-V.: Sensing service architecture for smart cities using social network platforms. Soft Comput. 1–10 (2016)
29. Ruan, Y., Durresi, A., Alfantoukh, L.: Trust management framework for internet of things. In: 2016 IEEE 30th International Conference on Advanced Information Networking and Applications (AINA). IEEE (2016)
30. Gu, L., Wang, J., Sun, B.: Trust management mechanism for internet of things. China Commun. **11**(2), 148–156 (2014)
31. Maddar, H., Kammoun, W., Youssef, H.: Effective distributed trust management model for internet of things. Procedia Comput. Sci. **126**, 321–334 (2018)
32. Fernandez-Gago, C., Moyano, F., Lopez, J.: Modelling trust dynamics in the internet of things. Inf. Sci. **396**, 72–82 (2017)
33. Caminha, J., Perkusich, A., Perkusich, M.: A smart trust management method to detect on-off attacks in the internet of things. Secur. Commun. Netw. **2018** (2018)
34. Liu, Y., Gong, X., Feng, Y.: Trust system based on node behavior detection in internet of things. Tongxin Xuebao J. Commun. **35**(5), 8–15 (2014)

35. Rehman, S.U., Manickam, S.: Rule-based mechanism to detect denial of service (DoS) attacks on duplicate address detection process in IPv6 link local communication. In: 2015 4th International Conference on Reliability, Infocom Technologies and Optimization (ICRITO) (Trends and Future Directions). IEEE (2015)
36. Rehman, S.U., Manickam, S.: Improved mechanism to prevent denial of service attack in IPv6 duplicate address detection process. Int. J. Adv. Comput. Sci. Appl. **8**(2), 63–70 (2017)
37. Google Cloud Platform. Online: https://cloud.google.com/solutions/iot/. Accessed Dec 2016
38. Lee, J.H., Kim, H.: Security and privacy challenges in the internet of things [security and privacy matters]. IEEE Consum. Electron. Mag. **6**(3), 134–136 (2017)

Review on Common IoT Communication Technologies for Both Long-Range Network (LPWAN) and Short-Range Network

Abdullah Ahmed Bahashwan, Mohammed Anbar, Nibras Abdullah, Tawfik Al-Hadhrami, and Sabri M. Hanshi

Abstract The Internet of Things refers to network of physical objects that use IP address for Internet connectivity and to communicate with other Internet-enabled devices and systems. The Internet of Things comes with a number of top-class applications and has also developed many things into smart devices. The enhanced objects would ultimately need to have better connectivity and wireless communication protocols with low power consumption. Hence, this chapter presents the survey which gives a picture of the current state of the art on conventional wireless technologies used in most IoT devices, as well as to select the most suitable common IoT communication technologies that can ensure uninterrupted connection and support real-time data transmission in energy efficient form. Also, this chapter contributed a comprehensive comparison for both long-range and short-range common IoT communication technologies that are used in different applications.

Keywords IoT communication technologies · Cellular network · LoRaWAN · SigFox 6LoWPAN · Zigbee · Bluetooth · NFC · RFID · Z-wave

---

A. A. Bahashwan · M. Anbar · N. Abdullah
National Advanced IPv6 Center (NAv6), Universiti Sains Malaysia (USM), 11800 Gelugor, Penang, Malaysia
e-mail: bahashwan@student.usm.my

M. Anbar
e-mail: anbar@usm.my

N. Abdullah
e-mail: nibras@usm.my

T. Al-Hadhrami (✉)
School of Science and Technology, Nottingham Trent University, Nottingham, United Kingdom
e-mail: tawfik.al-hadhrami@ntu.ac.uk

S. M. Hanshi
Seiyun Community College, Hadhramaut, Yemen
e-mail: mhanshi@ieee.org

© The Editor(s) (if applicable) and The Author(s), under exclusive license to Springer Nature Singapore Pte Ltd. 2021
F. Saeed et al. (eds.), *Advances on Smart and Soft Computing*, Advances in Intelligent Systems and Computing 1188, https://doi.org/10.1007/978-981-15-6048-4_30

1 Introduction

"Internet of the Thing" was attributed to Ashton, who was said to have coined it the term in 1999 [1]. In addition, contemporary development in cyber technologies has highlighted the attention given to the Internet of Thing. Presently, IoT plays a very significant role in the advancement of both manufacturing and society at large [2]. It is estimated that there are about 5 billion smart things which are connected to the cloud, and the number of the things is estimated to rise to about 25 billion [3]. Both Cisco and Ericsson have predicted that there would be about 50 billion things [4, 5] as shown in Fig. 1. It also shows how personal devices such as laptops, cell phones, smartwatches, tablets, and other devices like asset tracking, taking care of the elderly at home, smart energy, and so on.

Conversely, the Internet of the Things, which is already leading to a new technological revolution, is the pivotal change in the direction of a universal connection between everything. This has also lead to the gathering of many sciences and technologies such as wireless sensor network and data analytic. Furthermore, there are general applications which are derived from IoT like smart watches and which work very well as fitness tracker and smart appliances. It is clear that the legacy Internet technology has become grossly inadequate in addressing human needs. As such, IoT as the competitor object has the capacity to complement these new technological innovations of the Internet since it can allow communications between smart objects which can ultimately lead to the hitherto utopian vision where anything can communicate. For this, IoT should be seen as part and parcel of the global Internet [6–10].

Moreover, it is very important to show the most differences and contributions of this survey that compares to other existing ones. The survey by [6] presents a review on IoT communication protocols with their features for long-range and short-range standers. However, the drawback of this study is not including the most common

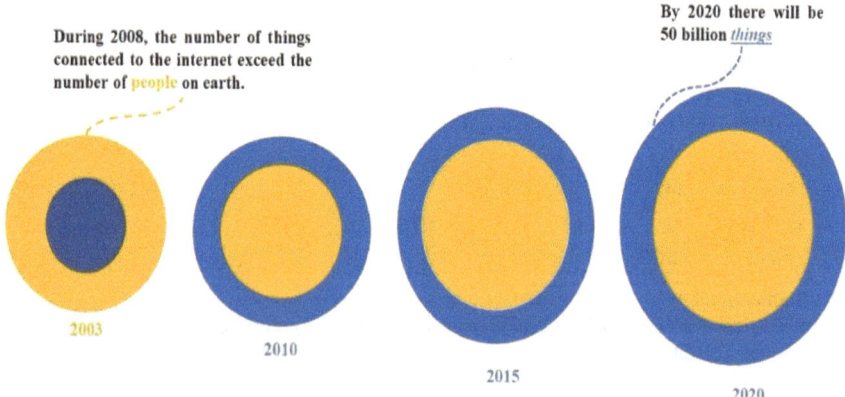

Fig. 1 IoT rising prediction [3]

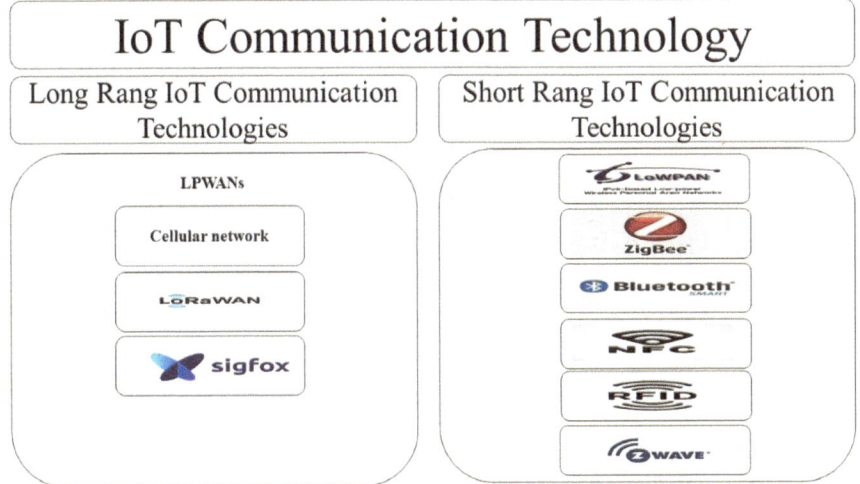

Fig. 2 Common IoT communication technologies

long-range technology such as LoRaWAN technology. Another survey by [16] just highlights the long-range technologies with their characteristics. However, the drawback of this survey mainly focus on to specific aspects of long-range technologies. One more survey by [4] just discusses the short-range low-power technologies of IoT. However, the drawback of this study just highlights the short-range technologies. Towards this end, this chapter is organized as follows. Section 2 covers the common IoT communication standards for long-range and short-range technologies. Section 3 presents the discussion, and finally, Sect. 4 illustrates the chapter conclusion (Fig. 2).

2 IoT Communication Standards

Generally, no restriction applies to the kind of technology that could be potentially be used in connecting the Internet of Things to the Internet or the most suitable solution for IoT service whose application is wireless communication standards. Any object that is connected is given dedicated proper protocol stacks, and the stacks are conceived to be convenient with the things, and it is then connected to the IP network (or gateway). Also, both applications and services are positioned on the connectivity type in two ways: locally or by using cloud computing services [11]. In particular, this section of this chapter discusses in details different communication technologies which are divided into long-range network (LPWAN) and short-range networks. Table 1 is abstracting comparison between long-range and short-range communication technologies.

Table 1 Comparison between long-range and short-range wireless standards of IoT [6, 7, 12, 27, 38, 39]

IoT communication standards

Parameter	Long-range IoT communication technologies			Short-range IoT communication technology					
	Cellular network	LoRaWAN	SigFox	6LoWPAN	Zigbee	Bluetooth LE	NFC	RFID	Z-wave
Network topology	NA	Star-of-stars	Star	Star and mesh	Star, mesh cluster network	Star, mesh, and P2P	P2P	P2P	Mesh
Data rate	NA	0.3–27 kbps	100 bps (UL), 600 bps (DL)	250 kbps	250 kbps	1 Mbps	424 kbps	4 Mbps	40 kbps
Network	WNAN	LPWAN	LPWAN	WPAN	WPAN	WPAN	P2P network	Proximity	WPAN
Frequency bands	Cellular band	Various, sub-gigahertz	868 or 902	2.4 GHz	2.4 GHz	2.4 GHz	860 MHz	902–928 MHz	868.42 MHz
Standard	3GPP and GSMA, GSM/GPRS/EDGE (2G), UMTS/HSPA LTE (4G)	LoRaWAN	SigFox	IEEE 802.15.4	IEEE 802.15.4	IEEE 802.15.4	ISO/IEC 14443 A&B, JIS X-6319-4	RFID	Z-wave
Coverage range (km)	5 km	1–10 km in rural, 1–3 km in urban	30–50 km in rural, 3–10 km in urban	10–10 m	10–100 m	15–30 m	0–10 cm	1.5 m	30 indoor 100 outdoor
Common application	M2M	Smart lighting pollution monitoring	Street lighting	Monitor and control over the Internet	Home monitoring and control	Many include a smartphone, headset	Smartphones E-ticket booking	Tracking inventory access	Home monitoring and control

(continued)

Table 1 (continued)

IoT communication standards

Parameter	Long-range IoT communication technologies			Short-range IoT communication technology					
	Cellular network	LoRaWAN	SigFox	6LoWPAN	Zigbee	Bluetooth LE	NFC	RFID	Z-wave
Features	High power consumption	Low power long range	Highly efficient energy and long range	Carry IPv6 datagrams on IEEE 802.15.4	Built on PHY and MAC layers of IEEE 802.15.4 and self-forming	Low power	Low power	Low power	Low power

2.1 Long-Range IoT Communication Technologies

Presently, wireless technologies are very crucial in interconnecting mobile devices as well as smart objects to each other or the cloud. Also, radio frequency wireless could be used to eliminate the wiring in view of these communication technologies since their features are suited to their speed, transmission type, security, cost, bandwidth reliability, and range [12]. This section is having common long-range IoT communications technologies with low-power wide area networks (LPWANs) for long-range transmission.

Cellular Network Cellular network technology is nothing but extremely high-speed transmission of data. By taking advantage of long-term evolution (LTE) mobile network, cellular network is used in deploying IoT device in long distance. Though this system consumes a lot of power, the interface gives a reliable communication and very high-speed access to the Internet. 100 km is the maximum of LTE. The technology could not be fitted into local area network or machine to machine (M2M). The cellular technology could be used for many applications, including mobile phone [6].

Nowadays, the standard cellular network technology is suitable to match M2M communication and IoT devices as well; it is otherwise known as long-term evolution advanced (LTE-A). Importantly, LTE-A is MAC layer access which uses orthogonal frequency-division multiple access (OFDMA) and which works to split the frequency into different bands, and each one is individually used. Three central elements made up the LTE-A architecture: the first one is radio access network (RAN) which is used for supervising the wireless communications and data planes. The next is the core network (CN) which manages the mobile devices and watch IPs [13]. The third one is the mobile nodes. The architecture of LTE-A is shown in Fig. 3.

LoRaWAN Stands for long-range wide area network (LoRaWAN) technology [14]. It is used in transmitting a few bytes of data over a long-range but with very low power consumption and the interval between 20 km in open places and 5 km in urban areas [15]. LoRa Alliances are the developers of LoRaWAN as well as its explicit criterion construction [16]. The technology gives middle access control and also allows to either communicate among themselves or to the gateways. LoRa's physical layer has some features such as long-range connecting, low data transmission, as well as lower power wireless connection. However, LoRaWAN is an unlicensed radio spectrum ISM band [17]. Therefore, star typology is the style of network used for LoRaWAN. Figure 4 shows the architecture of network.

As Fig. 4 shows, LoRaWAN architecture comprises of three central features [18]: End device: meaning any smart object can send and receive data. Gateway: which is the bridge between the smart objects and the network server. The portal is called referred to as modem or access point. The server is capable of analyzing the receiving date, routing, and forwarding data to an application server and managing the whole network. There are three classes of LoRaWAN: A, B, and C, and each

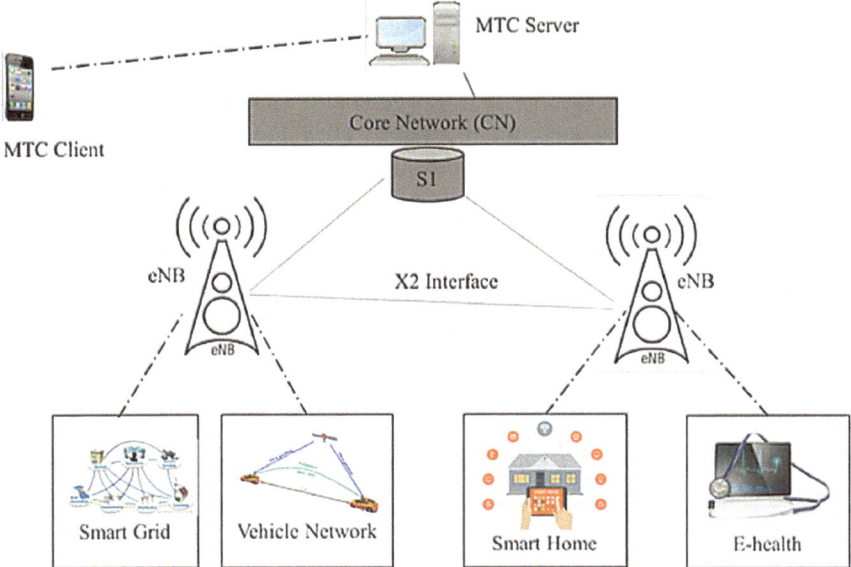

Fig. 3 LTE-A network architecture

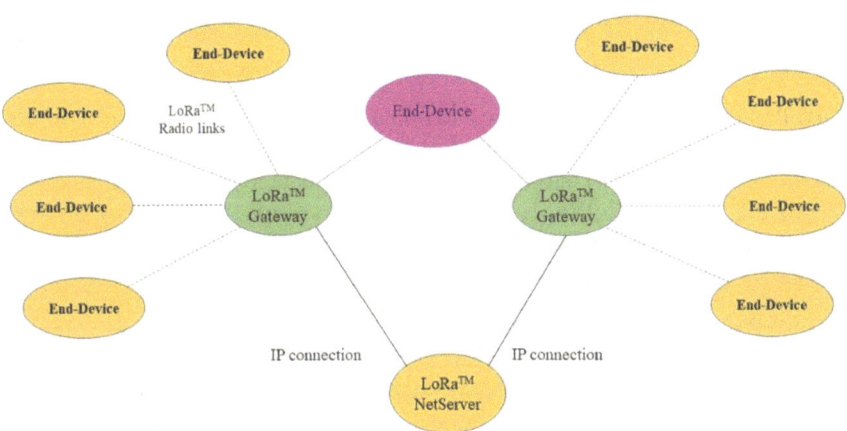

Fig. 4 LoRaWAN network

one of them have its skills. The devices in class A work on battery sensors and have two ways communication. The devices are very efficient because the end devices are in sleep modes often. Class B divides the up-link and down-link communications and decreases the latency. Devices under class B communicate with net server using beacon packets and listen to receive messages on a regular basis, while in class C, the end devices look for incoming messages and do not require latency for down-link transmission. However, in this class, power is not restricted, and this class is also used for real-time applications [19].

SigFox Technology Has been around since 2009 [20], and since then it has become very popular. In the beginning, the first model of SigFox only supports one-way communication, but now it supports bi-directional transmission. The technology can manage millions of smart objects with maximum long range 30–50 km which is contrary side and 3–10 km in a city. The technology is one of the low-power wide area wireless technologies since it can support many different applications such as M2M. Besides, its transmission rate is between 10 and 1000 bits per second, and its operates on battery while its style of network is star typology [6].

2.2 Short-Range IoT Communication Technology

The most significant block of the IoT model is the connectivity which is the same way communication technologies help us connect to smart objects and help us get premium services in the same way. On top of this, these technologies use very low-power network. In the next part of this chapter, we will highlight short-range IoT communications protocols including Pv6 over low-power wireless personal area network (6LoWPAN), Zigbee, Bluetooth, near-field communication (NFC), radio frequency identification (RFID), and Z-wave technology.

6LoWPAN Technology Before, most IoT data link frames are small and IP6 header is bigger, and there was an issue with that. Because of this issue, the Internet Engineering Task Force (IETF) introduced standard IPv6 datagrams for IoT applications which are known as 6LoWPAN [13, 21, 22]. With 6LoWPAN, we can use IPv6 packet in addition to low-power wireless network on IEEE 802.15.1. This is because IPv6 has a significant number of IPv6 address [23, 24].

6LoWPAN allows for the use of IPv6 packet in addition to low-power wireless network on IEEE 802.15.1 standard. The idea is done since the IPv6 has a significant number of IPv6 address. 6LoWPAN can support two types of network style such as star and mesh topology. Figure 5 shows the adaptation of 6LoWPAN layer between the data link and network layer and Zigbee protocol stack. Both 6LoWPAN and Zigbee use the same IEEE 802.15.4 protocol which is the most adversary [25]. It is very important to note that 6LoWPAN has some key features such low-power energy consumption and reliable transmission. Also 802.15.4 standard radio communication rate is 2.4 GHz which is very good for wireless communication, and the transmission rate is 250 kbps.

Zigbee Technology Zigbee Alliance is the developers of Zigbee technology [26]. The technology is a standard protocol for lower power wireless communication, and the radio frequency bands are 2.4 GHz, 915 MHz, and 868 MHz, while the transmission rate is 250 kbps. Also, the Zigbee objects work in low power or sleep mode and

Fig. 5. 6LoWPAN versus Zigbee protocol stack [18]

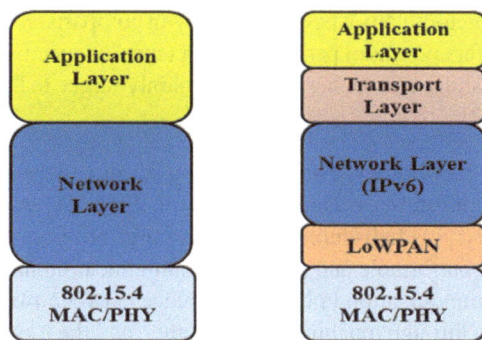

ZigBee Protocol Stack   6LoWPAN Protocol Stack

also work in batteries for a long period of time. Also, Zigbee is grounded on IEEE 802.15.4 standard to achieve high level of speed and energy efficient communication; it is also placed in the network layer. The technology is also suitable for personal area network with lower digital frequency. All in all, the Zigbee protocol supports different network style like star, tree, and mesh topology as illustrated in Fig. 6 [27, 28].

Bluetooth Technology A highly sophisticated technology which is integrated in many smart devices, Bluetooth, is used for short-range or in personal area network for sending and receiving data. Bluetooth ISM band is 2.400–2.485 GHz. Also, star topology is the network style for Bluetooth [29]. There are three types of Bluetooth:

- Bluetooth low energy (BLE) is popular technology which is used in IoT model as well in other smart devices in order to achieve low power energy, low latency, and low bandwidth [30]. Introduced in 2010, it is called Bluetooth smart [31].
- Typical Bluetooth, which is universal, is used in mobile phones with enough throughput. Bluetooth is also used for data stream application such as audio, and the limitation of standard Bluetooth can support seven slaves.

Fig. 6 Illustrated the Zigbee technology [25]

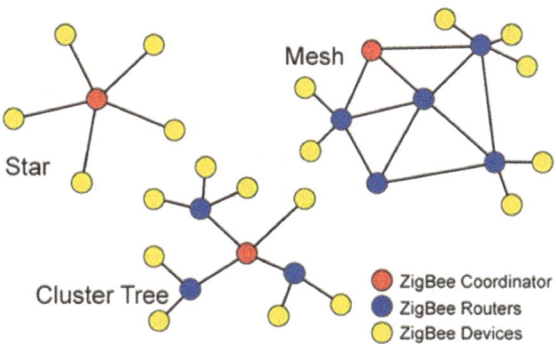

- Bluetooth 5 is a new technology in comparison with the older forms. Bluetooth 5 offers premium performance in terms of speed, range, and the capacity of broadcasting, and its new features mainly target IoT devices as well as smart phones [32].

NFC Technology Philips and Sony developed the near-field communication or NFC for connectionless communication in 2002 [33]. NFC is an extension of RFID technology; it is half-duplex and short-range wireless communication protocol. NFC is both comfortable and protected communication in a number of devices. The system of communication in NFC takes place between two nearby devices in a range of 4–10 cm through touching them together, and the topology is one to one. NFC operates in 13.56 MHz, and on the positive side, and it introduces a new type of services in mobile phones, in payment and smart homes or in accessing offices. Its protocol has two modes: the first being active where both devices use their power to transmit data, and both are active. The second is passive mode: one device is active, which is the initiator and produces the RF field, whereas the target is passive. There are also three modes in which NFC operate: reader/writer, card emulation, and peer-to-peer modes [34]. Also, the usability of NFC technology is fastly gaining ground because it is very simple to use. NFC has some key features: reduces cost by installing passive tags, low power consumption, and high flexibility.

RFID Technology Radio frequency identification or RFID consists of a reader, and the RF tag is small chip which uses wireless to transmit data. RFID tag sends the small size of data wireless to respond to the reader. RFID is a technology which is used for short-range communication standard. The technology is primarily used to identify objects and people automatically, and RFID tags are very well known since people used them always, for example, in poll payment, health care, agriculture, libraries, national security, as well as proximity cards. There are two categories of RFID tags: the active reader tag and the passive reader tag [35]; inactive tags are operated by batteries and use high radio frequencies, while passive tags are powerless and are used for fewer frequencies. RFID is critical in IoT applications, and the style of the RFID network is peer-to-peer topology [36].

Z-wave Technology Developed by Z-wave alliance [37], it is a wireless communication technology; it is very efficient in energy consumption. The technology can connect up to 30–50 devices which makes it an appropriate communication protocol for IoT objects. Specifically, it is very convenient for a light commercial environment and home automation. It is a reliable wireless transmission which is suitable for short messages sent from control unit, or between the other nodes within a network. It has five key layers for application, routing, transfer, medium access control (MAC), and physical layer (PHY). Moreover, its ISM band is 900 MHz and transmits at 9.6 and 40 kbps. Z-wave 400 series supports 2.4 GHz, and the bit rates are up to 200 kbps. The technology consists of two devices: the controller and the slaves; the controller guides the slaves while the slaves apply the commands which it gets from the controller. The controller keeps a routine table which shows the network topology in order to select the best route to the destination, and simultaneously, the slave's nodes work as router it passes the message to reach the other nodes [38].

3 Discussion

In the final analysis, the common IoT communication technologies are playing a vital role in activating IoT, connecting the sensor devices to IoT gateways and carry out data communication between these items. IoT communication technologies for both long range and short range are developed based on dissimilar standards to match the requirement of the applications, for example, coverage range, power consumption, data rate, and bandwidth. Therefore, there are a lot of communication technologies in IoT network, and every one associates with a unique features and benefits. This chapter contributed the main current and emerging IoT communication technologies for both long and short ranges to guide the researchers to select the right communication technology as their requirements.

4 Conclusion

In this chapter, the most common low power energy (LPWAN), as well as the short-range paradigm which are used for IoT wireless communication technologies, are discussed. These technologies are categorized based on their distance; they use low energy and perform efficiently. Many of the models are used for IoT since they are already verified and validated; each one of them has its own unique features and specifications. Therefore, there are many standards used for comparison between wireless communication protocols such as range, network topology, data rate, coverage range, and the power resources. All in all, this chapter is having the flexibility to be extended to many hopeful directions, for example, to review IoT communication technologies with their security issues and associated challenges.

References

1. Ashton, K.: That 'internet of things' thing. RFID J. 22(7), 97–114 (2010). https://www.rfidjournal.com/articles/view?4986
2. Bahashwan, A.O., Manickam, S.: A brief review of messaging protocol standards for internet of things (IoT). J. Cyber Secur. Mobil. 8(1), 1–14 (2018)
3. Kujur, P., Gautam, K.: Smart interaction of object on internet of things. Int. J. Comput. Sci. Eng. 3(2), 15–19 (2015)
4. Andersson, M.: Short range low power wireless devices and internet of things (IoT). Digi-Key Corporation R01, 1–16 (2015)

5. Alzubaidi, M., Anbar, M., Al-Saleem, S., Al-Sarawi, S., Alieyan, K.: Review on mechanisms for detecting sinkhole attacks on RPLs. In: 8th International Conference on Information Technology (ICIT), pp. 369–374. IEEE (2017)
6. Al-Sarawi, S., Anbar, M., Alieyan, K., Alzubaidi, M.: Internet of things (IoT) communication protocols. In: 8th International Conference on Information Technology (ICIT), pp. 685–690. IEEE (2017)
7. Amairah, A., Al-Tamimi, B.N., Anbar, M., Aloufi, K.: Cloud computing and internet of things integration systems: a review. In: International Conference of Reliable Information and Communication Technology, pp. 406–414. Springer, Cham (2018)
8. Al-Shalabi, M., Anbar, M., Wan, T.-C., Khasawneh, A.: Variants of the low-energy adaptive clustering hierarchy protocol: survey, issues and challenges. Electronics 7(8), 136 (2018)
9. Al-Fuqaha, A., Guizani, M., Mohammadi, M., Aledhari, M., Ayyash, M.: Internet of things: a survey on enabling technologies, protocols, and applications. IEEE Commun. Surv. Tutor. 17(4), 2347–2376 (2015)
10. Al-Shalabi, M., Anbar, M., Wan, T.-C., Alqattan, Z.: Energy efficient multi-hop path in wireless sensor networks using an enhanced genetic algorithm. Inf. Sci. 500, 259–273 (2019)
11. Pan, F., Li, L., Chen, X.: Decision making using a wavelet neural network based particle swarm optimization in stock transaction. ICIC Express Lett. 6(1), 9–14 (2012)
12. Saad, C., Cheikh, E.A., Mostafa, B., Abderrahmane, H.: Comparative performance analysis of wireless communication protocols for intelligent sensors and their applications. Int. J. Adv. Comput. Sci. Appl. 4(4), 76–85 (2014)
13. Salman, T., Jain, R.: Networking protocols and standards for internet of things. In: Internet of Things and Data Analytics Handbook, pp. 215–238 (2017)
14. LoRaWAN® Coverage | LoRa Alliance®: [Online]. Available: https://lora-alliance.org/lorawan-coverage. Accessed 15 Feb 2020
15. Basford, P.J., Bulot, F.M.J., Apetroaie-Cristea, M., Cox, S.J., Ossont, S.J.J.: LoRaWAN for smart city IoT deployments: a long term evaluation. Sensors 20(3), 648 (2020)
16. Home page | LoRa AllianceTM: [Online]. Available: https://lora-alliance.org. Accessed 03 Feb 2020
17. Foubert, B., Mitton, N.: Long-range wireless radio technologies: a survey. Future Internet 12(1), 13 (2020)
18. Vangelista, L., Zanella, A., Zorzi, M.: Long-range IoT technologies: the dawn of LoRa. In: Future Access Enablers for Ubiquitous and Intelligent Infrastructures, vol. 159, pp. 51–58. Springer, Cham (2015)
19. Navarro-Ortiz, J., Sendra, S., Ameigeiras, P., Lopez-Soler, J.M.: Integration of LoRaWAN and 4G/5G for the industrial internet of things. IEEE Commun. Mag. 56(2), 60–67 (2018)
20. Sigfox—The Global Communications Service Provider for the Internet of Things (IoT). [Online]. Available: https://www.sigfox.com/en. Accessed 03 Feb 2020
21. Alzubaidi, M., Anbar, M., Chong, Y.-W., Al-Sarawi, S.: Hybrid monitoring technique for detecting abnormal behaviour in RPL-based network. J. Commun. 13(5) (2018)
22. Alzubaidi, M., Anbar, M., Hanshi, S.M.: Neighbor-passive monitoring technique for detecting sinkhole attacks in RPL networks. In: Proceedings of the 2017 International Conference on Computer Science and Artificial Intelligence, pp. 173–182 (2017)
23. Mulligan, G.: The 6LoWPAN architecture. In: Proceedings of the 4th Workshop on Embedded Networked Sensors, pp. 78–82 (2007)
24. Alabsi, B.A., Anbar, M., Manickam, S., Elejla, O.E.: DDoS attack aware environment with secure clustering and routing based on RPL protocol operation. IET Circuits Devices Syst. 13(6), 748–755 (2019)
25. Lu, C.-W., Li, S.-C., Wu, Q.: Interconnecting ZigBee and 6LoWPAN wireless sensor networks for smart grid applications. In: 2011 Fifth International Conference on Sensing Technology, pp. 267–272. IEEE (2011)
26. Zigbee Alliance: [Online]. Available: https://www.zigbee.org. Accessed 03 Feb 2020
27. Samie, F., Bauer, L., Henkel, J.: IoT technologies for embedded computing: a survey. In: 2016 International Conference on Hardware/Software Codesign and System Synthesis, pp. 1–10. IEEE (2016)

28. Yang, Q., Huang, L.: Inside Radio: An Attack and Defense Guide, pp. 1–369. Springer (2018)
29. Celosia, G., Cunche, M.: Discontinued privacy: personal data leaks in Apple Bluetooth-low-energy continuity protocols. In: Proceedings on Privacy Enhancing Technologies 2020, no. 1, pp. 26–46 (2020)
30. Ghori, M.R., Wan, T.-C., Anbar, M., Sodhy, G.C., Rizwan, A.: Review on security in Bluetooth low energy mesh network in correlation with wireless mesh network security. In: 2019 IEEE Student Conference on Research and Development (SCOReD), pp. 219–224. IEEE (2019)
31. Fürst, J., Chen, K., Kim, H.S., Bonnet, P.: Evaluating Bluetooth low energy for IoT. In: 2018 IEEE Workshop on Benchmarking Cyber-Physical Networks and Systems (CPSBench), pp. 1–6. IEEE (2018)
32. Collotta, M., Pau, G., Talty, T., Tonguz, O.K.: Bluetooth 5: a concrete step forward toward the IoT. IEEE Commun. Mag. **56**(7), 125–131 (2018)
33. Home—NFC Forum I NFC Forum: [Online]. Available: https://nfc-forum.org/. Accessed 03 Feb 2020
34. Coskun, V., Ozdenizci, B., Ok, K.: A survey on near field communication (NFC) technology. Wireless Pers. Commun. **71**(3), 2259–2294 (2013)
35. Atlam, H.F., Alenezi, A., Alassafi, M.O., Alshdadi, A.A., Wills, G.B.: Security, cybercrime and digital forensics for IoT. In: Principles of Internet of Things (IoT) Ecosystem: Insight Paradigm, pp. 551–577. Springer, Cham (2020)
36. Sharma, M., Agrawal, P.C.: A research survey: RFID security & privacy issue. In: International Conference on Computer Science, Morelia, Mexico, pp. 255–261 (2013)
37. The Internet of Things is Powered by Z-Wave. [Online]. Available: https://z-wavealliance.org. Accessed 08 Feb 2020
38. Gomez, C., Paradells, J.: Wireless home automation networks: a survey of architectures and technologies. IEEE Commun. Mag. **48**(6), 92–101 (2010)
39. Motlagh, N.H., Mohammadrezaei, M., Hunt, J., Zakeri, B.: Internet of things (IoT) and the energy sector. Energies **13**(2), 1–27 (2020)

Information Security

An Overview of Blockchain Consensus Algorithms: Comparison, Challenges and Future Directions

Kebira Azbeg, Ouail Ouchetto, Said Jai Andaloussi, and Laila Fetjah

Abstract Like any other distributed system, the Blockchain technology relies on consensus algorithms in order to reach agreement and secure its network. Over the past few years, several kinds of consensus algorithms were created in the Blockchain ecosystem. In this paper, we present some main consensus algorithms used by the Blockchain technology. We also provide a comparison summary of their advantages and weaknesses as well as in which Blockchain type can be used. Furthermore, we present some challenges and possible directions related to the trendiest domains. With this work, for a given situation, we hope to help to choose the right algorithm to use. In addition, we hope to give directions to new possible algorithms propositions.

Keywords Blockchain · Distributed systems · Consensus algorithm

1 Introduction

Since the creation of Bitcoin 2008, which was the first implementation of the Blockchain technology by Nakamoto [1], the interest in this technology has been increased rapidly. Nowadays, Blockchain becomes one of the most interesting technologies in the world which witnessed an explosive growth during the past few years. Blockchain has started its popularity in the finance and banking sector [2], but it was quickly embraced by almost all life domains such as insurance [3], energy [4], health [5], government [6] and education [7]. The secret behind this success and the most

K. Azbeg (✉) · O. Ouchetto · S. Jai Andaloussi · L. Fetjah
Computer Science, Modeling Systems and Decision Support Laboratory, Faculty of Sciences
Ain-Chok, Hassan II University of Casablanca, Casablanca, Morocco
e-mail: azbegkebira@gmail.com

O. Ouchetto
e-mail: ouail.ouchetto@etude.univcasa.ma

S. Jai Andaloussi
e-mail: said.jaiandaloussi@etude.univcasa.ma

L. Fetjah
e-mail: leila.fetjah@etude.univcasa.ma

F. Saeed et al. (eds.), *Advances on Smart and Soft Computing*, Advances in Intelligent
Systems and Computing 1188, https://doi.org/10.1007/978-981-15-6048-4_31

357

attractive thing in the Blockchain are that it combines a set of interesting technologies and mechanisms that ensure its working. Blockchain can be defined as a decentralized encrypted database which is distributed across a peer-to-peer network without need of a central authority to control and secure it. The information stored in the Blockchain is protected and validated by a mechanism called consensus.

Consensus algorithms refer to the different mechanisms that are used in order to reach agreement and ensure the security in a distributed system. This kind of systems faces a fundamental problem similar to the Byzantine Generals Problem which relies on how to achieve consensus in the presence of a number of faulty and malicious participants. As it is based on a decentralized and distributed network, Blockchain needs also such algorithms to handle data and reach consensus. So that, every participant in the Blockchain network can have the same database state. There are several kinds of consensus algorithms implemented by different Blockchains. The most common algorithms are Proof of Work, which is the first consensus algorithm implemented by the first Blockchain (Bitcoin), Proof of Stake and Delegated Proof of Stake. Thus, we can say that the Proof of Work is the original consensus algorithm used by the Blockchain and all the other algorithms are just an alternative trying to overcome the weaknesses and limitations of the Proof of Work. In addition to these three algorithms, several other popular consensus algorithms are presented in this paper. The paper gives also a comparison summary between the different algorithms and some current challenges. This work aims to give insights that could lead to new directions for creating new possible algorithms. It could also help to choose the right algorithm to use according to each situation.

Our paper is organized as follows. Section 2 gives a brief explanation of the Blockchain technology. In Sect. 3, we describe the most popular consensus algorithms in the Blockchain world. Section 4 gives a brief comparison between the different treated consensus algorithms. Section 5 presents some challenges and possible directions, and Sect. 6 concludes the paper.

2 Blockchain

The Blockchain technology was firstly invented in 1991 by Haber and Stornetta [8] who proposed a solution to protect electronic documents. The solution is based on time stamping and linking hashes of documents in order to create a non-tampering with system. But it is only in 2008 that Blockchain had its popularity with its first application in Bitcoin. It was known as the technology behind Bitcoin, the first cryptocurrency, created by Satoshi Nakamoto. Blockchain is, then, a technology that provides both storage and data transmission. It is based on a peer-to-peer network to ensure communication between nodes in a transparent and secure way without need of a central control authority. Each node has a copy of the database called also ledger Fig. 1a. Data is grouped in blocks, and every block is linked with the previous one by a cryptographic hash Fig. 1b. Blockchain is basically composed of five components:

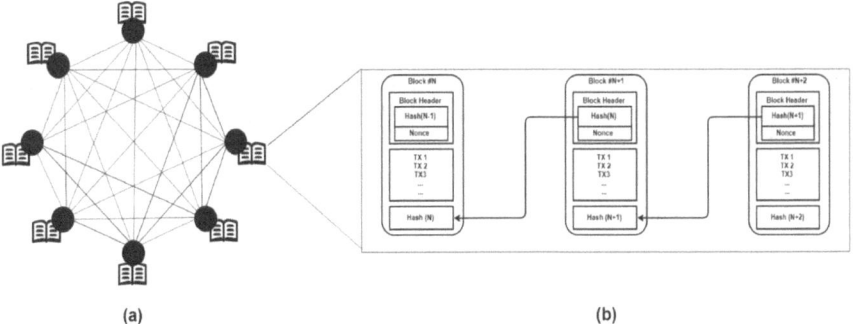

Fig. 1 Blockchain architecture

peer-to-peer network, a distributed ledger, cryptography, consensus mechanisms and smart contracts.

There are three types of Blockchains: public or permissionless Blockchain, private or permissioned Blockchain and consortium Blockchain. The main differences between the three types are the ability and the rights to read, write and validate a block. The public Blockchain type is the one used in cryptocurrency like Bitcoin and Ethereum [9]. It allows anyone to join the network, read and write to the ledger and participate in the consensus process. In the second type, a single authority is controlling access to the network and the consensus process. To join the network, nodes need permission from the authority. The last type is a hybrid between the public and the private one. In this case, we can have several organizations controlling the consensus process instead of a single one.

3 Consensus Algorithms

In a distributed system, the agreement is a fundamental problem, and it is generally illustrated by the Byzantine Generals Problem [10]. The problem is copied from the Generals of the Byzantine army who encamp around an enemy city and wish to attack it. They can communicate only using messengers and must establish a common battle plan. The idea is to coordinate an attack at a specific moment or to retreat safely from the city. The Generals share what they will do by sending the "attack" message to confirm the attack and "retreat" to cancel it. However, a number of these Generals may be traitors trying to confuse the military. Thus, they send the message "retreat" to convince some loyal Generals to retreat at the time of the attack and to cause defeat. The problem is to find a strategy to make sure that all loyal Generals can agree on a battle plan. Traitors will anyway retreat, but since their number is supposed to be small, the attack will be a success. To overcome this problem in a distributed system, it should implement a consensus algorithm in order to have Byzantine fault tolerance (BFT) which means to tolerate the faults caused by the Byzantine Generals Problem.

Like in any other distributed system where there is no trusted central authority that ensures the system security. Blockchain also uses consensus mechanisms to reach agreement and maintain the consistency of shared data. Thus, Blockchain is seen as a trust machine based on consensus algorithms that enhance the security among untrusted nodes. These algorithms can also protect it against malicious attacks by validating transactions and creating new blocks that will be broadcasted and adopted by all nodes participating which is the Blockchain network. We present below the most common mechanisms that are used to reach consensus in Blockchain.

3.1 Proof of Works

Proof of Work (PoW) is originally proposed by Dwork and Naor in 1993 [11] to combat junk mail and control access to shared resources. In 2008, Bitcoin integrates this idea in its protocol as a consensus algorithm to reach consensus on transactions and to prevent the double spending attack. Thus, PoW is the original and the first consensus algorithm used by Blockchain. Its main idea is to push miners to use their computing power in order to solve a complex puzzle based on a mathematical problem. Once the problem is solved, the miner broadcasts the block to the network so that the other miners can check if the solution is correct. If so, the block is added to the Blockchain. Technically speaking, the task of a miner is to find a random number called nonce, which is located in the block header. To do so, the miner increments the nonce until he finds a block hash which begins with a specific number of zero bits. The number of zero prefixing the hash is defined by a parameter called difficulty. The block hash is calculated by concatenating the data grouped in the block with the nonce found and passing the result to a hash function Fig. 2. The difficulty is adjusted to the total computing power used in the network; the more power is added to the network, the more difficulty is increased. In this way, a new block is generated in an average time. In Bitcoin, the average time spent to generate a new block is 10 min.

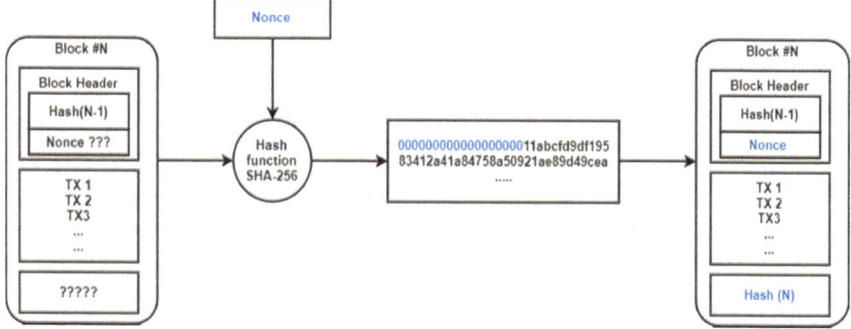

Fig. 2 PoW/nonce generation

In addition to Bitcoin, many other cryptocurrencies have adopted the PoW such as Ethereum [12] and Litecoin [13].

3.2 Proof of Stake

Proof of Stake (PoS) is another consensus algorithm which is firstly implemented by the Peercoin cryptocurrency proposed by King and Nadal in 2012 [14]. Instead of a competitive process that depends on energy consumption, PoS is based on a selection process that takes into account the validators' stake. The validators in a PoS system are the equivalent of miners in a PoW system. The selection process allows the network to choose the node that will validate the new block by proving its ownership of an amount of coins using the "coin age" metric. Roughly speaking, the coin age is the multiplication of currency amount by the duration of its possession. Once a validator is selected, he validates the block by placing a bet on it and he gets a reward proportionate to the bet he placed. Despite being the first cryptocurrency that implements the PoS, Peercoin does not use the PoS in 100%, and it also uses the PoW to facilitate the initial creation of the currency. Thus, Peercoin is a hybrid system combining both PoW and PoS. An example of a cryptocurrency that uses PoS in a purely way without combining it with PoW is Nxt [15]. We have also Ethereum which is planning to switch to the PoS in its 2.0 version which is called Casper protocol [16].

3.3 Delegated Proof of Stake

Delegated Proof of Stake (DPoS) is a decentralized consensus algorithm created by Larimer [17]. It implements a layer of technological democracy to offset the negative effects of centralization [18, 19]. It uses a reputation system, and it is based on an election process to reach a decentralized voting. The main idea is to appoint a set of delegates (known also as witnesses) to secure the network on the behalf of the other shareholders. Delegates take turns to create blocks in a random manner. If a delegate fails to create a block properly, it loses its reputation. Therefore, shareholders can pull back their votes in favor of that delegate and then replace it by another delegate. Various versions of DPoS are currently in use by many Blockchains like BitShares [20], EOS [21] and Lisk [22].

3.4   Proof of Authority

Proof of Authority (PoA) is a consensus mechanism that allows only predefined authorities to validate transactions and append blocks to the Blockchain. This algorithm consists of designating a set of nodes to be validators, and it gives them the authority to update the ledger and secure the Blockchain. This leads to a kind of agreement centralization at the hands of a small number of known actors. Thus, PoA is often used by private and consortium Blockchains and authorities are considered as trusted nodes. Instead of coins, PoA is based on identity and reputation as a stake. Because of their known identity, the authorities cannot behave maliciously to protect their reputation. Ethereum Blockchain has implemented this algorithm in two clients Parity [23] and Geth [24]. The Parity implementation is named Aura, and the Geth implementation is named Clique [25].

3.5   Proof of Capacity/Proof of Space

Proof of Capacity (PoC) also called Proof of Space (PoSpace) was first proposed on a paper titled "Proofs of Space" [26] as an alternative concept of PoW. The idea in this concept is to dedicate an amount of disk space for mining instead of computation power like in PoW. Thus, for example in BurstCoin [27] which is a cryptocurrency based on PoC rather than doing a lot of work in order to verify blocks, like we do in PoW, the work is done upfront using a process called plotting. During this process, the miner generates plots (files) storing a large number of hashes that are computed in advance using different nonces. These hashes can then be reused in the mining process. Other Blockchains that use PoSpace are SpaceMint [28] and Chia [29].

3.6   Practical Byzantine Fault Tolerance

The practical Byzantine fault tolerance (PBFT) is a BFT algorithm proposed by Castro and Liskov in 1999 [30]. It is destined for asynchronous systems, and it is based on several optimizations in order to improve the response time. With this algorithm, the system can handle f malicious nodes if we suppose that the system has $3f + 1$ nodes in total. The algorithm is based on state machine replication. The network is split into two types of nodes (primary and backups) in each PBFT consensus round (called view). Then, common patterns of decisions are looked for. The nodes are called replicas, and in each view, one node is primary and the others are backups. The algorithm consists of five phases as illustrated in Fig. 3:

- Request: The client submits a request to the primary node.
- Pre-prepare: The primary node assigns a sequence number to the request and broadcasts it to all backups.

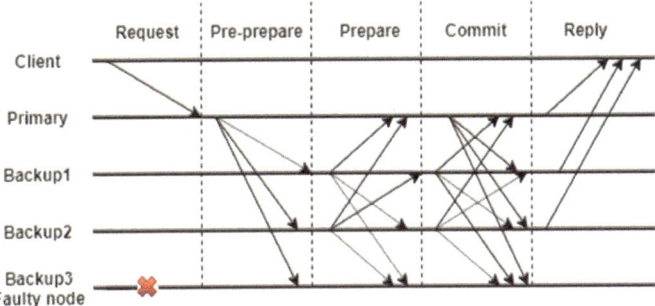

Fig. 3 PBFT phases

- Prepare: If a backup node accepts the request, it sends a prepare message to all other nodes. After collecting $2f + 1$ messages, if the majority accepts the request, nodes will start the commit phase.
- Commit: All nodes send a commit message. If a node receives $f + 1$ valid commit message, then they carry out the client request.
- Reply: Nodes send out the reply to the client, and then the client waits at least for $f + 1$ of the same reply to get the correct response.

There are many variants of this algorithm which is implemented by several Blockchains like Hyperledger [31].

3.7 Proof of Elapsed Time

The Proof of Elapsed Time (PoET) is another alternative way to achieve consensus that does not require high-energy consumption. The network assigns a random waiting time to each node. This time determines how long a node has to wait before being allowed to generate a new block [32]. The node with the shortest time wins first. The participating nodes in the network must run Software Guard Extensions (SGX) which is a trusted execution environment developed by Intel [33].

3.8 Proof of Burn

The Proof of Burn (PoB) is proposed firstly by Stewart [34, 35] in 2012 to be an alternative to PoW and PoS. The main idea in this algorithm is to burn coins in order to have the chance to mine a new block. Thus, nodes send coins to a burn address which allows only the reception of coins. These coins are unspendable, and we consider that they are destroyed or lost. The chance to get selected to mine increases with the amount of burned coins. This mechanism can also be used to bootstrap one cryptocurrency off of another. For example, one of the cryptocurrencies that are using the PoB is Counterparty [36], and it allows participants to send Bitcoins to an unspendable Bitcoin address in order to received Counterparty tokens (Known as XCP currency) in exchange.

3.9 Proof of Weight

The main idea in the Proof of weight (PoWeight) is to give a weight for each node based on some measures. The first example of this algorithm is Algorand cryptocurrency [37] which is based on a consensus model that consists of giving weight to users according to the amount of money that they have in their accounts. Filecoin [38] which is a decentralized storage network, for example, instead of giving weight based on the amount of coins, it uses the amount of IPFS data that each node stores in the network. Thus, the node having the highest weight (amount of stored data) has more chance to create a new block.

3.10 HotStuff

HotStuff is a leader-based Byzantine fault-tolerant replication protocol proposed by VMware Research [39] and destinated to the partially synchronous model. The algorithm considers that there are n replicas $n = 3f + 1$ where f is the number of replicas controlled by an adversary. It works in iterations of views, and each view has a single primary node. The algorithm is based on four phases:

- Prepare phase: The primary node starts by collecting new view messages from replicas, then sends a prepare message to all replicas.
- Pre-commit phase: Once the primary node receives $(n - f)$ prepare votes, he combines them in a prepare quorum certificate and broadcasts it in a pre-commit message. The replica sends, then, a pre-commit vote to the primary node.
- Commit phase: When the primary receives $(n - f)$ pre-commit votes, he prepares a pre-commit quorum certificate and sends a commit message to replicas. Replicas respond to the primary by a commit vote.
- Decide phase: Once the primary receives $(n - f)$ commit votes, he prepares a commit quorum certificate and broadcasts it to all replicas in a decide message. When a replica receives a decide message, he executes the commands for the current state transition in the committed branch and starts the next view.

This algorithm was the backbone behind the LibraBFT [40] consensus algorithm used by the Libra cryptocurrency created by Facebook.

4 Consensus Algorithms Comparison

In this section, we present a comparison of the most popular consensus algorithms used in Blockchain and discussed in the previous section. Table 1 gives summary about the advantages and weaknesses of each algorithm with the Blockchain type where it can be used the most. Thus, based on what we want to achieve in our Blockchain, we decide what kind of Blockchain network we need. Then, we choose the appropriate consensus protocol for our needs. For example, if we need to use

Table 1 Comparison of the most popular consensus algorithm in Blockchain

Consensus algorithm	Blockchain type	Advantages	Weaknesses/limitations
PoW	Public Blockchain	• Steady (since 2009)	• Energy consumption • Slow
PoS	Public Blockchain	• Fast • Energy-efficient	• Validators with large stakes could lead to centralization • Nothing at stake problem
DPoS	Public Blockchain	• Fast • Energy-efficient	• Concentrated voting power could lead to centralization
PoA	Private and consortium Blockchains	• Fast • Energy-efficient	• More centralized
PoC/PoSpace	Public Blockchain	• Flexibility in using any hard drive • Energy-efficient	• Possibility of malware attack • Possibility of a centralized storage
PBFT	Private and consortium Blockchain	• Energy-efficient	• Inefficient for large networks • Susceptible to Sybil attacks
PoET	Private and consortium Blockchains	• Energy-efficient	• Necessity of a specialized hardware security (reliance on Intel)
PoB	Public Blockchain	• Reduce power consumption	• Waste of energy (in the case of burning Bitcoins generated by PoW)
PoWeight	Public Blockchain	• Energy-efficient • Highly customizable	• Incentivization can be hard
HotStuff	Private and consortium Blockchain	• Energy-efficient	• Not fully decentralized

a public Blockchain, we can choose one of these algorithms: PoW, PoS, DPoS, PoC, PoB or PoWeight. The choice of the appropriate algorithm is based on our requirements and on the parameter that we need most. If our priority, for example, is to ensure a height security level and we have enough computation power, so in this case the appropriate algorithm could be the PoW. In other cases, the priority could be the speed of block validation. In these cases, we can use, for example, the PoS or DPoS and so forth.

5 Challenges and Possible Directions

Based on the weaknesses and limitations seen in the previous section, we can divide the challenges facing Blockchain consensus algorithms into several categories:

- Energy consumption: The mining process consumes a vast amount of power. For example, according to a study about the cost of cryptocurrencies mining [41] done between January 1, 2016, and June 30, 2018, they estimated that for the generation of one US$, Bitcoin consumed an average of 17 MJ, Ethereum 7 MJ and Litecoin 7 MJ. Comparatively, the generation of one US$ using mineral mining is consuming less energy than cryptocurrency mining. For example, the mining of copper, gold, platinum and rare earth oxides required 4, 5, 7 and 9 MJ, respectively. Several solutions were proposed to overcome this problem by either implementing alternative algorithms such as PoS or using alternative sources of power like solar and wind energy.
- Centralization: Despite that the Blockchain technology is decentralized by design, some consensus algorithms can lead to centralization. This could lead to security problems because of the monopoly of consensus and concentrating the Blockchain governance in limited nodes.
- Speed: The speed is related to the time needed for transaction validation process and the time spent for block creation. This depends on the algorithm used to choose the node which will create the block and how it proceeds to validate a transaction.
- Attacks: Each consensuses mechanism could have a vulnerability which could be exploited to attack the Blockchain. The most known attacks are 51% attack and selfish mining attack [42]. The 51% attack is possible when a mining node has a power which is greater than 50% of the total hashing power in the whole network. In the second attack, the selfish miner does not publish the block he mined to the rest of the network. He then continues to create the next block and maintains his lead to create his private branch so that he can create a fork. Once his branch is longer to the main one, he then publishes all blocks that he created and the honest miners must to adopt his branch instead of the main one.

In addition to these challenges, we can add the scalability issue; the scalability of a consensus algorithm is referring to its ability of scaling to large numbers of nodes (network size). It refers also to the transaction throughput that an algorithm offers. All these criteria should be taken into account when choosing or developing a consensus algorithm.

In addition to the financial sector, Blockchain has known a great interest in other domains like Internet of things (IoT) and artificial intelligence (AI). For example, many researches surveyed the challenges facing the integration of Blockchain with IoT [43]. One of the most challenges is the participation of these devices in the mining process, as we know, the Internet of things devices have limited resources and low power. That is making them unsuitable for participating in the consensus process. For that, researches should be directed to solve this problem by creating consensus algorithms specially designed for IoT devices. Another future direction is to use artificial intelligence for designing new consensus algorithms. For example, we can use artificial intelligence to choose the node which will create the new block or to verify transactions.

6 Conclusion

Blockchain is the technology behind the first cryptocurrency in the world "Bitcoin", created in 2008 by Satochi Nakamoto. It is seen as a technology that allows storage in a decentralized ledger and a distributed database and allows communication between nodes through a peer-to-peer network. The nature of its system pushed the Blockchain to use consensus mechanisms to secure and maintain its database. Over the past few years, Blockchain ecosystem has developed different consensus algorithms. In this paper, we presented a set of the most popular consensus algorithms used in the Blockchain technology. We started with an explanation of the Blockchain technology. Then, we provided a definition of each algorithm and how it works. In the end, we made a comparison between different algorithms by outlining the advantages, weaknesses and limitations of each one and in which Blockchain type can be used. Thus, the choice of the right consensus algorithm is depending on the Blockchain type and on the application nature. We also presented some challenges facing these algorithms in order to take them into consideration when developing new algorithms. For future possible directions, we considered the most trending domains, namely Internet of things and artificial intelligence.

Acknowledgements This research was supported by the National Center for Scientific and Technological Research (CNRST).

References

1. Nakamoto, S.: A peer-to-peer electronic cash system. Bitcoin. https://bitcoin.org/bitcoin.pdf (2008)
2. Moore, T.: The promise and perils of digital currencies. Int. J. Crit. Infrastruct. Prot. **6**, 147–149 (2013). https://doi.org/10.1016/j.ijcip.2013.08.002
3. Raikwar, M., Mazumdar, S., Ruj, S., Sen Gupta, S., Chattopadhyay, A., Lam, K.-Y.: A blockchain framework for insurance processes. In: 2018 9th IFIP International Conference on New Technologies, Mobility and Security (NTMS), pp. 1–4. IEEE, Paris (2018)
4. Li, Y., Yang, W., He, P., Chen, C., Wang, X.: Design and management of a distributed hybrid energy system through smart contract and blockchain. Appl. Energy **248**, 390–405 (2019). https://doi.org/10.1016/j.apenergy.2019.04.132
5. Mettler, M.: Blockchain technology in healthcare: the revolution starts here. In: 2016 IEEE 18th International Conference on e-Health Networking, Applications and Services (Healthcom), pp. 1–3. IEEE, Munich, Germany (2016)
6. Ølnes, S., Ubacht, J., Janssen, M.: Blockchain in government: benefits and implications of distributed ledger technology for information sharing. Gov. Inf. Q. **34**, 355–364 (2017). https://doi.org/10.1016/j.giq.2017.09.007
7. Han, M., Li, Z., He, J. (Selena), Wu, D., Xie, Y., Baba, A.: A novel blockchain-based education records verification solution. In: Proceedings of the 19th Annual SIG Conference on Information Technology Education—SIGITE'18, pp. 178–183. ACM Press, Fort Lauderdale, Florida, USA (2018). https://doi.org/10.1145/3241815.3241870
8. Haber, S., Stornetta, W.S.: How to time-stamp a digital document. In: Conference on the Theory and Application of Cryptography, pp. 437–455. Springer, Berlin, Heidelberg (1990)
9. Wood, G.: Ethereum: a secure decentralised generalised transaction ledger. Ethereum Project Yellow Paper 151 (2014)
10. Lamport, L., Shostak, R., Pease, M.: The Byzantine generals problem. ACM Trans. Program. Lang. Syst. **4**(3), 382–401 (1982)
11. Dwork, C., Naor, M.: Pricing via processing or combatting junk mail. In: Brickell, E.F. (ed.) Advances in Cryptology—CRYPTO'92, pp. 139–147. Springer Berlin Heidelberg, Berlin, Heidelberg (1993). https://doi.org/10.1007/3-540-48071-4_10
12. Ethereum. https://ethereum.org/. Accessed 06 Dec 2019
13. Litecoin—open source P2P digital currency. https://litecoin.org/. Accessed 06 Dec 2019
14. King, S., Nadal, S.: Ppcoin: peer-to-peer crypto-currency with proof-of-stake. Self-Published Paper, 19 Aug 2012
15. Nxt community: Nxt Whitepaper. https://whitepaper.io/document/62/nxt-whitepaper. Accessed 10 Feb 2020
16. Learn about Ethereum. https://ethereum.org/learn/#proof-of-work-and-mining. Accessed 06 Dec 2019
17. Schuh, F., Larimer, D.: Bitshares 2.0: Financial Smart Contract Platform. 12
18. BitShares Blockchain Foundation: The BitShares Blockchain. https://github.com/bitshares-foundation/bitshares.foundation/blob/master/download/articles/BitSharesBlockchain.pdf
19. Delegated Proof of Stake (DPOS)—BitShares Documentation. https://docs.bitshares.org/en/master/technology/dpos.html. Accessed 06 Dec 2019
20. Bitshares-foundation/bitshares.foundation. https://github.com/bitshares-foundation/bitshares.foundation. Accessed 06 Dec 2019
21. EOSIO—Blockchain Software Architecture. https://eos.io/. Accessed 06 Dec 2019
22. Slasheks/lisk-whitepaper. https://github.com/slasheks/lisk-whitepaper. Accessed 06 Dec 2019
23. Parity Technologies. https://www.parity.io/ethereum/. Accessed 06 Dec 2019
24. Go Ethereum. https://geth.ethereum.org/. Accessed 06 Dec 2019
25. Go Ethereum. https://geth.ethereum.org/docs/interface/private-network. Accessed 19 Dec 2019
26. Dziembowski, S., Faust, S., Kolmogorov, V., Pietrzak, K.: Proofs of space. In: Annual Cryptology Conference, pp. 585–605. Springer, Berlin, Heidelberg (2015)

27. Gauld, S., von Ancoina, F., Stadler, R.: The burst dymaxion (2017)
28. Park, S., Kwon, A., Fuchsbauer, G., Gaži, P., Alwen, J., Pietrzak, K.: SpaceMint: a cryptocurrency based on proofs of space. In: Meiklejohn, S., Sako, K. (eds.) Financial Cryptography and Data Security, pp. 480–499. Springer Berlin Heidelberg, Berlin, Heidelberg (2018). https://doi.org/10.1007/978-3-662-58387-6_26
29. Cohen, B., Pietrzak, K.: The Chia Network Blockchain. 44
30. Castro, M., Liskov, B.: Practical Byzantine fault tolerance. In: OSDI, vol. 99, no. 1999, pp. 173–186 (1999)
31. Seeley, L., Io, B.: Introduction to Sawtooth PBFT. https://www.hyperledger.org/blog/2019/02/13/introduction-to-sawtooth-pbft. Accessed 06 Dec 2019
32. Chen, L., Xu, L., Shah, N., Gao, Z., Lu, Y., Shi, W.: On security analysis of proof-of-elapsed-time (PoET). In: Spirakis, P., Tsigas, P. (eds.) Stabilization, Safety, and Security of Distributed Systems, pp. 282–297. Springer International Publishing, Cham (2017). https://doi.org/10.1007/978-3-319-69084-1_19
33. Intel® Software Guard Extensions (Intel® SGX). https://www.intel.com/content/www/fr/fr/architecture-and-technology/software-guard-extensions.html. Accessed 06 Dec 2019
34. User: Ids—Bitcoin Wiki. https://en.bitcoin.it/wiki/User:Ids. Accessed 06 Dec 2019
35. Proof of Burn—Bitcoin Wiki. https://en.bitcoin.it/wiki/Proof_of_burn. Accessed 06 Dec 2019
36. Counterparty. https://counterparty.io/. Accessed 06 Dec 2019
37. Gilad, Y., Hemo, R., Micali, S., Vlachos, G., Zeldovich, N.: Algorand: scaling Byzantine agreements for cryptocurrencies. In: Proceedings of the 26th Symposium on Operating Systems Principles—SOSP'17, pp. 51–68. ACM Press, Shanghai, China (2017). https://doi.org/10.1145/3132747.3132757
38. Protocol Labs: Filecoin: A Decentralized Storage Network (2017)
39. Yin, M., Malkhi, D., Reiter, M.K., Gueta, G.G., Abraham, I.: HotStuff: BFT consensus in the lens of blockchain. arXiv:1803.05069 (2019)
40. Consensus Libra: https://developers.libra.org/. Accessed 06 Dec 2019
41. Krause, M.J., Tolaymat, T.: Quantification of energy and carbon costs for mining cryptocurrencies. Nat. Sustain. 1, 711–718 (2018). https://doi.org/10.1038/s41893-018-0152-7
42. Li, X., Jiang, P., Chen, T., Luo, X., Wen, Q.: A survey on the security of blockchain systems. Future Gener. Comput. Syst. S0167739X17318332 (2017). https://doi.org/10.1016/j.future.2017.08.020
43. Reyna, A., Martín, C., Chen, J., Soler, E., Díaz, M.: On blockchain and its integration with IoT. Challenges and opportunities. Future Gener. Comput. Syst. 88, 173–190 (2018). https://doi.org/10.1016/j.future.2018.05.046

Critical Information Infrastructure Protection Requirement for the Malaysian Public Sector

Saiful Bahari Mohd Sabtu and Kamaruddin Malik Mohamad

Abstract Malaysian Public Sector (MPS) organizations span laterally within all ten critical sectors outlined in the National Cybersecurity Policy (NCSP). Critical Information Infrastructure Protection (CIIP) initiatives are mainly driven by the overarching NCSP and the National Cybersecurity Strategy. Down the hierarchy, CIIP initiatives become more focused at the sectoral level. However, a dedicated CIIP framework for the MPS is currently unavailable thus giving an opportunity for research in this area. This paper explores current CIIP requirements pertinent to MPS sectoral needs. The method used is comparative analysis. In this paper, analyzed resources include international organizations requirements, key national policy documents, published official directives, circulars, guidelines and tools related to the MPS CIIP. The study findings have shown that risk management and resilience are among the emerging themes. A total of 21 external strategic requirements and 26 available internal resources are identified. A comparison of MPS Cybersecurity Framework (RAKKSSA) against NIST Cybersecurity Framework is also established to highlight CIIP. For future work, five recommendations are proposed as guidelines for developing MPS CIIP Framework.

Keywords Critical Information Infrastructure Protection · Information security management · Cybersecurity

1 Introduction

Protecting Critical Information Infrastructure (CII) is vital to the national security and contributes to the preservation of a nation's sovereignty. Critical Information Infrastructure Protection (CIIP) can be defined as all activities aimed at ensuring the functionality, continuity, and integrity of CII to deter, mitigate, and neutralize

S. B. M. Sabtu (✉) · K. M. Mohamad
Universiti Tun Hussein Onn Malaysia, 86400 Parit Raja, Johor, Malaysia
e-mail: gi160011@siswa.uthm.edu.my

K. M. Mohamad
e-mail: malik@uthm.edu.my

F. Saeed et al. (eds.), *Advances on Smart and Soft Computing*, Advances in Intelligent Systems and Computing 1188, https://doi.org/10.1007/978-981-15-6048-4_32

a threat, risk, or vulnerability or minimize the impact of an incident [1]. Furthermore, the protection activities should balance and unite a broad range of people, processes, and technology-related activities. Exclusively focusing on cyber-threats that ignores important traditional physical threats is just as dangerous as the neglect of the virtual aspect of the problem [2]. CII can be described as assets that support normal functions of the modern society and the economy which includes essential services and utilities such as the electrical power grid, telecommunications, transportation, financial systems, and other essential services [3]. Moreover, the systems that help operate and control these services is widely known by their common names: Industrial Control Systems (ICS), Distributed Control Systems (DCS), Programmable Logic Controllers (PLC), Human Machine Interfaces (HMI), and Supervisory Control and Data Acquisition (SCADA) systems. All of the above can also collectively referred as operational technologies (OT). OT is hardware and software that detects or causes a change, through the direct monitoring and/or control of industrial equipment, assets, processes, and events [4]. Today's CII are connected through the Internet to allow remote monitoring, off site control, and management, making it increasingly vulnerable to cyber-related threats. The advent of Internet of Things (IoT) extends the protection parameter beyond the previous checkpoints.

Based on Global Cybersecurity Index (GCI) 2018 [5], Malaysia is ranked 8th in the top ten list. In a published report to the Malaysian government, Academy of Sciences Malaysia [6], proposed that action is required to shape a safer and more resilient cyberspace for public and businesses at large by creating greater resiliency within the critical services. The report also commented on the lack of guidelines and directives published at the sectoral level.

Malaysian CII is formally defined in the Malaysian National Cyber Security Policy (NCSP) [7, 8] and referred to as Critical National Information Infrastructure (CNII). According to the NCSP, CNII are assets (real and virtual), systems, and functions that are vital to the nations that their incapacity or destruction would have a devastating impact on national economic strength, image, defense and security, the government capability to functions, and the public health and safety. For the purpose of uniformity, CNII is referred as CII throughout this paper.

Malaysian CIIP management is spearheaded by the National Cyber Security Agency (NACSA) [9], an organization under the National Security Council. Previously, the role was mandated to CyberSecurity Malaysia [8]. Refocusing its roles, CyberSecurity Malaysia currently provides cybersecurity consultation, training and concentrates its expertise as a technical reference entity.

Henceforth, as a leading agency for CIIP, NACSA manages all the strategic initiatives under the NCSP and National Cybersecurity Strategy. For each sector, a nominated lead agency shall oversee the governance and its regulatory aspects. Furthermore, each sector has its own inner workings vocabulary, policies, legislative framework, and unique operational nature. As for the Malaysian Government Public Sector (MPS), the sector lead was initially under the purview of Malaysia Administrative, Modernization, and Planning Unit (MAMPU) until recently in 2019, the function has been transferred to NACSA as the single lead entity for the government [9, 10].

At the legislative level, ten acts are in force and provided cyber-related law enforcement agencies the mandate to act against cybercrime offenders. The acts are the Communications and Multimedia Act, Computer Crime Act, Digital Signatures Act, Copyright Act, Personal Data Protection Act, Electronic Government Act, Electronic Commerce Act, Telemedicine Act, Antifake News Act, and the Penal Code. Other related laws that apply to MPS CIIP include Official Secrets Act for protecting classified official information and Protected Areas and Protected Places Act for security protection of designated areas and places. However, discussion on the legislative aspect pertaining to CIIP is beyond the scope of this paper.

The organization of the rest of this paper is as follows. Section 2 explains the research methodology applied and its related conceptual framework, while Sect. 3 reports the current findings and discussion. Section 4 provides recommendations to be included in the establishment of a CIIP Framework for the MPS. The paper is then concluded in Sect. 5.

2 Methodology

This paper addresses two key questions regarding MPS CIIP which are:

- What are the external strategic requisites for developing a target protection profile for the MPS?
- What are the internal antecedents for CIIP management in the likes of resources and tools?

This paper establishes a qualitative inquiry in the direction of preparing a sectoral protection framework for MPS CII. The research is descriptive in nature. As resources related to national security are highly classified and restricted, document analysis and comparative study were carried out using publicly available resources such as key policy documents, related directives, circulars and guidelines pertaining to protection of CII within the MPS. Furthermore, NIST Cybersecurity Framework (NIST CSF) is compared to the existing Malaysian Public Sector Cybersecurity Framework. A conceptual framework for this paper is depicted in Fig. 1.

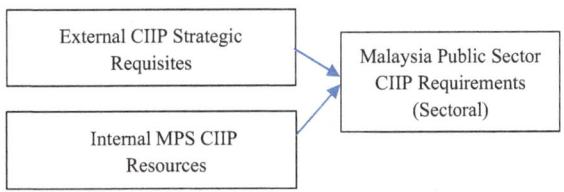

Fig. 1 Conceptual framework to define baseline requirements for MPS CIIP. Adapted from ENISA [11]

3 Findings and Discussion

3.1 External Strategic Requisites

In 2004, United Nations [12] has created a resolution on global culture of cyber-security and the protection of CII in which it recognizes that each country will determine its own critical information infrastructures (CII) and furthermore listed 11 elements for protecting CIIs. Other organizations such as G8 Nations [13], Organization for Economic Co-operation and Development (OECD) [14], International Telecommunications Union (ITU) [11, 15], European Union Agency for Cybersecurity (ENISA) [16–18] the Commonwealth Telecommunication Organization (CTO) [19], and Meridian Process [1, 20] also outlined important considerations when specifying CIIP requirements. Based on our review, a list of 21 strategic requirements were collected and analyzed (refer Table 1).

3.2 Existing Resources and Tools for MPS CII Management

In the context of MPS, agencies are required to conform to the Malaysian Public Sector Cybersecurity Framework, also known as RAKKSSA (Rangka Kerja Keselamatan Siber Sektor Awam) [21]. RAKKSSA consists of eight components: identify, protect, detect, response, recover, procurement, security audit, and enforcement. Compared to NIST CSF [22] which only comprises of five functions, RAKKSSA includes procurement, security audit, and enforcement as additional components. An earlier work [23] reviewing NIST CSF concurred that the framework is relevant in Malaysia and developing nations such as Malaysia could apply frameworks such as NIST CSF as a quick win effort to ensure a globally compliant CIIP. As the current reference framework for the public sector cybersecurity, RAKKSSA depends on previously published circulars and directives, for example: the Malaysian Public Sector Management of ICT Security Handbook (MyMIS) [24] and Malaysian Public Sector Information Security Risk Assessment Methodology (MyRAM) [25]. RAKKSSA is a directive-based guideline framework and contrary to the NIST CSF, and it is intended to be used by the public sector only. Even though visualized consisting of pillars and building blocks, RAKKSSA operates in a continuous cycle between its first and eight components. RAKKSSA also identified a prerequisite in terms of an enabling environment. The environment consisted of strategic partnerships of four lead agencies which are MAMPU, Chief of Government Security Office (CGSO), CyberSecurity Malaysia and MIMOS, Malaysia's National Applied Research and Development Center under the Ministry of International Trade and Industry. A comparison of NIST CSF and RAKKSSA is summarized in Table 2.

MPS organizations also have access to government security applications namely MyRAM, MyISMS, MyISGRC, and NC4 Web application. MyRAM is both a Web-based Malaysian risk assessment methodology and an information security risk

Table 1 CIIP requirements based on international agencies/organizations

	Requirements	G8	UN	ITU	OECD	ENISA	CTO	MP
1.	Emergency warning	•	•		•	•		•
2.	Stakeholder awareness	•	•	•			•	•
3.	Identify interdependencies	•	•	•	•	•		•
4.	Promote partnerships	•	•	•	•	•		•
5.	Crisis communication networks	•	•	•	•	•		•
6.	Incident response capability			•	•	•	•	•
7.	Policy to protect CII and information confidentiality	•	•	•	•	•	•	•
8.	Facilitate tracing of attacks	•	•					•
9.	Training exercises for continuity and contigency	•	•	•	•			•
10.	Capacity building			•			•	
11.	Cyber security culture			•	•		•	•
12.	Adequate law and personnel	•	•	•	•	•	•	•
13.	International cooperation	•	•	•	•	•	•	•
14.	Information sharing mechanism				•	•	•	•
15.	R&D, international standards	•	•	•	•		•	•
16.	Governance and organizational structures			•	•	•	•	•
17.	Technical procedural measures			•	•		•	
18.	National strategy			•	•	•	•	•
19.	Risk management approach			•	•	•	•	•
20.	Monitoring and continous improvement			•	•	•	•	•
21.	Resilience			•		•	•	•

assessment tool. MyISMS application provides a complete guide in implementing ISMS including tutorials, ISMS requirements list, guidelines, standards, and document templates are available for download by agencies to develop, adapt, and obtain ISO/IEC 27001: 2013 ISMS certification. MyISGRC [26] Web application measures governance, risk management, competency and culture, physical security, and operations security domains. MyISGRC compliance checklist is developed mainly from ISO/IEC 27001:2013 Information Security Management System Standard, COBIT 5: IT Governance and Management of Enterprise IT Framework, NIST SP800-100: Information Security Handbook: A Guide for Managers and a few other related references. National Cyber Coordination and Command Center (NC4) [27] is established in accordance with the NCSP and Directive No. 24: National Cyber Crisis Management Policy and Mechanism to assess the current cybersecurity level of preparedness in the face of threats and cyber-attacks at the national level. NC4 operationalized its portal as a focal point for referencing cyber-threats and crisis at the national

Table 2 Summary of NIST CSF and RAKKSSA comparison

	NIST CSF	RAKKSSA
Origin	NIST, US	MAMPU, CGSO, CSM, MIMOS
Version	1.1	1.0
Year published	First introduced 2014, updated 2018	2016
Intended audience	Public and private organizations	Malaysian Public Sector
Type	Voluntary framework, self-regulated guideline	Guideline framework
Depth	Flexible, high level to low level	High level only, requires agency adaptation for low level implementation
Process	Cyclic lifecycle	Cyclic lifecycle
Structure	Framework implementation tiers, framework core with five components (identify, protect, detect, response, recover), with categories and sub-categories, and framework profiles	Eight components (identify, protect, detect, response, recover, procure, audit, and enforce) with sub-components
Priority areas	Cybersecurity and CII. Applicable to many sectors	Mainly for Public Sector Cybersecurity, inclined toward project management and procurement, development, and maintenance processes. Not specific to CII
Implementation reference	Normative reference provided, i.e., COBIT, ISMS, NIST SP800-53	Not provided. Requires other supporting references to operationalize
Supporting tool	Tools widely available	MyISGRC Web application
Compatibility	Crosswalk and mappings with another framework and references are available	Not explicitly provided

level and ensures coordination and cooperation between CII agencies. The portal also hosts a Web application, which provides monitoring and centralized control of cyber-incidents. It is used to determine the level of threat and assess the impact of cybersecurity threats to the nation.

Both CyberSecurity Malaysia and CGSO serve as key technical experts and reference agencies. CyberSecurity Malaysia functions are as mentioned in the previous section. CGSO remains a focal point for managing government security involving confidential and official secrets protection [28], and physical protective security for critical installations such as critical infrastructures and prohibited places and areas [29].

A list of currently available CII management resources provided by NACSA [30], MAMPU [31], CyberSecurity Malaysia [32], and CGSO are inventoried in Table 3. Each resource is classified within six categories: policy, plan, directive, framework, guidelines, and tools applicable for sectoral reference required by CII owners, CII operators and CII policy makers. Overall, a total of 26 references are available to address MPS current sectoral requirements and provided antecedents for future development of a more holistic CIIP. The National Cybersecurity Policy and the National Cybersecurity Strategy under NACSA deliver the top-level reference for all CII organization while MAMPU maintains the highest number of references for MPS. CGSO provides two important documents which covers the fundamental management of classified information and physical security of critical installations.

4 Recommendations

Towards developing MPS CIIP in parallel with NCSP vision, a refined sectoral framework should consider these recommendations for establishing a solid sectoral protection planning:

- Identification of critical services and its dependencies. Based on the core critical technology roadmap in the NCSP, critical services that make use of secure communications, high availability systems, network surveillance, response and recovery, trust relationship, secure access, system integrity and enables forensics should be enlisted as critical identification criteria. Critical, essential, and supporting services should be classified and mapped to reflect intra/interdependencies for CII services.
- Reorganizing and preparing a substantive reference list, which provides stakeholders and CII operators detailed description of existing policy, regulations, standards, technical codes, guidelines, and best practices that can be implemented. Relevant standards includes the ISO/IEC 27000 series, e.g., ISO/IEC 27001 for information security management, ISO 22301 and ISO/IEC 27031 for business continuity, and ISO/IEC 31000 for risk management.
- Cross-knitting and adapting the 21 strategic requirements in the findings as principle considerations for developing sectoral CIIP planning and activities.
- Conducting gap analysis by comparing existing antecedents (as-is) enviroment and target protection profile (to-be). A starting point would be using the 26 CIIP-related resources in Table 3.
- Reviewing CII areas such as business continuity and disaster risk management alongside physical critical infrastructure protection for coherence and hollistic management as the Malaysian government [33–35] recognizes threat to CII as a new emerging dimension, hence re-enforcing the resilience strategy as per stated in NCSP vision.

Table 3 Existing MPS CII resources and tools

Agency	Category	Level	
		Strategic	Operational
NACSA	Policy	National Cybersecurity Policy (NCSP)	
		National Cybersecurity Strategy	
	Plan	National cyber crisis management plan	
	Directive	Directive No. 24: Policy and Mechanism of the National Cyber Crisis Management	
			Reporting of public sector ICT security officer and incident handling
	Tools		National cyber coordination command center (NC4) portal and Web application
MAMPU	Policy	Public Sector ICT Security Policy	
	Framework	Malaysian Public Sector Cybersecurity Framework (RAKKSSA)	
	Directive	Malaysian Public Sector Management of Information and Communications Technology Security (MyMIS) handbook	
			Malaysian Public Sector Information Security Risk Assessment Methodology (MyRAM)
			ICT security incident handling management
			Security posture assessment
			Log server activation for incident reporting
			Service continuity management
			ISO27001 ISMS requirement, implementation, and guideline
			Information security posture assessment
	Tools	Malaysian Information Security Governance, Risk Management, and Compliance (MyISGRC) Web-based app	

(continued)

Table 3 (continued)

Agency	Category	Level	
		Strategic	Operational
			MyRAM: Malaysian Public Sector Web-based Risk Assessment System version 2.0
			MyISMS: Information Security Management System (ISMS) online guide portal
CyberSecurity Malaysia	Guidelines	Malaysian guideline to determine information security professionals requirements for the CNII agencies/organizations	
		CyberSecurity Malaysia ISMS implementation guideline	
			Code of ethics of information security professionals
			Guide on computer security
			Third-party security assessment
CGSO	Directive		Security Directive
			Critical National Infrastructure Standing Order

5 Conclusion

This paper is part of ongoing research on CIIP framework development for the MPS. In this paper, work is mainly focused on identifying key requirements from external and internal resources. However, existing frameworks for the public sector such as RAKKSSA seemingly not CII centric, thus providing an opportunity to develop a more bespoke protection framework which incorporate CII management areas such as resilience as key attribute. Based on our findings, a sectoral CIIP consists of varying degree of implementations such as strategic, legislative, and operational levels. Resilience is an emerging theme, whereas it is linked to nation's sustainability and security but appears latent in the current MPS Public Sector documented operational level references. Generic CIIP requirements are extracted from seven international bodies as reference points while for the MPS, 26 officially published material consisting of policy documents, plans, frameworks, directives, guidelines, and tools by MPS leading agencies. Lastly, the recommendations serve as take away points for future directions in developing the MPS CIIP framework.

Acknowledgements We would like to express our outmost gratitude to the Public Service Department Malaysia, for the sponsorship of this research under the Federal Training Award Scholarship.

References

1. GFCE Global Good Practices Critical Information Infrastructure Protection (CIIP) (2017). https://www.thegfce.com/good-practices/documents/publications/2017/11/21/critical-information-infrastructure-protection-ciip. Accessed 31 Dec 2019
2. Dunn, M.: Understanding critical information infrastructures: an elusive quest. In: The International CIIP Handbook, pp. 27–53 (2006)
3. Wilson, C.: Cyber threats to critical information infrastructure. In: Cyberterrorism, pp. 123–136. Springer, New York (2014)
4. Gartner Glossary. https://www.gartner.com/en/information-technology/glossary/operational-technology-ot. Accessed 31 Dec 2019
5. International Communication Union Global Cybersecurity Index (GCI) (2018). https://www.itu.int/pub/D-STR-GCI.01. Accessed 31 Dec 2019
6. Malaysia Academy of Sciences: Cyber Security: Towards a Safe and Secure Cyber Environment (2018)
7. Abdullah, F., Mohamad, N.S., Yunos, Z.: Safeguarding Malaysia's cyberspace against cyber threats: contributions by cybersecurity Malaysia. OIC-CERT J. Cyber Secur. 1(1) (2018)
8. Hashim, S.: Malaysia's national cyber security policy towards an integrated approach for cyber-security and critical information infrastructure protection (CIIP). In: Proceedings: ITU Regional Cybersecurity Forum for Africa and Arab States, Tunis, Tunisia (2009)
9. National Cyber Security Agency (NACSA): Directive on Notification of Government Computer Emergency Response Team (GCERT) Management Function by NACSA. Putrajaya (2019)
10. National Cyber Security Agency (NACSA): Directive on Public Sector ICT Security Officer Notification. Putrajaya (2019)
11. ENISA: Stocktaking, Analysis and Recommendations on the Protection of CIIs. European Union (2016)
12. United Nations: UN Assembly Resolution A/RES/58/199: Critical Infrastructure (2004)
13. G8 Principles for Protecting Critical Information Infrastructures (2003)
14. OECD: OECD Recommendation of the Council on the Protection of Critical Information Infrastructures. Seoul, Korea (2008)
15. International Telecommunication Union (ITU), The World Bank, Commonwealth Secretariat (ComSec), The Commonwealth Telecommunications Organisation (CTO), NATO Cooperative Cyber Defence Centre of Excellence (NATO CCD COE): Guide to Developing a National Cybersecurity Strategy—Strategic Engagement in Cybersecurity (2018)
16. ENISA: Threat Landscape and Good Practice Guide for Internet Infrastructure. European Union (2015)
17. ENISA: Methodologies for the Identification of Critical Information Infrastructure Assets and Services. European Union (2014)
18. ENISA: Measurement Frameworks and Metrics for Resilient Networks and Services. European Union (2010)
19. Commonwealth Telecommunication Organization: Commonwealth Approach for Developing National Cybersecurity Strategies a Guide to Creating a Cohesive and Inclusive Approach to Delivering a Safe, Secure and Resilient Cyberspace. United Kingdom (2014)
20. GFCE-MERIDIAN Process: The GFCE-MERIDIAN Good Practice Guide on Critical Infor-mation Infrastructure Protection for Governmental Policy-Makers. TNO, Netherlands (2016)

21. Malaysia Administrative Modernization and Planning Unit (MAMPU): Public Sector Cyber Security Framework. In: Rangka Kerja Keselamatan Siber Sektor Awam (RAKKSSA). Cyberjaya (2016)
22. NIST Cybersecurity Framework Website. https://www.nist.gov/cyberframework. Accessed 31 Dec 2019
23. Teoh, C.S., Mahmood, A.K., Dzazali, S.: Is NIST CSF applicable for developing nations? A case study on government sector in Malaysia. In: 21st Pacific Asia Conference on Information Systems (PACIS), pp. 101–111. Association for Information Systems (AIS), Langkawi (2017)
24. Malaysia Administrative Modernization and Planning Unit (MAMPU): Malaysian Public Sector Management of Information & Communications Technology Security Handbook (MyMIS). Putrajaya (2002)
25. Malaysia Administrative Modernization and Planning Unit (MAMPU): MAMPU Director General Directive Letter Implementation of Information Security Risk Assessment Using MyRAM App 2.0 in the Public Sector Agencies. Putrajaya (2015)
26. Malaysia Information Security Governance, Risk Management and Compliance (MyISGRC) App. https://www.mampu.gov.my/en/myisgrc-bi. Accessed 31 Dec 2019
27. National Cyber Coordination and Command Center NC4 Portal. https://www.nc4.gov.my. Accessed 31 Dec 2019
28. Malaysia Government: Act 88 Official Secrets. Federal Gazette, Malaysia (1972)
29. Malaysia Government: Act 298 Protected Areas and Protected Places. Federal Gazette, Malaysia (1959)
30. National Cybersecurity Agency (NACSA) Website. https://www.nacsa.gov.my/governmen t.php. Accessed 31 Dec 2019
31. Malaysia Administrative Modernization and Planning Unit Website. https://www.mampu. gov.my. Accessed 31 Dec 2019
32. Cybersecurity Malaysia Resources Download. https://www.cybersecurity.my/en/knowledge_ bank/info_guiding/best_practices/main/detail/639/index.html. Accessed 31 Dec 2019
33. Malaysia National Security Council: National Security Policy. Putrajaya (2017)
34. Ministry of Home Affairs Malaysia: Public Safety and Security Policy. Putrajaya (2019)
35. Ministry of Defense Malaysia: Defense White Paper. Kuala Lumpur (2019)

Experimental Evaluation of the Obfuscation Techniques Against Reverse Engineering

**Mohammed H. Bin Shamlan, Alawi S. Alaidaroos,
Mansoor H. Bin Merdhah, Mohammed A. Bamatraf, and Adnan A. Zain**

Abstract Source code obfuscation is one of the techniques used by software developers to protect their software. Obfuscation techniques transform the original code to a new protected version which is harder for the attackers to understand but have the same function as the original one. Most of obfuscation techniques are not based on well-defined measurements to clarify their effectiveness in protecting the source code from reveres engineering attacks. This paper presents an experimental study to investigate the effectiveness of specific control flows obfuscation technique—A parameterized flattening—in software protection against human attacks. We conduct an experiment, where software developers participate to perform an attack task on two applications written in C sharp programming language where one of the applications is clear and the other one is obfuscated with control flow obfuscation technique. As a result of the statistical analysis used in this paper, it is shown that only the obfuscation treatment significantly affects the correctness of the attacker to perform a successful attack, where the obfuscation reduced the correctness of the attacker by 50%. The complexity of the application has no significant effect on the correctness of the understanding tasks. Also, neither obfuscation treatment nor the complexity of the application has any effect on the correctness of the modification attack tasks.

Keywords Obfuscation · Software protection · Control flow

M. H. Bin Shamlan (✉) · A. S. Alaidaroos · M. H. Bin Merdhah
Faculty of Computing and IT, University of Science and Technology, Sanaa, Yemen
e-mail: icdlmukalla@gmail.com

M. A. Bamatraf
Faculty of Computing and IT, Hadhramout University, Al Mukalla, Yemen
e-mail: mbamatraf1@yahoo.com

A. A. Zain
Faculty of Engineering, Aden University, Aden, Yemen
e-mail: adnan_zain2003@yahoo.com

© The Editor(s) (if applicable) and The Author(s), under exclusive license to Springer   383
Nature Singapore Pte Ltd. 2021
F. Saeed et al. (eds.), *Advances on Smart and Soft Computing*, Advances in Intelligent
Systems and Computing 1188, https://doi.org/10.1007/978-981-15-6048-4_33

1 Introduction

In software development, compiling is a process of converting the source code to machine language for purpose of generating executable application that can be interpreted and executed by the processor, unfortunately compiling process of many programming language like c sharp and java which are widely used transfers the source code to an intermediate language which is not a machine language, and the intermediate language makes it easy for the reverse engineering to analyze and recover the source code and extract valuable information such as algorithms or concepts and registration serial numbers or cryptographic keys available in the software. Software obfuscation is one of the techniques used by software developers to protect their source code against reverse engineering attacks. Obfuscation is usually used to prevent manual inspection of program internals. The most general strategies are renaming the variables and methods and flatting down the program structures to become more complex. It can conceal the location of a flaw in an obfuscated patch [1]. Obfuscation means that the original code is subjected to a series of transformations by changing its structure, so that it becomes more difficult to understand while preserving its original functionalities [2, 3]. Control flow obfuscation technique alters the flow of the original program, by creating conditional, branching, and iterative constructs which result in valid executable logic. A control variable that represents the program state is used to ensure the correct flow [4]. Most of the obfuscation techniques are not based on well-defined measurements to clarify their effectiveness in software protection. This paper presents an experimental study to investigate the effectiveness of the parameterized flattening control flows obfuscation technique to protect the software against reverse engineering. In this experimental study, software developers participate to perform an attack tasks. A binary logistic regression test was used to carry out the results.

The rest of the paper is organized as follows: In Sect. 2, a literature review on the assessment of obfuscation effectiveness is presented. Details of the experiment preparation are given in Sect. 3. Section 4 details the analysis method used. In Sect. 5, the results of the experiments are analyzed to confirm or reject the research hypothesis. Finally, Sect. 6 concludes.

2 Related Works

Various researches have been proposed to examine the effectiveness of obfuscation. The study at [5] presents some basic metrics to evaluate the obfuscated code as compared to the original code. The proposed metrics are complexity, execution time, and the nested if statements. Another large-scale study was proposed at [6]; in this study, the obfuscation was quantified in a sense of complexity and modularity. An experimental approach was presented. A quantitative model with a set of security measurements inferred from the well-known Kolmogorov complexity was displayed in [7]. At [8], the presented experiment measures the ability of a reverse engineer to accurately perform an attack against identifier renaming obfuscation techniques. The researchers at [9] conduct more than one empirical study to assess the rename of the identifiers in the program code. The research in [8, 9] were similar since they examine the identifier renaming obfuscation techniques. In [10], an experiment is carried to evaluate the obfuscation against attackers, where students pretend as attackers to attack programs written in C programming language. A recent large-scale study was presented in [11], where an online Android developer participated in an experiment for testing the affectivity of Android developers to use obfuscation for protecting their application. Many of the previously presented literature works used fundamental meters, which may not be considered as evidence of effectiveness for the obfuscation techniques. In addition to those works, most of them were theoretical study, and not experimental. For our study, we target the lack of empirical study, which considered more reliable. We investigate the effectiveness of a specific control flows obfuscation technique in C sharp programming language.

3 Experiment Design

3.1 Goal and Context

The goal of this experimental evaluating is to investigate the effectiveness of specific control flow obfuscation technique—A parameterized flattening [12]—in protecting the software against human attacks. The main objective is to estimate how this obfuscation technique prevents or limits the correctness of the attacker to alter or perform modification and understand the source code. The context of this experiment study consists of *subjects* who are 14 developers pretending as attackers and *objects* which are the applications to be attacked. The subjects are programmers working at Smart Vision Company for software development at Hadhramout governorate in Yemen. These developers are familiar with coding and developing Web and desktop applications, and the total number of subjects and the distribution of the tasks are shown in Table 1.

The objects in the experiment are two .exe files developed using C# programming language; the two applications have been programmed to be different in complexity. The control flow obfuscation technique is based on the Algorithm proposed in [12].

The first application named, Activation, is an application to simulate the software activation process. It requires the user to enter the registration number, After the user enters the number in the text box and clicks the activation button, the program checks

Table 1 Experimental design

Total subjects	Total tasks	Understanding tasks	Modification tasks
14	44	22	22

that number and if the activation number is correct, the program displays a message that informs the user that the activation is successful, and vice versa in case of wrong activation number.

The second application is named Login, and this application simulates the software login process. When the user runs the application, the application checks the availability of Internet service, and then, it accesses its company Web site to check for any available updates. Then, the application displays the login screen. The user enters the login credentials, a user name, and password and clicks on the login button after clicking by the user. In the case of the availability of new updates, the program displays a message to the user asking to update the program and opens the link of the new version on the browser.

3.2 Variables

In this experiment, we consider the correctness of an attacker to perform the task as dependent variable and two independent variables, obfuscation treatments, and application complexity. We try to analyze if there is an affection of the two independent variables on the correctness which is a dependent variable.

So, the experiment considers the following variables:

Correctness is considered as a dependent variable, where the correctness of the performed attack tasks is calculated as one if the subject made success in the task and equals zero if the subject failed in the task.

Treatment is the first independent variable that applied to the source code, i.e., whether obfuscation was applied to the code or not.

Application is the second independent variable which is used in a task; this can be used to understand how code complexity influences the correctness of the attack task.

3.3 Null Hypotheses

- **H01**: The parameterized control flow obfuscation treatment and the complexity of the application have no significant effect on the correctness of the understanding attack tasks.
- **H02**: The parameterized control flow obfuscation treatment and the complexity of the application have no significant effect on the correctness of the modification attack tasks.

3.4   Experiment Material and Procedure

Before the experiment starts, subjects were prepared and trained about obfuscation and performing understanding and modification tasks on the obfuscated and clearing source code. Subjects were provided with details and explanations on every task to be executed through the experiment. Subjects used a desktop computer with Microsoft visual studio development environment and dot net Reflector and Reflexil, which are the tools for debugging and editing source code. The following materials were distributed to subjects: A short textual description of the program (login and activation examples by C#) they have to attack, which contains running instructions. In two .exe programs (login and activation examples by C#), one is clear, and the other is obfuscated. Explanation slides of the overall procedure of the experiment.

The experiment has been done according to the following procedure:

1. Read the program description.
2. Run the program to be familiar with it.
3. For performing each task:

    (a) Ask the supervisor of the sheet that describes the task to be executed.
    (b) Note and write down the time to start and execute the task.
    (c) For the understanding tasks, an answer must be provided.
    (d) Finally, note the end time and submit the sheet with the maintained application to the supervisor.

4   Analysis Method

In this experiment, we consider the correctness of an attacker to perform the task as the dependent variable and two independent variables, obfuscation treatments and application complexity. To analyze the affection of the two independent variables on the correctness which is the dependent variable, we use a binary logistic regression of correctness versus the two independent variables. Such analysis is suitable for the dichotomous nature of the measure. The logistic regression is formulated according to the following model:

$$\text{Understanding Correctness} = \frac{1}{1 + e^{-(\beta 0 + \beta 1 \cdot x1 + \beta 2 \cdot x2)}} \qquad (1)$$

where $x1$ is variable for the obfuscation treatment and $x2$ is variable for the complexity of the application. The values of the tow variable as:

$$X1 = \begin{cases} 1 \text{ if Treatment} = \text{Obfuscated.} \\ 0 \text{ if Treatment} = \text{Clear.} \end{cases}$$

$$X2 = \begin{cases} 1 \text{ if the Application Complixity} = \text{Login.} \\ 0 \text{ if the Application complixity} = \text{Activation.} \end{cases}$$

The mentioned formula model (1) is also used for investigate the modification correctness, where the only difference is depending on the performed task at the experiment, where some tasks was about measuring the correctness of the understanding ability of the attacker and the other is for measuring the correctness of the modification ability of the attacker. The assessment of the statistical test results is carried out considering the significance at a 95% confidence level ($\alpha = 0.05$). So, we reject the null-hypothesis when p-value $< \alpha$. For analyzing the obtained results for this experiment, we used a MedCalc statistical analysis program.

5 Results

This section presents the results of this experimental study that aimed to assess specific control flow obfuscation technique—A Parameterized Flattening—in software protection against human attacks. Table 2 reports the correct and wrong tasks for understanding and modification tasks.

The results of the tests on the logistic regression for the understanding correctness are reported in Table 3. The P-value of the obfuscation treatment is (0.0189), which is less than (alpha $= 0.05$), based on the results; we cannot reject the null hypothesis H01: where only the obfuscation treatment significantly affects the correctness of the understanding attack tasks outcome. The application complexity has no significant effect on the correctness of the understanding task. Also, the odd ratio of the obfuscation treatment is equal to (OR $= 0.0530$) which means that the obfuscation reduced the correctness of the attacker by 50% to correctly perform understanding tasks. For the application treatment, the odds ratio (OR $= 2.134$), thus no important difference is found.

Table 2 Correct and wrong tasks

Treatment	Understanding		Modification		Total	
	Wrong	Correct	Wrong	Correct	Wrong	Correct
Clear	1	10	2	9	3	19
Obfuscate	7	4	5	6	12	10

Table 3 Logistic regression of understanding correctness

Variable	Coefficient	Std. err	Wald	(P) significance	Odds ratio
$\beta 0$ constant	1.25682	1.79720	0.4891	0.4844	
$\beta 1$ obfuscation treatment	−2.93793	1.25120	5.5135	**0.0189**	0.0530
$\beta 2$ application complexity	0.75799	1.11305	0.4638	0.4959	2.1340

Table 4 Logistic regression of modification correctness

Variable	Coefficient	Std. err	Wald	(P) significance
$\beta 0$ constant	1.26110	1.58433	0.6336	0.4260
$\beta 1$ obfuscation treatment	−1.32367	0.98971	1.7887	0.1811
$\beta 2$ application complexity	0.16858	0.96486	0.03053	0.8613

Modification correctness results are reported in Table 4. Based on the results, we can reject the null hypothesis H02: Neither obfuscation treatment nor application complexity has any effect on the modification correctness of the attack.

6 Conclusions

This paper presented an experiment aimed at assessing the effectiveness of specific control flow obfuscation technique—A parameterized flattening—in software protection against human attacks. The experiment involved 14 programmers. As a result of the statistical analysis used in this paper, it is shown that only the obfuscation treatment significantly affects the correctness of the attacker to perform a successful attack, where the obfuscation reduced the correctness of the attacker by 50%. The complexity of the application has no significant effect on the correctness of the understanding tasks. Also, neither obfuscation treatment nor the complexity of the application has any effect on the correctness of the modification attack tasks.

References

1. Savio, A., Saurabh, M., Paulami, S., Mudit, K.: A study & review on code obfuscation. In: World Conference on Futuristic Trends in Research and Innovation for Social Welfare, Tamil Nadu, India (2016)
2. Hui, X., Yangfan, Z., Yu, K., Michael, R.L.: On Secure and Usable Program Obfuscation: A Survey. Department of Computer Science, The Chinese University of Hong Kong (2017)
3. Chandan, K., Lalitha, B.: Different obfuscation techniques for code protection. In: 4th International Conference on Eco-friendly Computing and Communication Systems, ScienceDirect (2015)
4. Asish, K.D., Shakya, S.D., Sanjay, K.J.: A code obfuscation technique to prevent reverse engineering. In: 2017 International Conference on Wireless Communications, Signal Processing and Networking (2017)
5. Christian, C., Clark, T., Douglas, L.: Taxonomy of Obfuscating Transformations. Department of Computer Science, The University of Auckland, New Zealand (1997)
6. Mariano, C., Andrea, C., Paolo, F., Cornelia, B.: A Large Study on the Effect of Code Obfuscation on the Quality of Java Code, pp. 1–39. Empirical Software Engineering (2014)
7. Rabih, M.: Quantitative Measures for Code Obfuscation Security Imperial. Imperial College London, Department of Computing, London (2016)

8. Mariano, C., Massimiliano, D., Jasvir, N., Poalo, F.: Towards experimental evaluation of code obfuscation techniques. In: Proceedings of the 4th ACM Workshop on Quality of Protection, Alexandria (2008)
9. Mariano, C., Massimiliano, D., Jasvir, N.: The effectiveness of source code obfuscation: an experimental assessment. In: 2009 IEEE 17th International Conference on Program Comprehension (2009)
10. Alessio, V., Leonardo, R., Marco, T.: Assessment of source code obfuscation techniques. In: 2016 IEEE 16th International Working Conference on Source Code Analysis and Manipulation (SCAM), Department of Automatic and Informatics, Italy (2016)
11. Dominik, W., Nicolas, H., Yasemin, A., Brad, R., Patrick, T., Sascha, F.: A large scale investigation of obfuscation use in Google Play. In: Annual Computer Security Applications Conference (2018)
12. Zheheng, L., Wenlin, L., Jing, G., Deyu, Q., Jijun, Z.: A parameterized flattening control flow based obfuscation algorithm with opaque predicate for reduplicate obfuscation. In: International Conference on Progress in Informatics and Computing (PIC) (2017)

PassGAN for Honeywords: Evaluating the Defender and the Attacker Strategies

Muhammad Ali Fauzi, Bian Yang, and Edlira Martiri

Abstract The main challenge in a honeywords system is how to generate artificial passwords (honeywords) that are indistinguishable from the genuine password (sugarword). It is straightforward to consider the PassGAN for generating honeywords from the defender's perspective. In this work, we analyze a game situation between the defender and the attacker assuming the two parties exploit the PassGAN for their own competing advantage, i.e., the defender uses the generator model of PassGAN to generate honeywords, and the attacker uses the discriminator model of PassGAN to detect the sugarword. In this game, we investigate the feasibility of PassGAN as a honeywords generation strategy and the possible strategies that can be used by the defender and the attacker to reach their goal. The best strategy for the attacker is to use a large number of iterations and to use the same dataset as the defender. From the defender's point of view, the strategy of using many iterations is also beneficial to reduce the attacker's success rate.

Keywords Honeywords · PassGAN · Attacker · Defender · Strategy

1 Introduction

A password remains the most widely used identity authentication method due to its simplicity and familiarity to users and developers [1]. Unfortunately, there have been many password data leaks recently including data from well-known organizations such as Rock-you [2], Dropbox [3], and Yahoo [4]. Even worse, the leaks show that most users prefer poor passwords for their accounts that enable attackers to guess them easily using cracking techniques such as dictionary attack, despite that

M. A. Fauzi (✉) · B. Yang · E. Martiri
Norwegian University of Science and Technology, Gjøvik, Norway
e-mail: muhammad.a.fauzi@ntnu.no

B. Yang
e-mail: bian.yang@ntnu.no

E. Martiri
e-mail: edlira.martiri2@ntnu.no

F. Saeed et al. (eds.), *Advances on Smart and Soft Computing*, Advances in Intelligent
Systems and Computing 1188, https://doi.org/10.1007/978-981-15-6048-4_34

the passwords are saved in the hashed form [5, 6]. Most of these breaches were only discovered months or even years after they first happened. During the period between the breach and its discovery, the attacker surely had exploited most of these passwords, some had even published or sold them online. Therefore, it is important to have not only a mechanism to improve security but also a mechanism to detect password data leaks quickly [7].

Some approaches have been proposed to handle a password data breach [8, 9]. A promising one is introduced by Juels and Rivest called honeywords [10]. In this system, the mixture of both real passwords (sugarwords or sugars) and decoy passwords (honeywords or honeys) is stored in the password file. If an attacker manages to steal the file containing the password and successfully resolves all hash values in the file, she or he still has to be able to distinguish the real password from the artificial ones. When an attacker tries to log in using a honeyword, the system will detect it and provide an alarm to denote that there is an attack on the password file. This technique is more practical because it only needs a few changes on the system, on both the server and client sides [11].

The main challenge for the system administrator in this approach is how to generate honeywords that are difficult to distinguish from the real ones. Some previous methods for honeyword generation include random-replacement [10], utilizing existing passwords [12], questionnaire-based [13], text- and image-based non-realistic honeyword generation [14], and paired distance protocol (PDP) [15]. One of the newest methods that can show potential is PassGAN [16].

PassGAN is designed to be a password guessing method [16] that utilizes a generative adversarial network (GAN) [17] to learn from the real passwords as training data and generate password guesses that have a high level of similarity to the real passwords. PassGAN does not require prerequisite knowledge or intuition from experts about what type of passwords are often chosen by users since it will autonomously learn from the real passwords. Besides, PassGAN for honeywords generation is classified as a legacy-UI method because it does not require user intervention. This kind of method is preferred due to usability advantage [10]. Therefore, PassGAN can be a good tool for honeyword generation from the defender's perspective.

On the other hand, attackers are also becoming smarter and are trying to find ways to pick the right password. The attackers could use various password guessing tools and machine-learning techniques to distinguish the real passwords from the decoys. The worst case scenario for the defender is when the attacker also uses PassGAN to determine the correct password. This condition raises a question of how feasible PassGAN is as a honeyword generation strategy and what are the best strategies for both the defender and the attacker in this situation.

In this study, we assume that the attacker has managed to crack the hashed password file that is leaked from the defender and could use PassGAN to distinguish the sugarword from the honeywords. We will analyze the feasibility of PassGAN for the generation of honeywords in this case. We will also investigate strategies that can be used by the attacker to make this distinguishing process more accurate and strategies that can be used by the defender (e.g., the system administrator) to better secure their honeywords system against this kind of attack.

Most previous researches on honeywords (e.g., [10, 12, 13, 15].) only evaluate the honeywords generation strategies heuristically. In this study, we will conduct an evaluation of PassGAN for honeywords generation empirically with some near real-world scenarios and use datasets from some real systems (e.g., leaked password data).

2 Background and Related Works

2.1 Honeywords

Honeywords is a promising security mechanism introduced by Juels and Rivet [10]. The principal idea is to mix each real password with a number of fake passwords in order to confuse attackers. The real password is called a sugarword while the decoys are called honeywords. The combination of the two is often referred to as sweetwords. When the attacker succeeds in cracking passwords from the leaked password file, she or he still has to be able to choose which password is the genuine one out of all the cracked sweetwords. Once the attacker logs in using a honeyword, an alarm will be triggered to notify the defender that there is a malicious password attempt to the system and high confidence that the corresponding identity principal's shadow password file was leaked from the system and cracked.

Generally, the system works as follows. During the registration process, the user will choose a username and a password. Following that, $k - 1$ honeywords associated with the password will be generated, so that it has k sweetwords. The sweetwords, together with the username and the other information about the user, are then stored in the login server after which the order of the passwords is shuffled. Subsequently, the information about the user index and the index of the real password will be saved into a checker-sever called a honeychecker. Since the honeychecker only has information about user index and real password index, the honeychecker can be used only if we have access to user information in the login server. In this system, it is assumed that the attacker does not manage to steal both the login server and the honeychecker data in the same period.

During the login phase, the user enters his username and password. The system then looks for the matching record in the login server. If the username exists, after performing the hashing calculation to the entered password, the system then checks whether the password is in the sweetwords stored for the corresponding user. If the password is not found in the sweetwords, then the password entered is incorrect and login is denied. Otherwise, if the password is found, the system sends information about the user index and index of the entered password into the honeychecker. Honeychecker then makes a comparison between the entered password index and the real password index for the corresponding user in the database. If the two have the same index, then the login is successful. Alternatively, the honeychecker sends an alert to the system administrator about a password data breach or conducts other procedures in accordance with the policy that has been previously determined.

2.2   PassGAN

PassGAN is a deep learning-based password guessing method developed by Hitaj et al. [16]. PassGAN uses a generative adversarial network (GAN) to autonomously learn from the real password dataset and then generate passwords that have the similar distribution to the dataset. GAN is a deep learning approach for estimating generative models through an adversarial process invented by Goodfellow et al. [17].

GAN is comprised of two neural networks that are pitted against each other: a generative model G and a discriminative model D. G studies the distribution of training data and creates new data instances or samples with similar distributions, while D computes the probabilities whether each sample is from the real training data or generated by G. G and D are trained simultaneously where the goal of G is to maximize the probability of D making a mistake, while in contrast, D aims to successfully detect the fake samples. Competition in this game drives both G and D to improve their methods until the generated samples cannot be distinguished from the real data. Recently, there have been many attempts to improve GAN. One of the most promising ones is IWGAN [18] due to its ability to make training more stable. IWGAN has also been successfully implemented for text generation and achieves improvements in performance.

In PassGAN, a generator model is trained to learn about the characteristics and structures of passwords from training data (e.g., leaked password data). The generative model then generates some fake passwords based on the training model. At the same time, the discriminator model is simultaneously trained to distinguish the real password from the fake ones.

3   Our Work

3.1   Game Situation

In this work, we simulate a game between the defender and the attacker. As seen in Fig. 1, three datasets are used for the game: the defender's dataset for her or his PassGAN training, the attacker's dataset for her or his PassGAN training, and the system dataset as the real passwords (sugarwords). The game flowchart can be divided into two parts: the defender's work and the attacker's work.

The defender and the attacker train their own PassGAN model separately using their own dataset. The defender uses p iterations to train her or his PassGAN while the attacker uses q iterations to train her or his PassGAN. Both the attacker and

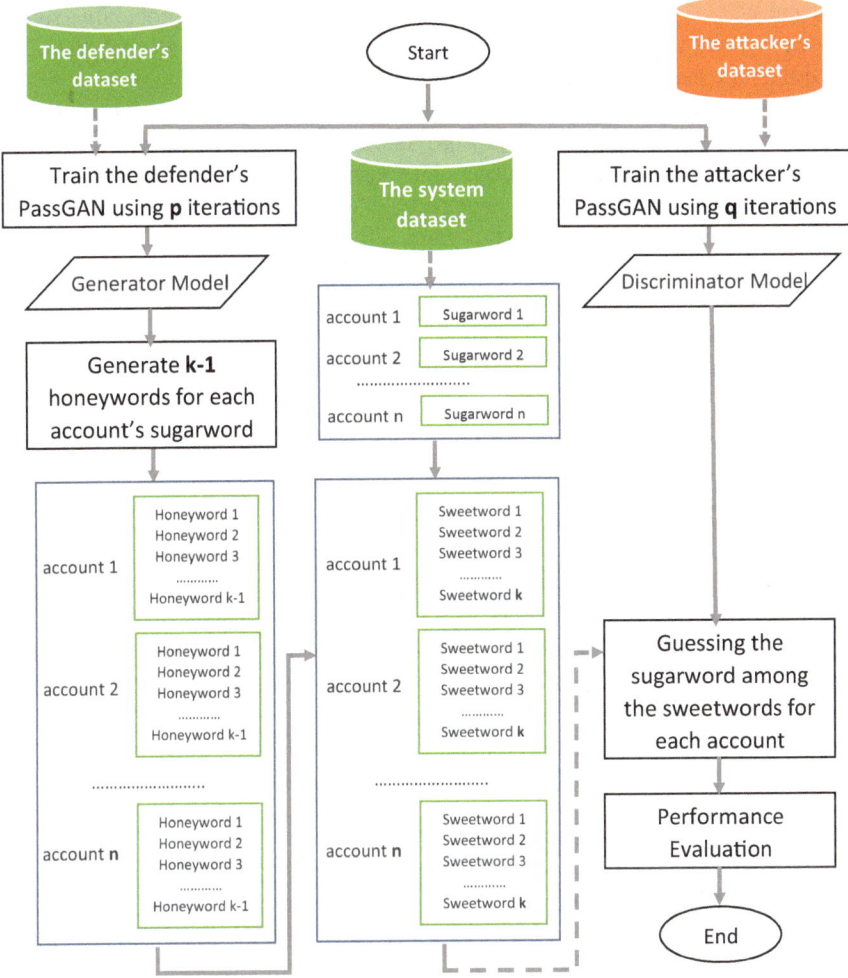

Fig. 1 Attacker and the defender game flow

the defender can use any number of iterations for their training process as it is one strategy that can be exploited by each party. After the training is complete, the defender takes the generator model of her or his PassGAN, which is originally comprised of generator and discriminator model, and then uses it to generate $k - 1$ honeywords for each account's sugarword in the system dataset. These generated honeywords are then combined with each sugarword from the system dataset to compose k sweetwords for each account.

Meanwhile, the attacker takes the discriminator model of her or his PassGAN and then uses it to determine the sugarword among the sweetwords for each account. The guessing process works as follows. For each account, the attacker uses the discriminator model to compute the probability of each sweetword becoming a sugarword.

Then, the sweetwords are sorted in descending order based on their probability value. The sweetword with the highest probability is then considered to be the real password of the account and will be submitted to the system by the attacker.

The guessed passwords are then evaluated. The attacker is considered successful if she or he can guess the correct password for each account. Otherwise, the attacker is considered to have failed. The success rate of the attacker will be used to evaluate both the attacker's and defender's strategies.

The objective of the defender in this game is to generate indistinguishable honeywords that can make the attacker's success rate low. On the contrary, the attacker aims to achieve the best discriminator performance thereby maximizing the probability of guessing a correct password and attain a high success rate. In this game situation, we analyze the best strategy that the two parties can exploit for their own competing advantage.

3.2 Dataset

Datasets used for this work are leaked passwords from Rock-you [19], Dropbox [20], and Linkedin [21]. The RockYou, Dropbox, and Linkedin datasets contain 21,315,673, 7,884,855, and 10,000 passwords, respectively. These three password datasets are quite similar because they were leaked from online media platforms that had similar users, from various professions who were mostly aged 18–34, and the password profile in these three datasets is very similar, e.g., there is no need for special characters or a combination of upper-case and lower-case letters.

3.3 Experimental Setup

In this experiment, we have made some assumptions as follows:

- The defender uses the generator model of PassGAN to generate honeywords.
- The defender uses a public password dataset (e.g., leaked passwords) as her or his training data.
- The attacker manages to steal the data and crack all the hashed passwords.
- The attacker uses the discriminator model of PassGAN to guess the sugarword.
- In the real world, the attacker most likely does not know and does not have access to the defender's training dataset. Therefore, the attacker is likely to use a different dataset from the defender's dataset for the PassGAN training. However, we also conduct an experiment when the attacker and the defender use the same dataset to simulate a worst case scenario for the defender.

The dataset used by the defender in this work is the RockYou (RY) dataset while the attacker uses the Dropbox (Db) dataset. In a scenario when the attacker and the defender use the same dataset, the RY dataset is used for the experiment. The

Linkedin (LI) dataset containing 10,000 accounts' real passwords is used as the system dataset's sugarwords ($n = 10,000$). For each account's sugarword, nine honeywords are generated, and then, these are combined to compose ten sweetwords ($k = 10$).

The following is the detail of the two scenarios:

Scenario 1: The defender and the attacker use the same number of iterations ($p = q$). A various number of iterations are used for the experiment ($p = q = \{5000, 10,000, …, 195,000\}$). In this scenario, two dataset variations (both parties use a different dataset and both parties use the same dataset) are also examined.

Scenario 2: The attacker and the defender use a different number of iterations. The attacker uses a fixed number of iterations ($p = 100,000$), while the defender uses several numbers of iterations ($q = \{5000, 10,000, …, 195,000\}$). In this sce-nario, two dataset variations (both parties use a different dataset and both par-ties use the same dataset) are also examined.

3.4 Performance Evaluation

Juels and Rivest [10] proposed flatness measurement to evaluate honeywords gener-ation strategies. A honeywords generation method is called ϵ-flat when the attacker has a maximum success rate ϵ given a one-time opportunity to choose a correct password. A generation strategy is called "perfectly flat" if the attacker's maximum success rate $\epsilon = 1/k$. If the success rate ϵ is not much greater than $1/k$, it is called "approximately flat." Therefore, the goal of both the defender and the attacker in this experiment is measured using the attacker's success rate. The formula is as follows:

$$\epsilon = \frac{\text{NoC}}{\text{NoA}} \tag{1}$$

where ϵ is the attacker's success rate, NoC is the number of accounts whose passwords have been guessed correctly, and NoA is the total number of accounts. In this experiment, since the number of sugarwords used is 10,000, then the number of accounts (NoA) is always 10,000. Besides, we will also calculate the attacker's success rate when the attacker is given the opportunity more than once to choose the correct password in case there are honeywords systems that allow more than one guess.

4   Experimental Results

4.1   Scenario 1

The experiment results from scenario 1 are shown in Fig. 2. Generally, the attacker's success rate in scenario 1 tends to be very low. This means that the generated honeywords are hard to distinguish from the sweetwords, so that the attacker discriminator model can only correctly guess a few passwords. Based on Fig. 2a, as expected, the attacker's success rate given only one guess when both parties use the same dataset is higher by almost 0.1 than when they use a different dataset. The use of the same dataset makes the attacker's discriminator model learn from the same defender's dataset, so that it can more successfully distinguish the honeywords. When the attacker and the defender use a different dataset, the success rate is relatively stable even though the number of iterations increases. The success rate only fluctuates by a very small margin, between 0.1 and 0.15. Meanwhile, when the attacker and the defender use the same dataset, the success rate increases as more iterations are used. The highest success rate (0.21) is obtained when both parties use 195,000 iterations.

Based on Fig. 2b, the success rate when both parties use the same dataset is also higher than when they use different datasets. However, the difference is relatively small, and the success rates in these two conditions are not much greater than the "perfectly flat" method. Therefore, this honeywords generation method can be considered as "approximately flat."

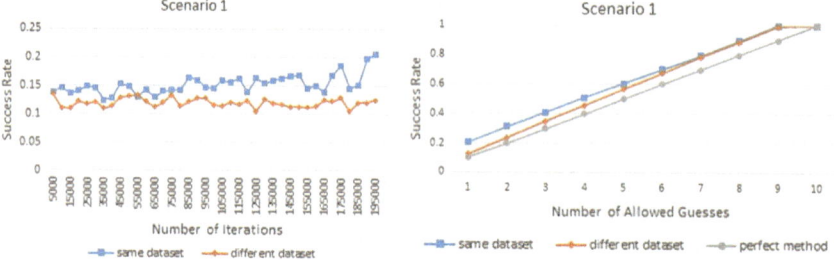

(a) The attacker's success rate given only 1 opportunity to guess the correct password in scenario 1. In this figure, the attacker and the defender use several iteration variations.

(b) The attacker's success rate given more than 1 opportunity to guess the correct password in scenario 1. In this figure, the defender and the attacker use 195000 iterations.

Fig. 2 Scenario 1 experiment results. **a** The attacker's success rate given only one opportunity to guess the correct password in scenario 1. In this figure, the attacker and the defender use several iteration variations. **b** The attacker's success rate given more than one opportunity to guess the correct password in scenario 1. In this figure, the defender and the attacker use 195,000 iterations

4.2 Scenario 2

The experiment results from scenario 2 are shown in Figs. 3 and 4. Generally, the use of a different number of iterations does not show significant differences as the attacker's success rate in scenario 2 also tends to be very low. Based on Fig. 3, the use of the same dataset also gives a higher success rate for the attacker than the use of a different dataset. In the same dataset condition, the success rate decreases, slightly, as the number of defender's iterations increases and becomes larger than the attacker's number of iterations. Meanwhile, when the attacker uses a different dataset from the

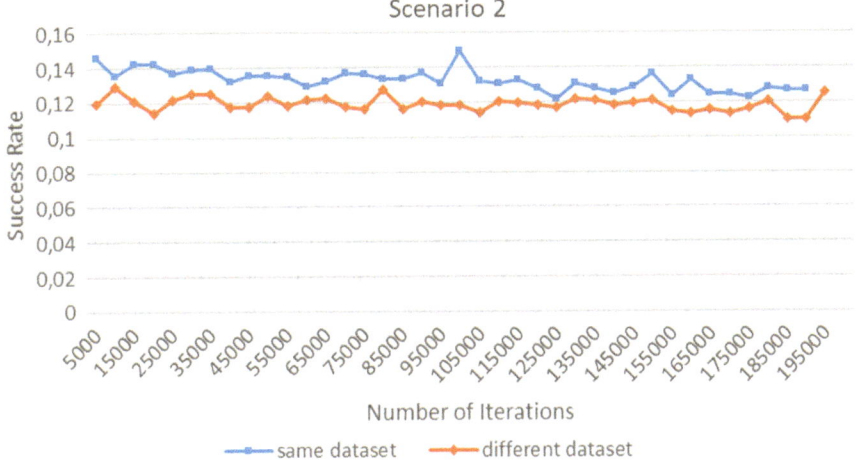

Fig. 3 Attacker's success rate given only one opportunity to guess the correct password in scenario 2. In this figure, the attacker uses a fixed number of iterations (100,000), while the defender uses several numbers of iterations

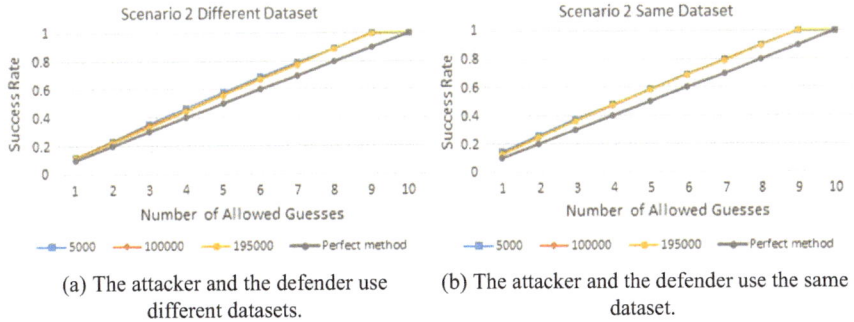

(a) The attacker and the defender use different datasets.

(b) The attacker and the defender use the same dataset.

Fig. 4 Attacker's success rate given more than one opportunity to guess the correct password in scenario 2. In these figures, the attacker uses a fixed number of iterations (100,000), while the defender uses several numbers of iterations. **a** The attacker and the defender use different datasets. **b** The attacker and the defender use the same dataset

defender's dataset, the success rate is relatively stable as it only fluctuated by a very small margin.

Based on Fig. 4a, b, the use of a different number of iterations by the defender does not show significant differences in success rate. The highest success rate is obtained when the defender uses 5000 iterations (fewer than the attacker's number of iterations), and the lowest success rate is obtained when the defender uses 195,000 iterations (more than the attacker's number of iterations). However, the difference is barely noticeable because it is very small.

5 Conclusion

In this work, we analyze the feasibility of PassGAN for a honeywords generation method. Furthermore, we also investigate some possible strategies that can be used by the attacker and the defender in a PassGAN-based honeywords system. The defender uses the generator model of PassGAN to generate high-quality fake passwords while it is assumed that the attacker can manage to steal and crack all the hashed password data and then uses the discriminator model of PassGAN to distinguish the real passwords from the fake ones. Based on the experiment results, PassGAN is feasible for use as a honeywords generation strategy. PassGAN is "approximately flat" even when the attacker also uses PassGAN to distinguish the sugarword from the honeywords.

The best strategy that the attacker can employ is to use the same dataset as the defender's dataset. In addition, the attacker can also use a large number of iterations as their strategy. A greater number of iterations are proven to enable the attacker to increase their success rate even though this increase is very small. From the defender's perspective, the use of many iterations is also beneficial to enable the defender to reduce the attacker's success rate.

In future works, several other strategies for both the attacker and the defender can also be examined. Despite its ability to generate high-quality artificial passwords, PassGAN is not a user context-aware method. If a user uses a password that is correlated to her or his data (e.g., birthday, pet name, etc.), the PassGAN-based generated honeywords would be easily distinguished. Therefore, the use of a targeted guessing attack to examine the PassGAN-based honeywords generation strategy would be relevant for future works.

References

1. Wang, D., Cheng, H., Wang, P., Yan, J., Huang, X.: A security analysis of honeywords. In: NDSS (2018)
2. Siegler, M.: One of the 32 million with a rockyou account? You may want to change all your passwords. Like now (2009). https://techcrunch.com/2009/12/14/rockyou-hacked/

3. Heim,P.: Resettingpasswordstokeepyourfilessafe(2016). https://blog.dropbox.com/topics/company/resetting-passwords-to-keep-yourfiles-safe.
4. Hackett, R.: Yahoo raises breach estimate to full 3 billion accounts, by far biggest known (2017). https://blog.dropbox.com/topics/company/resettingpasswords-to-keep-your-files-safe
5. Florencio, D., Herley, C.: A large-scale study of web password habits. In: Proceedings of the 16th international conference on World Wide Web, pp. 657–666. ACM (2007)
6. Furnell, S.M., Dowland, P., Illingworth, H., Reynolds, P.L.: Authentication and supervision: a survey of user attitudes. Comput. Secur. **19**(6), 529–539 (2000)
7. Flor^encio, D., Herley, C., Van Oorschot, P.C.: An administrator's guide to internet password research. In: 28th Large Installation System Administration Conference (LISA14), pp. 44–61 (2014)
8. Bojinov, H., Bursztein, E., Boyen, X., Boneh, D.: Kamouflage: Loss-resistant password management. In: European symposium on research in computer security, pp. 286–302. Springer (2010)
9. Rao, S.: Data and system security with failwords (2006). US Patent App.11/039,577
10. Juels, A., Rivest, R.L.: Honeywords: Making password-cracking detectable. In: Proceedings of the 2013 ACM SIGSAC Conference on Computer & communications security, pp. 145–160. ACM (2013)
11. Genc, Z.A., Karda¸s, S., Kiraz, M.S.: Examination of a new defense mechanism: Honeywords. In: IFIP International Conference on Information Security Theory and Practice, pp. 130–139. Springer (2017)
12. Erguler, I.: Achieving flatness: selecting the honeywords from existing user passwords. IEEE Trans. Dependable Secure Comput. **13**(2), 284–295 (2015)
13. Chakraborty, N., Singh, S., Mondal, S.: On designing a questionnaire based honeyword generation approach for achieving flatness. In: 2018 17th IEEE International Conference On Trust, Security And Privacy In Computing And Communications/12th IEEE International Conference On Big Data Science And Engineering (TrustCom/BigDataSE), pp. 444–455. IEEE (2018)
14. Akshaya, K., Dhanabal, S.: Achieving flatness from non-realistic honeywords. In: 2017 International Conference on Innovations in Information, Embedded and Communication Systems (ICIIECS), pp. 1–3. IEEE (2017)
15. Chakraborty, N., Mondal, S.: On designing a modified-ui based honeyword generation approach for overcoming the existing limitations. Comput. Secur. **66**, 155–168 (2017)
16. Hitaj, B., Gasti, P., Ateniese, G., Perez-Cruz, F.: Passgan: A deep learning approach for password guessing. In: International Conference on Applied Cryptography and Network Security, pp. 217–237. Springer (2019)
17. Goodfellow, I., Pouget-Abadie, J., Mirza, M., Xu, B., Warde-Farley, D., Ozair,S., Courville, A., Bengio, Y.: Generative adversarial nets. In: Advances in neural information processing systems, pp. 2672–2680 (2014)
18. Gulrajani, I., Ahmed, F., Arjovsky, M., Dumoulin, V., Courville, A.C.: Improved training of Wasserstein Gans. In: Advances in neural information processing systems, pp. 5767–5777 (2017)
19. Brannondorsey: Passsgan (2018), https://github.com/brannondorsey/PassGAN/releases/download/data/rockyou-train.txt
20. Leaks dropbox, https://hashes.org/leaks.php?id=91
21. Leaks linkedin, https://hashes.org/leaks.php?id=68

Measuring the Influence of Methods to Raise the E-Awareness of Cybersecurity for Medina Region Employees

Mahmood Abdulghani Alharbe

Abstract The use of computers has become inevitable for businesses, especially in the last decades. Whether in private or in public sectors, people have become extremely attached to the use of smart digital devices, such as smartphones, laptops, tablets, etc. However, the digitisation has made life easier and more comfortable, but this huge change in life has brought with it several challenges and raised the average of criminal activities known as cybercrimes. The observations showed that the main reason behind most people facing these crimes is the lack of awareness of cybersecurity and cybercrimes. This research sheds lights on cybersecurity and cybercrimes, measures the government's e-awareness of cybersecurity, and identifies the types of most effective material in raising the e-awareness of cybersecurity and cybercrimes. The study has divided the data into different categories, in consistent with (age, gender, education, and ethic). The sample size was 243 staff. The age of the staff was divided into three different groups, in relation to their IT capabilities. The suggestion was that the primary age group target was between (20 and 35) years and then between (36 and 50) as well as (51 and 60) years. Every participant within the sample was exposed to cybersecurity e-awareness by using different tools. The study indicates that directing sources to provide Arabic content through videos or audio clips, they are more effective than other methods, thus contributing to saving resources from the loss in textual awareness. Also, there is poor e-awareness among employees form an Arab origin, while the non-Arab have a much better awareness. The most effect cybersecurity measure is to enhance the e-awareness, which affects the employees based on the employee's level.

Keywords Cybersecurity · IT capability · E-awareness · E-listening · E-video

M. A. Alharbe (✉)
Department of Information Systems, Taibah University, Almadinah, Saudi Arabia
e-mail: mharbie@taibahu.edu.sa

© The Editor(s) (if applicable) and The Author(s), under exclusive license to Springer Nature Singapore Pte Ltd. 2021
F. Saeed et al. (eds.), *Advances on Smart and Soft Computing*, Advances in Intelligent Systems and Computing 1188, https://doi.org/10.1007/978-981-15-6048-4_35

403

1 Introduction

These days, a lot of people use the Internet believing it is a safe environment. Day after day, computers, smartphones, and tablets are being used in daily life. On the other hand, there are many attacks that are based on this daily use. Intrusions, attacks, and cyber-attacks are no longer just exceptions [14]. The costs of managing cybersecurity incidents are rising. Despite the fact that most cyber-attacks are considered to be harmless, number of these attacks have an enormous impact. The consequences of violations in cybersecurity vary from lack of impact, denial of distributed services, information manipulation, data theft, or perhaps the management of systems, and affecting the physical world [15]. People have a tendency asking, currently, why attackers attack? Herzberg et al. have answered this question. The attacks on the Internet are increasing. Attackers generally use straightforward techniques that attract users to a phishing web sites, and the use of spoofed emails reaching spoofed web sites. The purpose of this explicit method is to get users IDs, passwords, different personal and monetary data then misuse them for fraudulent activities. In line with a study by Gartner analysis of two million users saving this data in spoofed web sites, direct losses were calculated at one billion dollars and the annual damages are estimated at four hundred millions to one billion dollars [16].

2 Literature Review

McGuire et al. state that: "The motives of Internet-based crime, for the most part, supported the personal gain (for example, the utilisation of malware to access checking account details, it might also be a kind of protest and/or criminal guilti-ness, or web site distortion)". Motivations are often inferred by examining the operation tools that are used. Some analysis conjointly states that there are addi-tional non-traditional motives, such as satisfying intellectual curiosity challenge, general spite, revenge, establishing respect and power amongst online communities, or perhaps merely boredom" [17]. McCrohan et al. conducted a study on a staff relating to worker awareness of the necessity to launch sturdy passwords consis-tent with certain standards. In the beginning, they found that the staff did not have vital variations within the alternative of passwords. However, after a two-weeks education campaign, researchers found that the strength of passwords accumulated by end of the twelve months. One of the results of this study conducted that the behaviour of the staff can be modified by educating and coaching them on correct security practices to enhance the online security for themselves and for the corpora-tions they work for [18]. Albrechtsen et al. conjointly conducted analysis on a group of participants to ascertain the increase of awareness and whether it influences the development of participants' awareness and technology usage behaviour. An applied math analysis showed that the impact was vital enough to form an enormous distinc-tion to a large variety of participants in terms of awareness and behaviour via email messages, leaflets and posters, and short-run cluster collective thinking sessions" [19]. Kritzinger et al. wrote a research entitled: 'Cybersecurity for home users: a replacement means of protection through awareness', within which they discussed the utilisation of the Internet to profit people and organizations and showed that the internet is often helpful; however, it may bring new and dangerous risks due to cyber-attacks by third parties—unauthorised—attempting to hack data and access for private interests, which is unethical. It is conjointly necessary that every user of the Internet perceives the importance of securing their personal information and realising the implications they may face if this information was hacked. The authors say that users of UN agencies do not understand the risks of the Internet, and that they typically enter this world with no previous awareness. They stressed that there must be a compulsive model that is ought to force users of its web service to awareness programs explaining the risks that occur within the cyber-world and stress that data acquisition ought to be compulsive in order that the user will complete the browsing solely with awareness material [20]. Wells et al. stated that: "because of the technology progress, cybersecurity becomes at risk of a wider variety of attacks. These attacks cause a significant threat. Companies need to make sure that products changes and maintain the security for apparatus, employees, and customers". Authors mentioned conjointly the importance of analysis and development of cybersecurity tools, and they also ought to teach future industrial staff [21].

3 Research Methodology

The analysis basically relied on a quantitative data collection by employing a structured form. The sample of this study is staff of Madinah Region. A random sample technique was applied to select a sample. The participants were given three kinds of cybersecurity e-awareness related to IT capability: (a) e-text awareness, (b) e-listening awareness, and (c) e-video awareness, in order to measure the suitable method according to the Influence of employee's performance. There were many stages to achieve this study which started by asking the employees to complete a form, before being prepared or viewing course on e-awareness. The sample size was 243 staff. The age of the staff was divided into three different groups, in relation to their IT capabilities. The suggestion was that the primary age group target was between (20 and 35) years and then between (36 and 50) as well as (51 and 60) years. Every participant within the sample was exposed to cybersecurity e-awareness by using different tools.

3.1 Data Collocation

The data was randomly collected during this analysis by choosing a sample of staff and then dividing them into different categories, in consistence with (age, gender,

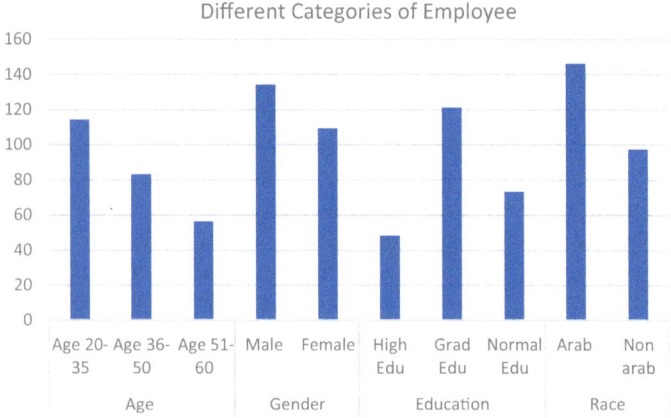

Fig. 1 The data dividing them into different categories

education, and ethnicity). The sample size included 243 staff members, who were classified into three categories in terms of their age in relation to IT capability: 20–35, 36–50, and 51–60 years. Every participant within the sample was exposed to cybersecurity e-awareness ways. Figure 1 explain how the data collocation were dividing into different categories.

3.2 Data Analysis

The data was gained and analysed in order to measure the improvement within the individual's cybersecurity e-awareness according to the Influence of employee's performance of these teams and to spot the ways that were the foremost helpful to them. First questionnaire was distributed to measure the participant's cybersecurity e-awareness regarding the safest ways they used to manage IT capabilities in their lives. Then, they were asked to use different e-awareness material; however, by a totally different means, the participant can see as well as hear a clip containing some directions and recommendations on an equivalent topic. Finally, testing the participant's e-awareness in the fields of cybersecurity to determine which of the previous methods was of most effectiveness for the participant.

4 Results and Discussion

In the results section, the study presents the outcomes obtained to measure the suitable method according to the influence of employee's performance by using some graphs that help the reader to understand the information more easily.

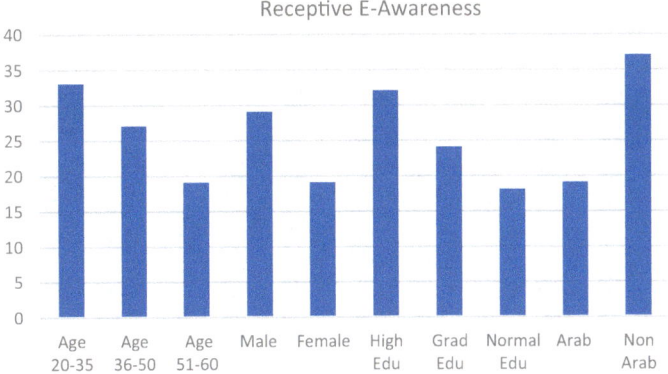

Fig. 2 Receptive E-awareness to the influence of employee's performance

4.1   Receptive E-Awareness

Before receiving any e-awareness materials in any platform and as in Fig. 2, we can see the age group 51–60, Arab, in normal education and females scores around 18 out of 40, they were the lowest level of all other categories, while the age group of 20–35 and non-Arab showed the highest level of awareness from all other categories around 35 out of 40. The difference between the highest and lowest degree of awareness was preceded by seven degrees, which represents 17% of the total score.

4.2   Subsequently E-Awareness

After receiving e-awareness materials by different platforms and as in Fig. 3, we can see people who are at the age 25–36 years, in high education and non-Arab score 40 out of 40; on the other hand, Arab and those in normal education get around 28 which are the lowest scores.

4.3   IT Capability and Categories

Cybersecurity centres as shown from the results in Fig. 4 are the best way to enhance e-awareness of its members to the required level. With 17.5%, e-video awareness methods are more effective than e-listening methods. The results show that the government agencies have an advantage to know where participants are less aware and are able to direct resources to address this vulnerability, rather than spending it on groups that may be at, a somewhat, reassuring level. Non-Arab employees have

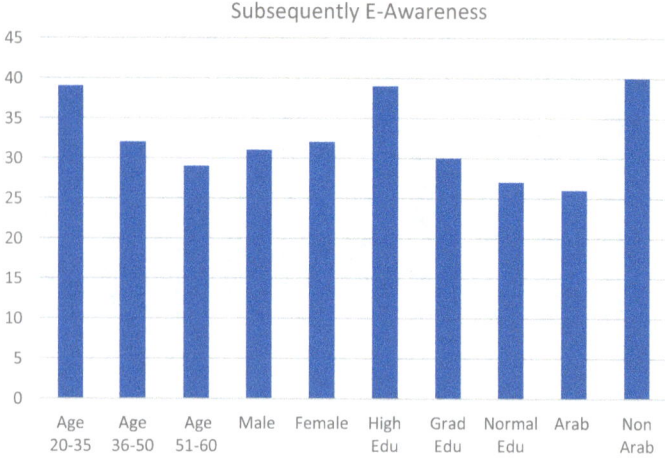

Fig. 3 Subsequently, e-awareness to the influence of employee's performance

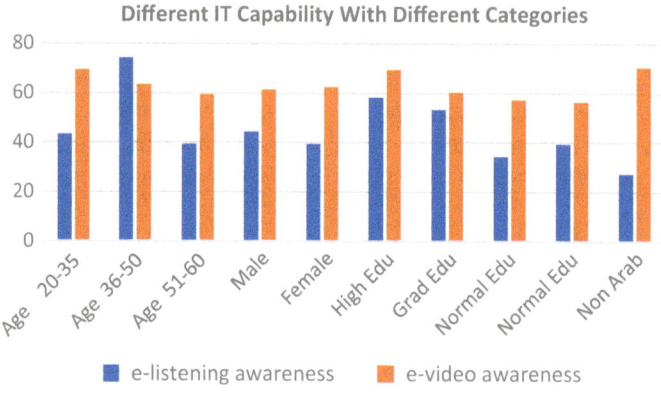

Fig. 4 Different IT capability with different categories

a higher level with 66.5% difference of e-awareness than other groups. The Foundations can focus on providing e-awareness materials in Arabic on an interim basis until further notice.

While in the second place, with 14% difference, the audio clips are more effective on the participants with high education. Video and audio clips achieved higher preference rate of 44% of the votes of participants by age (36–45). This gives the cybersecurity centres a great confidence that both e-awareness materials provided are effective and have a higher response when communicated to the participant.

The Foundations can focus on providing e-awareness materials in Arabic on an interim basis until further notice. Directing sources to provide Arabic content through videos or audio clips, they are more effective than other methods, thus contributing to saving resources from the loss in textual awareness. All categories and without exception have shown remarkable progress and improved the e-awareness outcomes, giving the officials more motivations to complete the process to reach an acceptable level of security.

5   Conclusion

Information centres in various governmental agencies can obtain their own employees' information by providing staff access to the system, completing procedures and reaching results. On the other hand, the cybersecurity centres in the governmental agencies can get other information by using different tools. The best ways for enhancing the e-awareness of cybersecurity are provided. These affect the employees based on their level of education, age, gender, race. The study provided an insight of the level of e-awareness in cybersecurity that employees have and gathered their IT capabilities with the safest method before exposing them to any kind of e-awareness.

6   Future Work

This study can be extended by adding different types of videos and other methods such as whiteboards, avatars, and motion-graphic. Also, studying the most influential techniques contribute to raising of e-awareness among employees. The investigation used in this study can be developed to include direct hands-on training on the areas where the employees have been engaged. It is also possible to study the effects of the male or female character's comment in the audio or video clips, the role to be in Arab and non-Arab clothes and so on.

References

1. Nazief, B., Yudatama, U., Nizar A.: Important factors in information technology governance awareness: an empirical survey of the expert's opinion in Indonesia Uky Yudatama et al. J. Comput. Sci. **15** (8), 1065–1073 (2019). https://doi.org/10.3844/jcssp.2019.1065.1073
2. Yadav, S., Uraon, A., Kumar Sinha, M.: Evaluating awareness and usage of library services among undergraduate students of Silchar Medical College, Silchar, Assam. Indian J. Inf. Sourc. Serv. **9**(1), 25–29 (2019). ISSN: 2231–6094
3. Zwilling, M., Klien, G., Lesjak, D.: Cyber security awareness, knowledge and behavior: a comparative study. J. Comput. Inf. Syst. (2020). https://doi.org/10.1080/08874417.2020.171 2269
4. Pusatli, T., Koyuncu, M.: Security Awareness Level of Smartphone Users: An Exploratory Case Study. Hindawi Mobile Information Systems vol 2019, Article ID 2786913, 11 pages. https://doi.org/10.1155/2019/2786913

5. Harmon, R.R., Auseklis, N.: Sustainable IT services: assessing the impact of green computing practices. In: Portland International Conference on Management of Engineering & Technology, 2009. PICMET 2009, pp. 1707–1717
6. Adela, J.W.C., et al.: Information systems and ecological sustainability. J. Syst. Inf. Technol. **10**, 186–201 (2008)
7. Guo, P.J., Kim, J., Rubin, R.: How video production affects student engagement: an empirical study of MOOC videos. In: Proceedings of the First ACM Conference on Learning@ Scale Conference (pp. 41–50). ACM (2014).
8. Abawajy, J.: User preference of cyber security awareness delivery methods. Behav. Inf. Technol. **33**(3), 237–248 (2014)
9. Andress, J.: The Basics of Information Security: Understanding the Fundamentals of InfoSec in Theory and Practice. Syngress, p. 240 (2014). ISBN 9780128008126.
10. Benitez-Amado, J., et al.: Information technology enabled innovativeness and green capabilities. J. Comput. Inf. Syst. **51**, 87–96 (2010)
11. Von Solms, R., & Van Niekerk, J. (2013). From information security to cyber security. Comput. Secur., **38**, 97–102
12. Finnerty, K., Motha, H., Shah, J., White, Y., Button, M., & Wang, V.: Cyber Security Breaches Survey 2018: Statistical Release (2018)
13. Wilson, M., Hash, J.: Building an information technology security awareness and training program. NIST Spec. Publ. **800**(50), 1–39 (2003)
14. Arora, A., Nandkumar, A., Telang, R.: Does information security attack frequency increase with vulnerability disclosure? An empirical analysis. Inf. Syst. Front. **8**(5), 350–362 (2006)
15. de Bruijn, H., Janssen, M.: Building cybersecurity awareness: the need for evidence-based framing strategies. Govern. Inf. Quart. **34**(1), 1–7 (2017)
16. Herzberg, A., Jbara, A.: Security and identification indicators for browsers against spoofing and phishing attacks. ACM Trans. Internet Technol. (TOIT) **8**(4), 16 (2008)
17. McGuire, M., Dowling, S.: Cyber crime: a review of the evidence, p. 75. summary of key findings and implications, Home Office Research report (2013)
18. McCrohan, K.F., Engel, K., Harvey, J.W.: Influence of awareness and training on cyber security. J. Internet Commer. **9**(1), 23–41 (2010)
19. Albrechtsen, E., Hovden, J.: Improving information security awareness and behaviour through dialogue, participation and collective reflection. An intervention study. Comput. Secur. **29**(4), 432–445 (2010)
20. Kritzinger, E., von Solms, S.H.: Cyber security for home users: a new way of protection through awareness enforcement. Comput. Secur. **29**(8), 840–847 (2010)
21. Wells, L.J., Camelio, J.A., Williams, C.B., White, J.: Cyber-physical security challenges in manufacturing systems. Manufact. Lett. **2**(2), 74–77 (2014)
22. Siponen, M.T.: Five dimensions of information security awareness. SIGCAS Comput. Soc. **31**(2), 24–29 (2001)
23. Kortjan, N., Von Solms, R.: A conceptual framework for cyber-security awareness and education in SA. South Afr. Comput. J. **52**(1), 29–41 (2014)

An Improvement of the Cryptographical Implementation of Coppersmith's Method

Sohaib Moussaid El Idrissi and Omar Khadir

Abstract In the late twentieth century, Don Coppersmith published a method to solve polynomial modular equations using lattice theory. This method was generalized by Alexander May in 2009. In 2014, Phi and Nguyen focused on the running time of the Coppersmith's algorithm and proposed an improvement on the underlying technique (L^2) to perform the speedup, then they exploit the matrix structure to reduce the running time even further. In our work, we propose a new implementation of the Coppersmith's method to improve the running time by finding the suitable lattice dimension experimentally.

Keywords Public key cryptography · Coppersmith's algorithm · Lattices · RSA cryptosystem

1 Introduction

In 1997, Coppersmith described a method for solving modular polynomial equations where, f is a monic polynomial of degree d and N a large number of unknown factorization. He showed a way to find all small solutions x_0 of $f(x) \equiv 0[N]$ such that $|x_0| \leq N^{\frac{1}{d}}$ in polynomial time. His work is considered as a strong tool for public key cryptography [3, 12].

The main idea of Coppersmith's algorithm is based on an article published by Lenstra-Lenstra-Lov´asz (LLL) [8] in 1982. The LLL technique has also proven to be efficient for the cryptanalysis of several cryptosystems, see Nguyen [6, 12].

First, we recall the example given by Coppersmith [3] to show the utility of this method. Let $N = pq$ be an integer with unknown large factors p, q. It is well known that to encrypt a message M using the Rivest, Shamir and Adleman (RSA)

S. Moussaid El Idrissi (✉) · O. Khadir
Laboratory of Mathematics, Cryptography, Mechanics, and Numerical Analysis Fstm, University Hassan II of Casablanca, Casablanca, Morocco
e-mail: moussaid.i.sohaib@gmail.com

O. Khadir
e-mail: khadir@hotmail.com

F. Saeed et al. (eds.), *Advances on Smart and Soft Computing*, Advances in Intelligent Systems and Computing 1188, https://doi.org/10.1007/978-981-15-6048-4_36

method we compute $C \equiv M^e[N]$, with (e, N) the public key. Assuming that we know enough information about M, we use (e, N) and C to compute the unknown part regarding M. For the last forty years, the RSA cryptosystem has been proven to be robust and secure, as demonstrated by Boneh [2] gathering multiple attacks on RSA cryptosystem over 20 years of its creation, then 20 years after by Mumtaz [11] who collected a brief history of recent attacks.

In practice, to reduce computation, we use $e = 3$. So, if we apply Coppersmith's method on 2/3 bits of the plaintext M and the ciphertext C, then we can obtain the 1/3 unknown part. This task can be performed by building the polynomial $f(x) = (M + X)^3 - C$ and finding all possible solutions.

When applying Coppersmith's theorem, the main difficulty faced is due to the lattice dimension n. In 2005, May [10] showed that the complexity upper bound of Coppersmith's method is $O\big(d^5 \log^9(N)\big)$. The lattice dimension n grows in the function of N, since $n = \log(N)$. In order to improve the running time, many researchers [10, 15, 16] looked for generalizations and new variant. The approach presented by Nguyen and Bi in 2016 uses a rounding technique. Their contribution reduces the size of the matrices entries.

Coppersmith's method is constructed in two parts to achieve the upper bound $N^{\frac{1}{d}}$ on the roots. The first part enables us to find all small roots x_0 such that $|x_0| \leq \frac{1}{2} N^{\frac{1}{d} - \epsilon}$ with $\epsilon > 0$. The second part ensures that the solutions x_0 are $|x_0| \leq \frac{1}{4} N^{\frac{1}{d}}$. To calculate all roots $|x_0| \leq \frac{1}{4} N^{\frac{1}{d}}$, the interval $[-N^{\frac{1}{d}}, N^{\frac{1}{d}}]$ is subdivided into four segments and the first step of Coppersmith's method is performed on each one. A recent work in 2018 by Lu et al. [9] discusses many techniques to speed up Coppersmith's technique, but these ideas are targeted to the RSA problem. The lack of papers in the literature addressing the structure of Coppersmith's algorithm motivated us to reach out for a new approach, where the lattice dimension is our key parameter.

In this work, we suggest a new way to apply Coppersmith's theorem if we subdivide the interval $[-N^{\frac{1}{d}}, N^{\frac{1}{d}}]$ into more than four segments. An immediate consequence of our result will be that the lattice dimension will become smaller. The LLL running time will also decrease, improving the performance of Coppersmith's method.

This article is arranged as follows. In Sect. 2, we give a general overview of the mathematical aspect behind lattice theory. In Sect. 3, we present Coppersmith's method with theorems and algorithmic implementations. In Sect. 4, we present our main idea provided with numerical comparisons between our approach and Coppersmith's algorithm. Finally, we provide a brief summary of the work in Sect. 5.

Notation: Let p, q be prime numbers. The set of integers is $\mathbb{Z}$. The set of real numbers is $\mathbb{R}$. $\|.\|$ denote the Euclidean norm. f is a monic polynomial of degree d, $\|f\| = \sqrt{\sum_{i=1}^{n} |a_i|^2}$. $x \in \mathbb{R}$, we use $\lfloor x \rceil$ to specify the closest integer to x. Let $a, b, c \in \mathbb{Z}$. Let $\log_b(x)$ be the logarithm of a positive real number, with a respect to a fixed base $b \in N$ (if $b = 2$, we use $\log(x)$). The notation $a \equiv b[c]$ is used when c divides a-b.

2 Methods

2.1 Lattice Theory

We will now provide the important notions needed in lattice theory to understand Coppersmith's theorem.

A lattice is a subset of R^m.

Definition 1 ([5]) *Let $\{b_1, \ldots, b_n\} \in \mathbb{R}^m$ be a linearly independent set of vectors. The lattice generated by $\{b_1, \ldots, b_n\}$ is defined as*

$$\mathcal{L} = \left\{ \sum_{i=1}^{n} a_i b_i \,|\, a_i \in \mathbb{Z} \right\}$$

the set of all integer linear combinations of $\{b_1, \ldots, b_n\}$.

We call the vectors $b_1, \ldots, b_n$ a lattice basis. The lattice rank is n and m is the lattice dimension. In case of $n = m$ the lattice $\mathcal{L}$ is called a full rank lattice. We define a basis matrix B formed by taking the vectors $b_1, \ldots, b_n$ as rows.

In what follows we consider only full rank lattices. We note both the lattice basis and its respective matrix as B.

Definition 2 (c) *The determinant of a lattice L with a matrix basis B is $|det(B)|$. Also noted as lattice volume noted $vol(\mathcal{L})$.*

The choice of the basis matrix is independent of the determinant of the lattice.

For example taking $\mathcal{L} = \mathbb{Z}^2$, we can see that $B_1 = \{(0, 1), (1, 0)\}$ and $B_2 = \{(1, 1), (2, 1)\}$ are both valid basis. In both cases, the determinant value is ± 1.

The notion of a lattice provides us with many computational problems. In what follows we list the most well known of these. We consider the lattices $\mathcal{L}$ in $\mathbb{Z}^n$.

Shortest Vector Problem (SVP): Given a basis matrix B of a lattice $\mathcal{L}$, find the shortest nonzero vector $v \in \mathcal{L}(B)$.

Closest Vector Problem (CVP): Given a basis matrix B of a lattice $\mathcal{L}$, and $v \in \mathbb{R}^n$, find a vector x in $\mathcal{L}$ such that $||x - v||$ is minimal.

See Nguyen [13] for a thorough explication regarding over their complexity and the algorithms used to solve these problems.

Lattice reduction Lattice reduction is a set of methods to find a 'good' lattice basis. Computing this will enable the problems defined previously to be solved easily. A 'good' lattice basis is defined as a set of vectors that are short and nearly orthogonal.

The LLL algorithm is a generalization of the Lagrange–Gauss [14] algorithm using Gram–Schmidt Orthogonalization. Performing Gram–Schmidt Orthogonalization alone does not always yield a lattice basis. Therefore LLL utilizes the Gram–Schmidt basis as a reference while ensuring that the new vectors lie within the lattice [8]. LLL is an algorithm that given a basis $\{b_1, \ldots, b_n\}$ of a lattice $\mathcal{L}$, outputs a lattice basis which is nearly orthogonal in polynomial time $O\left(d^5 \log^9(N)\right)$, where $b = \max_{0 \le i \le n} \log(b_i)$. By construction, the first vector of the new basis b_1 is considered practically a short vector $\|b_i\| \le 2^{\frac{n-1}{4}} vol(\mathcal{L})$. LLL is used to solve the SVP.

2.2 Coppersmith's Method

In 1997, Don Coppersmith [3] showed a way to find the bounded roots of a modular polynomial equation. In this section, we will present Coppersmith's theorem with demonstrations and we will explain and discuss all aspects of this method.

The main idea of the Coppersmith algorithm is as follow:

Let the problem be $f(x) \equiv 0[N]$, with f an integer polynomial and N an integer number with unknown factorization.

It is easy to solve $f(x) = 0$ over Z using Newton's method, yet the hardness of the problem resides in its modular aspect. To solve this problem, we need to build a new family of polynomials g from f such that $g(x_0) = 0$ over $\mathbb{Z}$, then apply lattice reduction to find a new polynomial with relatively small coefficients. For the proof of the main theorem, we need the next result of Howgrave-Graham [7]. Let $g(x)$ be a univariate polynomial with $d + 1$ monomials.

Let X and N be positive integers. We suppose that $\|g(xX)\| \le \frac{N}{\sqrt{d+1}}$.

If $|x_0| < X$ satisfies $g(x_0) \equiv 0[N]$, then $g(x_0) = 0$ over $\mathbb{Z}$.

Proof By using Cauchy–Schwarz inequality [17] $\sum_{i=0}^{n} x_i y_i \le \left(\sum_{i=0}^{n} x_i^2\right)\left(\sum_{i=0}^{n} y_i^2\right)$ for $x_i, y_i \in \mathbb{R}$.

If $x_i \ge 0$ and $y_i = 1$ for $0 \le i \le d$, we can write:

$$\sum_{i=0}^{d} x_i \le \sqrt{(d+1)\sum_{i=0}^{d} x_i^2}$$

Now, we have:

$|g(x_0)|$

$$= \left|\sum_{i=0}^{d} a_i x_0^i\right| \le \sum_{i=0}^{d} |a_i||x_0|^i \le \sum_{i=0}^{d} |a_i|X^i \le \sqrt{d+1}\|g(xX)\| \le \sqrt{d+1}\frac{N}{\sqrt{d+1}}$$

Then $g(x_0) = 0$ over $\mathbb{Z}$.

Now, we give the statement provided by Coppersmith [3]. Let f be a monic integer polynomial of degree d, $\varepsilon > 0$, and an integer N with unknown factorization. We can find all roots x_0 to $f(x_0) \equiv 0[N]$ such that $x_0 \le \frac{1}{2}N^{\frac{1}{d}-\epsilon}$ in polynomial time in $\log(N)$ and d.

Proof See Chap. 4 in [5] for the complete proof.

An algorithmic version of this theorem is elaborated by Nguyen [15]. This part is considered the first step for finding the solutions. Its complexity is mainly bounded by the LLL [8] running time. As the lattice dimension grows, the value X approaches the upper bound $N^{\frac{1}{d}}$.

Algorithm 1 Coppersmith's First Step

Input: N large integer with unknown factorization and f a monic polynomial of degree d, n lattice dimension, X an upper bound over the solutions.

   Output: All solutions such that $f(x0) \equiv [N]$ and $|x_0| \leq X$

1. Set $h = \lceil \frac{n}{d} \rceil$
2. Build a matrix $n \times n$, B, where the rows are the coefficients of $g_{i,j}(x) = N^{h-1-j} f(x)^j x^i$ with $h > 1$ and for $0 \leq i < d, 0 \leq j < h$.
3. Apply the LLL algorithm to B.
4. Compute the equivalent polynomial $p(xX)$ of the first vector of the new basis.
5. Find all roots x_0 of $p(x)$ over $\mathbb{Z}$.

   Return all roots such that $f(x_0) \equiv 0[N]$ and $|x_0| \leq X$
   The term ϵ can be removed to achieve the upper bound $N^{\frac{1}{d}}$ by taking $\epsilon = \frac{1}{\log_2(N)}$ the new upper bound over the roots will be $\frac{1}{4}N^{\frac{1}{d}}$. May [10] proposed subdividing the interval into four segments. Applying algorithm 1 on each one enables us to find all the bounded roots, if they exist. More precisely. Coppersmith [3], 1997 Let f be a monic polynomial of degree d and an integer N with unknown factorization. We can find all roots x_0 to $f(x_0) \equiv [N]$ such that $|x_0| \leq N^{\frac{1}{d}}$ in polynomial time in $\log(N)$ and d. A recent version of Coppersmith's method was published in 2013 by Bi and Nguyen [1]. Their method implements a faster LLL algorithm, L^2 and uses a rounding technique over the entries of the lattice matrix.

Algorithm 2 Coppersmith's Second Step

Input: N large integer with unknown factorization and f a monic polynomial of degree d.

   Output: All solutions such that $f(x_0) \equiv [N]$ and $|x_0| \leq X$

1. Set $n = \lceil \log(N) \rceil$, $X = \lfloor \frac{1}{4}N^{\frac{1}{d}} \rfloor$, $t = \lfloor -N^{\frac{1}{d}} \rfloor + X$
2. **while** $t \leq N^{\frac{1}{d}}$ **do**

Call Algorithm 1: $N, f(x - t), n, X$
Store $x_0 + t$ for each solution from Coppersmith's step 1.

$$t = t + 2X$$

Return All solutions x_0 stored.
This method is deterministic, with a running time of $O(d^5 \log^9(N))$ [1].

3 Improved Implementation

We will now present our contribution for improving Coppersmith's method.

Finding all solutions $f(x) \equiv [N]$ with $|x_0| \leq N^{\frac{1}{d}}$ can be achieved in a polynomial time $O(d^5 \log^9(N))$. In Coppersmith's method section, it was shown in the demonstration that $n \approx \log(N)$ with n as the lattice dimension. We safely assume that the complexity can be expressed as $O(d^5 n^9)$. Based on this observation, we can try to compute the same solutions using a different approach. The first step of Coppersmith's method find all solutions such that $x_0 \leq \frac{1}{2} N^{\frac{1}{d} - \epsilon}$, and by taking $\epsilon = \frac{1}{\log_2(N)}$ we find an upper bound $|x_0| \leq \frac{1}{4} N^{\frac{1}{d}}$.

The second step of Coppersmith's method is a consequence of the factor $\frac{1}{4}$, by subdividing the interval $[-N^{\frac{1}{d}}, N^{\frac{1}{d}}]$ into four subintervals and performing the first step of Coppersmith's method on each one.

Our amelioration goes as follows, by taking $\epsilon = \frac{1}{\log_a(N)}$ we find a new upper bound $|x_0| \leq \frac{1}{2a} N^{\frac{1}{d}}$ over the roots x_0. We consider now a new variable 'number of subintervals' in the problem . This will allow us to relieve the pressure on the dominant factor in the complexity by calling the first step of Coppersmith's method multiple times, since $n \approx \log_a(N)$ will get smaller as a grows.

Intuitively, if we suggest calling Coppersmith's algorithm multiple times, it will only add more in the running time. In our approach, we fortify our claim by using the fact that LLL algorithm complexity $O(n^6 \log^3(b))$ uses, b, the greatest length of b_i under the Euclidean norm and n the lattice dimension. Both parameters will be reduced using our approach. Since the new lattice dimension is $n \approx \log_a(N)$, this implies that the matrix entries are bounded by a new upper bound $\frac{1}{2a} N^{\frac{1}{d}}$.

Algorithm 3 Amelioration of Coppersmith's Second Step
Input: N large integer with unknown factorization and f a monic polynomial of degree d, a number of sub-intervals.
   Output: All solutions such that $f(x_0) \equiv [N]$ and $|x_0| \leq X$

1. Set $n = \lfloor \log_a(N) \rfloor$, $X = \lfloor \frac{1}{2} N^{(\frac{1}{d} - \frac{1}{n})} \rfloor$, $t = \lfloor -N^{\frac{1}{d}} + X \rfloor$.
2. **while** $t \leq N^{\frac{1}{d}}$ **do**

   Call Coppersmith's First Step: N, $f(x - t)$, n, X
   Store $x_0 + t$ for each solution from Coppersmith's First Step.

$$t = t + 2X$$

   Return All solutions x_0 stored.
   If we take $a = 2$, we find Coppersmith's original approach to be a worst-case scenario.

4 Experimental Results

We will now study the amelioration over the running time compared with the usual approach.

We use monic cubic polynomials with solutions near the upper bound as test examples: $f(x) = (x - (x_0 - 1))(x - x_0)(x - (x_0 + 1))$ with $x_0 \approx N^{\frac{1}{3}}$.

These examples are considered a worst case scenario, since the solutions are exactly $X - 1, X$, and $X + 1$. We avoid using any speed-up method for calculation that will affect Coppersmith's theorem. We also use *fplll* [4] to perform lattice reductions, since it is considered the fastest LLL method used in practice.

All our tests are performed on polynomials with solutions around the upper bound $N^{\frac{1}{3}}$. We assume that a message was encrypted and sent using a certain public key (e, N), where N varies in size. We also possess an amount of information about the plain text, with $e = 3$ we need to know at least $\frac{2}{3}$ of the message M. Let l and l' be respectively the length of M and M' the known part of the message and C the intercepted cryptogram. We build the polynomial as follows: $f(x) = \left(M' \cdot 2^{l-l'} + x\right)^e - C$ (Table 1).

In this table, we can see that for N of size 256, it took on average six days using Coppersmith's approach. We managed to reduce that time to 30 min. Of course, the number of subintervals cannot exceed a certain threshold, since the ideal situation is where the complexity is balanced between the variable a and the lattice dimension $n \approx log_a(N)$. To find this balance, we provide the following theorem. Let f be a monic polynomial of degree d and an integer N with unknown factorization of size n bits. Let $\alpha \in \mathbb{N}$ such that $\alpha < n$, Algorithm 1 can find all roots x_0 to $f(x_0) \equiv 0[N]$ such that $|x_0| \leq N^{\frac{1}{d}}$ in $2^{\alpha+1}$ repetitions at least.

Proof We know for $a = 2$ and $\epsilon = \frac{1}{log_a(N)} = \frac{1}{n}$, Algorithm 1 can find all solutions in 4 repetitions.

For $a = 2^\alpha$ and $\epsilon = \frac{1}{log_a(N)} = \frac{1}{n}$, we find:

$$|x_0| \leq \frac{1}{2}N^{\frac{1}{d}-\frac{1}{n}} = \frac{1}{2}N^{\frac{1}{d}-\frac{1}{log_a(N)}} = \frac{1}{2a}N^{\frac{1}{d}} = \frac{1}{2^{\alpha+1}}N^{\frac{1}{d}}$$

On the other hand, $log_a(N) = \frac{\ln(N)}{\ln(2^\alpha)} = \frac{log_2(N)}{\alpha} = \frac{n}{\alpha}$. By doing so, the new lattice dimension will be $\frac{n}{\alpha}$ and it will need at least $2^{\alpha+1}$ repetitions of Coppersmith's step 1 to find all roots x_0 such that $|x_0| \leq N^{\frac{1}{d}}$.

The optimal choice of α would be maximal, such that the lattice dimension is minimal. For one polynomial f, we have created $log_2(N)$ polynomials g such that $g(x_0) = 0$ over $\mathbb{Z}$. Since we know that the number of repetitions isn't restricted only to four times. We will apply the same principal, by repeating Coppersmith's step 1 $log_2(N)$ time. We find:

$$2^{\alpha+1} = log_2(N)$$

Table 1 Calculating running time in seconds for different size of N in bits

Bits	a								
	2	3	4	5	6	7	8	9	10
32	0.215	0.085	0.066	0.072	0.08	0.84	0.082	0.135	0.155
64	6.24	0.77	0.56	0.33	0.26	0.27	0.25	0.24	0.25
128	3858.46	203.08	54.18	19.93	15.8	10.71	8.28	7.77	6.61
256	523,656 ≈6 days	52,029.4 ≈14 h	21,074.5 ≈5 h	10,787.6 ≈3 h	5880.62 ≈ 1 h38min	3620.21 ≈1 h	3066.03 ≈50 min	2351.29 ≈35 min	1816.13 ≈30 min

Table 2 Calculating running time in seconds for different size of N in bits

Bits	$a = 2^\alpha$						
	2^4	2^5	2^6	2^7	2^8	2^9	2^{10}
128	3.8	2.47	2.73	3.71	7.8	7.5	26.97
256	692.53 $\approx$10 min	350.8 $\approx$5 min	225.44 $\approx$3 min	179.55 $\approx$3 min	176.26 $\approx$3 min	176.41 $\approx$3 min	194.49 $\approx$3.2 min

Table 3 Calculating running time in seconds for N of 512 bits

$a = 2^\alpha$ Bits	2^8	2^9	2^{10}
512	37,264.3 $\approx$10 h	31,964.5 $\approx$8.7 h	30,761.3 $\approx$8.5 h

$$\alpha = log_2(log_2(N)) - 1$$

The new lattice dimension will be $\frac{log_2(N)}{log_2(log_2(N))-1}$.

Performing the same tests on the new choice of a yields the next table (Table 2).

From these experimental tests, we notice that the running time starts to increase when $a = 2^\alpha > log_2(N)$, and the optimal value of α is not too far from $\alpha \approx log_2(log_2(N))$.

Applying our result to N of 512 bits right around the optimal choice of α gives us conclusive results as shown in Table 3.

5   Conclusion

In this work, we have presented a new way to ameliorate Coppersmith's method. Our technique uses more than four subintervals as in the original version of Coppersmith's paper. Our results have been confirmed by experimental tests on computers.

References

1. Bi, J., Nguyen, P.Q.: Rounding LLL: finding faster small roots of univariate polynomial congruences. Public-Key Cryptogr.-PKC **2014**, 185–202 (2014)
2. Boneh, D.: Twenty years of attacks on the rsa cryptosystem. Not. AMS **46**(2), 203–213 (1999)
3. Coppersmith, D.: Finding a small root of a univariate modular equation. Adv. Cryptol. Eurocrypt **96**, 155–165 (1997)
4. Development team, T. F. fplll, a lattice reduction library. Available at 2016
5. Galbraith, S.D.: Mathematics of Public Key Cryptography. Cambridge University Press (2012)
6. Goldreich, O., Goldwasser, S., Halevi, S.: Public-key cryptosystems from lattice reduction problems. In: Advances in Cryptology-CRYPTO'97: 17th Annual International Cryptology

Conference, Santa Barbara, California, USA, August 1997. Proceedings, p. 112. Springer (1997)

7. Howgrave-Graham, N.: Approximate integer common divisors. In: CaLC, vol. 1, pp. 51–66. Springer (2001)
8. Lenstra, A., Lenstra, H., Lovasz, L.: Factoring polynomials with rational coefficients. Math. Ann. **261**(4), 515–534 (1982)
9. Lu, Y., Peng, L., Kunihiro, N.: Recent progress on coppersmiths lattice based method: a survey. In: Mathematical Modelling for Next-Generation Cryptography, pp. 297–312. Springer (2018)
10. May, A.: Using LLL-reduction for solving RSA and factorization problems. Inform. Secur. Cryptogr., 315–348 (2009)
11. Mumtaz, M., Ping, L.: Forty years of attacks on the rsa cryptosystem: A brief survey. J. Discrete Math. Sci. Cryptogr. **22**(1), 9–29 (2019)
12. Nguyen, P.Q.: Cryptanalysis of the Goldreich-Goldwasser-Halevi cryptosystem from crypto97. In: Annual International Cryptology Conference, pp. 288–304. Springer (1999)
13. Nguyen, P.Q.: Public-key cryptanalysis. Recent Trends Cryptogr. Contemp. Math. **477**, 67–120 (2009)
14. Nguyen, P.Q., Stehle, D.: Low-dimensional lattice basis reduction revisited. In: ANTS, vol. 3076, pp. 338–357. Springer (2004)
15. Nguyen, P.Q., Stehle, D.: An LLL algorithm with quadratic complexity. SIAM J. Comput. **39**(3), 874–903 (2009)
16. Nguyen, P.Q., Stehle, D.: Low-dimensional lattice basis reduction revisited. ACM Trans. Algorithms (TALG) **5**(4), 46 (2009)
17. Steele, J.M.: The Cauchy-Schwarz Master Class: An Introduction to the Art of Mathematical Inequalities. Cambridge University Press, Cambridge (2004)

A Conceptual Model for Dynamic Access Control in Hadoop Ecosystem

Hafsa Ait idar, Hicham Belhadaoui, and Reda Filali

Abstract Big data, huge and varied collections of data, is a blanket concept that is greatly used in recent years. Hadoop has imposed itself as the principal big data platform for storing and processing multiple data types. It is widely used in various sectors which may contain a huge amount of sensitive data. Hadoop provides a highly available, cost-effective, and fault-tolerant ecosystem to deal with sensitive data with high speed. Due to the distributed nature of this platform, privacy and security threats have become a critical concern within Hadoop. For this reason, the proposed Dynamic Data Sensitivity Access Control (D2SAC) framework in [1] aims to protect sensitive data contained in Hadoop platform. In this paper, we aim to keep enhancing this framework by presenting a conceptual model of the D2SAC framework, then we provide a formal definition of each component of this conceptual model. In addition, we give a detailed description of the Access Enforcement Module that is responsible for providing the access decision about the user request.

Keywords Sensitive data · Access control · Authorization · Hadoop · Big data

1 Introduction

Over recent years, the amount of data has exploded around us. Data are generated from numerous sources, mainly from social media, server logs, business transactions, and sensors [2]. Such massive data which cannot be stored, processed, and analyzed using traditional Relational Database Management System (RDBMS) are typically referred to as the blanket term Big Data [3].

H. Ait idar (✉) · H. Belhadaoui · R. Filali
RITM, CED Engineering Science, National High School for Electricity and Mechanics (ENSEM), Hassan II University, Casablanca, Morocco
e-mail: hafsa.aitidar93@gmail.com

H. Belhadaoui
e-mail: belhadaoui_hicham@yahoo.fr

R. Filali
e-mail: filalihilalireda@gmail.com

F. Saeed et al. (eds.), *Advances on Smart and Soft Computing*, Advances in Intelligent Systems and Computing 1188, https://doi.org/10.1007/978-981-15-6048-4_37

The term big data includes all sorts of data we generate today. With volume, variety, velocity, and veracity of data, a scalable cluster with high processing power and huge storage capabilities is required to treat this reams of data [2].

Hadoop [4] has emerged as an important open platform to deal with the main big data challenges. Distributed storage, scalability, reliability, high availability, and cost-effective are the major features of this platform [5]. Hadoop is specially designed to handle large volumes of data in a highly reliable and timely manner. This is mostly due to several Hadoop ecosystem components like HDFS, MapReduce, YARN, Hive, HBase, Ambari, and so on.

Hadoop is widely used in numerous sectors like health care, financial services, manufacturing systems, higher education, telecommunications, etc. The information generated by these sources may contain a huge amount of sensitive data [3]. Such data can reveal a person's identity or a company's trade secrets, which could present a serious danger if recognized by undesired individuals. Hence, protecting sensitive data against illegitimate users and maintaining secure access became a necessary requirement for Hadoop.

Considering the amount and diversity of sensitive data, on one hand, the distributed nature and scalability of Hadoop ecosystem, on the other hand, sensitive data are even more vulnerable to many attacks (denial of service, man in the middle, impersonation, etc.) [6]. For this reason, securing data within Hadoop is becoming increasingly difficult, which requires appropriate authentication, authorization, and access control models to overcome these security breaches [5].

Apache Sentry [7] and Apache Ranger [8] are the most used security projects to ensure fine-grained access control for Hadoop cluster. In our previous paper, we proposed the Dynamic Data Sensitivity Access Control (D2SAC) framework in order to enhance the protection of sensitive data located in Hadoop ecosystem [1]. In this paper, our main purpose is to present the conception of the D2SAC framework and detail each component of the conceptual model. Furthermore, we focus on the Access Enforcement Module that is responsible for providing the access decision within the D2SAC in order to provide an automated, strong, and dynamic framework.

The remainder of this paper is as follows: Sect. 2 contains an overview of access control models in Hadoop. We present the proposed D2SAC framework and its components in Sect. 3. In Sect. 4, we introduce the conceptual model of D2SAC, followed by the description of the Access Enforcement Module in Sect. 5. Section 6 gives the conclusion of our work.

2 Access Control for Hadoop Ecosystem

Access control is considered one of the major requirements to keep sensitive information protected against unauthorized access and illegitimate users within Hadoop ecosystem [9].

For a while, the role-based access control (RBAC) model was regarded as a mature and natural approach for controlling resources access within organizations [10]. The idea behind RBAC is that every user is assigned a particular role and every role is associated with a set of permissions, which means that a user may only access the organization resources if his role has the relevant privileges [9].

Using roles minimizes the administrative tasks for companies, facilitates the management of users' rights by adding or revoking roles, thereby enabling organizations to determine their constraints with more flexibility [10]. However, limitations and difficulties are also many especially for large companies with thousands of employees [9]. Defining the right roles and assigning them to appropriate users is becoming increasingly cumbersome to handle. Also, considering all permissions a user will need without granting too many privileges could result in creating a hundred of roles, making the implementation and deployment of the RBAC model in the context of big data more challenging.

Attribute-based access control (ABAC) model is used to provide great flexibility in making access decisions and to prevent the role explosion problem [11]. ABAC model was proposed as a solution to protect sensitive information from deliberate abuse in various sectors [11, 12].

Issues related to security and privacy in big data are handled in [5, 6, 13]. There have been several papers [14–16], which discuss the main security challenges in Hadoop in order to propose suitable solutions. A security system called Vigiles to ensure fine-grained access control for MapReduce systems is proposed in [17]. Reddy in [18] proposed an access control framework to protect sensitive data in HDFS. This model depends entirely on the data owner's guidelines. A fine-grained access control policy is presented in [19] to ensure big data security in the cloud. In this approach, the authorization to access data is only dependent on the data owners.

Authors in [3] proposed a content sensitivity-based access control (CSBAC) framework to estimate the sensitivity of data with minimal intervention of user. The CSBAC uses the dataset itself with no base set to compute its sensitivity, unlike the work presented in [20] where it is necessary to define a base set. In this framework, the data sensitivity estimator estimates the sensitivity of data item according to its information gain. However, the domain expert is involved when the data have not been previously encountered in order to train the neural network.

3 Proposed Dynamic Data Sensitivity Access Control Framework

Considering the distributed nature of Hadoop ecosystem, sensitive data located in Hadoop like Personally Identifiable Information (PII), financial data, payment data, and many others may suffer from multiple threats. Distributed denial of service (DDoS), data leakage, data breach, and data theft represent the common attacks that target the contained data in Hadoop. Thus, securing sensitive data in Hadoop becomes absolutely the primary concern and the major challenge for many organizations.

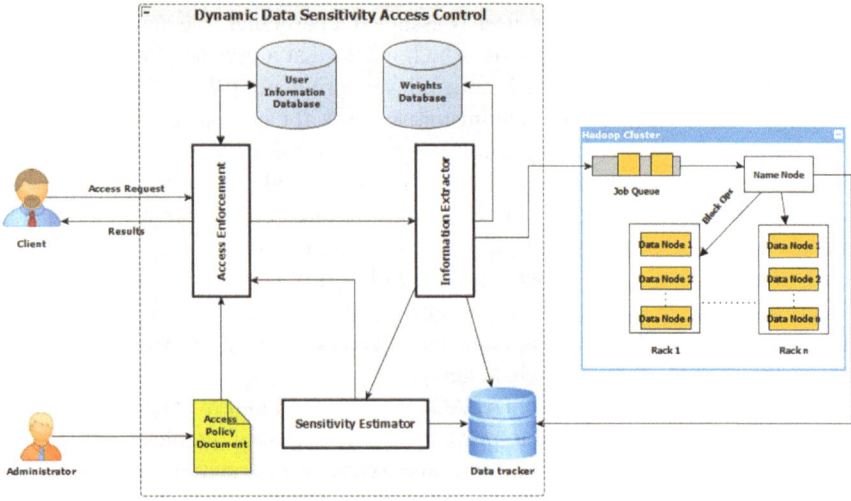

Fig. 1 Proposed D2SAC framework

To address this issue and keep sensitive data secured, we proposed in our previous work the Dynamic Data Sensitivity Access Control (D2SAC) framework, as shown in Fig.1.

The D2SAC consists of a set of components, each one has its proper functionalities in order to propose a strong access control framework. The main components of D2SAC framework [1] are:

- The Access Enforcement Module (AEM) receives the requester's request and makes access control decisions. A description of this component is given in Sect. 5.
- The Sensitivity Estimator Module (SEM) calculates the sensitivity value of data using a mathematical model that is based on the Analytic Hierarchy Process method.
- The Information Extractor Module (IEM) receives the criteria used in the calculation of data sensitivity and extracts their corresponding weights from the weights database. Later, it submits these weights to the Sensitivity Estimator Module.

In this paper, we aim firstly to concentrate on the conception of the D2SAC framework and to present a formal definition of each element of the conceptual model. Then, we give details about the Access Enforcement Model.

4 Conceptual Model of D2SAC

The Dynamic Data Sensitivity Access Control (D2SAC) framework aims to protect sensitive data within Hadoop cluster, and for this reason, we calculate the sensitivity value for each object or data stored in HDFS using a mathematical model. This value is represented as a numerical value and it is safeguarded by our framework. On the other hand, users are classified based on their actual jobs, thus, each user will be assigned the most needed and relevant role to accomplish his job. In the D2SAC framework, every role has a weight that is indicated in the Access Policy Document (APD). This weight is also expressed as a numerical value and aims to specify the access level that users can reach. Consequently, based on the actual role of user, he may or not be authorized to access the sensitive data.

In this section, we present the conceptual model for the D2SAC framework, as shown in Fig. 2, followed by formal definitions of each component.

1. The essential components of the D2SAC include:

   - Users (U), a user is an individual who interacts with computer to access data and services within Hadoop platform.
   - Groups (G), a group is a set of users with similar job requirements.
   - Subjects (S), a subject refers to a process executed on behalf of the user to fulfill operations in the cluster. The subject is always executed with full privileges of its creator.
   - Roles (R), a role is a set of permissions assigned to users in the system.
   - Hadoop services (HS), Hadoop ecosystem consists of a set of components, like HDFS, HBase, Hive, Yarn, Sqoop, etc. A Hadoop service is considered as Hadoop component which allows access to data and objects inside Hadoop.
   - Objects (OB), an object is considered as a resource of the system that needs permissions to get access.
   - Operations (OP), an operation is considered as an action that will be executed upon objects (e.g., read, write, edit, etc.).

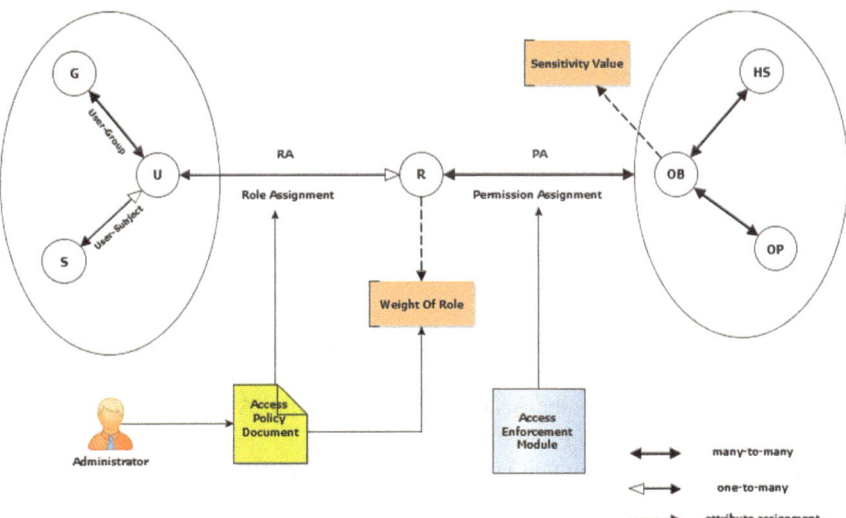

Fig. 2 Conceptual model of D2SAC

- Sensitivity value is an attribute attached to each object in the system. This value is calculated by the Sensitivity Estimator Module in an automated way within our framework, and it is expressed as a numerical value in order to limit access to sensitive data.
- Role weight: the weight of role is defined in the APD and it is attributed to each role. When a user attempts to reach an organization's object, the weight of its role must dominate the sensitivity value of the requested object before accessing it.

2. In this model, each user (U) can have only one role (R) at the same time in order to eliminate the risk of granting too many rights to one user. This is illustrated by one-to-many relation between user and role. Similarly, the user–subject relationship is also one-to-many.
3. The rest of the relationships in our model are many-to-many:

- A user can be assigned to multiple groups, and a group may contain many members which explain the many-to-many relation.
- Hadoop services defined as Hadoop components can deal with a variety of data, files, etc. Also, several operations can be performed on these objects, thus depicting the many-to-many relationships.

4. There are two types of assignments in our model:

- Role assignment (RA): based on the information presented in the Access Policy Document, the adequate role is assigned to the user.
- Permission assignment (PA): permissions represent the access rights to the organization's resources. Based on the weight of the user's role and the sensitivity value of the object, access can be granted or not to the requesting user.

In the next section, we provide more detail on these assignments.

5   Analysis of Access Enforcement Module

The Access Enforcement Module (AEM) is an important component of the D2SAC framework, it receives the user request and returns the result if this user is authorized to access the requested data.

After receiving the user request, the AEM verifies that the input data path exists in the HDFS, then it uses the user information database to check if the submitted information is correct. If so, the Access Policy Document (APD) is invoked in order to retrieve the corresponding role of this user.

The Access Policy Document is extremely important to ensure the proper functioning of our framework. It contains information related to each role within the company. Figure 3 depicts an example of the Access Policy Document.xml file. As

```xml
<?xml version="1.0"?>
    <Roles>
        <Role id="101" name="Chief Executive Officer">
            <SecurityLevel>high</SecurityLevel>
            <WeightOfRole>0.8</WeightOfRole>
            <Category>Administrator</Category>
            <Users>
                <User id="1011"/>
                <User id="1012"/>
            </Users>
        </Role>
        <Role id="201" name="Financial Analyst">
            <SecurityLevel>medium</SecurityLevel>
            <WeightOfRole>0.07</WeightOfRole>
            <Category>Staff</Category>
            <Users>
                <User id="2011"/>
                <User id="2012"/>
                <User id="2013"/>
            </Users>
        </Role>
    </Roles>
```

Fig. 3 Example of APD.xml file

shown in this figure, the information includes the security level, weight of the role, category, and users assigned to this role.

The APD indicates the user role and the associated data sensitivity score that this user can access. For example, the first role "Chief Executive Officer" is assigned to two users. This role has a weight of 0.8, which means that the attributed users can access all the data with a sensitivity value less than 0.8. Considering the importance of the APD, the administrator is the only one capable of editing this document, if necessary.

Algorithm 2 shown in Fig. 5 is invoked whenever a user tries to access a data stored in the HDFS. This algorithm accepts the user name and the required data and returns the access decision (either true or false). Algorithm 2 calls Algorithm 1 (Fig. 4) that returns the weight of the user role defined in the APD. Then, it obtains the sensitivity value of data computed by the SEM. Based on these values, if the weight of the user role is greater than or equal to the sensitivity value of data, then the user can be able to access the demanded data. Otherwise, the user is not authorized to access it. Furthermore, the data tracker receives every time all the information about the requester and the sensitivity of data after accessing this data.

With:

UR: the role of user.
ID: the user identifier.
W_r: the weight of the user role.
S_d: the sensitivity of data.

Fig. 4 Algorithm 1

> **Algorithm 1: Retrieving the weight of user role**
>
> **Input:** ID: User identifier
>
> **Output:** W_r: Weight of user role
>
> Begin
>
>   If ID exists in APD then
>
>   UR ←getUserRole(ID)
>
> W_r ← weightOfRole(UR)
>
>   Return W_r
>
>   Else
>
>   Return null
>
>   End If
>
> End

Fig. 5 Algorithm 2

> **Algorithm 2: Returning the decision**
>
> **Input:** D: Input data, U: User name
>
> **Output:** Access Decision
>
> Begin
>
>   ID ← getUserID(U)
>
> W_r ← Algorithm_1(ID)
>
> S_d ← getDataSensitivity(D)
>
>   If $W_r >= S_d$ then
>
>   Return true
>
>   Else
>
>   Return   false
>
>   End If
>
> End

6 Conclusion

Because Big Data technology is evolving at a fast pace, traditional means to implement security systems are now insufficient to overcome security issues introduced by this technology. Hadoop is considered as the most popular framework to deal with massive data through the use of a cluster scale-out environment. Sensitive data contained in Hadoop are exposed to countless security threats. In order to improve access to sensitive data, we have proposed the D2SAC framework in our previous work with the aim to protect and secure sensitive data within Hadoop.

In this paper, we first present related works of access control in Hadoop. Then, we introduce our proposed D2SAC framework and its principal components. We further propose the conception of the D2SAC and we detail each component of the conceptual model. We additionally highlight the functioning of the Access Enforcement Module which plays a crucial role in strengthening access control of the D2SAC. For future work, we plan to enhance our framework by providing access control policies to protect sensitive data depending on their degree of sensitivity. Also, we aim to continue developing our framework in order to test the efficiency, flexibility, and rapidity of D2SAC.

References

1. Ait Idar, H., Aissaoui, K., Belhadaoui, H., Hilali, R.F.: Dynamic data sensitivity access control in Hadoop Platform. In: 2018 IEEE 5th International Congress on Information Science and Technology (CiSt), pp. 105–109. IEEE, Marrakech (2018)
2. Gupta, M., Patwa, F., Sandhu, R.: Object-tagged RBAC model for the Hadoop ecosystem. In: Livraga, G., Zhu, S. (eds.) Data and applications security and privacy XXXI, pp. 63–81. Springer International Publishing, Cham (2017)
3. Ashwin Kumar, T.K., Liu, H., Thomas, J.P., Hou, X.: Content sensitivity based access control framework for Hadoop. Digit. Commun. Netw. **3**, 213–225 (2017)
4. Apache Hadoop, https://hadoop.apache.org/
5. Bhathal, G.S., Singh, A.: Big Data: Hadoop framework vulnerabilities, security issues and attacks. Array. **1–2**, 100002 (2019)
6. Behera, M., Rasool, A.: Big data security threats and prevention measures in cloud and Hadoop. In: Balas, V.E., Sharma, N., Chakrabarti, A. (eds.) Data Management, Analytics and Innovation, pp. 143–156. Springer Singapore, Singapore (2019)
7. Apache Sentry, https://sentry.apache.org/
8. Apache Ranger, https://ranger.apache.org/
9. Ait Idar, H., Aissaoui, K., Hicham, B., Filali Hilali, R.: A Survey on Semantic Access Control in Cloud Computing (2018)
10. Younis, A., Kifayat, Y., Merabti, K.M.: An access control model for cloud computing. J. Inf. Secur. Appl. **19**, 45–60 (2014)
11. Cavoukian, A., Chibba, M., Williamson, G., Ferguson, A.: The importance of ABAC: attribute-based access control to big data: privacy and context. Priv. Big Data Inst. Ryerson Univ. Tor. Can. (2015).
12. Krishnan, R.: Access control and privacy policy challenges in big data. In: NSF (National Science Foundation) workshop on big data security and privacy, p. 2 (2014)

13. Inukollu, V.N., Arsi, S., Rao Ravuri, S.: Security issues associated with big data in cloud computing. Int. J. Netw. Secur. Its Appl. **6**, 45–56 (2014)
14. Das, D., O'Malley, O., Radia, S., Zhang, K.: Adding security to apache hadoop. Hortoworks IBM. 26–36 (2011)
15. Gupta, M., Patwa, F., Sandhu, R.: POSTER: Access control model for the Hadoop ecosystem. In: Proceedings of the 22nd ACM on Symposium on Access Control Models and Technologies-SACMAT '17 Abstracts, pp. 125–127. ACM Press, Indianapolis, Indiana, USA (2017)
16. Sharma, P.P., Navdeti, C.P.: Securing big data Hadoop: a review of security issues, threats and solution. Int J Comput Sci Inf Technol. **5**, 2126–2131 (2014)
17. Ulusoy, H., Kantarcioglu, M., Pattuk, E., Hamlen, K.: Vigiles: fine-grained access control for mapreduce systems. In: 2014 IEEE International Congress on Big Data, pp. 40–47. IEEE (2014)
18. B. Reddy, Y.: Access control for sensitive data in Hadoop distributed file systems. In: The Third International Conference on Advanced Communications and Computation, pp. 17–22 (2013)
19. Yuan, Q., Ma, C., Lin, J.: Fine-grained access control for big data based on CP-ABE in cloud computing. In: International Conference of Young Computer Scientists, Engineers and Educators, pp. 344–352. Springer (2015)
20. Zeng, W., Yang, Y., Luo, B.: Access control for big data using data content. In: 2013 IEEE International Conference on Big Data, pp. 45–47. IEEE (2013)

Towards a Blockchain and Smart Contracts-Based Privacy Framework for Decentralized Data Processing

Marc Gallofré Ocaña, Abdo Ali Al-Wosabi, and Tareq Al-Moslmi

Abstract Analysing and processing data are not exempt of privacy policies, particularly when talking about personal data. This problem becomes more complex and
cumbersome when sharing the data between parties, entities or even across borders.
Different policies, implications and interests may apply, and the risks of losing control
and privacy protection arise. Yet, new forms of analysing, protecting and sharing data
should be considered in order to safeguard the privacy. This paper introduces a general
framework for privacy assurance in data processing applications. This framework is
based on blockchain and smart contract technology that have received increased
attention recently. Our solution reduces the need for a credible centralized authority
and maintains the privacy of data and information for data processing applications.

Keywords Data processing · Blockchain · Privacy · Smart contracts

1 Introduction

With the development of technology, an abundance of data became available on the
Internet making the benefit of it an urgent need for many companies, institutions and
governments, and hence, the need and interest of data processing applications using
artificial intelligence (AI), natural language processing (NLP) [1], machine learning
(ML) [2] or data analysis.

With the advancement in data processing applications, it has become essential
to consider aspects like data privacy. Recently, blockchain technology has become
more popular for applications that require strong privacy and security. Examples

M. Gallofré Ocaña (✉) · T. Al-Moslmi
University of Bergen, Bergen, Norway
e-mail: Marc.Gallofre@uib.no

T. Al-Moslmi
e-mail: Tareq.al-moslmi@uib.no

A. A. Al-Wosabi
BIT Group Sdn Bhd, Cyberjaya, Malaysia
e-mail: abdoali.abdullah@bit.com.my

© The Editor(s) (if applicable) and The Author(s), under exclusive license to Springer   431
Nature Singapore Pte Ltd. 2021
F. Saeed et al. (eds.), *Advances on Smart and Soft Computing*, Advances in Intelligent
Systems and Computing 1188, https://doi.org/10.1007/978-981-15-6048-4_38

are Bitcoin [3], Ethereum [4, 5] and other cryptocurrencies which allow people to make safe and trustworthy purchases and transfers anonymously. Blockchain [6] is a shared repository that tracks all transactions (blocks) as a sequence from the start and secures the data from unauthorized access using encryption and consensus algorithms. A blockchain network can be either a public network where all encrypted transactions are public or a private network where only authorized users can read and write blocks. The entire network, as opposed to a single agency, such as a bank, hospital or a state, is constantly checking its credibility [7]. In this way, users do not have to trust a single authority, rather security is assured by the vigour and computing power of the entire network involved in the blockchain network which validates all encrypted transactions on it by consensus. Some blockchains, like Ethereum, have implemented the concept of smart contracts which are computer programs that directly control the transaction between users and can define obligations, rules or sanctions, whereas the blockchain ledger contains information about the present and past state of a collection of business items (such as healthcare data and online news), a smart contract determines the functional logic that generates new facts that are added to the distributed ledger. Chain code is usually utilized by administrators to assemble similar smart contracts for deployment. In fact, the smart contract specifies the laws for various organizations in the executable code. When a smart contract is executed by users, it produces a transaction if the stabilized conditions are fulfilled. Apps rely on a smart contract to produce transactions that are registered on the distributed ledger [8].

This paper presents the blockchain-based framework for data processing applications that utilizes blockchain technology and smart contracts to guarantee data privacy between participants. The proposed solution eliminates the need for a reliable centralized authority and ensures privacy of sensitive data and information for data processing apps. The rest of this paper is organized as follows: Sect. 2 presents the related works. Section 3 introduces the proposed framework, and Sect. 4 concludes the paper and outlines plans for future work.

2 Related Works

Analysing and processing data are not exempt from privacy discussions either, particularly when talking about personal data. This problem becomes more complex and cumbersome when sharing the data between parties, entities or even across borders. Different policies, implications and interests may apply, and the risks of losing control and privacy protection arise. Yet, new forms of analysing, protecting and sharing data should be considered in order to safeguard control over data and privacy.

For instance, clinical and medical sectors are used to deal with personal data, not surprisingly, they are becoming one of the first adopters of new ways of analysing data while maintaining such privacy. In [9], authors proposed an infrastructure for decentralized data analysis over personal health data. The infrastructure involves three different actors: personal health owners, a trusted entity and researchers. The trusted secured and independent entity is in charge of executing the analysis designed by the researchers and accessing the data owned by different entities. This trusted party does not contain data by itself, it only accesses data required for the analysis. Finally, results (e.g. plots, models, statistics without) without personal or sensitive data are sent back from the trusted entity to the researchers. The proposed infrastructure neither requires to transfer the original data (it can be transformed, anonymized or pseudonymized and encrypted before being shared), nor sent the whole data to a central party that does the analysis too; and data owners can decide whatever they make available for processing, and no other parties can see the data of each other's.

As mentioned in the previous work, there is still an issue how to ensure, manage and control agreements and consensus between the different parties, enhancing security when encryption keys are shared between parties and how to provide a complete transparency and auditability over data that is being collected, ownership and agreements. To do so, blockchain technologies are being positioned as a possible solution, as [10] proposed for enhancing control over personal data, ownership and manage access permissions. Authors proposed a decentralized platform using blockchain to address and protect personal data ownership, transparency, auditability and access control for mobile user applications and services which want to process their personal data. Users own their personal data which is stored in the platform and accessed through the blockchain. Through the blockchain, users can agree, modify and revoke data access permissions to services. Both users and services can access the data only if their digital signature and permissions are verified by the blockchain.

Recently, a new paradigm has been proposed: the re-decentralization of the Web by its creator Tim Berners-Lee. This new approach aims to bring the data back to the owners (users) using SOLID[1] pods and semantic and linked data. Yet, using the SOLID pods, the user can store the data in a safe place, keeps its ownership while using semantic vocabularies and ontologies that facilitate its sharing and linking between data sets. As seen in the previous work, the decentralization of the data and the improved privacy and data access control can be enhanced using decentralized technologies like blockchain which can be combined and improved by using SOLID pods like in [11, 12].

Considering the literature, it makes sense to examine the possibility of using blockchain and decentralized infrastructures for data processing applications, which can lead to performance benefits while maintaining data privacy, security and ownership and mitigating the chances of data leaks and misuses as well.

3 The Framework

In this section, we explain our proposed framework, the blockchain-based privacy framework for data processing applications, which uses smart contracts to provide

---

[1] SOLID stands for: "Social linked data". https://solid.mit.edu

data privacy in data processing. Our approach removes the need for a reliable central-ized authority and guarantees high integrity, transparency and confidentiality of data and information for data processing applications.

3.1 General Framework Overview

Using blockchain technology and smart contracts has the potential to transform the protection of the intended data set and applications into an automated distributed system that guarantees the quality of the data and the outcome set provided to the end-user. Our suggested framework would concentrate on the use of smart contracts implemented separately on the open-source blockchain platform. The execution of the smart contract functions will be carried out by selected peers (nodes) and will be authenticated (approved) by a number of validators nodes that are globally spread based on the designed consensus protocol (agreement). Smart contracts will accept pre-defined requests in the form of method calls and will also cause events to allow the contributing validators to actively observe, monitor and receive relevant warnings when deviations occur. In doing so, it ultimately aims to return the targeted data and application to the optimum and to react to deviations that arise within the data chain.

Figure 1 illustrates a general overview of the proposed framework. As shown in the figure, the framework integrates four entities, the participating user, the blockchain and smart contracts platform, data processing applications and the targeted data.

Fig. 1 General overview of the proposed framework

3.2 Framework Design

Each participating entity has a role, association and interactions with other entities. There are four main participating entities, and their roles (function) are summarized as follows:

- **User**: The participating user is the end-user who requests and receives the manipulated data from the data processing application. Each user has a well-defined access control (permissions) in order to retrieve the desired data and get the generated information (knowledge). Users need to have specific permissions and unique private key in order to retrieve the processed data or results. Thus, users (i.e. system analyst or researcher) would be able to access and request encrypted results data by using his public key.

- **Data processing application**: The data processing application collects the required data from the data pool (data set), process the data (e.g. classifies and clusters the collected data, extracts and do mining the required knowledge) or execute applications contained in dockers[2] sent by the users. All the data to be processed is encrypted and stored in the blockchain using hash values. Thus, all data can be available to the intended application based on the previous approval given by the data owner using smart contracts and after being encrypted using a public key. Then, intended application can retrieve the data for its processing, which will be logically correlated with a specific owner using unique identifier which is not conveying personality information of that person.

- **Data**: The targeted data is generally a non-correlated data that is produced from different systems (or applications) and needs to be processed for the participating users. Personal data will be stored in the blockchain as digested values using and encrypted using unique keys. The privacy factor of data that could be classified as private would be guaranteed by our framework as all related data (either user data or generated information) will be stored in blockchain network as encrypted hash values. Thus, data records cannot be read or processed unless specific permissions are granted from the owner of that data using smart contracts. Hence, private data (such as health data records) will be protected and prevented from unwanted potential access (i.e. not approved). Once the needed permissions are given and data is legally revealed, the intended application can process that data without revealing the personality information and produce the necessary analysis.

- **Blockchain platform**: The blockchain and smart contract platform is responsible for constantly securing data from unauthorized (illegal) entry, maintaining anonymity by avoiding any sensitive (private) data and tracking and validating the availability of data based on user authentication and privileges. All conducted processes by those intended applications will be signed by the private key that belongs to end-user (i.e. system analyst or researcher). And thus, the generated analysis reports will not be appended to the blockchain network unless they are authenticated by validators nodes based on the defined consensus protocol,

---

[2]https://www.docker.com.

where all intended nodes approve the submitted data record to be appended to the blockchain network, and a consensus protocol guarantees that the blockchain nodes decide on the particular sequence in which the entries are appended [13, 14].

Essentially, before data processing may deal, each participant will specify a common set of contracts addressing common terms, details, laws, meanings of concepts and procedures. Taken collectively, these arrangements lay out a business model that controls all relations between the participants to the deal. A smart contract, along with a distributed ledger, constitutes the core of this decentralized blockchain framework.

3.3 Framework Implementation

In this part, we overview some of the processes intended for the implementation of the proposed framework.

- **Data collection**: Targeted data (e.g. personal healthcare details or news) is collected by participating organizations (e.g. clinics, hospitals, and online news websites). To synchronize those data to the cloud for proper access and further processing, an administrator/end-user (i.e. an authorized person) may first enrol for an online account to the designed cloud service with adequate storage.
- **Data integrity and privacy**: Intended data records come from several of sources throughout the day, leading to a massive number of data records. The designed framework utilized a proprietary blockchain layer (called channel) that facilitates data separation and secrecy. The channel-specific record is exchanged among peers in the channel and the transacting participants must be correctly authenticated to a channel in order to communicate with it. Thus, the private info is sent by peer-to-peer through a gossip protocol to only the organization(s) that are allowed to access it. Such records are maintained in a private state database of peers of approved entities (also known as "side" database, or "SideDB") that can be retrieved by these designated peers from the chain code. Once the designed application submits the transaction to the ordering service, private data hashes transfers are included as usual in chains. And the private data hashes block is spread to all peers. In this case, all channel peers will verify transactions with private data hashes in a coherent way without understanding the underlying private data.
- **Data processing and sharing**: A data centre can share data with authorized participants only (e.g. an authorized end-user or researcher) to process the data and produce related analysis and reports. Whenever data sharing is detected in the framework, there will be an event generated to record the data access request. The event record will be recorded into history records (logging records). That data record held in a data centre and all associated operations will be sent to a blockchain network accompanied by several measures to turn a collection of records into a transaction. A collection of transactions is then used to construct a block, and the block will be checked by validator nodes on the blockchain network. After a set of steps, the integrity of the ledger can be maintained, and

subsequent verification of the record and the transaction relevant to this data can be made accessible. That when a personal data procedure is carried out, a log will be recorded in the blockchain. It means that all acts pertaining to personal records are responsible. We will enforce the access control (CA) system by using the Hyperledger Fabric membership service feature and the channel system [15]. The CA is responsible for membership enrolment by providing digital certificates for involved participants in the blockchain network. A different type of access can be defined in the credential, such as querying and modifying chain code execution operations in the network. In the case of privacy concerns, the system can specifically exchange confidential data with the data requester on the basis of the need to include data to aid the intended service. In order to issue a certificate, the data owner can clearly indicate in the certificate in which category of personal data is allowed to access, whether read-only or read–write access is allowed.

3.4 Framework Use Case

To illustrate our framework, we exemplify the case of Saraswati (user), a researcher who wants to train a model for analysing social exclusion patterns, and Mímir (Data), is a governmental agency who owns data about income, families, welfare, health and so forth. Using a smart contract, Saraswati signs an agreement with Mímir about the data terms of usage and processing limitations which will be registered in the blockchain. Then, using a new smart contract, Saraswati sends the dockerized program for training a model for analysing social exclusion patterns to the Fuxi (the data processing application). At the same time, Mímir sends the data to the Fuxi using another smart contract. When both the data and dockerized application are collected from Fuxi, the docker is run, and the results are sent back through the blockchain. Finally, Saraswati can collect the results from the blockchain.

The framework allows Mímir to share the data with Saraswati without letting Saraswati to see the personal data or transferring the data to Saraswati. Yet, both data and personal data are safeguarded and blockchain platform is provided with security and traceability of all transaction. At any time, both Saraswati and Mímir can verify the process and results by checking the blockchain transactions.

4 Conclusions

This paper presented a blockchain-based general framework for data processing applications which avoids sharing data between different parties for its processing and as a enhance privacy assurance. The proposed framework integrates four entities, the participating user, the blockchain and smart contracts platform, data processing applications and the targeted data. Our solution reduces the need for a trusted centralized authority and maintains privacy of data and information for data processing applications by applying newer technologies like Docker and Blockchain. All participants of blockchain can see every transaction as its contents, thus it is easy to audit and validate them, making the system transparent; at the same time, all transactions are encrypted which can improve their security. As every transaction must be agreed by its participants and validated by consensus in blockchain, it adds an extra level of protection allowing to ensure its posterior fulfilment and possible retractions. It also adds a traceability property to the system inherited from the blockchain, as every transaction is recorded in the blockchain, it makes possible to validate the result or process and replicate them.

In our future work, we are planning to validate the proposed framework and make it publicly available on the Internet.

Acknowledgements This work was supported by the News Angler project funded by the Norwegian Research Council's IKTPLUSS programme as project 275872.

References

1. Young, T., Hazarika, D., Poria, S., Cambria, E.: Recent trends in deep learning based natural language processing. IEEE Comput. Intell. Mag. **13**(3), 55–75 (2018). https://doi.org/10.1109/MCI.2018.2840738
2. Simeone, O.: A brief introduction to machine learning for engineers. Now Found. Trends (2018). https://doi.org/10.1561/2000000102
3. Nakamoto, S.: Bitcoin: A peer-to-peer electronic cash system. White paper (2008). https://bitcoin.org/bitcoin.pdf
4. Buterin, V., et al.: A next-generation smart contract and decentralized application platform. White paper (2014)
5. Wood, G., et al.: Ethereum: a secure decentralised generalised transaction ledger. Ethereum Project Yellow Paper **151**(2014), 1–32 (2014)
6. Drescher, D.: Blockchain Basics: A Non-Technical Introduction in 25 Steps. Apress (2017)
7. Alidin, A.A., Ali-Wosabi, A.A.A., Yusoff, Z.: Overview of blockchain implementation on islamic finance: Saadiqin experience. In: 2018 Cyber Resilience Conference (CRC). pp. 1–2 (2018). https://doi.org/10.1109/CR.2018.8626822
8. Hyperledger Fabric: Smart contracts and chaincode (2019), https://hyperledger-fabric.readthedocs.io/en/release-1.4/smartcontract/smartcontract.html Accessed: 20 Dec. 2019
9. Sun, C., Ippel, L., Soest, J., Wouters, B., Malic, A., Adekunle, O., Berg, B., Mussmann, O., Koster, A., Kallen, C., Oppen, C., Townend, D., Dekker, A., Dumontier, M.: A privacy-preserving infrastructure for analyzing personal health data in a vertically partitioned scenario. In: the 17th World Congress on Medical and Health Informatics (MEDINFO 2019). vol. 264, pp. 373–377 (2019). https://doi.org/10.3233/SHTI190246
10. Zyskind, G., Zekrifa, D., Alex, P., Nathan, O.: Decentralizing privacy: using blockchain to protect personal data. In: 2015 IEEE Security and Privacy Workshops. pp. 180–184 (2015). https://doi.org/10.1109/SPW.2015.27
11. Third, A., Domingue, J.: Linkchains: Trusted personal linked data. In: Blockchain-enabled Semantic Web (2019). https://oro.open.ac.uk/id/eprint/68229
12. Bucur, C.I., Ciroku, F., Makhalova, T., Rizza, E., Thanapalasingam, T., Varanka, D., Wolowyk, M., Domingue, J.: A decentralized approach to validating personal data using a combination of blockchains and linked data. Tech. rep., Linked OpenData Validity—Technical Report from

the International Semantic Web Research Summer School (ISWS) 2018 (2019). https://arxiv.org/abs/1903.12554

13. Cachin, C., et al.: Architecture of the hyperledger blockchain fabric. In: Workshop on distributed cryptocurrencies and consensus ledgers, vol. 310, p. 4 (2016). https://pdfs.semanticscholar.org/f852/c5f3fe649f8a17ded391df0796677a59927f.pdf

14. Cachin, C., Vukolic, M.: Blockchain consensus protocols in the wild. In: 31st International Symposium on Distributed Computing (DISC 2017), vol. 91 (2017). https://doi.org/10.4230/LIPIcs.DISC.2017.1

15. Hyperledger Fabric: Membership service providers (2019). https://hyperledger-fabric.readthedocs.io/en/release-1.4/msp.html Accessed: 20 Dec. 2019

Cloud Computing and Networking

VOIP in MANETs Based on the Routing Protocols OLSR and TORA

Hamza Zemrane, Youssef Baddi, and Abderrahim Hasbi

Abstract Mobile ad hoc network are new generation of networks based on no pre-existing infrastructure, these type of networks are used a lot for collaboration between staff members in train stations, airports, and even in military applications, the MANETs have a lot of routing protocols, and choosing the most suitable for the network becomes a serious problem. Our works consists of applying the vocie over IP protocol to the MANETs and studying the performance of the network when it uses the OLSR and the TORA protocols.

Keywords Mobile ad hoc network (MANET) · VOIP protocol · Optimized link state routing (OLSR) · Temporary ordering routing algorithm (TORA) · SIP protocol

1 Introduction

The recent evolution of technology in the field of wireless communication and the emergence of portable computing units, today push researchers to make efforts to achieve the goal of networks access to information anywhere and anytime. Mobile ad hoc networks (MANETs) belong to a category of wireless networks, this does not need any infrastructure, and each node is playing the role of the host as well as the router. Mobile terminals in these networks can move randomly and at any speed. In this context, challenges or characteristics must be taken into account when deploying ad hoc mobile networks. Our work consists of applying the VOIP protocol in the MANETs. Voice over IP is an emerging voice communication technology. It is part

H. Zemrane (✉) · A. Hasbi (✉)
Lab RIME, Mohammadia School of Engineering, UM5, Rabat, Morocco
e-mail: zemranehamza93@gmail.com

A. Hasbi
e-mail: ahasbi@gmail.com

Y. Baddi
Lab STIC, High School of Technology Sidi Bennour, UCD, El Jadida, Morocco
e-mail: baddi.y@ucd.ac.ma

© The Editor(s) (if applicable) and The Author(s), under exclusive license to Springer
Nature Singapore Pte Ltd. 2021
F. Saeed et al. (eds.), *Advances on Smart and Soft Computing*, Advances in Intelligent
Systems and Computing 1188, https://doi.org/10.1007/978-981-15-6048-4_39

of a turning point in the world of communication. The big number of the MANETs routing protocols makes the choice of the most adapted routing protocol to route the voice over IP packets in the MANETs a serious problem. In our simulation scenario, we compared the behavior of the MANETs when it uses the optimized link state routing protocol (OLSR) and the temporally ordered routing algorithm (TORA) protocols for routing the voice data. In Sect. 2, we talk about the ad hoc mobile networks, and we give a definition of ad hoc networks, the advantages of mobile ad hoc network, and the QoS models. In Sect. 3, we talk about the voice over IP standards: the H323 protocol, the SIP protocol, the RTP and RTCP transport, and the audio standard. In Sect. 4, we make a simulation of the VOIP protocol based on the MANETs, and we compared the performance of the OLSR and TORA routing protocols.

2   Ad Hoc Mobile Networks

2.1   Definition of Ad Hoc Networks

An ad hoc mobile network [1] generally called mobile ad hoc network (MANET) consists of a large population, relatively dense, mobile units that move in any territory. The only means of communication is the use of "radio waves" that spread between different mobile nodes that can be wireless communication technology such as: ZigBee [2], WiFi [2], WiMAX [3], NB-IoT [4], and others, without using a pre-existing infrastructure or centralized administration.

2.2   Advantages of Mobile Ad Hoc Networks

The benefits of the mobile ad hoc [5] network technology are inherent in the fact that there is no need for pre-existing infrastructure:

- Mobile ad hoc networks [5] can be deployed in any environment and can be used in Internet of Things [6] application such as Smart Homes [7, 8] and Ehealth [4, 9].
- The cost of operating the network is low: No infrastructure is to be put in place initially, and especially, no maintenance is to be expected.
- The deployment of an ad hoc network is simple: It does not require any prerequisites since it is enough to have a certain number of terminals in a space to create an ad hoc network and fast since it is immediately functional as long as the terminals are here.
- The flexibility of use: This is a very important parameter since the only elements that can break down are the terminals themselves. In other words, there is no

breakdown "penalizing" overall (a station that is used for routing can be replaced by another if it fails).

2.3 QOS Models

In telecommunication networks, the objective of quality of service is to achieve better communication behavior, so that the content of the latter is correctly routed, and the resources of the network are used in an optimal manner. A quality of service model defines what types of service can be provided in a network and certain mechanisms used to provide these services.

In other words, a quality of service model describes a set of end-to-end services, which allow customers to select a number of guarantees that govern properties such as time, scheduling, and reliability. The quality of service model specifies the architecture that must take into account the challenges imposed by ad hoc networks, such as the change in topology and the constraints of delay and reliability.

FQMM (flexible quality of service model for MANETs) The FQMM model was the first quality of service model proposed for ad hoc networks in 2000. It is a hybrid model combining the properties of the IntServ and Di Serv models but suitable for small- or medium-sized ad hoc networks (approximately 50 knots).

SWAN (service differentiation in stateless wireless ad hoc networks) The SWAN model implements packet admission control. A packet is accepted if the bandwidth of the route it must carry is sufficient to ensure its transit without causing network congestion.

iMAQ (an integrated mobile ad hoc QoS framework) The iMAQ model provides a quality of service solution for the transfer of multimedia data in ad hoc networks.

3 Voice Over IP Protocol Standards

3.1 H323 Protocol

With the development of multimedia over networks, it has become necessary to create protocols that support these new features, H.323 is a communication protocol encompassing a set of standards used for sending audio and video data over the Internet. Network component defined by H.323:

- The terminals: Two types of H.323 terminals are available today. An IP telephone set connected directly to the company's Ethernet network. A multimedia PC on which an H.323 compatible application is installed.
- Gateway: They provide the interconnection between an IP network and the telephone network, the latter being either the public telephone network or a corporate Pabx.

- Gatekeeper: Their role is to perform address translation (phone number—IP address) and authorization management.
- Multipoint control unit: It allows clients to connect to data conferencing sessions. Multipoint control units can communicate with each other to exchange conference information.

3.2 SIP Protocol

SIP is a signalling protocol belonging to the application layer of the OSI model. Its role is to open, modify, and release sessions. The opening of these sessions allows the realization of audio or videoconferencing, distance education, voice (telephony), and multimedia broadcasting on IP essentially.

Operation mode SIP intervenes at the different phases of the call:

- Location of the corresponding terminal.
- Analysis of the recipient's profile and resources.
- Negotiation of the type of media (voice, video, data ...) and communication parameters.
- Availability of the correspondent, determines if the called party wishes to communicate, and authorizes the caller to contact him.
- Establishing and monitoring the call, notifying the calling and called parties of the login request, call forwarding and closing management.
- Management of advanced functions: encryption, error feedback, ...

3.3 RTP and RTCP Transport

Real-time transport protocol (RTP): The RTP protocol, standardized in 1996, aims to organize the packets at the entrance of the network and control them at the output. This is in order to reform the flows with its starting characteristics. RTP is managed at the application level so does not require the implementation of a kernel or libraries. RTP is a protocol for applications with real-time properties. It allows to:

- Reconstitute the flow time base (packet time stamp).
- Set up sequencing of packets by dialling to allow detection of lost packets.
- Identify the content of the data to associate them with secure transport.
- The identification of the source, i.e., the identification of the sender of the packet.
- Transport audio and video applications in frames (with dimensions that are dependent on the codecs that perform the scan).

Real-time transport control protocol: The RTCP protocol [10] is based on periodic transmission of control packets to all participants in a session. It is the UDP protocol (for example) that allows the multiplexing of RTCP control packets. The

RTP protocol uses the RTCP, which carries the following additional information for session management:

- The receivers use RTCP [10] to send a QoS report back to the transmitters. These reports include the number of packets lost, the parameter indicating the variance of a distribution, and the round-trip delay.
- An additional synchronization between the media. Multimedia applications are often transported by separate streams. For example, voice, image, or even scanned applications on multiple hierarchical levels can see managed streams follow different paths.
- Identification because RTCP packets contain address information, such as the address of an e-mail message, a phone number, or the name of a participant in a conference call.
- The control of the session, because RTCP allows participants to indicate their departure from a conference call or simply to provide an indication of their behavior.

3.4 Audio

The transport of voice over an IP network requires all or some of the following steps:

- Digitalization: In the case where the telephone signals to be transmitted are in analog form, the latter must first be converted into digital form according to the pulse code modulation (PCM) format at 64 kbps. If the telephone interface is digital (ISDN access, for example), this function is omitted.
- Compression: The 64 kbps PCM digital signal is compressed in one of the codec formats (compression/decompression) and then inserted into IP packets. The codec function is most often performed by a digital signal processor (DSP).
- Decompression: On the receiving side, the received information is uncompressed. To do this, it is necessary to use the same codec as for compression and then converted back to the appropriate format for the recipient (analogue, PCM 64 kbps, etc.). The goal of a codec is to achieve good voice quality with the lowest possible bit rate and compression delay.

4 Simulation of the VOIP in MANET Using OLSR and TORA

4.1 Operating Mode of the MANET Routing Protocols

Optimized link state routing (OLSR) [11]: As the name suggests, the OLSR is link state protocol. OLSR offers optimal routes in terms of the number of hops in the network. In a link state protocol, each node declares its direct links with its neighbors to the entire network. In the case of OLSR, the nodes only declare a subset of their neighborhood. All declared neighbors are chosen so that they can reach the whole neighborhood at two jumps. This set is called the set of multipoint relays. Multipoint relays are used for the purpose of minimizing traffic due to the broadcast of control messages in the network. Roads are built from multipoint relays, to keep the information necessary for the choice of multipoint relays and the calculation of the routing table up to date, and the OLSR nodes need to exchange information periodically. To inquire about the near neighborhood, OLSR nodes send HELLOS messages containing their list of neighbors. By this message, the neighborhood subsets are periodically declared in the network, using these same multipoint relays. This information provides a network map containing all the nodes and a partial set of links, but sufficient for the construction of the routing table. The routing table is built at all nodes, and the data routing is done jump by jump without the intervention of OLSR whose role stops at the update of the routing table of the IP stack.

Temporary ordering routing algorithm (TORA) [12]: The TORA algorithm is applied to directed acyclic graphs (DAGs) modeling the network. He is trying to maintain a property called "destination orientation" of the DAGs. The graph becomes undirected if a link becomes faulty. The acyclic-oriented graph uses a notion of node height. Each node communicates its height to all its direct neighbors, so that the link orientation between two neighboring nodes is from the highest node to the smallest one. This height represents the fundamental routing metric of this protocol. It determines several routes to join a destination, so that each node has a copy of the algorithm in order to run it locally. It uses a routing technique called "reverse link." This technique allows the protocol to respond quickly if a source-recipient link fails, reversing that link and establishing a new path.

4.2 Simulation Scenario of the MANET Network

The scenario simulate the behavior of 15 nodes in motion that exchange voice information using a voice server, over a MANET network based on the 802.11g standard, and we compare the performance of the network when it use the OLSR and the TORA protocols, for the routing of the voice information. The simulation is done by the RiverBed Modeler Academic Edition 17.5, and it takes a period of 50 min (Fig. 1).

4.3 Comparison in the Data Link Layer: 802.11g

The RTP delay: This records the difference in time at which the application packet was time stamped at the source node to the time at which the packet was received by RTP at the destination node. The curve that represents the AODV protocol starts

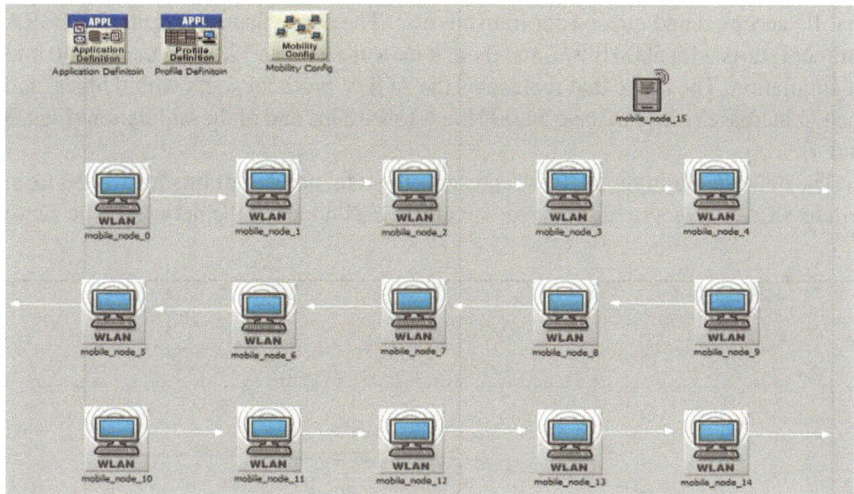

Fig. 1 Simulation scenario of the MANET

with 0.000043 s, and then, it makes a slow decrease that goes to 0.0000425 s, toward the end of the simulation. The curve that represents the TORA protocol starts with 0.013 s, and then, it makes a decrease that goes to 0 s toward the end of the simulation (Fig. 2).

The network load for BSS:1: The voice server and all the mobile nodes are configured with the same BSS identifier (1) to communicate. The graph represents the total data traffic received by the entire WLAN BSS:1 from the higher layers of the MACs

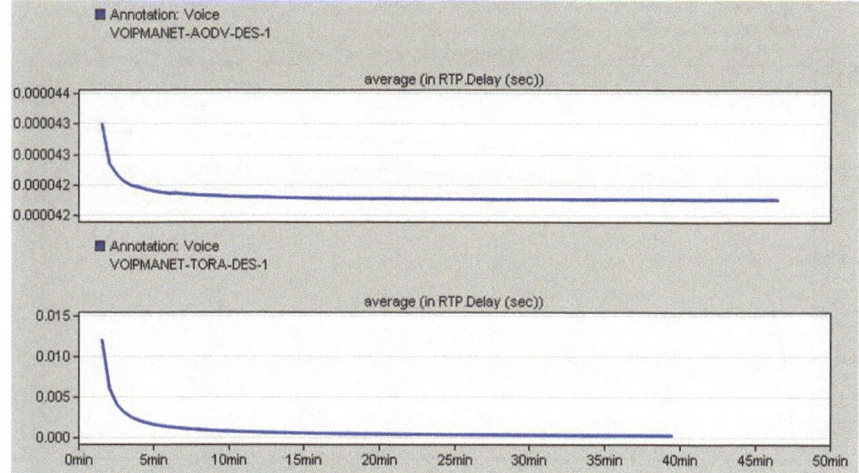

Fig. 2 Average of the RTP Delay

that ID accepted and queued for transmission. The curve that represents the TORA protocol starts with 5000 bits/s, and then, it increase to have 58,000 bit/s after 40 min of simulation. The curve that represents the AODV protocol starts with 0 bits/s, and then, it increase slowly to have 54,000 bits/s toward the end of the simulation (Figs. 3 and 4).

The network throughput: This represents the total number of bits forwarded from wireless LAN layers to higher layers in all WLAN nodes of the network. The curve

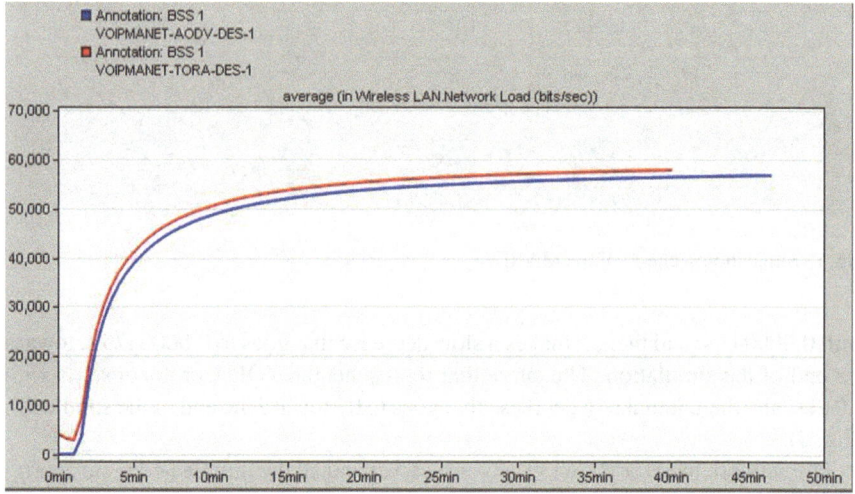

Fig. 3 Average of the network load for BSS:1

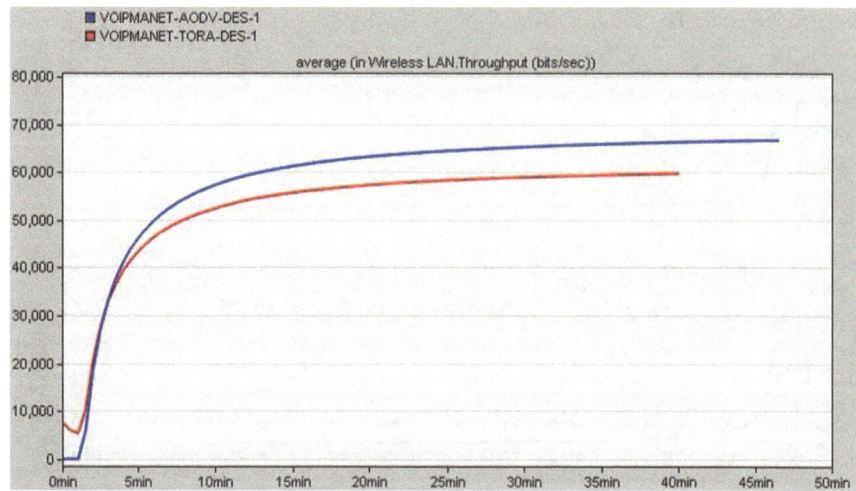

Fig. 4 Average of the network throughput

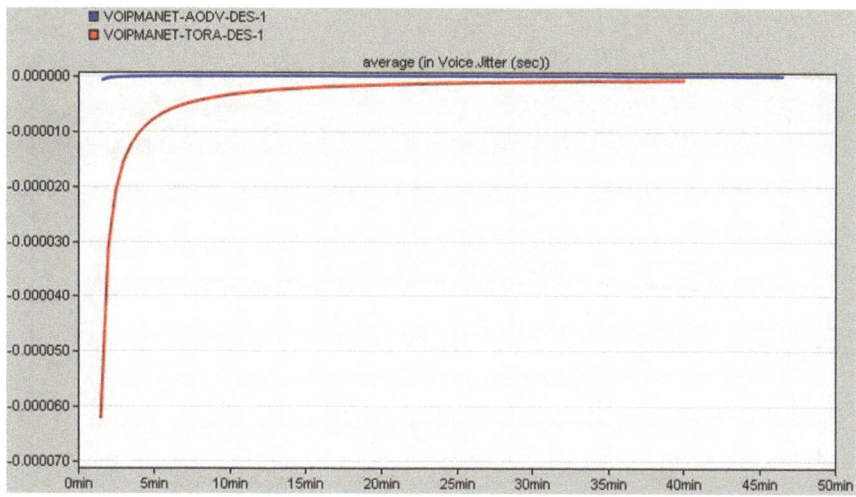

Fig. 5 Average of the voice jitter

that represent the TORA protocol starts with 8000 bits/s, then it increase to have 50,000 bits/s after 7 min of simulation, and after it continues the increase slowly to have 60,000 bits/s toward the end of the simulation. The curve that represent the AODV protocol starts with 0 bits/s, then it increase to have 60,000 bits/s after 10 min, and then, it continues the increase slowly that takes 67,000 bits/s toward the end of the simulation (Fig. 5).

4.4 Comparison in the Application Layer: Voice Application

Voice jitter: If two consecutive voice packets leave the source node with time stamps t_1 t_2 and are played back at the destination node at time t_3 t_4, then: Jitter $= (t_4 - t_3)$ $(t_2 - t_1)$. The curve that represents the AODV protocol has a constant evolution that takes a little bit less than 0 s until the end of the simulation. The curve that represents the TORA protocol starts with -0.000062 s, then it makes a fast increase to have -0.00001 after 4 min of simulation, and then, it continues the increase slowly to have -0.000001 until the end of the simulation.

Voice packet end to end delay: This is the time at which the sender node gave the packet to RTP to the time the receiver got it from RTP. The curve that represents the TORA protocol starts with 0.125 s, and then, it makes a decrease that goes to 0.123 s toward the end of the simulation. The curve that represents AODV protocol starts with 0.123 toward the end of the simulation (Fig. 6).

MOS is the mean opinion score: It is a measure of the quality used in telephony to express the quality of the voice applications.

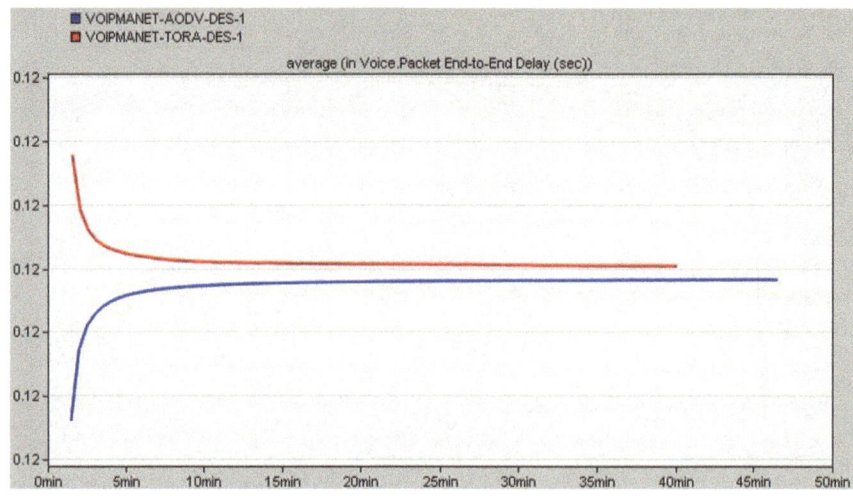

Fig. 6 Average of the voice end to end delay

The AODV protocol has a constant evolution, and it takes 4.35 until the end of the simulation. The curve that represent the TORA protocol starts with 4.33, then, it makes a fast increase that takes 4.346 after 7 min of simulation, and then, it continues the increase slowly to have 4.47 toward the end of the simulation (Fig. 7).

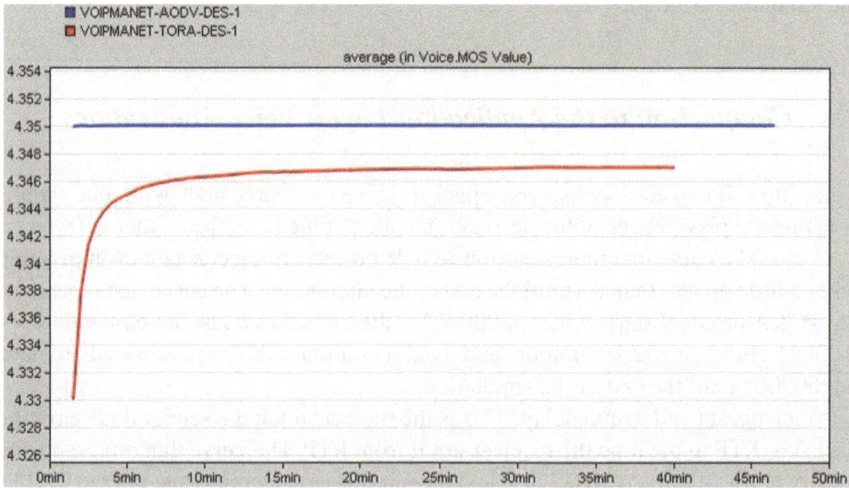

Fig. 7 Average of the voice MOS

5   Conclusion

Voice over IP refers to broadcasting the voice stream over Internet networks, instead of traditional PSTN telephone networks. The Internet protocol (IP) was originally designed for managing data networks, and after its success, the protocol was adapted to voice management, transforming, and transmitting information into an IP data packet. The latest network technology is based on no existing communication infrastructure known under the name MANTEs, and the MANETs are very useful for communication and collaboration between staff at airports for example. It exists several routing protocols that adapt MANET networks for finding the best one to adapt for voice transmission which become a difficult thing. Our simulation allows us to study the performance of our MANET network when it transmits voice over IP using the OLSR and TORA routing protocols.

References

1. Truong, N.B., Lee, G.M., Ghamri-Doudane, Y.: Software defined networking- based vehicular adhoc network with fog computing. In: 2015 IFIP/IEEE International Symposium on Integrated Network Management (IM). pp. 1202–1207. IEEE (2015)
2. Zemrane, H., Baddi, Y., Hasbi, A.: Comparison between IOT protocols: Zigbee and wifi using the opnet simulator. In: Proceedings of the 12th International Conference on Intelligent Systems: Theories and Applications, p. 22. ACM (2018)
3. Zemrane, H., Abbou, A.N., Baddi, Y., Hasbi, A.: Wireless sensor networks as part of IOT: performance study of wimax-mobil protocol. In: 2018 4th International Conference on Cloud Computing Technologies and Applications (Cloudtech), pp. 1–8. IEEE (2018)
4. Zemrane, H., Baddi, Y., Hasbi, A.: Ehealth smart application of WSN on wwan. In: Proceedings of the 2nd International Conference on Networking, Information Systems Security, p. 26. ACM (2019)
5. Zemrane, H., Baddi, Y., Hasbi, A.: Mobile adhoc networks for intelligent transportation system: comparative analysis of the routing protocols. Procedia Comput. Sci. **160**, 758–765 (2019)
6. Zemrane, H., Baddi, Y., Hasbi, A.: Sdn-based solutions to improve IOT: survey. In: 2018 IEEE 5th International Congress on Information Science and Technology (CiSt), pp. 588–593. IEEE (2018)
7. Zemrane, H., Baddi, Y., Hasbi, A.: Internet of things smart home ecosystem. In: Emerging Technologies for Connected Internet of Vehicles and Intelligent Transportation System Networks, pp. 101–125. Springer (2020)
8. Zemrane, H., Baddi, Y., Hasbi, A.: IOT smart home ecosystem: Architecture and communication protocols. In: 2019 International Conference of Computer-Science and Renewable Energies (ICCSRE), pp. 1–8. IEEE (2019)
9. Zemrane, H., Baddi, Y., Hasbi, A.: Improve IOT ehealth ecosystem with sdn. In: Proceedings of the 4th International Conference on Smart City Applications, pp. 1–8 (2019)
10. Reddappagari, P.J., Li, N.: Embedded RTCP packets (2019). uS Patent 10,469,630
11. Prakash, K., Philip, P.C., Paulus, R., Kumar, A.: A packet fluctuation-based OLSR and efficient parameters-based OLSR routing protocols for urban vehicular ad hoc networks. In: Recent Trends in Communication and Intelligent Systems, pp. 79–87. Springer (2020)
12. Liu, S., Zhang, D., Liu, X., Zhang, T., Wu, H.: Adaptive repair algorithm fortora routing protocol based on food control strategy. Comput. Commun. (2020)

A Survey on Network on Chip Routing Algorithms Criteria

Muhammad Kaleem and Ismail Fauzi Bin Isnin

Abstract As number of components on the semi-conductor industry is growing at a healthy rate, results in an increase in number of cores integrating on a chip.. Demand for core interconnections brings network on chip (NoC) under consideration. Instead of using traditional on chip interconnection, NoC is proved to be a promising solution. Communicating nodes require routing algorithm for successful transmission of packets. In this paper, we highlight the recent challenges of routing algorithms and classification of routing algorithm techniques in different criteria such as aging-aware, thermal-aware, congestion-aware, fault-aware, resilient, and energy efficient. It was challenging to design a NoC routing algorithm capable of providing less congested paths, better energy efficiency, and high scalability. Now, it is even more challenging to design routing algorithm to deal with other issues such as aging of network, hotspot in the network due to high temperature, high temperatures can produce faults in the system, enhance power efficiency and resilient to enhance tolerance. Routing algorithm has a significant impact on latency and throughput of network on chip. In the end, we discussed performance matrices commonly used for evaluation of NoC routing algorithms. To the best of our knowledge, this is the only study which is focusing on modern criteria of NoC routing algorithms.

Keywords Network on chip · Routing algorithm

1 Introduction

Network on chip is promising solution that helps to facilitates system on chip (SoC) limitations [1]. Network on chip (NoC) is communication infrastructure for hundreds of microprocessors and components on chip. In conventional NoC architecture, each

M. Kaleem (✉) · I. F. B. Isnin
Faculty of Engineering, School of Computing, Universiti Tecknologi Malaysia, Johor Bahru, Malaysia
e-mail: kaleem.muhammad@graduate.utm.my

M. Kaleem
Department of Computer Science and IT, University of Sargodha, Sargodha, Pakistan

© The Editor(s) (if applicable) and The Author(s), under exclusive license to Springer Nature Singapore Pte Ltd. 2021
F. Saeed et al. (eds.), *Advances on Smart and Soft Computing*, Advances in Intelligent Systems and Computing 1188, https://doi.org/10.1007/978-981-15-6048-4_40

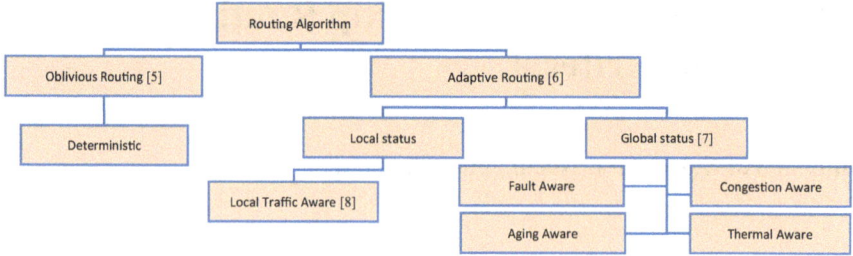

Fig. 1 Taxonomy of routing algorithm

core comprises of processing element (PE), network interface, and a router [2]. In NoC, most widely adopted topology is mesh topology for actual implementation [3] due to its short wire length and simplicity of physical layout. Once topology for network on chip is settled, the main other concerning factors are its routing algorithm and flow control mechanisms. NoC routing algorithm is responsible for delivery of packets from source to destination within the network. Routing algorithm selects path that a packet will take to reach to its destination. Routing algorithm must determine three different conditions: how path is defined; where the routing decisions are taken; and total path length.

An ingenious routing algorithm should possess the properties of low average latency, good load-balancing in various workloads, reasonable routing lengths, higher network throughput, deadlocks freedom, livelocks freedom [4], starvation freedom, and resolving network transfer errors particularly. Hence, routing algorithms are thereby became the point of interest form NoC inception. In this paper, we are focusing on imminent challenges for routing algorithms in mesh topology. Routing algorithms are categorized in two broader terms such as oblivious routing and adaptive routing, Fig. 1. Oblivious routing is a state-of-the-art routing algorithms meant for providing seamless communication without exhibiting deadlocks, livelocks with better throughput, and less latency but unable to cope with modern routing demands [5]. Traditionally, adaptive routing is known for dynamically adjustment of routes in congestion situation and fault aware scenarios [6]. Recently, researches are not only focusing on fault awareness and congestion situations in routing but also on age awareness, thermal awareness [7], providing resilience, and energy efficiency under extreme stress. Adaptive routing has two major classes. Local adaptive routing algorithms take decision with the help of neighboring nodes which are capable of delivering information relevant to its close neighbor only [8]. It is challenging to build an immaculate map of the whole network with extremely limited information. Whereas in global status, routing immaculate information of the whole network is eminent, hence better decisions could be made but it is challenging to enhance performance. To the best of our knowledge, this is the only study which is focusing on modern demands of NoC routing algorithms such as ageing aware, thermal aware, and fault aware routing algorithms at this time.

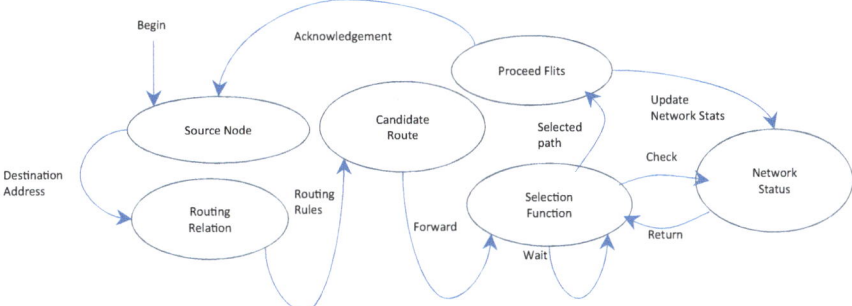

Fig. 2 Routing algorithm state diagram

Decisions for flits trying to traverse from source *A* to the destination *B* took place in various steps. Process model for routing algorithm Fig. 2 source nodes pass the address of the destination node to its local router, local router find out the routing relations, assessment of candidate route under the routing rules [9]. Selection function will check the current network state to decide or select an appropriate route for the flits to proceed with the selected route. Acknowledgement of task accomplishment given to the source node and incase of any update learnt from the transmission is updated with network status.

2 Challenges of Network on Chip Routing Algorithms

It is challenging to design a routing algorithm capable of providing less congested path, energy efficient, scalability with less latency, and high throughput due to the amount of load need to be carried out in network on chip. One of the main issues of routing algorithm is performance. Routing algorithm must be properly defined to achieve high performance. Below are the challenges that must be considered during designing and development of routing algorithm for NoC:

- If multiple packets/flits are trying to find its way to the destination node, without having any other option of traversing, this will cause congestion on links. Congestion in NoC results in low throughput and high latency. Both latency and throughput result in reducing performance [7].
- When a set of resources are required by nodes and due to cyclic dependency no progressive target is achieved. If a packet holds its position in a buffer of one router and another packet is waiting in another routers buffer to take its place, all packets waiting for each other to release a resource leaving the entire packets in a deadlock [4].
- When packets circulate within a network without making its movement toward its target. Consider a packet that is meant to traverse in east direction but due to congestion situation or to avoid deadlock conditions, it is directed towards other

Fig. 3 Key features of
routing algorithm

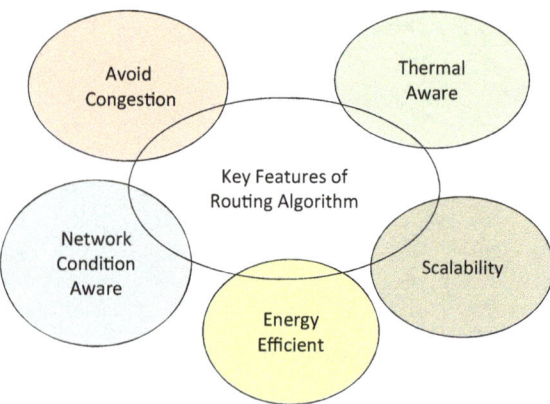

Fig. 3 Key features of
routing algorithm

directions. It will be circulating around in the entire network and will not only create congestion but also waste routing resources [5].

- If a packet already occupies output channel of router and a packet already present in a buffer requests for an output channel, hence, the packet requesting an output channel is blocked. Due to blockage, no packet will move and will be starved to end up sitting in the buffer [10].
- Low latency is an amount of time taken by the flit to traverse the system. Latency is calculated in terms of cycles it has required to traverse through the system [11].

A successful routing algorithm must possess some essential features Fig. 3. A routing algorithm must be aware of latest network condition, thermal aware to reduce temporary and permanent fault ratio, and energy efficient path selection. Similarly, it must adapt according to the size of a network on chip topology. Routing must be equally flexible in variable topology sizes and lengths.

3   Routing Algorithm Criteria

Routing algorithms plays an important role in providing communication among the nodes in high speed NoC architectures. In the recent work, researchers have tried to manage congestion situations and temporary or permanent faults of NoC. NoC switching technique determines the resource allocation method when messages through the network. Packets switching can be used to utilize network resources better. Nowadays, researchers are also considering the aging of network, hotspot in the network due to high temperature, high temperatures can produce faults in the system, enhance power efficiency, and resilient to enhance tolerance.

Aging is an emerging reliability concern, which degrades the system performance and causes timing failure eventually. Aging happens when a transistor is under stress and temperature is high. AROMa [12], an aging-aware deadlock-free adaptive routing algorithm online aging monitoring system for 3D NoCs. It obtains routers aging rates for each layer of 3D NoCs periodically. AROMa, constantly switches between different paths, is suggested by the online age monitoring system. To distribute traffic evenly throughout the network is also beneficial in avoiding aging factor [13]. Oblivious routing techniques have no information about the congestion situations and only route packets in a deterministic way. On the contrary, oblivious routing is low cost with low latency and high throughput. LAXY [14] is an adaptive routing technique using the benefits of oblivious and adaptive routing. It is also temperature aware. ETW [15] is a routing algorithm for 3D partially connected NoC. Partially connected NoC use TSV through silicon Vias to communicate between the layers of 3D.

As the technology scales, temperature effects become more significant and designing for performance becomes more difficult. It is crucial for designers to understand the impact of thermal variations on these systems to reduce hotspots and maintain performance [16]. Thermal issues are an increasing concern in microelectronics due to increased power density as well as the increasing vulnerability of the system to temperature effects (delay, leakage, reliability) [17].In 3D NoC, it is challenging to tackle heat dissipation problem due to absence of heat sink among the layers and poor traffic distribution [18]. LEAD [10], ATAR [7] uniformly distribute load among the layers and nodes to protect transistor biasing from overheating. INT [19] uses immediate neighborhood temperature to tackle this problem. In [20], it uses coolest path algorithm to find such a path which is at low heat dissipation level (Table 1).

Faulty node, faulty router, or faulty links are common in network on chip. Faults may be of permanent nature or of temporary. Permanent faults are eternal; NoC is unable to recover from such faults. Temporary faults are time bound. It may occur due to over heat dissipation or high energy consumption. HRA [21], heuristic bases routing algorithm, is based on a bi-objective weighted sum cost function. In case of a TSV experiences a fault, enhanced first last [22] is capable of handling at a low cost. By prohibiting many turns, fault awareness can be achieved [23, 31]. EFuNN [24] algorithm examines network traffic status, link faults of the area, and distance to the destination by obtaining the distance matric. It uses fuzzy neural networks to find the best path. In order to avoid faults [25, 27] uses LBFT and bypass routing path algorithms use detouring technique at the faulty area. Topology agnostic greedy protocols [26] use spanning tree for greedy routing in network on chip. Key features are: fully distributive, no routing table required, fault aware, deadlock free, no virtual channel required, multiple path search between the source and destination. Virtual TSV [29] is an online adaptive algorithm to assist instantly work around the newly defected area without using redundancies. PDA-FTR [30] works efficiently to compute flexibility of the routing function and also achieves fault resilient data transfer and improve quality of selection to balance traffic load. PDA-HyPAR [32] sets of the rules for routing, only successful routes that fulfill the requirements are allowed to transfer on the prescribed path. Topology agnostic [33] is applied at runtime to the packets and it does not require partitioning of virtual channels. This allows it to fully exploit them to reduce packet blocking and boost performance (Table 2).

Table 1 Classification of routing algorithms

Criteria	Algorithm full name	Routing algorithm	Citations
Aging aware	Aging aware routing for online monitoring	AROMa	[12]
	Workload aware routing	WAR	[13]
	Location-based age resilient	LAXY	[14]
	East then West	ETW	[15]
Thermal aware	Floating XY-YX	Floating XY-YX	[18]
	Longitudinal exclusively adaptive or deterministic	LEAD	[10]
	Aging aware routing for online monitoring	AROMa	[12]
	Adaptive thermal aware routing	ATAR	[7]
	Immediate neighbor temperature	INT	[19]
	Coolest path	CP	[20]
	Location-based age resilient	LAXY	[14]
	East then West	ETW	[15]
Fault aware	Heuristic-based routing algorithm	HRA	[21]
	Enhanced first last	Enhanced first last	[22]
	Longitudinal exclusively adaptive or deterministic	LEAD	[10]
	Lattice-based routing algorithm	LBRA	[23]
	Evolving fuzzy neural network	EFuNN	[24]
	Load balancing fault-tolerant	LBFT	[25]
	Topology agnostic greedy protocol	Topology agnostic greedy protocol	[26]
	Bypass path routing	Bypass path routing	[27]
	Fully adaptive congestion aware routing scheme	FACARS	[28]
	Virtual through silicon via	Virtual TSV	[29]
	Path diversity aware-fault tolerant routing	PDA-FTR	[30]
	Repetitive turn model	RTM	[31]
	Path diversity aware hybrid planar adaptive routing	PDA-HyPAR	[32]
	Topology agnostic	Topology agnostic	[33]
	East then West	ETW	[15]

(continued)

Table 1 (continued)

Criteria	Algorithm full name	Routing algorithm	Citations
Resilience	Heuristic-based routing algorithm	HRA	[21]
	Load balancing fault-tolerant	LBFT	[25]
	Workload aware routing	WAR	[13]
	Topology agnostic greedy protocol	Topology agnostic greedy protocol	[26]
	Fully adaptive congestion aware routing scheme	FACARS	[28]
	Virtual through silicon via	Virtual TSV	[29]
	Path diversity aware-fault tolerant routing	PDA-FTR	[30]
	Topology agnostic	Topology agnostic	[33]
	East then West	ETW	[15]
Power efficient	Aging aware routing for online monitoring	AROMa	[12]
	Adaptive thermal aware routing	ATAR	[7]
	Load balancing fault-tolerant	LBFT	[25]
	Path diversity aware-fault tolerant routing	PDA-FTR	[30]
	Immediate neighbor temperature	INT	[19]

Resilience is an ability to resist and withstand under stress and errors. The system must enforce strict integrity to meet real-time requirements. All possible random hardware errors and possible effects must be taken into account. Along with other criteria, researchers are considering the resilience in routing algorithm such as in [12] is age-aware as well as resilient. Power efficiency is a generalized property; it does not matter from which genre the routing algorithm is; it should be power efficient. List of power efficient routing algorithms are as follows [7, 12, 19, 25, 30].

4 Routing Algorithm Performance Matrices

To compare and contrast different NoC routing algorithms, a set of performance metrics is commonly used [34]. Most commonly performance assessment is conducted on two basic parameters, i.e., throughput and packet latency. Throughput is the rate of traffic at which packets can be sent across the network. For a packet within a system, throughput can be defined as follows:

Table 2 Tools and topologies in recent routing algorithms

Routing algorithm	Simulation tool used	Simulation tool availability	Topology mesh	Virtual channels
HRA [21]	Nirgam	Open source	2D regular	No
Floating XY-YX [18]	Access Noxim	Open source	3D regular	No
Enhanced first last [22]	Built in house	Limited access	3D partially connected	Yes
LEAD [10]	Access Noxim	Open source	3D partially connected	Yes
AROMa [12]	Gem5	Open source	3D regular	Yes
LBRA [23]	Noxim	Open source	2D regular	No
ATAR [7]	Access Noxim	Open source	3D regular	No
EFuNN [24]	HNOC	Open source	2D regular	No
LBFT [25]	Booksim 2.0	Open source	2D regular	No
WAR [13]	Gem5	Open source	2D regular	No
Bypass path routing [27]	Noxim	Open source	2D regular	No
FACARS [28]	Noxim	Open source	2D regular	No
Virtual TSV [29]	Verilog-HDL	Open source	3D partially connected	Yes
PDA-FTR [30]	Noxim	Open source	2D regular	Larger buffer size
INT [19]	Ctherm with NagaM	Open source	3D regular	No
RTM [31]	Noxim	Open source	2D regular	No
CP [20]	Noxim	Open source	3D regular	No
LAXY [14]	Booksim 2.0	Open source	2D regular	No
PDA-HyPAR[32]	Noxim	Open source	3D regular	No
Topology agnostic [33]	Built in house	Limited access	3D partially connected	Yes
ETW [15]	Access Noxim	Open source	3D partially connected	Less

$$\text{Throughput(TP)} = \frac{(\text{Total packets completed}) \times (\text{Packet Lenght})}{(\text{Number of nodes}) \times (\text{Total time})} \qquad (1)$$

where, total packets completed are the number of complete packets successfully reached their destination node [35]. Packet length is measured in flits. Number of nodes is the amount of functional nodes it jumps during the communication. Total time is the time taken by the packets to traverse in the network from the beginning to the last packet reception. Throughput is the degree of physically handling of the maximum load. Throughput is measured in flits/cycle/node.

Packet latency is the time lapse (in cycles) from tail flit occurrence at the destination node to the time head flit was released at the source node. A packet has to travel multiple interconnections and router depending upon the topology and the routing algorithm [11]. Each packet may have a different latencies. Consider packet P_i, the latency L_i is defined as:

$$L_i = \text{recieving time(tail flit } P_i) - \text{sending time(head flit of } P_i) \qquad (2)$$

Consider F flits are considered to reach its destination node successfully where L_i is the latency of each packet P_i. Where i range from 1 to F. The average latency L_{avg} can be calculated as follows

$$L_{avg} = \frac{\sum_{i=1}^{F} L_i}{F} \qquad (3)$$

Finally, for overall network latency L, after calculating all packet latencies from source to destinations which is represented by $L_{s,d}$, the overall latency of the network (considering all source destination pairs) is found by the following waited averaging: Where $x_{s,d}$ packet generation rate, $L_{s,d}$ is the average latency between source and destination.

$$L = \sum_{s,d} \frac{x_{s,d}}{\sum_{s',d'} x_{s',d'}} L_{s,d} \qquad (4)$$

Performance of network is usually evaluated in terms of packet latency, overall network latency, and throughput. In order to make reliable conclusion, researchers preferred synthesis inputs and benchmarks to inject to the network under evaluation [36]. Synthesis modes are extracted from application behavior. Most popular synthesis modes are shuffle, uniform, neighbor, transpose, and many more.

Workload models are also used for the evaluation of routing algorithms of NoC. Parameters for workload model for simulations are injection rate and packet length. The injection rate of these synthetic modes can be precisely controlled to test the algorithms under different network stress. For testing routing algorithms, synthesis mode can be applied. To evaluate routing algorithm, injection rate and packet length can be changed. Benchmarks such as SPEC CPU2000, Splash-2, or PARSEC are used to achieve more convinced and sophisticated results [37]. Behavior of different benchmark applications has verity of variable effects on algorithm under observation. For example, some computing dependent application benchmarks generate extremely low traffic in the network. On the other hand, some memory dependent applications work in an opposite way.

5 Future Directions

NoC routing algorithm has three primary issues which directly affect performance parameters such as latency and throughput. If congestion in NoC increases, the latency of system also increases. Even when there is no congestion, system can experience a high latency which is due to vague routing algorithm. If a routing algorithm has low throughput, it will reduce the overall performance of the network on chip architecture. Congestion in the network is one of the main causes of low throughput. Hence, while designing a routing algorithm, designer must keep track of features necessary to perform an efficient routing mechanism which is also capable of handling its problem without exceeding cost of system. It is challenging to design a NoC routing algorithm capable of providing less congested paths, better energy efficiency, and high scalability. Now, it is even more challenging to design routing algorithm to deal with other issues such as aging of network, hotspot in the network due to high temperature, high temperatures can produce faults in the system, enhance power efficiency, and resilient to enhance tolerance along with its local issues. It is challenging to distinguish between temporary and permanent faults. In case of a fault, it is challenging to declare link as dead especially in case of temporary fault. It is challenging to overcome thermal hotspots in 3D NoC which usually appears initially in the center of top layer. As far as aging aware, thermal aware, and fault aware are concerned, it is easy to declare any path unavailable for future communication if it encounters any issue but it is extremely challenging to make it available and once issue is solved.

6 Conclusion

In this paper, we have introduced taxonomy of routing algorithm and concepts for mesh topology. Recent routing algorithms with its criteria are also presented and discussed recent trends in network on chip. Oblivious routing algorithms are cheap with high throughput and low latency but under stress, its performance suffers. Researchers are using adaptive routing, usually modify oblivious algorithm to make them adaptive. Multi-disciplinary algorithms are commonly applied to achieve NoC routing and power efficiently. This is the only study which is focusing on modern criteria of NoC routing algorithms such as ageing aware, thermal aware, fault aware, resilience, and power efficiency routing algorithms at the same time.

References

1. Kadri, N., Koudil, M.: A survey on fault-tolerant application mapping techniques for network-on-chip. J. Syst. Archit. **92**, 39–52 (2019)
2. Wang, J., Ebrahimi, M., Huang, L., et al.: Efficient design-for-test approach for networks-on-chip. IEEE Trans. Comput. **68**, 198–213 (2019)
3. Meng, H., Yang, L., Liu, Z., Wang, D.: A high-throughput network on-chip in full-mesh architecture. IEICE Electron Express **15**, 20180635 (2018)
4. Muhammad, S.T., Saad, M., El-Moursy, A.A., et al.: CFPA: Congestion aware, fault tolerant and process variation aware adaptive routing algorithm for asynchronous Networks-on-Chip. J. Parallel Distrib. Comput. (2019)
5. Rezaei-Ravari, M., Sattari-Naeini, V.: Dynamic clustering-based routing scheme for 2D-mesh networks-on-chip. Microelectronics J. **81**, 123–136 (2018a)
6. Zulkefli, F.W., Ehkan, P., Warip, M.N.M., et al.: A Comparative Review of adaptive routing approach for network-on-chip router architecture. In: International Conference of Reliable Information and Communication Technology, pp 247–254. Springer (2017)
7. Dash, R., Majumdar, A., Pangracious, V., et al.: ATAR: An adaptive thermal-aware routing algorithm for 3-D network-on-chip systems. IEEE Trans. Compon. Packag. Manuf. Technol. 1–8 (2018)
8. Akbar, R., Safaei, F.: A novel adaptive congestion-aware and load-balanced routing algorithm in networks-on-chip. Comput. Electr. Eng. **71**, 60–76 (2018)
9. Gabis, A.B., Bomel, P., Sevaux, M.: Application-aware multi-objective routing based on genetic algorithm for 2d network-on-chip. Microprocess Microsyst. **61**, 135–153 (2018)
10. Salamat, R., Khayambashi, M., Ebrahimi, M., Bagherzadeh, N.: LEAD: An adaptive 3D-NoC routing algorithm with queuing-theory based analytical verification. IEEE Trans. Comput. **1** (2018)
11. Maqsood, T., Bilal, K., Madani, S.A.: Congestion-aware core mapping for network-on-chip based systems using betweenness centrality. Futur. Gener. Comput. Syst. **82**, 459–471 (2018)
12. Ghaderi, Z., Alqahtani, A., Bagherzadeh, N.: AROMa: aging-aware deadlock-free adaptive routing algorithm and online monitoring in 3D NoCs. IEEE Trans. Parallel Distrib. Syst. **29**, 772–788 (2018)
13. Pano, V., Lerner, S., Yilmaz, I., et al.: Workload-aware routing (WAR) for network-on-chip lifetime improvement. In: 2018 IEEE International Symposium on Circuits and Systems (ISCAS), pp 1–5. IEEE (2018)
14. Rohbani, N., Shirmohammadi, Z., Zare, M., Miremadi, S.G.: LAXY: a location-based aging-resilient Xy-Yx routing algorithm for network on chip. IEEE Trans. Comput. Des. Integr. Circ. Syst. **36**, 1725–1738 (2017)
15. Salamat, R., Khayambashi, M., Ebrahimi, M., Bagherzadeh, N.: A resilient routing algorithm with formal reliability analysis for partially connected 3D-NoCs. IEEE Trans. Comput. **65**, 3265–3279 (2016)
16. Lee, D., Das, S., Pande, P.P.: Analyzing power-thermal-performance trade-offs in a high-performance 3D NoC architecture. Integration **65**, 282–292 (2019)
17. Lee, D., Das, S., Doppa, J.R., et al.: Performance and thermal tradeoffs for energy-efficient monolithic 3D network-on-chip. ACM Trans. Des Autom. Electron Syst. **23**, 60 (2018)
18. Safari, M., Shirmohammadi, Z., Rohbani, N., Farbeh, H.: WiP: Floating XY-YX: An efficient thermal management routing algorithm for 3D NoCs. In: 2018 IEEE 16th International Conference on Dependable, Autonomic and Secure Computing, 16th International Conference on Pervasive Intelligence and Computing, 4th International Conference on Big Data Intelligence and Computing and Cyber Science and Technology Congress (DASC/PiCom/DataCom/CyberSciTech, pp 736–741. IEEE (2018)
19. Kumar, S.S., Zjajo, A., van Leuken, R.: Immediate neighborhood temperature adaptive routing for dynamically throttled 3-D networks-on-chip. IEEE Trans. Circ. Syst. II Express Briefs **64**, 782–786 (2017)

20. Fu, Y., Li, L., Wang, K., Zhang, C.: Kalman predictor-based proactive dynamic thermal management for 3-D NoC systems with noisy thermal sensors. IEEE Trans. Comput. Des Integr. Circ. Syst. **36**, 1869–1882 (2017)
21. Gabis AB, Bomel P, Sevaux M (2018) Bi-Objective Cost Function for Adaptive Routing in Network-on-Chip. IEEE Trans Multi-Scale Comput Syst
22. Charif, A., Coelho, A., Ebrahimi, M., et al.: First-last: a cost-effective adaptive routing solution for TSV-based three-dimensional networks-on-chip. IEEE Trans. Comput. (2018)
23. Fusella, E., Cilardo, A.: Lattice-based turn model for adaptive routing. IEEE Trans. Parallel Distrib. Syst. **1** (2018)
24. Rezaei-Ravari, M., Sattari-Naeini, V.: Reliable congestion-aware path prediction mechanism in 2D NoCs based on EFuNN. J. Supercomput. **74**, 6102–6125 (2018b)
25. Xie, R., Cai, J., Xin, X., Yang, B.: LBFT: a fault-tolerant routing algorithm for load-balancing network-on-chip based on odd–even turn model. J. Supercomput. **74**, 3726–3747 (2018)
26. Stroobant, P., Abadal, S., Tavernier, W., et al.: A general, fault tolerant, adaptive, deadlock-free routing protocol for network-on-chip. In: 2018 11th International Workshop on Network on Chip Architectures (NoCArc), pp 1–6. IEEE (2018)
27. Priya, S., Agarwal, S., Kapoor, H.K.: Fault Tolerance in Network on Chip Using Bypass Path Establishing Packets. In: 2018 31st International Conference on VLSI Design and 2018 17th International Conference on Embedded Systems (VLSID), pp. 457–458. IEEE (2018)
28. Touati, H.C., Boutekkouk, F.: FACARS: a novel fully adaptive congestion aware routing scheme for network on chip. In: 2018 7th Mediterranean Conference on Embedded Computing (MECO), pp 1–6. IEEE (2018)
29. Dang, K.N., Ben, A.A., Okuyama, Y., Ben, A.A.: Scalable design methodology and online algorithm for TSV-cluster defects recovery in highly reliable 3D-NoC systems. IEEE Trans. Emerg. Top Comput. (2017)
30. Chen, Y.Y., Chang, E.J., Hsin, H.K., et al.: Path-diversity-aware fault-tolerant routing algorithm for network-on-chip systems. IEEE Trans. Parallel Distrib. Syst. **28**, 838–849 (2017)
31. Tang, M., Lin, X., Palesi, M.: The repetitive turn model for adaptive routing. IEEE Trans. Comput. **66**, 138–146 (2017)
32. Dai, J., Jiang, X., Huang, H., et al.: HyPAR: hybrid planar adaptive routing algorithm for 3D NoCs. In: 回路とシステムワークショップ論文集 Workshop on Circuits and Systems, pp 186–191. [電子情報通信学会] (2017)
33. Charif, A., Coelho, A., Zergainoh, N.-E., Nicolaidis, M.: A dynamic sufficient condition of deadlock-freedom for high-performance fault-tolerant routing in networks-on-chips. IEEE Trans. Emerg. Top Comput. (2017)
34. Aswale, S., Ghorpade, V.R.: Geographic multipath routing based on triangle link quality metric with minimum inter-path interference for wireless multimedia sensor networks. J. King Saud Univ. Inf. Sci. (2018)
35. Blessington, T.P., Bhaskara, B., Basha, F.N.: A perspective on collaboration with interconnects & routing in network on chip architectures. Procedia Comput. Sci. **89**, 180–186 (2016)
36. Alonso, M.G., Flich, J.: Prosa: Protocol-driven network on chip architecture. IEEE Trans. Parallel Distrib. Syst. **29**, 1560–1574 (2018)
37. Boraten, T., Kodi, A.K.: Runtime techniques to mitigate soft errors in network-on-chip (NoC) architectures. IEEE Trans. Comput. Des. Integr. Circ. Syst. **37**, 682–695 (2018)

Modeling and Developing a Geomatics Solution for the Management of Sanitation Networks in Urban Areas: A Case Study of El Fida District of Greater Casablanca

Kaoutar El Bennani, Hicham Mouncif, and ELMostafa Bachaoui

Abstract This paper aims to model the sanitation network and to develop a smart geomatics solution to renovate the management of this network by utilizing a powerful tool, namely Lizmap, an open-source advanced software for developing a geoportal (GIS), with the latest versions of QGIS desktop and QGIS server. Quantitative documents and data are important to the management of the sanitation network as previous difficulties have been experienced due to the use of classic archaic methods of managing data, for example, manual archiving, plans, etc. The rapid growth of the population of the city of Casablanca combined with technical installation problems related to sanitation equipment has generated many ideas for the use of modern resolutions based on developing modern Web-mapping applications (GIS) that are highly suitable for the management of these networks; the city of Casablanca could join other international smart cities by using these kinds of applications. In the interests of efficiency, conceptual object models for various equipment of the sanitation network are realized using a specialized software for spatial modeling, namely Argo-CASEGEO. This modeling is necessary to create a spatial and descriptive sanitation network database. The open source geoportal developed will facilitate managing the sanitation network for engineers, technicians, field workers, and concerned citizens; the spatial information and new tech are key elements for supporting management in every service activity of the various networks related to urban areas in Morocco.

Keywords Lizmap · Smart cities · Geoportal · Sanitation network · Modeling · GIS

K. El Bennani (✉) · E. Bachaoui
Remote Sensing and Geographical Information Systems Applied To the Geosciences and the Environment, Faculty of Sciences and Technology, University Sultan Moulay Slimane, Beni Mellal, Morocco
e-mail: k.elbennani@usms.ma

E. Bachaoui
e-mail: m.bachaoui@usms.ma

H. Mouncif
Multidisciplinary Faculty, University Sultan Moulay Slimane, Beni Mellal, Morocco
e-mail: m.mouncif@usms.ma

© The Editor(s) (if applicable) and The Author(s), under exclusive license to Springer   467
Nature Singapore Pte Ltd. 2021
F. Saeed et al. (eds.), *Advances on Smart and Soft Computing*, Advances in Intelligent
Systems and Computing 1188, https://doi.org/10.1007/978-981-15-6048-4_41

1 Introduction

The sanitation sector in Morocco has experienced major problems in recent years, related to malfunctioning sanitation networks; unlimited spills, for example, increase the level of pollution. Malfunctioning urban networks are often caused by runoff rates and climate change coupled with the noticeable wear and lack of maintenance on the sanitation systems. A first step to solve these problems should start with upstream studies of sanitation networks. These types of systems will significantly contribute to geographic intelligence fueled by modern geographic information system (GIS) technology; improvements consist of improving the quality of rainwater collection services and wastewater recycling and ensuring adequate real time and solid long-term sanitation network management.

GIS enables the modern administration and management of territory based on business applications developed by specialized service providers. Business applications are turn-key tools that easily integrate the basic features of the GIS that can be applied by agents in their daily management tasks [1]. They enable, among other things, the management of sanitation networks. There are equally specialized spatial analyses and modeling tools for resource management. From this starting point, we have designed our own geoportal for the management and modeling of sanitation networks. The conceptualized geomatic solution for the management and modeling of sanitation networks aims to solve future equipment problems and has become a key factor in modern city planning. Modern and smart city policies both participate to improve the quality of the life of the citizens of a city [2, 3]

The purpose of the research presented in this paper is to establish a GIS-based sanitation data model for the integration of a multi-representation and multi-modal urban sanitation network [4, 5] The project is validated in the context of an urban sanitation GIS in the urban region of El Fida, Greater Casablanca which was chosen as the study area. In this large Moroccan city, Web-mapping applications of sanitation networks have been developed to offer multiple management functions that include inserting, editing, deleting, geolocation, printing of plans, measurement, statistics, and various features of the attribute table, using modern GIS by creating a Web-mapping application with Lizmap.

2 Methodology

2.1 The Sanitation Network Conceptual Object Model

In this section, we present the sanitation network conceptual object model used to develop the first geoportal for the management of sanitation networks in the Web-mapping domain. In the object model, sanitation network data are presented as stereotypes: conventional objects, geographic objects, and network objects. Figure 1

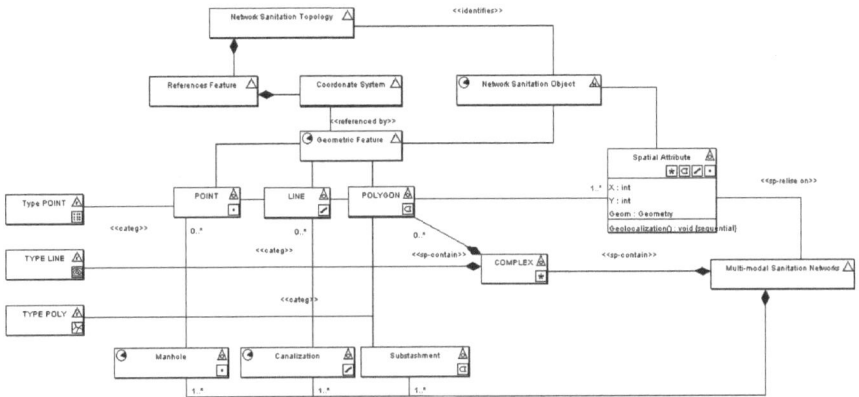

Fig. 1 Sanitation network conceptual model

illustrates the conceptual object model of the sanitation network using a UML-GeoFrame conceptual model schema. The concept of a sanitation network object can be referred to as a feature representing a real-world phenomenon providing a view of and checking the correct functioning of the sanitation network [6]. It may embody one or more spatiotemporal properties that are valid for a period of time. Spatial properties are represented in geometric features and are referenced as spatial referencing systems. For geometric objects, a spatial object can have four methods for abstracting its spatial component—punctual, polygonal, linear, and complex—while non-spatial properties are used to describe the characteristics of the parts of the sanitation network, namely manholes, canalization, and subcatchment. These are specified by geographic object stereotypes. As the properties of the sanitation object, such as its spatiality or attributes, may change over time, sanitation network objects can only be valid for a period or even an instant. However, different sanitation transmission modes for storm water, wastewater, and industrial water are commonly applied to the utility of various equipment such as manholes, sections, and subcatchments. Finally, the stereotype "Categ" is used to characterize a particular type of association that occurs when modeling specific fields for each geometric feature [4, 7, 8].

2.2 The Installation Steps of a Sanitation Network

The modeling of the sanitation network begins with the creation of a node; we can create a watershed because each watershed requires a single node. To create a canalization, we need an upstream node and a downstream node; moreover, to build a pumping station, it is requisite to have a node and two or three sections; with three sections minimum and a node, we can construct a spillway. In the case of a water

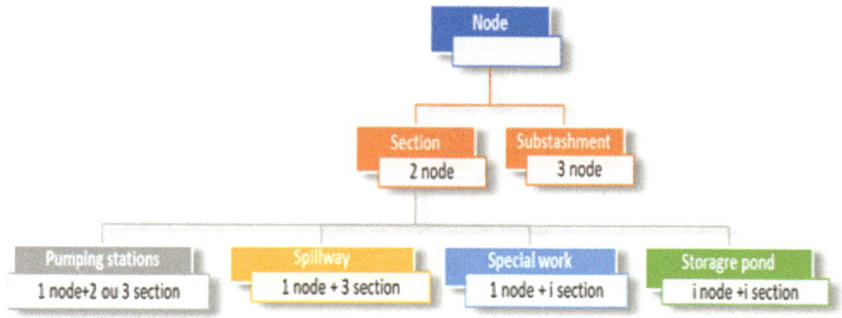

Fig. 2 Installation steps of a sanitation network

works, we need a node and maybe three or four sections depending on the complexity of the work. Figure 2 shows the installation steps of sanitation networks [9].

2.3 Architecture

This geoportal was developed using the latest version of Lizmap 3.2.2 designed by 3Liz, an open-source software that consent the development of Web-mapping applications in QGIS Desktop and Lizmap, written in JavaScript and Python that allows users to share and edit geospatial data. This makes it easier for the publication of a map intended for the general public to valorize geographic information; the database has been implemented on PostgreSQL using the extension PostGIS 2.0.4. PostGIS is an extension that joints support for geographic objects to the PostgreSQL object-relational database, Open Layers, Google Maps, Bing Maps, and the IGN geoportal displaying the dynamic maps. One of the runtime libraries used was the Geospatial Data Abstraction Library (GDAL), whishes a Web cartographic application GIS that stocking the storage and management of spatial and non-spatial data of relative information of a sanitation network. The developer of the geoportal configures the options and the functionalities for each map that will be available within the map application in QGIS desktop, with the Lizmap plugin installed [10].

The Lizmap Web client is installed on a map server with QGIS Server, and the transfer of the data between Lizmap and QGIS Server is made by FTP protocol with FileZilla. The developer prepares data and projects in QGIS desktop and the configuration of Lizmap in the QGIS server. Users can access our geoportal applications using their Web browser and their smartphones or tablets. The geoportal development took shape in Laravel which is a PHP framework for building scalable, standards-based Web applications following the model-view-controller (MVC), an architectural pattern. The Lizmap architecture is shown in Fig. 3.

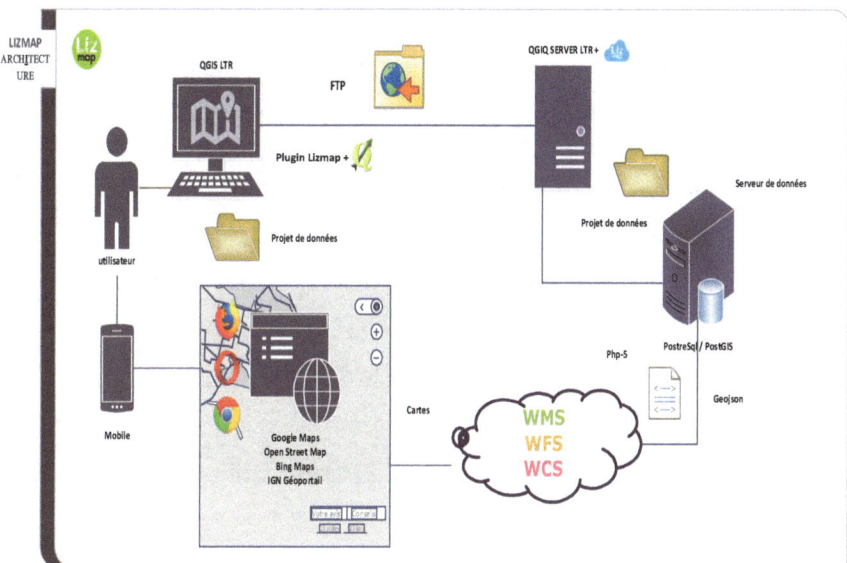

Fig. 3 Lizmap architecture

2.4 The Reason for Selecting Lizmap

Lizmap, is an open-source software, that may be installed on different platforms including Linux and Windows. Sadly, setting up and maintaining a cartographic server "QGIS Server" effectively is time-consuming and can be more complex to install in Windows 10. Installing the latest version of Lizmap requires manual installation and configuration for various software:

- QGIS Server LTR 3.4.6 (installed via OSGeo4W ×64)
- QGIS Desktop (installed via OSGeo4W ×64)
- Lizmap Web Plugin (installed via QGIS desktop plugin manager)
- etc.

The various software cited above must be on the same machine, the manual installation of Lizmap, QGIS Server, and Apache are passed step by step, then we need to properly configure the virtual hosts. The latest version of Lizmap has more difficulties in configuring the virtual hosts and QGIS server because it depends on a variable when configuring GRASS (@grasspath@) specifying the exact version of GRASS, for example, (grass-6.4.3), and it will be complicated to discover the configuration problem between the latest version and the previous version of Lizmap as it takes a long time to configure. The correct configuration of QGIS-Ltr server is shown in Fig. 4 [10, 11].

```
C:\OSGeo4W64\httpd.d\httpd_qgis-ltr.conf - Sublime Text (UNREGISTERED)
File  Edit  Selection  Find  View  Goto  Tools  Project  Preferences  Help

    httpd_qgis-ltr.conf

 1   LoadModule fcgid_module modules/mod_fcgid.so
 2
 3   DefaultInitEnv O4W_QT_PREFIX "C:\OSGeo4W64/apps/Qt5"
 4   DefaultInitEnv O4W_QT_BINARIES "C:\OSGeo4W64/apps/Qt5/bin"
 5   DefaultInitEnv O4W_QT_PLUGINS "C:\OSGeo4W64/apps/Qt5/plugins"
 6   DefaultInitEnv O4W_QT_LIBRARIES "C:\OSGeo4W64/apps/Qt5/lib"
 7   DefaultInitEnv O4W_QT_TRANSLATIONS "C:\OSGeo4W64/apps/Qt5/translations"
 8   DefaultInitEnv O4W_QT_HEADERS "C:\OSGeo4W64/apps/Qt5/include"
 9   DefaultInitEnv O4W_QT_DOC "C:\OSGeo4W64/apps/Qt5/doc"
10
11   DefaultInitEnv PATH "C:\OSGeo4W64\apps\qt5\bin;C:\OSGeo4W64\bin;C:\OSGeo4W64\apps\qgis-ltr\bin;C:\OSGeo4W64
     \apps\grass\@grasspath@\bin;C:\OSGeo4W64\apps\grass\@grasspath@\lib;C:\Windows\system32;C:\Windows;C:\Windo
     ws\System32\Wbem"
12   DefaultInitEnv QGIS_PREFIX_PATH "C:\OSGeo4W64\apps\qgis-ltr"
13   DefaultInitEnv QT_PLUGIN_PATH "C:\OSGeo4W64\apps\qgis-ltr\qtplugins;C:\OSGeo4W64\apps\qt5\plugins"
14   DefaultInitEnv TEMP "C:\Users\kouter\AppData\Local\Temp"
15   DefaultInitEnv PYTHONHOME "C:\OSGeo4W64\apps\Python37"
16   DefaultInitEnv PYTHONPATH "C:\OSGeo4W64\apps\Python37;C:\OSGeo4W64\apps\Python37\Scripts"
17
18   Alias /qgis-ltr/ C:\OSGeo4W64/apps/qgis-ltr/bin/
19   <Directory "C:\OSGeo4W64/apps/qgis-ltr/bin/">
20       SetHandler fcgid-script
21       Options ExecCGI
22       Order allow,deny
23       Allow from all
24   </Directory>
```

Fig. 4 Configuring of virtual host

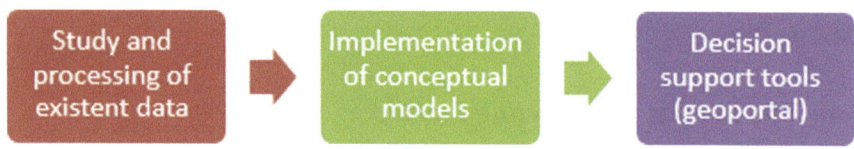

Fig. 5 Steps of development geoportal

3   Results and Discussion

3.1   Geoportal for Management of Sanitation Network

To develop this geoportal, we have carried out the following steps (Fig. 5):

3.2   Development of the Geoportal

Poor management of wastewater and stormwater collection systems can have significant environmental and economic impacts: high frequency of human interventions (clogged systems, overflows) or untreated wastewater discharged directly into the environment. The developed Web-mapping application, referred to as the geoportal in the article, encompasses a variety of features. It offers a default Web mapping of sanitation networks with the following practical features: pan, zoom on any selected equipment of the sanitation network (canalization, sub catchment, manhole), zoom in, zoom out, selecting a zoom level with a scale bar, scale display, numerical zoom, and panning. These latest functionalities are positioned on the right toolbar of the map. To zoom in on a geographic zone, the user simply selects the zoom rectangle and drags to draw a rectangle, defining the area to reach. Selection boxes enable the user to hide and display the proposed layers such as canalization, manholes, and subcatchments. To fully use the map, the user can hide the management layers panel. The administrator of the geoportal allows the user to add some features, all depending on their needs, including:

- selecting various kinds of map background
- locating by layer (spatial research)
- calculus distance, area, and perimeter of equipment sanitation networks
- printing the plan of sanitation networks
- editing spatial and descriptive data of the sanitation network's equipment
- show the statistics (e.g., classification of canalization by diameter, average flow).

The window of our developed geoportal is very useful as it offers a variety of advanced features for managing the sanitation network, but administrators must confirm their identity to be able to access a few of the map's functionalities such as editing spatial and descriptive data of various equipment of sanitation networks. The button connects are situated at the top left for authentication. After authentication, the administrator easily gets full access to editing. The user can access a number of features some of which we refer to: selecting a base map, locating by layers, distance, area, and perimeter measurements, printing the plan of the sanitation network, atlas and statistics, etc. All latest functionalities are situated on the right part of the map. What is more, users can also use our geoportal on their smartphones and tablets [10, 12].

3.3 Features of Geoportal

3.3.1 Spatial Research

Spatial research is a default feature that can be easily used by a simple activation from the map users, it is located above the layer management panel and it appears in the form of a list, some lists require entering some characters (e.g., research of subcatchment by area, locate sections by length, locate manholes by type …) before inputting each location. The user is required to simply choose an emplacement from the list to be able to zoom the desired geographic location [12, 13] (Figs. 6 and 7).

3.3.2 Editing Data of Sanitation Networks

The administrator of our geoportal is permitted not only to edit but also to modify descriptive and spatial data of the sanitation network. The administrators of our geoportal can limit the possible changes desired of descriptive and spatial data, as

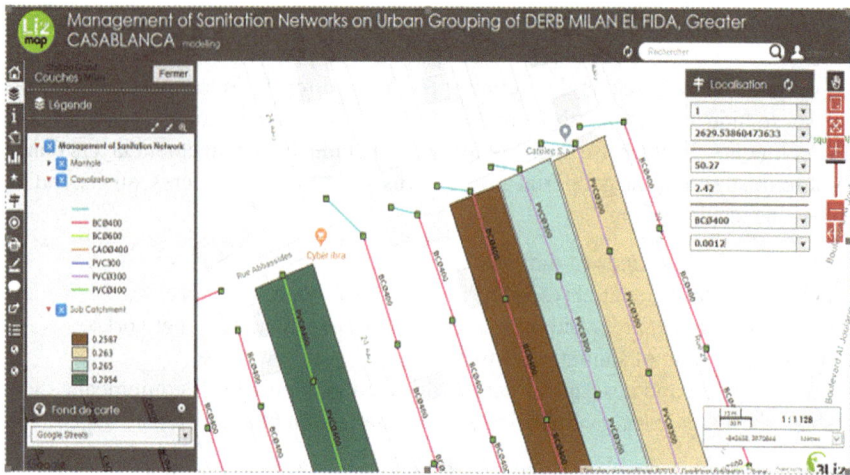

Fig. 6 Spatial research

Fig. 7 Geoportal windows interface

well by adding spatial objects (subcatchment, manholes, and canalization), making geometric as well as field modifications, or deleting spatial objects. This feature appears in the geoportal toolbar. The edit menu allows the administrator to select the data (subcatchment, manhole, and canalization) they want to update. Once the layer is selected, the edit panel is viewed.

This varies depending on the configuration of the geoportal desired by the administrator. If any changes are inaccessible, the administrator can decide between adding new equipment of sanitation networks (subcatchment, manholes, and canalization) or selecting one. If the user selects Add, they will be asked to draw a simple form that

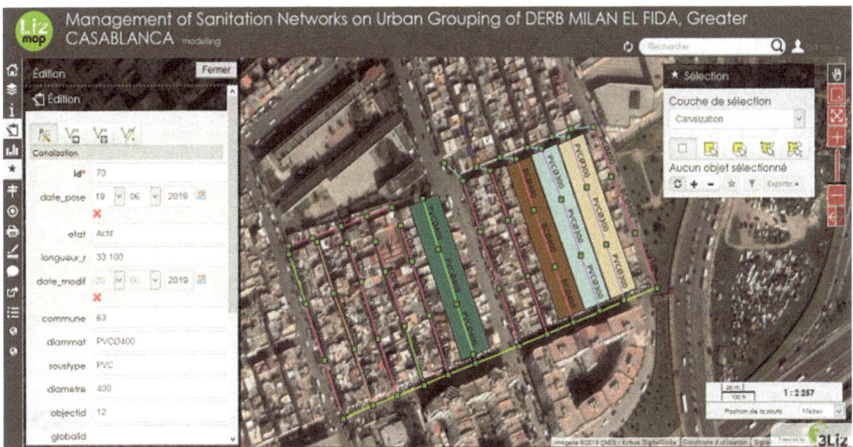

Fig. 8 Add new canalization

depends on the selected data layer: we have selected sections (canalization) in the form of a line. To finish the line, the administrator needs to add the last canalization by double-clicking on the desired location. Once the drawing is finished, an editing form for the fields pops up. If the user wants to restart drawing the geometry, they should click cancel. The administrator of the geoportal has provided all the required information of the canalization. If the geometry is appointed, the user may click on the Save button. The new canalization will be added, and this new object can be updated any time by simply selecting it.

To update data rather than an object of a sanitation network (subcatchment), the administrator needs to click on it on the map, then click on the Edit button. The selected object appears on the map and its geometry may be changed immediately. To validate the geometry modifications or simply access the editing form for fields, the user needs to click Save. A dialog box comprising the editing form for field shows up. The Save button saves the geometry and the attributed changes (area, name, slope, runoff coefficient, impervious area). If the user wants to remove the selected object, a simple click on Del will do this [12, 13] (Fig. 8).

3.3.3 Some Queries for Managing Sanitation Network in Our Geoportal

The poor management of various sanitation networks (wastewater and stormwater) can have significant problems, both environmental and economic. For this reason, the best incentive for developing a geoportal of the management of sanitation networks, using Lizmap, is to respond to a large number of requests from engineers, technicians, and fieldworkers who cite poor management as the cause of the many problems [14, 15].

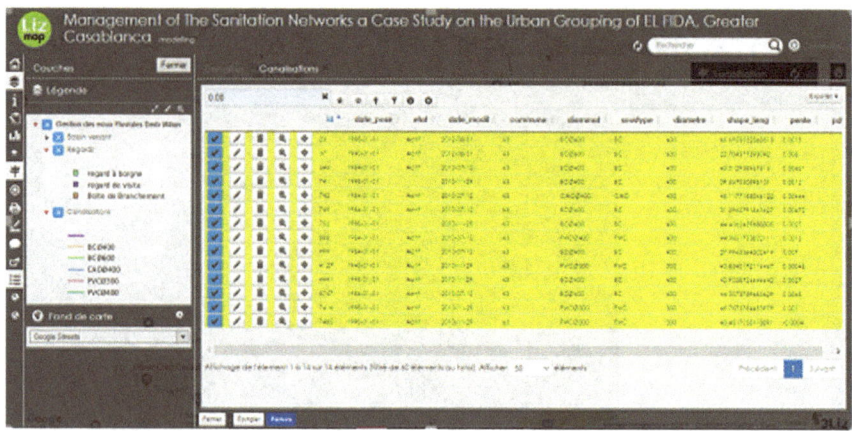

Fig. 9 Result of query 1

3.3.4 Query 1: Give Canalizations that Are at Risk of Being Blocked Quickly

The user can answer this question using the feature attribute layers of our geoportal, by entering condition, for example, (slope under 0.03) the result of the query is represented in Fig. 9. The user can also export the canalizations that are slopes under 0.03 that has a risk of blocking in GeoJSON format.

3.3.5 *Group the Canalizations According to Their Diameter and Slope (Diameter ≤ 300 mmand Slope < 0.03)*

See Fig. 10.

3.3.6 Give the Canalizations Made of Constrained Concrete BC

See Fig. 11.

3.3.7 Distribute the Canalization Diameters According to Their Classification (primary, Secondary, and Tertiary)

In the latest version of Lizmap, a way to show charts is implemented (DataViz). The user will be able to create a few kinds of graphs (scatter, pie, histogram, box, bar histogram 2d, polar) with only a few clicks. This function may replay all questions; in this case, it represents diameter classification [12] (Fig. 12).

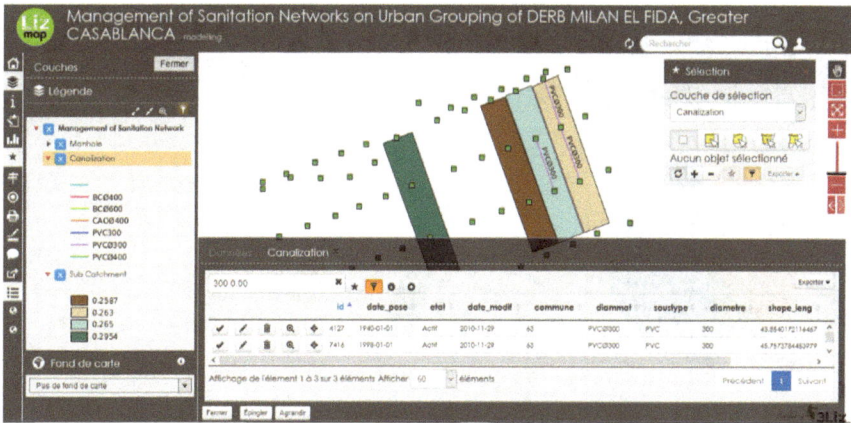

Fig. 10 Result of query 2

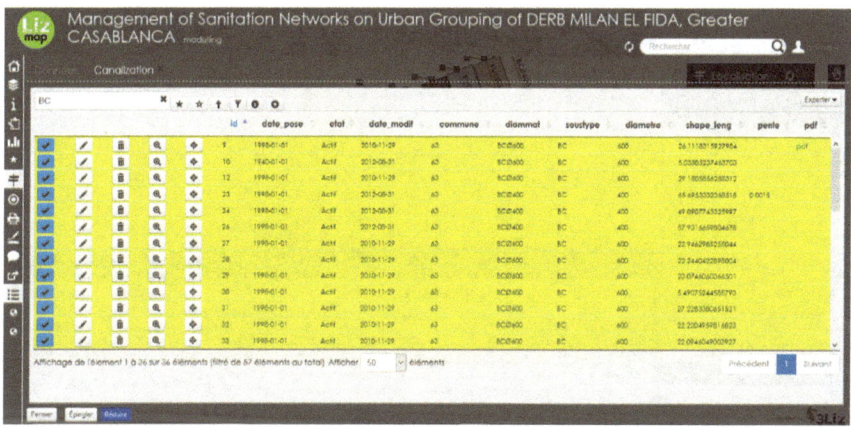

Fig. 11 Result of query 3

3.3.8 Distribute the Subcatchment Area Less Than or Equal to 0.26 Hectares

According to some needs (requests) of the management of sanitation networks, we can conclude:

The sanitation network of Urban Grouping DERB MILAN is entirely gravity-fed.

- 54.1% of canalizations have a diameter equable to 400 mm (Fig. 13);

- 37. 9% of canalizations have a diameter equal to 600 mm;
- 8.11% of canalization have a diameter equal to 300 mm, resulting in under sizing problems.

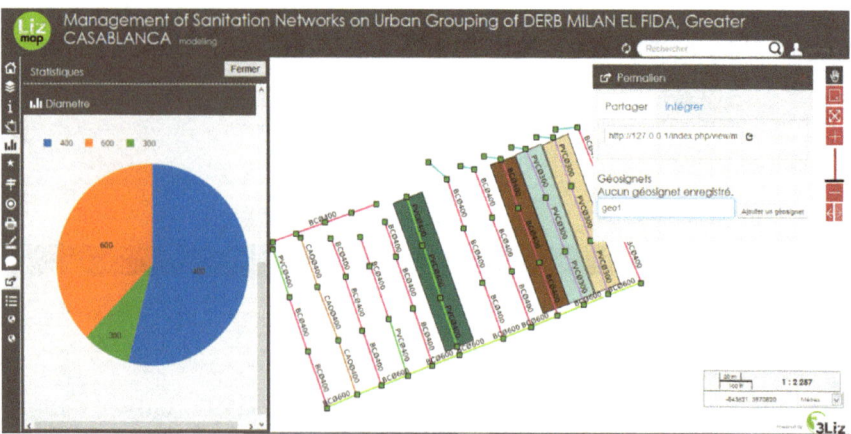

Fig. 12 Result of query 4

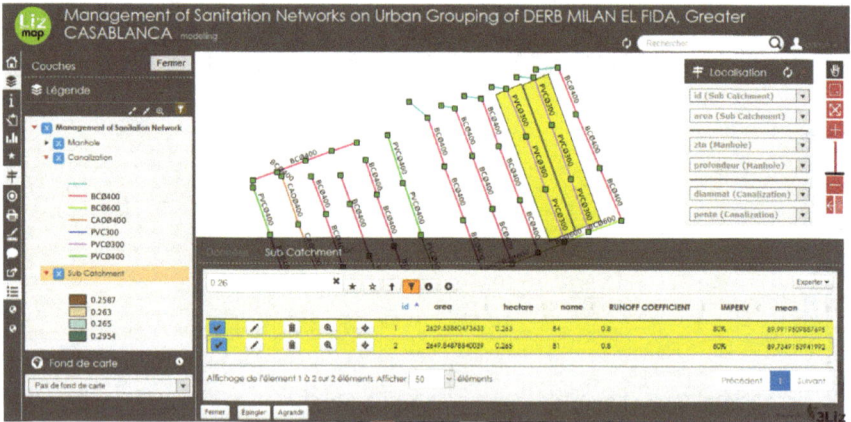

Fig. 13 Result of query 5

- it is strongly advisable to obtain at least a 3% longitudinal slope to avoid any deposit formation.
- 88.14% of the canalizations have a slope less than 0.03, which presents a risk of deposits due to the low slope.
- The sanitation network canalizations of the Urban Group of El Fida are entirely made of concrete; these canalizations have a sudden break.

The requests offered show various possible applications. The flexibility of our geoportal has allowed the formulation of several queries on the database that are difficult to answer using traditional management methods. The results obtained made it possible to detect and represent anomalies in the sanitation network [14, 16].

4 Conclusion

Specialized researchers, engineers, technicians, field workers, and concerned citizens can easily use this geoportal. As information and communication technologies are key elements for supporting management planning in every service activity of a sanitation network, this GIS application can be used to support any network related to urban areas. This genre of applications can help the city of Greater Casablanca to be among the smart international cities. Our geoportal can help urban managers exploit sanitation network equipment data to allow fieldworkers to provide better environmental services to citizens. Benefits such as improving the resolution of leaks, sending a list of intervention orders to fieldworkers, monitoring and evaluating leaks, and signaling leaks are among the aims of future work of developing a mobile GIS application for fieldworkers. This specific geoportal includes open-source code that can be upgraded in the future when new management techniques emerge. It is adjustable to diverse software platforms so that in the future, data may be broadly distributed. Finally, using new technology applications such as the Internet, iPads, and iPhones to gather information and communicate instantaneously with multiple users make this application easily transmitted and operated. In conclusion, this GIS application is an appropriate tool for municipality administration services, and used in cooperation with researchers, engineers, fieldworkers, technicians, and concerned citizens; it can make a serious contribution to the better management of sanitation networks, thus providing citizens with better living conditions.

References

1. Jinal, P., Gayatri, D.: Exploring enterprise resource planning (ERP) and geographic information system (GIS) integration. In: Applied Geoinformatics for Society and Environment conference, p. 7 (2013)
2. Pierre-Alain, A., Sophie, S.: Systèmes d'Information Géographique: outil d'aide à la gestion territoriale. In: Techniques de l'ingénieure (2009)
3. Martin, S.E., Reynard, R., Pellitero, O., Ghiraldi, L.: Multi-scale web mapping for geoheritage visualisation and promotion. Geoheritage 6(2), 141–48 (2014)
4. Jugurta, L., Maurício, F., Jaudete, D.: ArgoCASEGEO - an open source CASE tool for Geographic Information Systems modelling using the UML-GeoFrame model. In semantic scholar, p. 8 (2016)
5. Shaopei, C., Jianjun, T., Christophe, C.: Multi-scale and multi-modal GIS-T data model. J. Transp. Geogr. 19(1) (2011)
6. Shashi, S., Hui .X.: Encyclopedia of GIS, p. 54, (2008)
7. Christiaan, L., Peter, O., Rod, H.: The modelling of spatial units (Parcels) in the land administration domain model (LADM). In: FIG Congress Facing the Challenges – Building the Capacity Sydney, p. 28 (2010)
8. Victor de Freitas, S., Lisboa, F., Marcus, V.: Improving productivity and quality of GIS databases design using an analysis pattern catalog. In: Second Asia-Pacific Conference, Conceptual Modelling (APCCM2005), p. 9. Newcastle, NSW (2005)
9. ENGEES—École Nationale du Génie de l'Eau et de l'Environnement de Strasbourg, Généralités sur la modélisation. https://engees.unistra.fr. Last Accessed 2010

10. Lizmap installation: https://docs.lizmap.com/current/fr/install/upgrade.html. Last Accessed 2018
11. Lizmap description: https://www.3liz.com/en/lizmap.html,last. Last Accessed 2018
12. Lizmap documentation: https://docs.lizmap.com/current/fr/user/index.html. Last Accessed 2018
13. Lizmap services: https://www.3liz.com/en/services.html. Last Accessed 2019
14. Abdelbaki, C., Zerouali, M. : Modélisation d'un réseau d'assainissement et contribution à sa gestion à l'aide d'un système d'information géographique - cas du chef-lieu de commune de chetouane-wilaya de Tlemcen Algérie. Larhyss Journal, ISSN 1112-3680, n° 10, pp. 101–113 (2012)
15. Chikh, M., Trache, M. : Modélisation des données d'un réseau d'assainissement dans une base de données de type S.I.G. In : 2ème Colloque Maghrébin sur l'Hydraulique, pp. 56–63 (2012)
16. Mousty, P., Chartier, B., Barrere, S., Montaner, J., Delacour, J., Pauline, D., Pauline, D.: Système d'information géographique et télégestion d'un réseau d'assainissement. TSM. **9**, 431–438 (1990)

Networked Expertise Based on Multi-agent Systems and Ontology Design

Yman Chemlal and Anass El Haddadi

Abstract Performance of competitive intelligence needs network expert which is composed by various profiles and skills to analyze and evaluate information collected from internal or external sources. Unfortunately, different studies reveal that is difficult for a network; human expert disposes a limited time to acquire and process a huge quantity of information. The core idea of our research paper is to develop an intelligent approach based on ontology and multi-agent system to provide a platform able to analyze and treat information as a network human expert and produce new knowledge useful for decision making based on SWOT analysis method.

Keywords Networked expertise · Collective intelligence · Multi-agent system · Ontology · Rules mining

1 Introduction and Related Work

Competitive intelligence forms an integral part of company's strategic planning process [1] to gather and analyze information about its industry, business environment, competitors, competitive products, and services [2]. The information gathering and analysis process can help a company to develop its strategy or to identify competitive gaps [2]. In this respect, many method and tools are used like PESTEEL, PEST analysis, SWOT, etc. .... This paper focused on SWOT analyze method because SWOT has been widely applied in strategic planning for enterprises, market demand analysis, industry analysis and it is generally seen as a practical and effective approach [3]. Principally, it used to evaluate the internal strengths, weaknesses, opportunities, and external threats involved of a specific domain [4]. By identifying the factors in these four fields, the organization can recognize its core competencies for decision making, planning, and building strategies [1].

Human expert is a backbone of competitive intelligence process; his role was confined to process and analyze uncertain and incomplete information from various

Y. Chemlal (✉) · A. El Haddadi
Data Science and Competitive Intelligence Team, National School of Applied Sciences of Alhoceima (ENSA), Alhoceima, Morocco
e-mail: Chemlal_yman@yahoo.fr

© The Editor(s) (if applicable) and The Author(s), under exclusive license to Springer Nature Singapore Pte Ltd. 2021
F. Saeed et al. (eds.), *Advances on Smart and Soft Computing*, Advances in Intelligent Systems and Computing 1188, https://doi.org/10.1007/978-981-15-6048-4_42

sources and create new knowledge useful for decision making. Furthermore, human expert collaborates each other within a network called networked expert, to overcome a huge quantity of uncertain/incomplete information and provide efficiency intelligent knowledge. This collaboration creates collective intelligence; collective intelligence is defined as shared intelligence emerging from a group of people when they work on the same tasks that could result in more innovative outcomes than when individuals work alone [5, 6]. Unfortunately, various field studies [7, 8] show that generally network is difficult to be implemented and managed successfully within a company and that due to a number of factors such:

- Technological factors: company does not adopt a new technology to facilitate communication between experts.
- Organizational factors: lack of management of expert network's structure. It must include, at least, institutions of subject supervisors, moderators, and experts [7].
- Personal-related factors: lack of experts involvement in sharing experience and expertise.
- Motivational factors: expert must have a time and financial conditions to be engaged in decision-making process.
  All these factors and others motivated our research to develop an intelligent network as networked expertise. Our contribution allows creating collective intelligence and supporting a decision making. Furthermore, this framework must fully achieve the requirement specification of a human expert network such as:
- Self organization: Self organization favors fast revelation of original ideas, factors, goals, and proposals, as well as appreciable reduction of risks and prevention of the negative consequences of managerial decisions. Self-organized environments bear networked leaders and identify talented moderators and experts [9].
- Specific methods of expert information acquisition and processing: Among the most important method to analyze and treat information is the SWOT analysis method (strengths, weaknesses, opportunities, and Threats), and this method can reveal both the external and internal factors that affect the company. The internal factors are categorized as either strengths, weaknesses, while the external factors are either opportunities or threats [1].
  Thus, the design of an artificial network involves the design of a complex system. This complexity lies in reasoning about uncertain/ incomplete information. Our approach, therefore, based on multi-agent system (MAS) technologies [10], composed of autonomous entities called agents. Each agent reasons about uncertain and incomplete information and agents self-organized locally to produce the desired global function [3] which is the creation of intelligence knowledge useful for decision-making process. However, to guarantee mutual understandings "Interoperability" between agents and facilitate agent's reasoning behavior to process and treat information with SWOT analysis method, a specific ontology is designed and implemented. Ontologies offer significant benefits to multi-agent systems: interoperability, reusability, support for multi-agent system (MAS) development activities (such as system analysis and agent knowledge

modeling), and support for MAS operation (such as agent communication and reasoning) [4].

There are many scholars at home and abroad have already developed a large architecture of MAS that integrated ontology; Lavbic [11], proposed a model of ontology to make agents able to reason about unstructured information found on the Web and information available in several internal data sources; Espinasse et al. [12] conducted an interesting research on the collection and extraction of relevant information on the Web from software agents and ontologies. However, there are few theories as a guidance of methodology for ontology-based MAS development. This paper will benefit from MOBMAS [4] methodology and Gaurino [13] to develop a specific Model of Ontology-based SWOT analysis method to be integrated in a core of our agents. This methodology comprised of three steps:

1. Architecture of MAS dedicated to collective intelligence.
2. Construction of Ontology based on SWOT analysis method for competitive intelligence.
3. Construction of MAS application ontology and rules mining.

This paper presents experimental results obtained with MAS designed to create networked expertise systems, and applied to the information processing based on SWOT analysis method for competitive intelligence system. Section 2 introduces architecture of MAS dedicated to collective intelligence. Section 3 gives a construction of ontology model based on SWOT analysis method for competitive intelligence system and Sect. 4 presents implementation of our proposition. Section 5 concludes with our perspectives.

2 Methods

2.1 Architecture of MAS Dedicated to Collective Intelligence

The first step is to design an architecture of MAS which is composed of four agent groups: (Fig. 1).

- Information transformation and extraction: responsible for transforming all the data captured into a formal, shareable language, and usable by the next layer. Data received in different forms by collector agent is transformed to OWL descriptions by transformative agent. This operation converted the raw data into XML documents. These XML documents will then be automatically transformed into semantic form.
- Management: Monitor agent consolidated all information received and dispatched it to appropriate cluster in "Information processing and decision making" group.

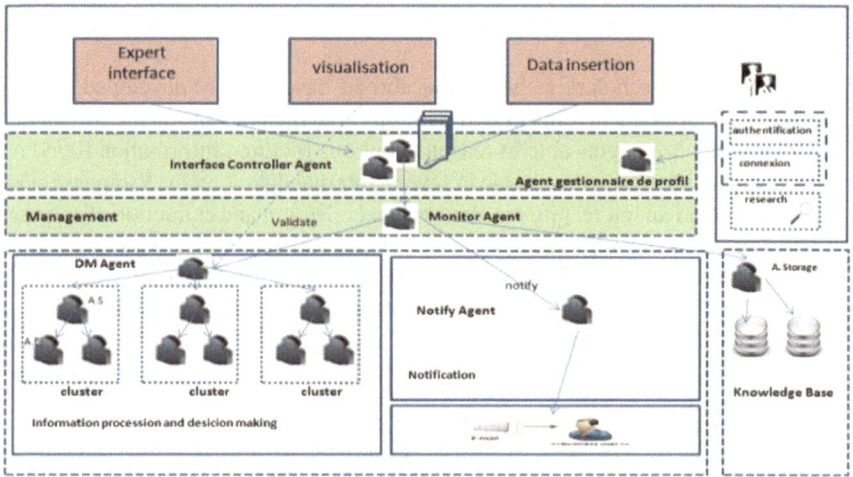

Fig. 1 Architecture of MAS dedicated to collective intelligence

– Information processing and decision making: is composed by different clusters of
  expert agents and each cluster is reliable to a specific subject; each cluster evaluated and validated the information sent by monitor agent. Decision maker agent
  consolidated all information treated by each cluster and inferred new knowledge
  useful for decision making.
– Notification: Notify agent shared information processed and/or produced with the
  appropriate decision makers.

2.2 Construction of Ontology Based on SWOT Analysis Method for Competitive Intelligence System-Case Study: Banking Sector

The second step is a deep understanding the strategy of the banking sector to confront
competition using SWOT method. This step based on theories of strategy's potential profitability for banking organization to create the core concept terminology.
This paper proposes to divide these cores into two parts: the internal context of the
bank and the external context of the bank. The first term set represents opportunities
and threats that can anticipate and/or influence future developments of a bank. The
internal context defines two main sub-concepts: Macro environment element and
micro environment element. Thus, the core concept terminology of "macro environment element" will be created with the PESTEL model [14] (Political, Economic,
Social, Technological, Ecological, and Legal) to get its sub-concepts. The concept
"micro-environment element" describes competitive intensity with the model of
five competitive forces of Porter [15] (Competitor—Customer—Suppliers—New

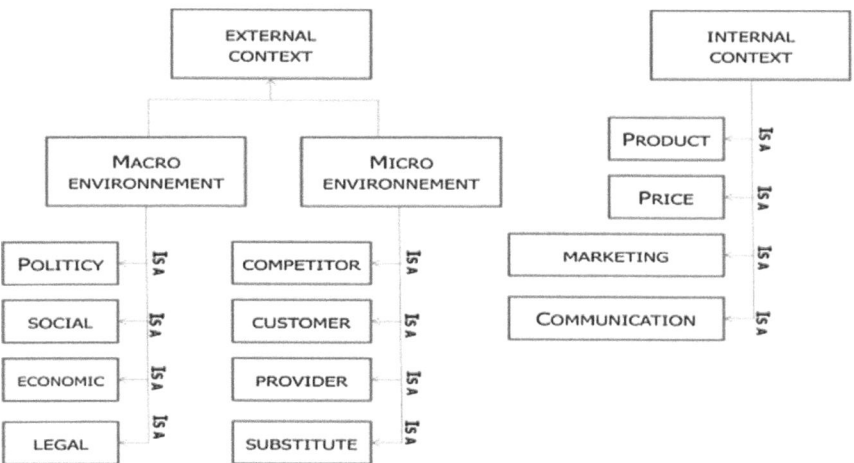

Fig. 2 Extract of ontology based on SWOT

entry—Substitutes) to get its sub-concepts. The second-term set represents a source of strengths/weaknesses, to identify the specificities of the organization and see how to use them to build a competitive advantage. In our case study, this concept divided on the following concepts: Product, price, marketing, communication, human resources. All this concepts and sub concepts are linked with a hierarchical relationship "is a" as shown in Fig. 2.

Construction of MAS application ontology: MAS application ontology is an essential phase in our research work; it inspired from MOBMAS methodology [4] to construct a new level able to integrate ontology in reasoning ability of MAS. It is the level that defines the adaptive behaviors of the agents through a specific ontology and logical rules. This paper proposed a specific ontology that describes the content model of the communication and interaction protocols between agents. This paper proposed the concept "communication" (Fig. 3) which is represented by FIPA message as a sub-concept; this one is a super concept of "content" that divided in Agent Identifier (AID) describes the identification of an agent and "Concepts" that divided into sub-concepts indicating how agent can reason about information received and how group of agents can create new knowledge (Figs. 4 and 5). It is: "Finding" present received information that includes all attributes related to the fields of banking sector. "Issues_competitive_intelligence", its sub-concepts "quantitative issues" and "qualitative issues" are used to describe the nature of processing elements that may be quantitative (e.g., competitors' pricing policy, the cost of their products and services, etc.) or qualitative (e.g., competitors' products and services, the types of their customers and suppliers, etc.). "Result" describes the results of the processing information received and related to "Finding", it is characterized by all attributes of the "Finding" plus the attribute "type" to indicate the four aspects of SWOT (strengths, weaknesses,

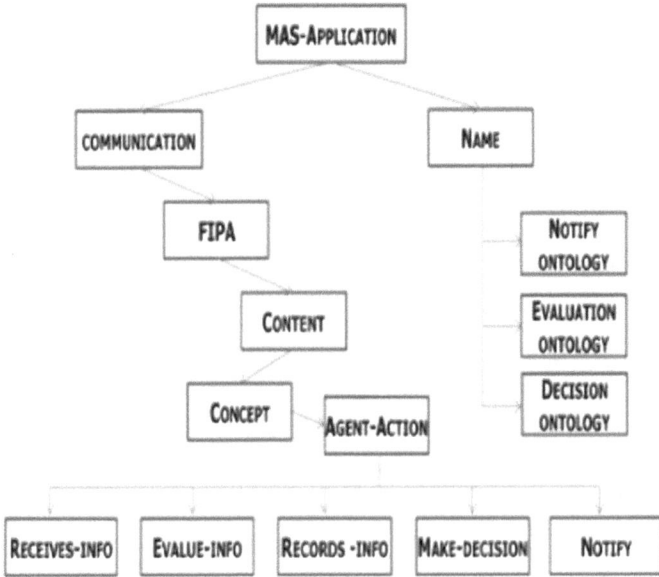

Fig. 3 Extract of MAS-application

opportunities, and threats). This attribute "type" is generated after executing logical rules by agents. "StrategicModel" describe strategies policy of the decision (Fig. 5).

In this study, four strategic options are considered as defined in SWOT matrix (Fig. 6): Opportunity/strength, opportunity/weakness, threat/strengths, threat/weakness, which are defined as a sub-concept of the "Strategic-Model" concept. "Agent Action" a special sub-concepts indicating what actions are required by agents, it is: "Receive-info" agent action used by CI agent to receive information to treat. "Evalue-info" enables expert agent to treat information (the instance of the Finding concept). Expert agent executes a logical rules dedicated to information processing and defined in section below to classify information in the four aspects of SWOT (strengths, weaknesses, oportunities, and threats). "MakeDecision" Agent Action", allows decision maker agent to consolidate all information treated by expert agent and execute strategic logical rules defined in section bellow to infer new knowledge useful for decision making.

Competitive intelligence analysis and rule mining: The basic definition of competitive intelligence association inference rules is described as follows:

- **Definition 1**. Result set is R, R = Result ($R1$), Result ($R2$), ..., Result (Rm), and m is a number of Result inferred by expert agent.
- **Definition 2**. Opportunites_Result set is O, O = Opportunity ($O1$), opportunity ($O2$), ..., opportunity (On) and n is the number of Opportunites_Result.
- **Definition 3**. Weakness_Result set is W, W = Weakness ($W1$), weakness ($W2$), ..., weakness (Wn) and n is the number of Weakness_Result.

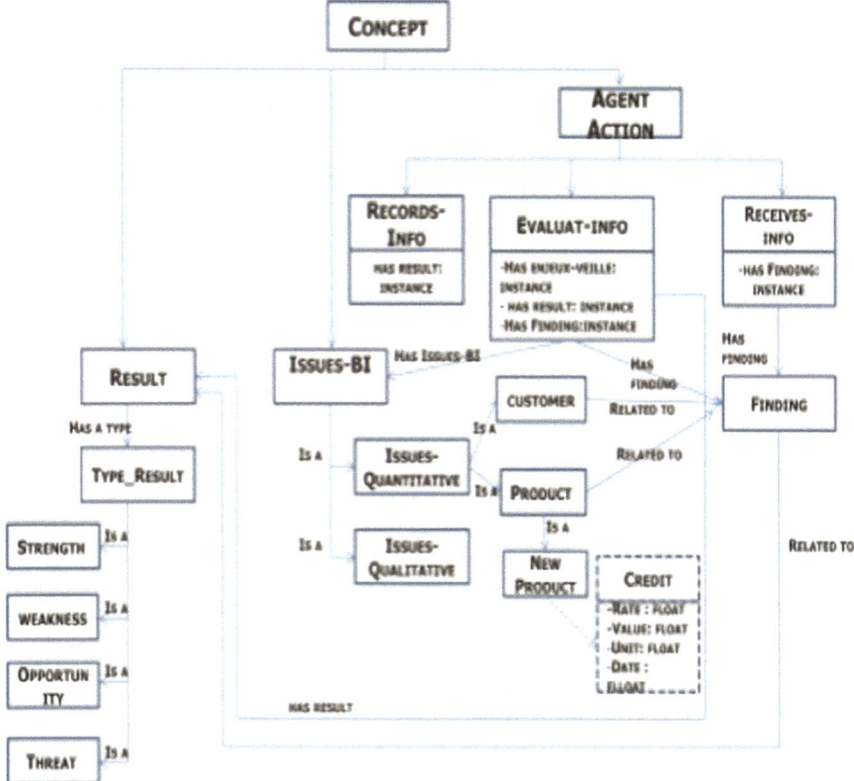

Fig. 4 Extract of Evalue_Info ontology

- **Definition 4**. Threat_Result set is T, $T =$ Threat $(T1)$, threat $(T2)$, ..., threat (Tn) and n is the number of Threat_Result.
- **Definition 5**. Strengths_Result set is S, $S =$ Strengths $(S1)$, strengths $(S2)$, ..., strengths (Sn) and n is the number of Strengths_Result.
- The object of this paper is analyzing and making decision using SWOT method. Therefore, we propose to further mining association inspiring by Apriori algorithm of association rules mining conditions. Apriori algorithm is the most classic association rule mining algorithm. Its main idea is to derive other high-frequency data item sets from known high-frequency data item sets. According to the association rules between data items, according to the occurrence of some items in a transaction, it can be deduced that other items also appear in the same transaction [16]. The transaction defined in Apriori algorithm can correspond to the result executed by agent. The data items defined can be corresponding to type of result in each transaction. Thus, this paper defined strategic rules as:

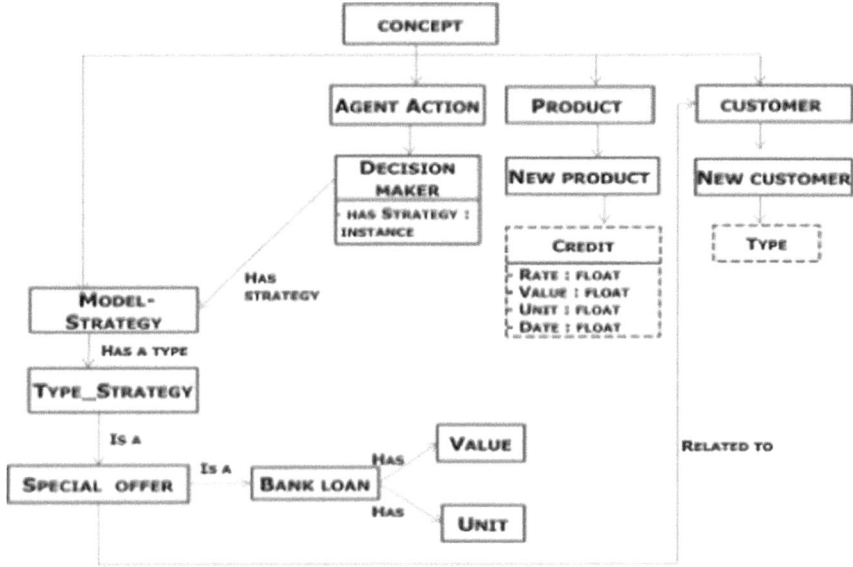

Fig. 5 Extract of decision-making ontology

| | | internal factors | |
		Strength	Weakness
External factors	Opportunity	strategic option	strategic option
		use forces to seize opportunities	minimize weaknesses to seize opportunities
	Threat	strategic option	strategic option
		uses forces to avoid threats	minimize weaknesses to avoid threats

Fig. 6 SWOT analyze method

- **Definition 6**. (Strategic rules): All types of result defined in (2), (3), (4), (5) and inferred by expert agent. Example: If "finding1: a competitor lowered their property rate by 4.5%" then "Type_Result==weakness_Result"".

The basic definition of mining association rules is described as follows:

- **Definition 7**. Let $I = \{...\}$ be a set of "n" binary attributes called items. Let $D = \{...\}$ be set of transaction called database. Each transaction in D has a unique transaction ID and contains a subset of the items in I.
- **Definition 8**. (Frequent Itemset Mining): Support (Opportunites_Result) = (number of Result in which type = opportunity)/total number of Result.

  - minSup is a threshold minimum indicating the percentage of texts where Opportunites_Result appears.
  - Definition 8 is employed with Weakness_Result, Threat_Result and Strengths_Result.

- **Definition 9**. (Making decision rules) these rules reflect a SWOT analyzes method, Example: If "Support (Opportunites_Result) >=minSup (Opportunites_Result) ^ Support (Strengths _Result) >=minSup (Strengths_Result)" then "Strategic Option is Opportunity/Strength" (see Fig. 6), this rule executed by decision maker agent and its result shared to appropriate maker decision by Notify agent.

3   Results

The implementation was done by OWL and Protected Editor 5.20; the implementation of the reasoning rules of this research was carried out by SWRL rules [17, 18]. However, the SWRL rules cannot be directly executed in a rule engine. The SWRL syntax must be converted into understandable syntax in a rule engine, such as Java Expert System Shell (JESS), to enable rules in the agent body. Semantic reasoning agents are software agents developed with the Jess inference engine. Their behaviors are mostly specified using inference rules and they use ontologies through JessTab, a Protégé component. We chose the JADE platform, to implement our prototype. Figure 7 illustrates the implementation of our ontology (concepts and properties) by OWL.

When information is received by monitor agent, this one combines all information founded "finding", and transmit it to appropriate cluster to evaluate it by executing strategic rules. Example: "If "finding 1 ==" Competitor XXX lowered their property rate by 4.5%" then Type_Result ==Weakness_Result", this example is represented as an excerpt from ontology and is depicted in Fig. 8. Figure 9 illustrates edition of strategic rules by SWRL in protégé. After evaluating all findings inserted in our framework as *strength/weakness/opportunity/threat*, decision maker agent consolidates all Results and calculates minSup of each type of result to execute strategic rules defined in *Definition* 6. See Table 1. In our case study, minSup is 40%, the system calculate that support (Opportunites_Result) ==60% and support (Weakness_Result) ==40% and according to SWOT method (Fig. 6), Decision maker agent infers that a company must minimize weakness to seize opportunities and subsequently the Decision maker agent executes a rule below by using inference engine knowledge asserted in ontology: "*Reduce fixed credit rates by 0.25 points*

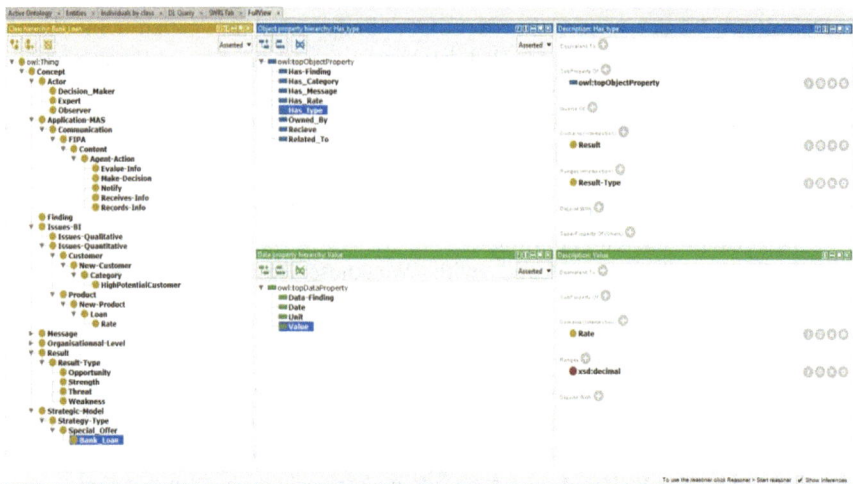

Fig. 7 Implementation of our ontology (concepts and properties) by OWL

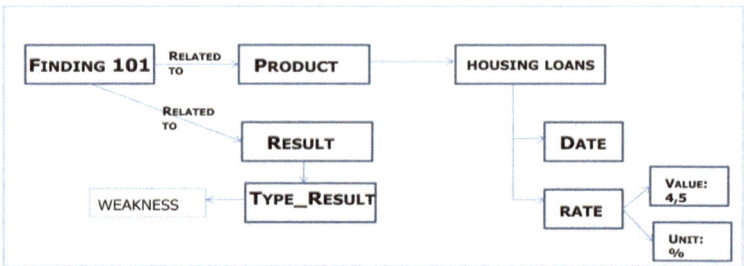

Fig. 8 Example of finding and executing rules

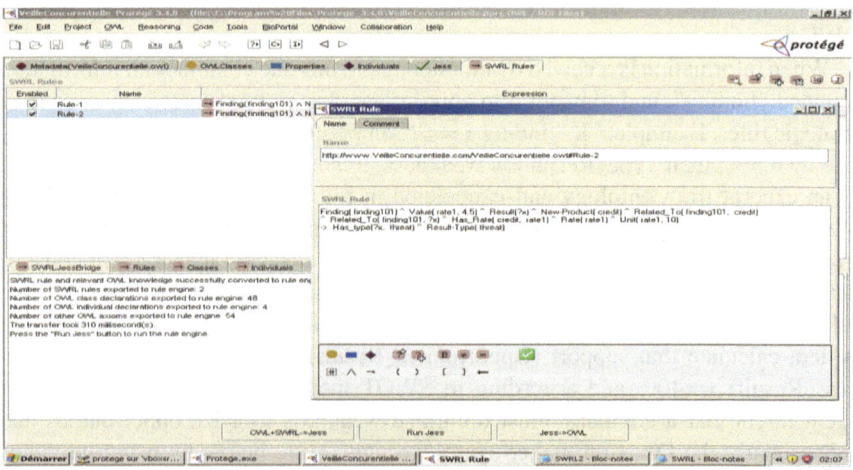

Fig. 9 Edition of strategic rules by SWRL in protégé

Table. 1. Example of executing strategic rules by expert agent

Id finding	Finding	Opportunites_Result	Weakness_Result	Force_Result	Threat_Result
F1	Competitor XXX lowered their property rate by 4.5%	0	1	0	0
F2	The survey revealed a satisfaction rate of 61% for online service	1	0	0	0
F3	Limited banking rate of 54% offering growth potential for banks	1	0	0	0
F4	Significant growth potential in Africa for Moroccan banks	1	0	0	0
F5	Market activities concentrated on a few local operators	0	1	0	0

for high-potential customers (receiving a monthly salary of more than 50,000 DH)".
This rule represented in Fig. 10.

4   Conclusion

This review proposed a platform based on a multi-agents system (MAS) whose agents reason as a human actor and collaborate with them to create a collective intelligence in network expert. This paper proposes also ontology model to ensure interoperability between agents and define the reasoning behaviors of agents based on SWOT analysis method. The prototype was implemented by the JADE platform

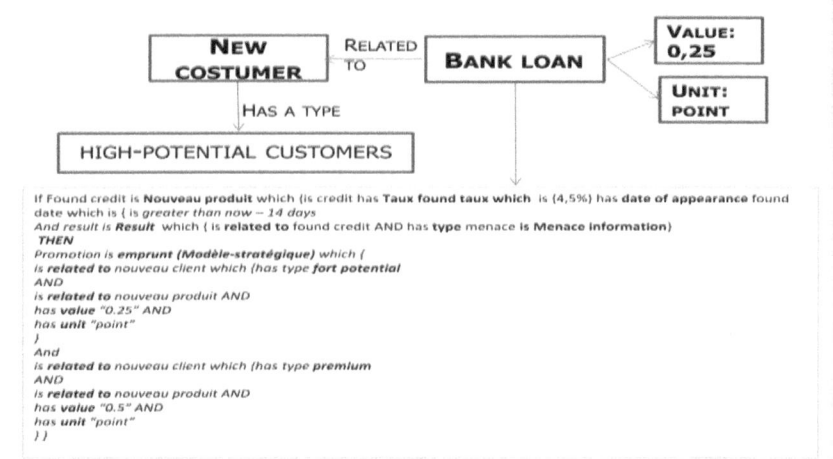

Fig. 10 Example of executing making decision rule

and ontological knowledge organization method was implemented with Protégé and JESS. The future work of this paper is mainly reproduced in the following two points: (1) the development of a graphical interface to allow actor to insert their knowledge directly into the knowledge base of agents, and subsequently modify the behavior of agents and (2) use datamining and machine learning technology with MAS to ensure self learning.

References

1. Phadermrod, B., Crowder, R.M., Wills, G.B.: Importance-performance analysis based SWOT analysis. Int. J. Inf. Manag., 1–10 (2016). https://doi.org/10.1016/j.ijinfomgt.2016.03.009
2. Pai, M.-Y., Chu, H.-C., Wang, S.-C., Chen, Y.-M.: Ontology-based SWOT analysis method for electronic word-of-mouth (2013). https://doi.org/10.1016/j.knosys.2013.06.009
3. Ferber, J.: Les systemes multi-agents. Vers une intelligence collective (1997)
4. Quynh-Nhu, N., Low, G.: MOBMAS: a methodology for ontology-based multi-agent systems development (2008). https://doi.org/10.1016/j.infsof.2007.07.005
5. Sieg, J.H., Beucheler, T., Pfeifer, R., Feuchslin, R.: Crowdsourcing, open innovation and collective intelligence in the scientific method: a research agenda and operational framework. In: Artificial Life. International Conference on the Synthesis and Simulation of Living Systems. Odense, Denmark (2010)
6. El Haddani, M.: Les bons eleves de la veille strategique au maroc (2014)
7. El Mabrouki, N.: La pratique de l'intelligence économique dans les grandes entreprises: voyage au cœur d'un système non univoque. In: XVI International Conference of AIMS, Montreal, 6–9 June 2007
8. Gubanov, D., Korgin, N., Novikov, D., Raikov, A.: E-expertise: modern collective intelligence. Springer (2014). ISBN: 978-3-319-06770-4. https://doi.org/10.1007/978-3-319-06770-4
9. Phadermrod, B., Crowdera, R.M., Willsa, G.B.: Importance-performance analysis based SWOT analysis. Int. J. Inf. Manag. **44**, 194–203 (2019)

10. Boes, J., Migeon, F.: Self organizing multi-agent systems for the control of complex systems. (2017). https://doi.org/10.1016/j.ieri.2014.08.006
11. Lavbič, D.: Knowledge Management with multi-agent system in BI systems integration. E Business—Applications and Global Acceptance (2012). ISBN 978-953-51-0081-2
12. Espinasse, B., Fournier, S., Albitar, S., Lima, R.: AGHATHE-2: an adaptive, ontology-based information gathering Multi-agent system for restricted web domains (2010)
13. Guarino, N., Oberle, D., Steffen, S.: What is an ontology? (2009). https://doi.org/10.1007/978-3-540-92673-3_0
14. PESTEL. https://www.professionalacademy.com/blogs-and-advice/marketing-theories---pestel-analysis
15. Porter. https://www.mindtools.com/pages/article/newTMC_08.htm
16. Yang, B., Wang, X.: Construction of logistics financial security risk ontology model based on risk association and machine learning. In: International Conference on Modeling, Simulation, Optimization and Numerical Techniques (2019)
17. Horrocks, I., Patel-Schneider, P.F., Boley, H., Tabet, S., Grosof, B., Dean, M.: SWRL: a semantic web rule language combining OWL and RuleML. W3C Memb. Submiss. **21**, 79 (2004)
18. Beimel, D., Peleg, M.: Using OWL and SWRL to represent and reason with situation-based access control policies. Data Knowl. Eng. **70**, 596–615 (2011)

Overview of Mobility Management in Autonomous and Centralized Wi-Fi Architecture

Hind Sounni, Najib Elkamoun, and Fatima Lakrami

Abstract Wi-Fi has become an optimal solution for interconnecting and managing wireless communication devices. For a reliable service, a higher security level, and seamless roaming for supporting mobile devices, a single management point is required. To this end, it is necessary to deploy a centralized controller to manage the network configuration. In this paper, we present an experimental study of mobility management in autonomous and centralized Wi-Fi architecture using a traditional Wi-Fi controller and SDN controller. The current study reveals analytically and experimentally the advantages of using centralized architecture in a Wi-Fi network. The main objective is to analyze and compare the impact of introducing the studied controllers on mobility management in wireless networks to highlight the high reliable scheme that enhances network performances. Obtained results prove that the deployment of an SDN controller provides better results compared to the traditional Wi-Fi controller.

Keywords Wi-Fi · Autonomous architecture · Centralized architecture · Software-defined network · Mobility management · Performance evaluation

1 Introduction

The popularity of wireless local area networks (WLANs) especially Wi-Fi has recently increased considerably because of its capacity to provide high mobility support, ease of use, and flexibility plus reducing installation and maintenance costs. Wi-Fi networks can be divided into two architectures: autonomous and centralized. The APs in autonomous architecture are stand-alone APs with complete integrated

H. Sounni (✉) · N. Elkamoun · F. Lakrami
STIC Laboratory, Chouaib Doukkali University, El Jadida, Morocco
e-mail: sounni.h@ucd.ac.ma

N. Elkamoun
e-mail: elkamoun.n@ucd.ac.ma

F. Lakrami
e-mail: lakrami.f@ucd.ac.ma

© The Editor(s) (if applicable) and The Author(s), under exclusive license to Springer   495
Nature Singapore Pte Ltd. 2021
F. Saeed et al. (eds.), *Advances on Smart and Soft Computing*, Advances in Intelligent
Systems and Computing 1188, https://doi.org/10.1007/978-981-15-6048-4_43

intelligence; this architecture is least expensive and fastest to procure. Still, it is useful only when having few access points to manage. In a dense Wi-Fi network, the significant number of devices connected to the Wi-Fi network leads to the necessity for more APs to cover the entire network. In this case, the managing of the wireless network is a very complicated and time-consuming task, requiring network studies and planning before physical installation and configuration, while ensuring network efficiency, performance, and security. Consolidating around a central control structure using a WLAN controller allows greater automation and makes rolling out new services faster. Also, it allows to fulfill dynamically to policy changes, and the network administrator is freed from the manual, which is a time-consuming task. Centralized Wi-Fi architectures based on WLAN controllers and access points were designed to decouple the data and control planes. The vendors implemented their models of WLAN controllers, and they became very proprietary and were closed to extensions to meet business needs. The introduction of software-defined network simplifies the provisioning over multi-vendor networks and also offers features which can only be achieved using SDN controller in Wi-Fi such as managing unified policy, network visualization, and application of SLAs through the unified network, which allows users to have a uniform experience, regardless of their access method, and finally, location-based services by leaving the dynamic decision-making to the IT using the physical localization of the user.

In a wireless network, it is quite conceivable during a movement to end up on another IP network than the original one. The mobility in Wi-Fi network is the capability of a mobile node to switch between access points without losing an effective network connection. The mobility has become an essential requirement for existing applications over wireless IP-based networks. Despite the number of researches that work on mobility management [1, 2], both between heterogeneous and homogeneous networks, to be on local or extended locations, mobility management remains one of the key challenges for the deployment of such wireless networks.

This paper provides a mobility management overview in autonomous and centralized architecture. A performance evaluation is done to analyze the impact of introducing centralized architecture while using a traditional Wi-Fi controller, or the SDN controller, which one gives better performance in an enterprise Wi-Fi network. The paper is structured as follows. Section 2 presents an overview of autonomous and centralized architecture. Section 3 presents and gives details about our experimental test. Section 4 provides the obtained results. Section 5 contains conclusion and future works.

2 Overview

2.1 Autonomous Architecture and Mobility Management

In an autonomous architecture, access points are stand-alone, containing the features and capabilities needed to operate without depending on another device. APs in autonomous architecture can be linked to the wired infrastructure. Multiple stand-alone APs in the same infrastructure form an extended service set (ESS). A stand-alone AP operates at all three network levels: management, control, and data. In a mobile Wi-Fi network [3], two types of mobility are involved, horizontal and vertical mobility [4]. The first refers to mobility within the same access technology and the second to mobility between different access technologies. The horizontal evokes several mobility levels. Most research on WLAN mobility focuses on mobility at level 2 and level 3. Level 2 mobility is performed when a station is re-associated to a new AP in the same subnetwork. If the new AP arrives from a different network, it is classified as a level 3 mobility [5].

In 802.11 standard [6], when a station enters a BSS/ESS, it chooses an AP to which it associates. The association of a station to an AP is done according to a certain number of criteria: received signal strength indication, packet error rate, and network load. When the station moves into a BSS or ESS, it must remain synchronized to communicate. To keep synchronization, tagged frames called beacon frames are sent periodically by the AP, containing the AP clock value. When these frames are received, the stations update their clocks to ensure synchronization with the AP. The station starts searching for another suitable AP to associate with, if it notices that the transmission power of the old AP is too weak, or when detecting transmission failures. Two listening ways are possible: passive and active. In passive listening, the station waits to receive a beacon frame from the AP announcing its presence which is emitted periodically. For active listening, once the station has located the most appropriate AP, it sends it an reassociation request directly via a probe request frame and waits for a response from the AP to reassociate with it with probe response frames. When the new AP accepts the station, an authentication process is performed. Following successful authentication, the new AP sends a reassociation response to the mobile station which is the last step to finalize and complete the communication transfer so that the exchange between the station and AP can begin. Periodically, the station monitors all channels in the network to detect if another AP has better performance than the old one. When a station changes its original AP, it is either because it has detected a decrease in signal strength or the network traffic is too high on the original AP.

Mobile IP [7] is defined according to several standards. The protocol allows maintaining the existing communications between the mobile node and its correspondents, even during movements to a different network. Also, it seeks to optimize direct routing between the mobile node and its correspondents. Besides, it supports

routing of the multipoint packets between the mobile(s) and the rest of the participants in group communication. To fulfill all these functions, the IETF has identified
two mobility protocols: mobile IPv4 and mobile IPv6 [8].

2.2 Mobility Limitations in Autonomous Architecture

The autonomous Wi-Fi infrastructure is considered as unstructured, leading to significant issue concerning mobility management. The current structure limits the ability
to integrate network applications, innovative services, and policies, since it is coupled
between the control plane and the data plan. This makes the implementation of intelligent mobility management functions based on real-time network status difficult [9].
As mentioned in the section above, for the association procedure, the nodes search for
new APs in the radio range, receive a list of SSIDs, send the authentication probe to the
AP with a strong RSSI, and start the re-association process to establish a connection
with the selected AP. Deploying multiple APs offer a better network coverage and a
seamless connectivity to the node during mobility. When a user wants to establish a
connection, it selects the SSID of an AP on the basis of the most reliable received
signal strength indication (RSSI). If the node associates with an overloaded access
point, this results in a low throughput. The load balancing mechanism is not provided
by IEEE 802.11 standard; this leads to a degradation of the throughput of the entire
network at the expense of the packet loss ratio [9], and therefore a network performance degradation. The authentication during handover is a time-limited task, in the
case of time-out, the network will lose the connection. These problems restrict the
perfect use of Wi-Fi, leading to inefficient mobility management, and as a result the
Wi-Fi network becomes unable to meet the needs of future networks. Centralized
Wi-Fi architecture comes to solve these limitations.

2.3 Centralized Architecture and Mobility Management

Today, users are confronted with a unique username, password, and access method
for the wireless network as well as for the wired network. With the introduction of
centralized architecture using a controller, a unique profile and access method can be
assigned for users in different types of networks, giving them the proper rights and
ensures the necessary security, while keeping the same login credentials. Also, it can
provide the appropriate quality of service through traffic prioritization, the controller
is capable of identifying and differentiating between different types of data packets
and prioritizing critical traffic over the wireless network infrastructure, which is
essential for real-time wireless traffic such as voice and video. Solving problems
such as interference is done by keeping adjacent access points operating in different
non-overlapping channels, by this way, it minimizes packet loss due to interference
in a dense network. Consolidation around a centralized structure ensures greater

automation, makes the network more dynamic and flexible to policy changes and traffic loads, and releases the network administrator from time-consuming manual tasks. Therefore, a prospective WLAN infrastructure is built, making it possible to introduce new mobility management functions and a smooth handover activity.

Traditional centralized architecture is designed to decouple the data and control planes using a Wi-Fi controller. Lightweight wireless access point protocol (LWAPP) [10] is the protocol used by the majority of traditional Wi-Fi controller to configure and manage the AP. Technically, a controlled AP is known as the lightweight access point (LWAP). With the rise of WLAN controllers, vendors have developed their solution and have closed extensions to meet commercial needs. This leads to losing visibility of the unified network (wired and wireless) and impedes troubleshooting, capacity planning, traffic optimization, and other critical functions. The reason why researchers are looking for other new solutions which simplify the provisioning over multi-vendor networks and resolve visibility issues.

The software-defined network (SDN) [11] concept is a very promising alternative in wireless network customization which allows transparency despite the techniques used by the constructor equipment. It introduces the programmability of the control plan and data plan. These features improve WLAN efficiency. SDN controls and configures the AP by monitoring traffic flows in the network and specifying how the traffic must be routed, and which application flows should have priority access to the RF. In this way, the SDN can provide assurance that users and applications get the connectivity they need. The network visualization through the unified network is enabled using SDN Wi-Fi. Also, it allows dynamic decisions to be made using the user's physical location. Users have a seamless experience regardless of the access method they use. In SDN architecture, the SDN controller and the switches are the principal elements. They exchange information via the OpenFlow protocol [12] which is responsible for defining the control messages and for establishing and installing the flows on the switches. The OpenFlow switch has a flow table, and each entry in this table contains a set of actions and packet fields. All packets without a corresponding entry in the OpenFlow switch's flow table are sent to the controller, and then it determines how to treat the packet (delete/add) and decide how to forward other similar packets in the future. Policies are used to define rules and actions, and it can simply and rapidly be changed and modified through the installation of new software controller. In the wireless context, the OpenFlow switches are located at the APs. Table 1 provides a comparison of controllers in a centralized architecture.

In a centralized architecture, different types of mobility can occur. If the mobile station roams between APs on the same controller, it is called intra-controller mobility [13]. If the mobile station roams between two APs registered to two different controllers, then it is an inter-controller L2 roaming [13]. In this paper, we focus on intra-controller mobility. Stations mobility in centralized architecture is managed by the controller. The first step is the dissociation between the AP and the mobile station. It is performed either by the AP based on the controller's request or by the mobile station if it detects a deterioration in RSSI quality. Then, the station sends Probrequest message to associate with a new AP, the information within this message are sent to the controller, the controller accepts the association request if this station

Table 1 SDN controller versus traditional Wi-Fi controller

	SDN controller	Traditional Wi-Fi controller
Protocol	OpenFlow	LWAPP/CAPWAP
Programmability	Yes	No
Network visualization	Yes	No
Load balancing	Yes	Yes
Transparency over multi-vendor	Yes	No
Mobility management	Yes	Yes

doesn't figure in the reject list and send to this AP all information about the station and its communication state. Once the station is associated (or dissociated), the controller sends this information to all APs to update their routing flow table. The APs periodically send the list of associated stations to the controller. The exchanges between the AP and the controller are managed by the LWAPP protocol [14] in the case of traditional centralized architecture, and by the Openflow protocol in the case of a centralized architecture based on the SDN [15].

3 Experimental Test

The centralized architecture offers a better mobility management compared to the autonomous architecture, especially in a dense network. Our objective is to determinate which centralized topology is more appropriate for managing the mobility of stations in a small network. To that end, an experimental test bed is done to emulate, evaluate, and compare the performance of the proposed scenarios, which are presented in Fig. 1.

The two centralized scenarios, Fig. 1a, b, consist of one controller, several APs, and a mobile station roaming from one AP to another in a straight line with a mobility speed of 1 m/s, and the MS is exchanging UDP traffic with a fixed node (server). Iperf is used as a traffic generator which is very suitable for the end-to-end delay, throughput, and packet loss measurements [16]. The distance from one AP to another is 15 m, with an overlapping area. The client part is represented by one mobile computer sending and receiving the generated data. This station has i5, 2.7 GHz processor, and 8 GB RAM, and uses Intel Dual Band Wireless-N cards. The SDN controller is running on a desktop computer with 64-bit Ubuntu 14.04 operating system. This desktop computer has i5, 2.7 GHz processor, and 16 GB RAM. The HP Van version 2.6 is used as the SDN controller. The used APs LWAPP and Openflow protocols. In the autonomous architecture, the test is performed using the same topology, without requiring the use of controllers.

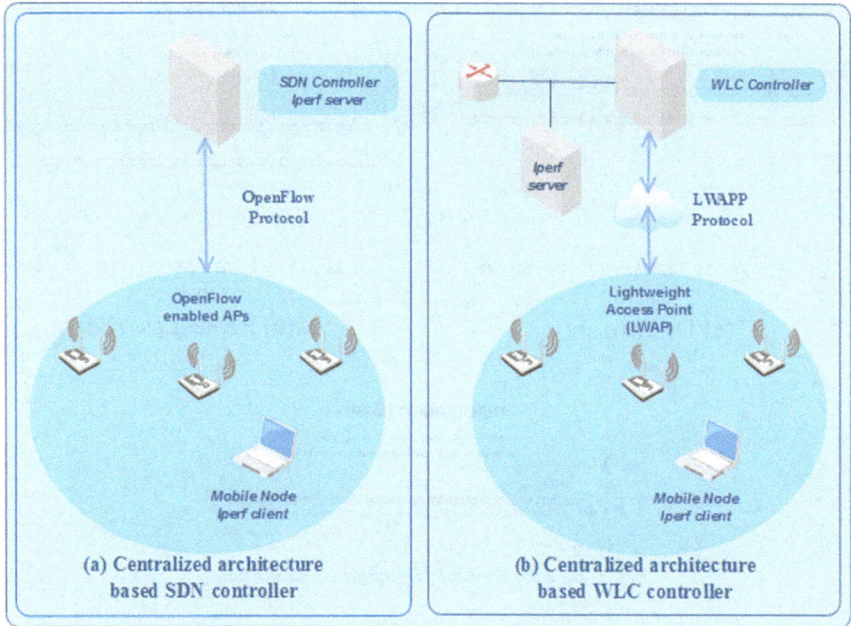

Fig. 1 Centralized architecture scenarios

4   Obtained Results

The evaluation criteria are delay: time it takes for a node to send a packet to another; loss rate: quantity of non-received traffic; and throughput: average rate of successful data transmission over a channel (bps). Figures below illustrate the delay, packet loss, and throughput for the proposed scenarios in the case of centralized architecture using the traditional Wi-Fi controller (Fig. 2) and SDN controller (Fig. 3). The results of the two scenarios were compared to the autonomous architecture, and the improvement percentage in each case was calculated. For centralized architecture using the traditional Wi-Fi controller, the delay is enhanced by a percentage of 18%, the packet loss of 29%, and the throughput of 15%. When using the SDN controller, the delay enhancement amounts to 20%, the packet loss to 34%, and the throughput to 17%.

5   Conclusion

The deployment of centralized architecture is more suitable for a high density network. However, it can also be used to enhance the performance of small networks by solving the problem of mobility management encountered in the

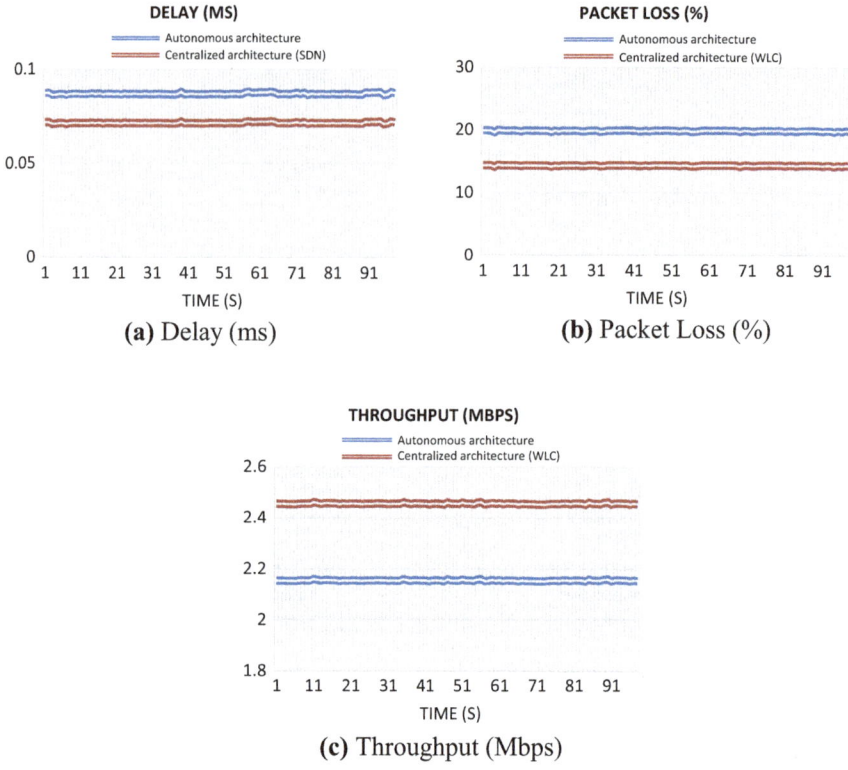

Fig. 2 Autonomous versus centralized (traditional Wi-Fi controller)

autonomous architecture. The adoption of such architecture leads to the need for controllers to manage the network. This paper provides an overview on the different controllers that can be used in a Wi-Fi network. An experimental study was conducted to evaluate the behavior of traditional Wi-Fi controller versus the SDN controller. The results prove that an SDN-based architecture is more suitable than traditional Wi-Fi controller, especially for managing node mobility in a Wi-Fi network, in addition to its ability to provide a seamless network despite the equipment used from a particular vendor. Also, it allows easy implementation of new algorithms through the programming interface. In the future work, we aim to develop a new algorithm to improve the performance of the mobile Wi-Fi network, and the implementation will be based on the SDN concept.

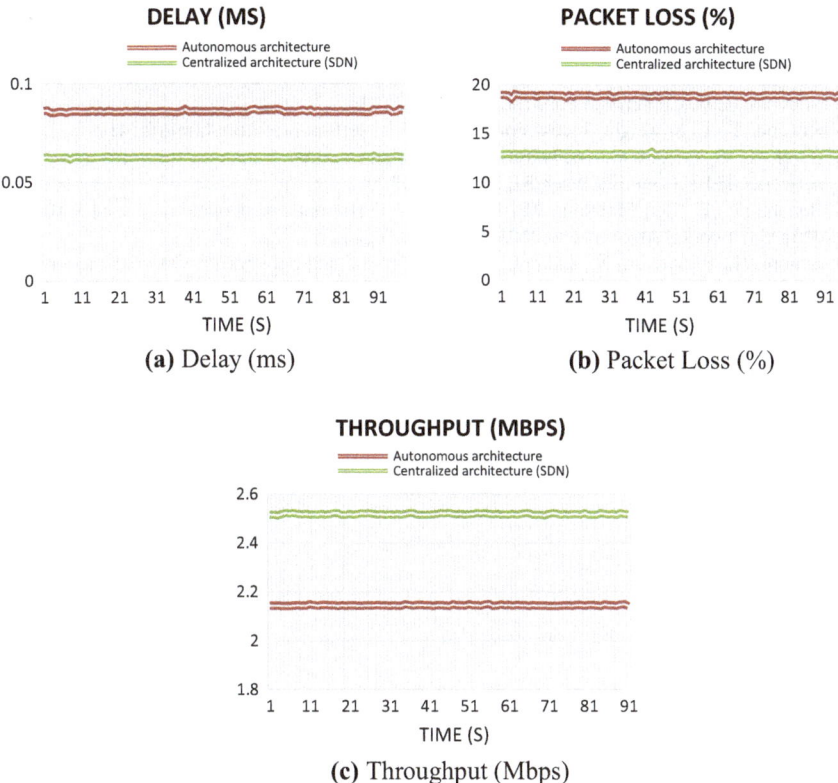

Fig. 3 Autonomous versus centralized (SDN)

References

1. Bin, L., Qi, Z., Weiqiang, T., Hongbo, Z.: Congestion-optimal WiFi offloading with user mobility management in smart communications. Hindawi Wirel. Commun. Mob. Comput., 15 (2018). Article ID 9297536
2. Gilani, S.M.M., Hong, T., Jin, W., Zhao, G., Heang, H.M., Xu, C.: Mobility management in IEEE 802.11 WLAN using SDN/NFV technologies. EURASIP J. Wirel. Commun. Netw. **2017**(1) (2017)
3. Aijaz, A., Aghvami, H., Amani, M.: A survey on mobile data offloading: technical and business perspectives. IEEE Wirel. Commun. **20**, 104–112 (2013)
4. Payaswini, P., Manjaiah, D.H.: Simulation and performance analysis of vertical handoff between WiFi and WiMAX using media independent handover services. IJCA **87**, 14–20 (2014)
5. Andersson, K.: Interworking techniques and architectures for heterogeneous wireless networks. J. Int. Serv. Inf. Secur., 27 (2012)
6. Siddiqui, F., Zeadally, S., Salah, K.: Gigabit wireless networking with IEEE 802.11ac: technical overview and challenges. JNW 10, 164–171 (2015)
7. Jora, N.: Mobile IP and comparison between mobile IPv4 and IPv6. J. Netw. Commun. Emerg. Technol. **2**, 7 (2015)
8. Anandakumar, H., Umamaheswari, K., Arulmurugan, R.: A study on mobile IPv6 handover in cognitive radio networks. In: Smys, S., Bestak, R., Chen, J.I.Z., Kotuliak, I. (eds.) International

Conference on Computer Networks and Communication Technologies, vol. 15, pp. 399–408. Springer, Singapore (2019)

9. Sounni, H., Elkamoun, N., Lakrami, F.: Towards QoS enhancement in wireless network through SDN. In: IEEE Conference Publication. IEEE (2019)
10. RFC 5412—Lightweight Access Point Protocol. https://tools.ietf.org/html/rfc5412.
11. Sezer, S., et al.: Are we ready for SDN? Implementation challenges for software-defined networks. IEEE Commun. Mag. **51**, 36–43 (2013)
12. Benamrane, F., Ben Mamoun, M., Benaini, R.: Performances of openflow-based software-defined networks. An overview. JNW **10**, 329–337 (2015)
13. Sounni, H., Elkamoun, N., Lakrami, F.: Software-defined network for QoS enhancement in mobile Wi-Fi network. IJRTE **8**, 4863–4868 (2019)
14. Fontes, R.D.R., Mahfoudi, M., Dabbous, W., Turletti, T., Rothenberg, C.: How far can we go? Towards realistic software-defined wireless networking experiments. Comput. J. **60**(10), 1458–1471 (2017)
15. Verma, R., Jose, S., Janakiraman, R., Jayarajan, S.: Identification for roaming mobile clients (2015)
16. Mishra, S., Sonavane, S., Gupta, A.: Study of traffic generation tools. Int. J. Adv. Res. Comput. Commun. Eng. **4**, 4 (2015)

Wireless Technologies and Applications for Industrial Internet of Things: A Review

Nisrine Bahri, Safa Saadaoui, Mohamed Tabaa, Mohamed Sadik, and Hicham Medromi

Abstract Industrial automation has developed a lot thanks to the concept of IoT and this is the reason why many industries have been working on new emerging technologies to ensure the concept of connectivity, such as the most popular ones: Low-power wide-area networks (LPWANs) designed to solve many problems such as scalability and range encountered by traditional cellular networks and short-range communication technologies, in order to also address the application of IoTs in the industrial context. The objective of this document is to provide an overview in one of the main pillars of Industry 4.0, contextualizing the IoT in Industry 4.0, discussing the most relevant technologies in the IoT: wireless technologies highlighting the major differences between the two ecosystems IoT and IIoT based on their different criteria and applications, and then based on this projecting existing technologies to study the appropriate one, although these technologies have their own drawbacks in the external and industrial environment.

Keywords Industry 4.0 · Industrial Internet of things (IIoT) · Wireless technologies · LPWAN · Industrial context

1 Introduction

Industry 4.0, also known as the factory of the future, is fundamentally distinguished by the integration of new technologies and intelligent automation in the company's value chain. This new digital transformation that is changing the manufacturing

N. Bahri (✉) · H. Medromi
Foundation for Research, Development and Innovation in Engineering Sciences, Casablanca, Morocco
e-mail: nissrinebahri@gmail.com

S. Saadaoui · M. Tabaa
Pluridisciplinary Laboratory of Research and Innovation (LPRI), EMSI Casablanca, Casablanca, Morocco

N. Bahri · M. Sadik
RTSE Research Team, Laboratory of Research Engineering, ENSEM, Hassan II University, Casablanca, Morocco

© The Editor(s) (if applicable) and The Author(s), under exclusive license to Springer Nature Singapore Pte Ltd. 2021
F. Saeed et al. (eds.), *Advances on Smart and Soft Computing*, Advances in Intelligent Systems and Computing 1188, https://doi.org/10.1007/978-981-15-6048-4_44

business by making radical changes not only to systems and processes but also to management methods and the business models. Industry 4.0 has been built over a strong backbone: big data and analytics, Internet of things, augmented reality, cybersecurity, the cloud platform, and the autonomous robot.

One of the major key technologies in Industry 4.0 is the industrial IoT, destined to play a disparate role in this digitized world. This category has emerged the world of manufacturing so fast previously (Microsoft, Schneider Electric, Cisco Systems, Siemens, IBM, Huawei Technologies, etc.) with several innovative ideas to achieve the same goals: find new ways of working or automate processes and maintenance, to reduce asset maintenance costs, enable finer operations and optimize logistics lead times and costs.

Connectivity is probably the most fundamental element of the IIoT architecture where it ensures the criteria and requirement of IIoT application. Thus, several solutions have been used to meet the various demands of IIoT applications based on existing wireless technologies. However, some recent survey statistics suggest that low-power WANs (LPWANs) are fast-becoming the most widespread communication networks used in several industrial IoT applications, but given that most of those technologies are implemented in the saturated ISM band (industrial, scientific and medical), it is difficult to choose the suitable one.

In this paper, we give an overview of IoT in the context of Industry 4.0 by focusing on the applications of IoT in the industrial domain to add intelligence through wireless communication to industrial systems. This paper will be organized as follows: The evolution of the industry will be the subject of the second section. A state of the art on wireless connectivity and differences between IoT and IIoT will be presented in the third section. Connectivity in an industrial context will be presented in Sect. 4. A discussion on the implementation of wireless technologies will be presented in Sect. 5. Towards the end, a conclusion.

2 The Evolution of Industry

The goal of the industry was not only the self-improvement and the race to meet industrial needs but also the improvement of society's mass level and making life easier for the consumer. Consequently, this economic growth should always go hand in hand during each industrial revolution. The development of industrialism is dealt with in four main phases (dates can be changed depending on the country).

From 1760 to 1880, the first industrial revolution was initiated by James Watt in England by developing a steam engine. From 1880 to 1950, the use of new energy sources as gas, electricity, and oil contributed to the second industrial revolution. From the fifties to the advancement of the industry is due to electronics with the arrival of transistors and microprocessors but also telecommunications and information technology. This revolution has led to the development of the aerospace sector and many others. Since 2010, and after three successive industrial revolutions (mechanization, mass production, and digitization), the time has come for smart factories (industry 4.0).

The term Industry 4.0 was invented to define a system that evolved from a computer-controlled automated facility (Industry 3.0), into a system that processes floor data to provide intelligent decisions in an automated approach [1]. The new initiative adopted by the German government was first used in a 2011 paper and mentioned for the first time in public at the international trade fair [2].

Considering the industrial IoT, cloud, cybersecurity, autonomous robot and artificial intelligence (AI) mainly the fuel of the industry 4.0. These technologies are attending today the more important phase of development. Therefore, the main challenge is to develop the appropriate technology that will immediately meet market requirements and to have flexibility in multiple environments. Looking at the growing interest in the fourth industrial revolution, currently this evolution is witnessing initial exploratory stages and academics still struggle to properly define the approach. However, few other searchers began to forward the scientific model of industry 5.0 (impact of robotics in human life) [3] and biology [4]. Modern theories suggest that economic growth depends on the diffusion and absorption of new technologies, but they require skills, not only to create them but also to apply them so eventually, the relation is mutual. In this context, Japan has defined the term Society 5.0 that refers to a modern society that effectively uses technology, robotics, computers, and communications systems, with the aim of improving the world [5], where Industry 4.0 and organizations overall will be major components in Society 5.0.

3 Wireless Connectivity in IoT

In 1999, Ashton expert in RFID technology from MIT coined the term Internet of things (IoT) to define a network that connects not only people but also the objects that surround them. There will be approximately 50 billion connected devices worldwide by 2020 according to Cisco's Internet Business Solution Group. Also, Ericsson states that connected wide-area IoT devices will grow from 11 billion in 2014 to 30 billion in 2022. Eventually, the amount of research in IoT field had an exponential evolution in parallel and the market size evolution (see Fig. 1).

Generic electrical coined the name industrial Internet as their term for IoT's industrial application. However, it is important to differentiate between IoT and IIoT because they have different target audiences, technical requirements, and strategies. IoT technologies have been widely used in industrial systems to monitor the manufacturing environment and to control production lines.

The industrial IoT system can collect and process data and generate services for production decisions. However, the IIoT system still faces the challenge of maintaining quality and quantity of data collected on sensor networks the industrial IoT system can collect and process data and generate services for production decisions.

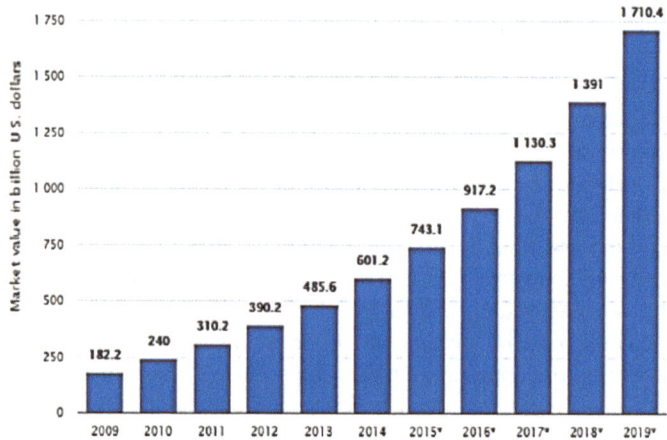

Fig. 1 The evolution of IoT's market value between 2009 and 2019

However, the IIoT system still faces the challenge of maintaining quality and quantity of data collected on sensor networks.

In terms of transmission modes, communication technologies can be divided into wired and wireless technologies [6]. The first one mainly includes Fieldbus and Ethernet technologies. Wireless communications technologies mainly include wireless local area networks (WLAN) and low-power wide-area networks.

(LPWAN), gradually replaced conventional technologies. However, the most significant comparison should be based on the system requirement which is the IIoT criteria (see Fig. 2) presents an overview of the involved technologies.

3.1 Wireless Local Area Networks (WLAN)

Bluetooth

Launched in 1999, Bluetooth technology is a wireless technology known for its ability to transfer data over very short distances in personal networks. Bluetooth low-power consumption (BLTE) is a recent addition to Bluetooth technology (by Nokia in 2006) and consumes about half the power of a conventional Bluetooth device and allows a throughput of the same order 1 Mbit/s; it is deliberately inexpensive and energy-efficient in design. BLE currently supports full mesh topology (several to several, up to 32,767 devices) in direct competition with the IEEE 802.15.4 communication battery types [7]. In fact, BLE has the biggest advantage in personal area network.

WiFi

The IEEE 802.11 (WiFi) standards are still evolving in several directions to enhance certain operational characteristics in different usage scenarios using direct-sequence spread spectrum (DSSS) and orthogonal frequency-division multiplexing (OFDM)

Standart

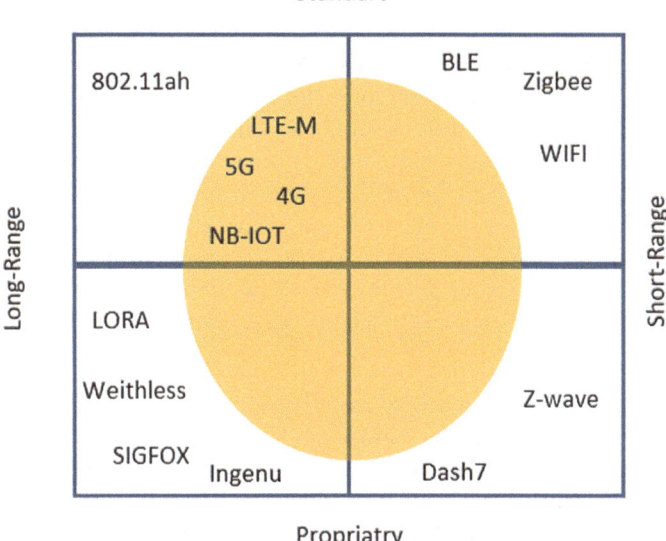

Fig. 2 Different wireless technologies

modulations. A widely adopted version is the IEEE 802.11n standard, which was specifically designed to boost the throughput of WiFi. So obviously WiFi's big advantage is the data rate but (depending on application) the range, power consumption and cost are the biggest downsides. Relatively high costs and limited range mean it would not find a place in many large-scale IoT implementations; however, 802.11ah is designed with less over-head to meet IoT requirements in a large range.

ZigBee

ZigBee is a low-power, low-cost, wireless mesh network standard. This technology provides small amounts of data, with most connected devices running off of a battery. This emerging wireless network technology has developed rapidly in recent years and is widely used for indoor navigation and positioning (less than 10 m). Based on a direct-sequence spread radio signal in the 868 MHz band (Europe, North America), or the 2.4 GHz ISM (worldwide) this technique that convert bits to chips, and transfer 20 kbps (868 MHz band), 40 kbps (915 MHz band) and 250 kbps (2450 MHz band) ensures a long battery life comparing to Bluetooth and WIFI. The features of the ZigBee standard have inspired researchers to integrate it into the new Internet trend of things. However, many other technologies have entered the world of IoT to take over and offer many opportunities giving that Zigbee's coverage is limited and hence cannot be used as an outdoor wireless communication system.

3.2 Cellular Network

Long-Term Evolution for Machines (LTE-M)

Has been designed for the IoT in terms of performance, throughput is high compared to traditional LPWAN technologies since we are on rates evaluated as going as low as 375 kbps and even as high as 1 Mbps by some operators. The latency times are also very good since they are less than 100 ms and there is no limit to the number of messages sent per day. In addition, as with traditional cellular technologies (2G/3G/4G), the handover is also provided, which is not negligible for objects in mobility. The LTE-M also optimizes the energy consumption of these objects through power saving mode. It is also a secure solution since it uses SIM card authentication.

Narrowband IoT (NB-IoT)

NB-IoT based on narrow-band radio technology, this LPWAN technology is standardized by the Third-Generation Partnership Project (3GPP). NB-IoT can coexist as LTE or GSM in licensed frequency bands. This technology occupies a frequency band of 200 kHz, which can be deployed within an LTE carrier [8].

NB-IoT provides connectivity of over 100 K devices per cell and could be increased by operating multiple NB-IoT carriers. It uses (quadrature phase shift keying) QPSK modulation, frequency-division multiple access on the uplink, and orthogonal FDMA (OFMA) on the downlink.

5G

LTE Cat M and NB-IoT are integrated into the 5G specifications, first to guarantee the interoperability with existing deployment, then to benefit from the 5G bands. It can also handle a large number of IoT-compatible devices. The term 5G includes massive-input massive-output (MIMOs), which allow higher network capacities than the current 4G LTE, and also "small cells," which allow a more condensed network infrastructure. One of the main features of 5G is "network splitting," which allows companies to reach different levels of connectivity to satisfy multiple use cases. 5G will be based on three main use case scenarios: enhanced mobile broadband (eMBB): higher bandwidth, ultra-reliable low-latency communication (uRLLC).

critical communications, and the **massive Internet of things** (mIoT).

Low-power wide area networks (LPWANs) are promising new wireless technologies planning to support 3 billion IoT connections by 2025. With its long range, deep penetration and ultra-low-power consumption, LPWAN brings reliable connectivity to previously unachievable industrial site.

3.3 Low-Power Wide-Area Networks

LORAWan

In 2012, Semtech has announced the acquisition of Cycleo SAS technology, created by French engineers in 2009. Semtech seems to project the industry's requirements in 2012 with Lora technology before the IoT's big exponential broadcast.

It is a radio modulation that uses a regional ISM radio band that offers the possibility to connect any object to the Internet through unlicensed bands with higher coverage. Simultaneously, it operates with the lowest data rate. Lora chirp spread spectrum modulation (CSS) insure complete bidirectional communication, and the resulting signal has low noise levels, allows high resistance to interference.

Sigfox

Created in 2009 is positioned as the first low-speed cellular operator dedicated to the "machine to machine" and the Internet of things. Sigfox uses 192 KHz of the publicly accessible band to send and receive messages by air. The modulation is the ultra-narrow band; every message has a width of 100 Hz and is sent at a rate of 100 or 600 bits per second depending on the region. SIGFOX reports that each gateway can manage up to one million connected objects, with a coverage area of 30–50 km in rural areas and 3–10 km in urban areas. Currently, Eutelsat is planning to launch 25 nanosatellites in a partnership with wireless company Sigfox to serve Internet of things (IoT) applications.

Ingenu

Ingenu was created in 2008 and initially focused on utilities, oil and gas applications. It proposed a proprietary LPWAN technology, which unlike most other unlicensed LPWA technology works in the ISM band of 2.4 GHz [9]. This solution uses random phase multiple access technology RPMA designed and built by Ingenu for IoT and M2M applications, and it enables data rates in the hundreds of thousands of bits per second. It also uses channel coding Viterbi algorithm to ensure data delivery and provide a high quality of service. RPMA, therefore, has clear advantages over competing Lora and Sigfox technologies, an Ingenu access point would cover the equivalent of 18 Lora antennas, 70 Sigfox antennas, and 30 NB-IoT accesses on cellular networks, reports expert blogger Bob Emmerson on IoT Global Network.

Weightless

Weightless SIG proposed a set of types in IoT techniques. There are three different norms for weightless: the weightless P is a bi-directional transmission technique used for the secure exchange of data for industrial applications, based on a frequency hopping procedure in a time slot, the weightless W using frequency hopping time division duplex operation it increase range and accommodate low-power devices for machine-to-machine communication, and weightless N based on ultra-narrow band modulation which follows unidirectional communication and which can provide long battery life and a wide coverage area [10].

DASH7

DASH7 Alliance Protocol (D7A) is an open-source wireless sensor and actuator network protocol, which operates in the 433, 868 and 915 MHz unlicensed ISM band/SRD band D7A complies with the ISO/IEC 18000-7. Which is an open standard for the license-free 433 MHz. It offers several years of battery life, a range of up to 5 km, an indoor location accurate to 1 m, low latency for connection to moving objects, a very small stack of open-source protocols, 128-bit AES shared key encryption support and data transfer up to 166 kbps and it uses asynchronous method communication which response to low-power wake up.

For over many years, short-distance multipoint communication technologies such as ZigBee and Bluetooth have been considered a suitable way to implement IoT services. Although these standards are characterized by power consumption, which is a key issue for many IoT devices, their limited coverage is a major constraint, especially when the application scenario concerns services that require urban or wide area coverage such as in industrial applications for the smart city.

The smart city applications contain and mix the two concepts: **IoT (consumer/commercial) and industrial IoT**, however the two ecosystems are largely different in many areas: information flow, data-volume, data-rate, interoperability, lifecycle, security and cost.

The first class of applications is the consumer IoT, these applications are characterized by the interaction between a human and a device, in case of failure, a human is there to recover or restart the application (exp: Google's Home and Apple's HomePod, smart watches etc.), the second one are commercial IoT, targets our daily environment outside the home (Consumer IoT), there is a range of applications that can be deployed in places we frequent, such as commercial office buildings, supermarkets, stores, hotels, health facilities or entertainment venues. Finally, the industrial IOT, this class of applications uses client/server communications and sends smaller amounts of data, IIoT sensors are often installed to measure parameters in remote and difficult to access infrastructures, for examples: Secure, health care, energy, aerospace, maintenance, management of oil, Gaz, material tracking, etc.

4   Industrial Connectivity

The IIoT connectivity is now based on wireless technologies that are used to provide access to data from terminal devices. It must be noted that solutions already existing and explicitly conceived for the industrial market have certain limitations. They are based on IEEE 802.15.4 compliant radio and are not designed to connect a large number of equipment, such as in typical IIoT applications [11]. On the other side, Zigbee designed to be useful in a wide variety of applications, including industrial command and control, public safety and remote sensing, but it is a little far to meet IIoT's demand comparing to many emerging technologies. However, the low-power wide-area network (LPWAN) solutions have been gaining popularity in recent years to occupy the bottom two levels of the protocol stack, with many competing technologies available or under development. Multiple standardization efforts by various standardization organizations aimed to develop wireless IIoT where LPWA connectivity technologies attracted a tremendous amount of interest as it caters to the requirements of a wide range of wireless IIoT applications. Unlike other connectivity methods, LPWA connectivity solutions are developed on the basis of a simple system, although the needs are many and diverse and they meet the requirements of the IIoT:

efficient signaling and channel access protocols to support high connection densities, extreme energy efficiency to prolong the operation of a battery-powered device for up to ten years, ultra-low cost to enable widespread adoption in an economically viable manner, and broad coverage to enable the deployment of a variety of devices with high reliability. These benefits make it easy for LPWA to be the most emerged technologies that will respond to the IIOT requirements [12]. LPWAN solutions are indeed examples of short-range devices with cellular-type ranges of about 10–15 km in rural areas and 2–5 km in urban areas. This is possible thanks to a radically new design of the physical layer, aiming at a very high sensitivity of the receiver.

As an example, LoraWAN respond directly to the category of industrial IoT such as asset tracking, smart agriculture, intelligent building, facilities, and management. Where also NB-IOT plays a major role but in different applications [13].

5   Discussion

As seen, different technologies may respond to a category of IoT application besides each one has its requirements and challenges. Before moving to a single technology, there are many issues and limitations of the work environment that need to be considered. Thus, moving toward the suitable technology is with regard to these questions (see Fig. 3).

Low-power wide area networks (LPWANs) have made progress in many areas. On one hand, the low-power, unlicensed communications technologies, some are highly proprietary and application-specific, and on the other hand, the large telecoms companies that have adopted low-power versions as extensions to their cellular the networks, with the major cellular protocols, NB-IoT and LTE, now wrapped up into the 5G standard.

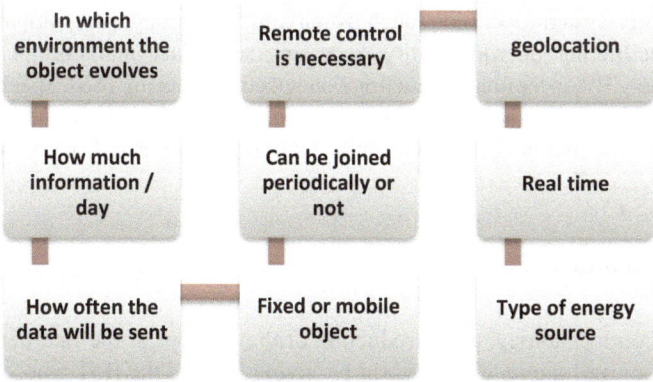

Fig. 3   The main questions to converge into a suitable technology

On one side if we need very low latency (1–100 ms) in industrial application with a large coverage, it is far away to be done with the proprietary Lora wan or Sigfox; however, the cellular extension is the solution (NB-IoT). Also having a mobile object or tracking it is not achievable with the NB-IOT due to mobility issues. On the other side, LPWANs usually function in the sub-GHz area, which allows good coverage but is often limited by cyclic transmission time side. The use of unlicensed spectrum has caused some reliability problems, as there is no assurance that it will be reliable (5G and satellite may overcome those technologies), the reason that led many researchers to conduct indoor and outdoor experiments [14] where some claimed that Lora wan can provide very robust communication even high noise level presence [15] and others begin to explore the physical layer (which is harder due to its corporate property nature). In most LPWAN technologies, two major alternative communication techniques, ultra-narrow band (UNB) and spread spectrum (SS) are used at the physical layer. However, the biggest challenge is choosing the most appropriate UNB and SS technique for LPWAN in the industry. Some research has argued that UNB could achieve moderately better results in terms of capacity and coexistence due to its flawless implementation, but also that SS could offer better safety, speed, and accuracy characteristics [16]. However, even if those technologies are the only suitable one to IIoT requirement till now, they are still in a study phase as one of the main challenges is industrial noise as well as interferences between unlicensed bands and they can be seen as one of the major problems faced by LPWANs deployed for IIoTs. Presently, it constitutes a serious challenge to a better spectrum usage for effective and quality data and digital communications.

In recent years, the cognitive radio for LPWAN began to attract the attention of researchers to help overcome some of the challenges faced by the conventional wireless networks especially the unlicensed band where, interference, coexistence, and scalability problem exist more.

The cognitive radio (CR) is defined by the Federal Communications Commission (FCC) as: "A radio or system that senses its operational electromagnetic environment and can dynamically and autonomously adjust its radio operating parameters to modify system operation, such as maximizing throughput, mitigating interference, facilitating interoperability, accessing secondary markets" [17]. Others resumed the CR concept basic functions on spectrum sensing (SS), spectrum decision making, spectrum access, and spectrum mobility [18, 19] and some began to test the implementation of CR spectrum detection technique on Lora network.

6   Conclusion

In automation technology, the introduction of the concept of the Internet of things has revolutionized the industrial world by giving birth to the IIoT whose objective is to provide industrial connectivity for intelligent factories. In this paper, we have presented a state of the art of the so-called IoT&IIoT systems and their applications in the industrial revolution Industry 4.0 and discussed the main LPWANs and their major future challenge in a noisy environment; in this context, many future work can be achieved.

References

1. Horacio, A.-G., Kurfess, T.: A brief discussion on the trends of habilitating technologies for Industry 4.0 and smart manufacturing. Manufact. Lett. **15**, 60–63 (2018)
2. Pfeiffer, S.: The vision of "Industrie 4.0" in the making a case of future told, tamed, and traded. Nanoethics **11**(1), 107–121 (2017)
3. Demir, K.A., Döven, G., Sezen, B.: Industry 5.0 and human-robot co-working. Proc. Comput. Sci. **158**, 688–695 (2019)
4. Sachsenmeier, P.: Industry 5.0—the relevance and implications of bionics and synthetic biology. Engineering **2**(2), 225–229 (2016)
5. Fathi, M.: Optimization in Large Scale Problems: Industry 4.0 and Society 5.0 Applications. Springer Nature (2019)
6. Cheng, J., et al.: Industrial IoT in 5G environment towards smart manufacturing. J. Ind. Inform. Integr. **10**, 10–19 (2018)
7. Tsiatsis, V., et al.: Technology fundamentals. Internet Things, 67–126 (2019)
8. Vejlgaard, B., Lauridsen, M., Nguyen, H., Kovacs, I.Z., Mogensen, P., Sorensen, M.: Coverage and capacity analysis of Sigfox, LoRa, GPRS, and NB-IoT (2017)
9. Bánáti, A., László, E., Kozlovszky, M.: Óbuda University, John von Neumann Faculty, Budapest, Hungary Security Survey of Dedicated IoT Networks in the Unlicensed ISM Bands Eszter Kail
10. Roy, M.: Survey of the interference sources in several IoT technologies information technology. Masters Thesis, May 2019
11. Sisinni, E., et al.: Industrial Internet of Things: challenges, opportunities, and directions. IEEE Trans. Ind. Inform. **14**(11), 4724–4734 (2018)
12. Mumtaz, S., et al.: Massive Internet of Things for industrial applications: addressing wireless IIoT connectivity challenges and ecosystem fragmentation. IEEE Ind. Electron. **11**(1), 28–33 (2017)
13. Sinha, R.S., Wei, Y., Hwang, S.-H.: A survey on LPWA technology: LoRa and NB-IoT. ICT Exp. **3**(1), 14–21 (2017)
14. Liando, J.C., et al.: Known and unknown facts of LoRa: experiences from a large-scale measurement study. ACM Trans. Sens. Netw. (TOSN) **15**(2), 16 (2019)
15. Angrisani, L., et al.: LoRa protocol performance assessment in critical noise conditions. In: 2017 IEEE 3rd International Forum on Research and Technologies for Society and Industry (RTSI). IEEE (2017)
16. Nitin, N.: LPWAN technologies for IoT systems: choice between ultra-narrow band and spread spectrum. In: 2018 IEEE International Systems Engineering Symposium (ISSE) (2018)
17. Yucek, T., Arslan, H.: A survey of spectrum sensing algorithms for cognitive radio applications. IEEE Commun. Surv. Tutor. **11**(1), 116–130 (2009)
18. Onumanyi, A.J., Abu-Mahfouz, A.M., Hancke, G.P.: Cognitive radio in low power wide area network for IoT applications: recent approaches, benefits and challenges. IEEE Trans. Ind. Inform. (2019)

19. Roncancio, G., et al.: Spectral sensing method in the radio cognitive context for IoT applications. In: 2017 IEEE International Conference on Internet of Things (iThings) and IEEE Green Computing and Communications (GreenCom) and IEEE Cyber, Physical and Social Computing (CPSCom) and IEEE Smart Data (SmartData). IEEE (2017)
20. Tendeng, R., Lee, Y.D., Koo, I.: Implementation and measurement of spectrum sensing for cognitive radio networks based on LoRa and GNU radio. Int. Natl. J. Adv. Smart Conver. **3**, 23–36 (2018)

Visual Vehicle Tracking via Deep Learning and Particle Filter

Hamd Ait Abdelali, Omar Bourja, Rajae Haouari, Hatim Derrouz, Yahya Zennayi, François Bourzex, and Rachid Oulad Haj Thami

Abstract Visual vehicle tracking is one of the most challenging research topics in computer vision. In this paper, we propose a novel and efficient approach based on the particle filter technique and deep learning for multiple vehicle tracking, where the main focus is to associate vehicles efficiently for online and real-time applications. Experimental results illustrate the effectiveness of the system we are proposing.

Keywords Computer vision · Vehicles tracking · Computer vision · Deep learning · YOLO · Bhattacharyya kernel · Histogram-based · Particle filter · IoU metric

1 Introduction

Visual vehicle tracking has received much attention in the past decade since it becomes a vital component in intelligent transportation surveillance systems. Visual vehicle tracking approaches aim to associate the target vehicles in consecutive frames of video. The core issue of those approaches is their dependency on the target vehicle behavior: In the case where the vehicle's speed is relatively higher, then the frame rate or the orientation changing during the tracking process can have a severe impact on the model's performance. A typical visual vehicle tracking system consists of five

H. Ait Abdelali (✉) · O. Bourja · R. Haouari · H. Derrouz · Y. Zennayi · F. Bourzex
Embedded System and IA Department, MASciR, 10100 Rabat, Morocco
e-mail: h.aitabdelali@mascir.com

O. Bourja
RIME Laboratory, Mohammadia School of Engineers, Mohammed V University, 10000 Rabat, Morocco

R. Haouari · H. Derrouz · R. Oulad Haj Thami
ADMIR Laboratory, Rabat IT Center, IRDA Team, ENSIAS, Mohammed V University, 10000 Rabat, Morocco

Y. Zennayi
Smart Communications Research Team, Department of Electrical Engineering at Mohammadia School of Engineers, Mohamed V University, 10000 Rabat, Morocco

© The Editor(s) (if applicable) and The Author(s), under exclusive license to Springer Nature Singapore Pte Ltd. 2021
F. Saeed et al. (eds.), *Advances on Smart and Soft Computing*, Advances in Intelligent Systems and Computing 1188, https://doi.org/10.1007/978-981-15-6048-4_45

essential components: object representation, dynamic model, object location, search mechanism, and data association. Based on the implementation of those components, visual vehicle tracking approaches can be categorized in different manners. The object representation component is responsible for the process of matching the target appearance under different influencing factors and determine the objective function to be used for finding the object in the frames. A dynamic model is used to predict the possible object. Since it is difficult to adapt an efficient dynamic model for fast movements and due to faster processors, the most current tracking algorithms use a random walk model to predict likely states. The main ideas of this work are: We leverage the power of the deep learning approach You Only Look Once (YOLO) [1]-based detection in the context of multiple vehicle tracking, a pragmatic tracking approach based on the particle filter [2–5], and the Bhattacharyya [6–8] kernel is presented and evaluated.

The rest of the work is organized as follows: Sect. 2 describes related and existing work in this field. Sections 3.2 and 3.3 presents basic particle filter algorithm and Bhattacharyya kernel method. Section 4 introduces the proposed approach. The experimental results are presented in Sect. 5. Finally, Sect. 6 concludes the paper.

2   Related Work

The Visual object tracking problem is formulated as a sequential recursive estimation of the probability distribution of the object in the previous frame. In the literature, researchers have proposed a variety of tracking model-based. In recent years, approaches based on the particle filter algorithm start gaining increasing interest, especially for real-time applications, such as traffic monitoring. This interest could be explained by the fact that the particle filter algorithm has shown a significant capability of nonlinear, non-Gaussian, and multi-model data processing. In [9], the authors presented the integration of color distributions in particle filter. In this work, for the measure of similarity between two distributions of color, the Bhattacharyya coefficient is applied. This approach has the ability to work in real time, but the processing time depends on the size of the region and the number of samples. The authors of [10] presented a color-based method for object tracking and developed an efficient tracking algorithm based on particle filter. To accelerate the weight computation of each particle, the integral images are used to compute the histogram. The main weakness of this approach is the large storage capacity. Similarly, Li et al. [11] also used a particle filter in combination with an adaptive fusion method for visual object tracking. In this work, the proposed method is more robust to illumination changes, particle occlusions, pose variations, cluttered backgrounds, and camera motion. However, the used dataset is limited comparing to other works. The research in [12] describes a novel approach that tackles the particle filter sample impoverishment problem. The main limitation of this approach is its dependency on the object speed since its performance decreases in the case where the speed of the moving object is higher than the frame rate. In [13], the authors used a hybrid estimation algorithm based on the particle filter method. This algorithm used the standard color of the particle filter in two steps: In the first step, the proposed algorithm handled the non-rigid and deformation of object, and in the second, it clutters the background. The major defect of this algorithm is that the number of particles increases with the increase of the number of tracked objects, which impacts the capability of the algorithm in real time, since the more significant the size of the state vector, the higher the competition time needed. The authors in [14] suggested the use of an immune genetic algorithm (IGA) to optimize the particle set. This optimization aims to describe the state of the target more significantly. Moreover, the use of this optimization technique increases the number of important particles, improves particle diversity, and decreases the error of state estimation and video tracking. The authors in [2] also proposed an optimized algorithm based on a continuously adaptive mean-shift (CamShift) method. The main idea of this work is improving the tracking accuracy by exploring the interaction between particle filtering and CamShift outputs. The use of the CamShift method helps improve the performance of the particle filter in sampling the position and scale space and reduce computing and execution time. Another optimization approach is presented in [15]. In this work, the authors used a particle filter with a posterior mode tracker to track target objects in the presence of illumination variations in the video frames. Similarly, in [16], the authors developed an object tacking approach that handled not only the illumination variations, but also the pose changes and the partial occlusions, as well. In [3], a Feature-driven (FD) motion-based model for object tracking is proposed. The proposed model uses features from the accelerated segment test (FAST) and matches them between the video sequences frames. This model was mainly developed to handle the abrupt motion-tracking problems that considered one of the significant challenges of tracking approaches. The main downside of this model is that it only uses the location cues for guiding particle propagation.

Recently, researchers start using deep learning techniques [17–19], to extract competent deep characteristics. In many complicated tasks relied on handcrafted features such as monitoring, location, traffic crowd detection, identification, self-stabilization, crash avoidance, obstacle, and vehicle tracking, deep learning algorithms have exhibited impressive results. With the smart video surveillance, autonomous vehicles, and numerous people-counting applications, facial recognition, the demand for rapid and accurate object detection systems is becoming a necessity. Such systems require not only the classification and identification but also the localization of each object in the image, which makes objects identification a much harder task than their conventional counterparts in computer vision.

3 Deep Learning

3.1 YOLO Detector

YOLO stands for You Only Look Once [1]. This technique watches at the image just for once and processes it simultaneously. Rather than of a large softmax, YOLO uses several softmax as a hierarchical tree; each softmax determines for a similar group of objects. YOLO is considered to be one of the fastest and most used object detection algorithms to predict several bounding boxes and their category probabilities at the same time YOLO use a single convolutional network.

3.2 Particle Filter

Particle filter is a sequential Monte Carlo for estimating the posterior distribution. It includes two parts: observation model and system model. See Fig. 1:

To predict the possible position of the targets, we use a system model given by:

$$X_t = g(X_{t-1}) + U_t \tag{1}$$

where $g(.)$ is the transition function, X_t presents the state of targets at time t, and U_t is the system noise. The observation model is used to determine the locations of the targets and has the form:

$$Y_t = f(X_t) + V_t \tag{2}$$

where $f(.)$ is a measurement function and V_t system noise. Let Y_t and X_t measurements and the hidden state of the object of interest at discrete time t, respectively. For tracking, the filtering distribution of interest is presented by $p(X_t|Y_{1:t})$, where $Y_{1:t} = (Y_1, \ldots, Y_t)$ is the observations in the current time. The estimation of filtering distribution content two-step recursion:

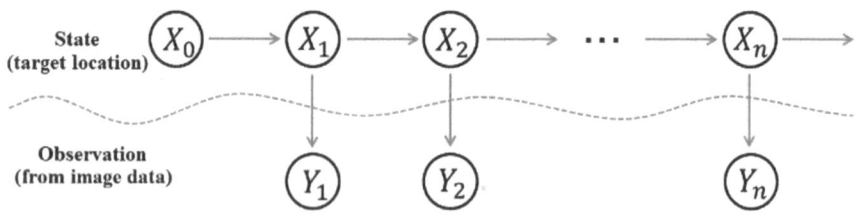

Fig. 1 Observation model and system model

$$p(X_t|Y_{1:t-1}) = \int p(X_t|X_{1:t-1})p(X_{t-1}|Y_{1:t-1})dX_{t-1} \tag{3}$$

$$p(X_t|Y_{1:t}) = p(Y_t|X_{1:t})p(X_t|Y_{1:t-1}) \tag{4}$$

where $p(X_t|Y_{1:t-1})$ presented the state evolution and $p(Y_t|X_{1:t})$ is the current observation. To this end, let given a weighted set of samples $\left\{X_{t-1}^{(i)}, W_{t-1}^{(i)}\right\}_{i=t}^{N}$ approximately according to $p(X_{t-1}|Y_{1:t-1})$, generated form a proposal distribution, $X_t^{(i)} \sim q(X_t|X_{t-1}, Y_t)$, $i = 1, \ldots, N$, the new weights are set to:

$$W_t^{(i)} = W_{t-1}^{(i)} \frac{p(Y_t|X_t^{(i)})p(X_t^{(i)}|X_{t-1}^{(i)})}{q(X_t|X_{t-1}^{(i)}, Y_t)}, \quad \sum_{i=1}^{N} W_t^{(i)} = 1. \tag{5}$$

The new particle set $\left\{X_t^{(i)}, W_t^{(i)}\right\}_{i=t}^{N}$ is then approximately distributed according to $p(X_t|Y_{1:t})$.

3.3 Bhattacharyya Kernels

The Bhattacharyya kernel is a perfect similarity measure for discrete distributions. Let h_V be the histogram of vehicle tracking, and the number of pixels inside the targets V as $|V|$, which is also equal to $|V| = \sum_k h_V(k)$ the sum over bins. Let q be the normalized histogram obtained in the frames and p be the normalized version of the targets h_V defend by $p = \frac{h_V}{|V|}$, and $\delta[n]$ is the Kronecker delta:

$$h_V(k) = \sum_{x \in V} \delta[b(x) - k], \tag{6}$$

where $\delta[n] = 1$ if $n = 0$, and $\delta[n] = 0$ otherwise. The Bhattacharyya kernel can be expressed as:

$$\begin{aligned} \text{Bhat}(p, q) &= \sum_k \sqrt{p(k)q(k)} \\ &= \sum_k \sqrt{p(k)\left(\frac{1}{|V|}\sum_{x \in V}\delta[b(x) - k]\right)} \\ &= \frac{1}{\sqrt{|V|}}\sum_{x \in V}\sum_k \sqrt{p(k)\delta[b(x) - k]} \\ &= \frac{1}{\sqrt{|V|}}\sum_{x \in V}\sqrt{p(b(x))}. \end{aligned} \tag{7}$$

where $b(x)$ is a map function a pixel x to its corresponding bin index, the computation of the Bhattacharyya kernel taking the sum of values of $\sqrt{p(b(x))}$ within the target V.

4 Proposed Approach

The proposed system is a filtering method based on recursive Bayesian estimation. The core idea of this work is that the density distribution is presented using random sampling of particles, with a nonlinear problem. The working mechanism of the proposed system has four major steps, namely selection step, prediction step, likelihood step, and measurement step. These steps are shown in Fig. 2:

Firstly, we generate N particles in an area that contains the possible locations of the targets. This step named selection or initialization. During the prediction step, each particle is transformed according to the state model by insertion random noise to emulate the impact of noise on the state. In the likelihood step, the weight of apiece particle is calculated, and the similarity (using Bhattacharyya kernel) between the target model and candidate regions is described. Furthermore, in this step, the weights are normalized to ensure that the sum of the weights is equal to 1. Finally, in the measurement step, the weight of each particle is re-examined based on the new sate. We assess the likelihood probability and re-sampling the particles for the next iteration. The steps of the proposed approach are presented in Fig. 3:

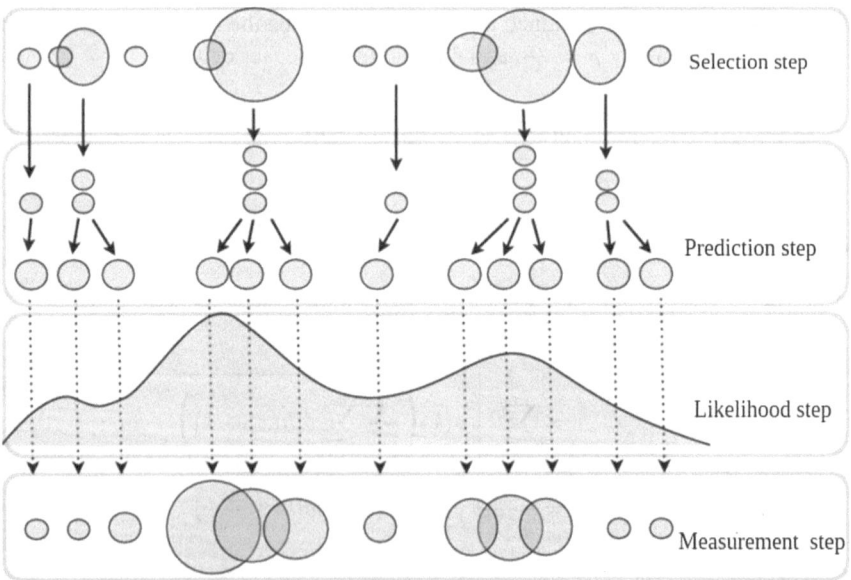

Fig. 2 Methodology of particle filter for vehicles tracking

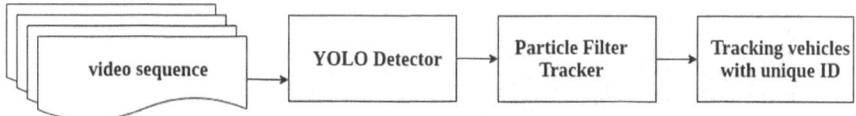

Fig. 3 Flowchart representation for visual vehicles detection and tracking

---

The Algorithm of the proposed method can be explained as follows:

Step 1. For each frame, the regions of targets (vehicles) are detected by the YOLO detector, and the sample set of N particles.

Step 2. Prediction the localisation's of the targets using second-order auto-regressive dynamics, for evaluate how the state of an targets will change by feeding it through a observation and system model, using Eqs. (1) and (2).

Step 3. Compute the Likelihood of each particle, and calculate the similarity between targets model and candidates regions in the current frame, using Eqs. (5) and (7).

Step 4. Measurement the weight of each particle is re-examine created on the new frame, we estimate the Likelihood, and Resampling $\left\{X_n^{(i)}, W_n^{(i)}\right\}_{i=n}^{N}$, and then approximately distributed according to $p(X_n | Y_{1:n})$.

Step 5. Draw trajectory of the targets by ID in the current frame, and go to step 2.

---

5 Experiment Result

This section shows the tracking results using the proposed method. The proposed approach components are run at 3.70 GHz single core of an Intel(R) Core (TM) i7-8700 K machine with 16 GB memory, NVIDIA GeForce GTX 1080 Ti, and Operating System Linux 64-bit. The program was implemented using C++, Qt Framework, and the Open Source Computer Vision (OpenCV) library. To check the performance effectiveness of the proposed approach, we use Intersection over Union (IoU) metric [20], also known as the Jaccard index, to compare the similarity between two arbitrary shapes (the predicted bounding box and the ground truth). Intersection over Union encodes the shape properties of the objects under comparison, e.g., the widths, heights, and locations of two bounding boxes. In the experiment, we use three videos; each has its characteristics.

We first use a road video (where the frame rate is 15 fps, the resolution is 2456 × 2054, and the number of frames is 1225) to check the effectiveness of the proposed approach. As shown in Fig. 4, the external bounding box red color represents the real

Fig. 4 Tracking results of the road sequence. Frames 1 and 78 are displayed. The predicted bounding box is drawn in blue, while the ground-truth bounding box is drawn in red. The metric IoU is higher than 0.95

trajectory of the vehicle (the ground truth, i.e., hand labeled), and the blue bounding box represents the trajectory of the vehicle by the proposed approach.

The second test is an Intersection$_1$ video (where the frame rate is 15 fps, the resolution is 2456 × 2054, and the number of frames is 1106), used to check the effectiveness of the proposed approach on a more complicated situation Fig. 5. The vehicles exhibit large-scale changes with partial occlusion. However, the proposed approach performs better in estimating the orientation and the scale of the targets, especially when an occlusion occurs.

The last experiment is an Intersection$_2$ video (where the frame rate is 15 fps, the resolution is 2456 × 2054, and the number of frames is 1124) presented in Fig. 6. The results show that the proposed method estimates the good accuracy of orientations and the scales of the targets.

The results show that the proposed approach is robust to detect and track the trajectory of the vehicles in different situations (pause, scale variation, occlusion, and rotation).

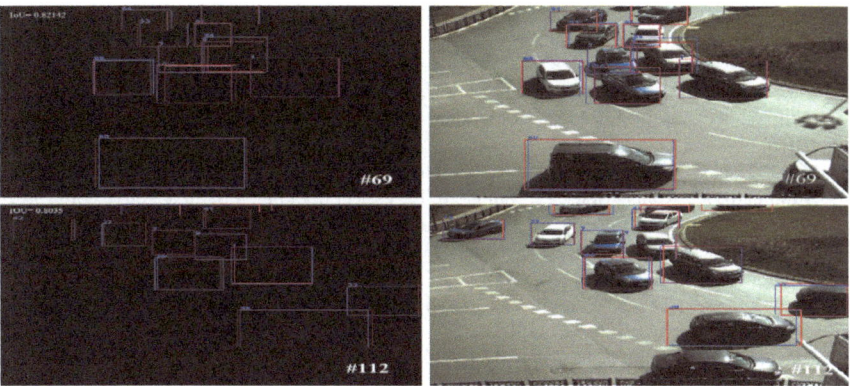

Fig. 5 Tracking results of the Intersection$_1$ sequence. Frames 1 and 172 are displayed. The predicted bounding box is drawn in blue, while the ground-truth bounding box is drawn in red. The metric IoU is higher than 0.80

Fig. 6 Tracking results of the Intersection$_1$ sequence. Frames 69, and 112 are displayed. The predicted bounding box is drawn in blue, while the ground-truth bounding box is drawn in red. The metric IoU is higher than 0.88

6 Conclusion

In this paper, a novel visual vehicle tracking algorithm has been presented. The detection of moving vehicles is done using the YOLO detector and the particle filter algorithm for tracking vehicles in consecutive frames of a video sequence. In this method, we combine the Bhattacharyya kernels and the particle filter. Numerous experiments include challenging video when the proposed algorithm achieves excellent results.

References

1. Liu, W., Anguelov, D., Erhan, D., Szegedy, C., Reed, S., Fu, C.Y., Berg, A.C.: SSD: single shot multibox detector. In: European Conference on Computer Vision, pp. 21–37. Springer, Cham (2016)
2. Ma, J., Han, C., Chen, Y.: Efficient visual tracking using particle filter. Inform. Fusion (2007)
3. Nummiaro, K., Koller-Meier, E., Van Gool, L.: Object tracking with an adaptive color-based particle filter. Pattern Recogn., 353–360 (2002)
4. Abdalmoaty, M.: Identification of stochastic nonlinear dynamical models using estimating functions. Dissertation. KTH Royal Institute of Technology (2019)
5. Li, A., Jing, Z., Hu, S.: Particle filter based visual tracking with multi-cue adaptive fusion. Chin. Opt. Lett. 3(6), 326–329 (2005)
6. Tony, J., Risi, K.: Bhattacharyya expected likelihood kernels. Paper Presented at the Meeting of the COLT (2003)
7. Tony, J., Risi, K., Andrew, H.: Probability product kernels. J. Mach. Learn. Res. 5, 819–844 (2004)
8. Chang, H.-W., Chen, H.-T.: A square-root sampling approach to fast histogram-based search. In: 2010 IEEE Conference on Computer Vision and Pattern Recognition (CVPR), pp. 3043–3049, 13–18 June 2010
9. Vadakkepat, P., Jing, L.: Improved particle filter in sensor fusion for tracking randomly moving object. IEEE Trans. Instrum. Meas. 55(5), 1823–1832 (2006)
10. Czyz, J., Ristic, B., Macq, B.: A particle filter for joint detection and tracking of colour objects. Image Vis. Comput. 25, 1271–1281 (2007)
11. Han, H., Ding, Y., Hao, K., Liang, X.: An evolutionary particle filter with the immune genetic algorithm for intelligent videotarget tracking. Comput. Math. Appl. 62(7), 2685–2695 (2011)
12. Samarjit Das, N.V., Kale, A.: Particle filter with a mode tracker for visual tracking across illumination changes. IEEE Trans. Image Process. 21(4), 2340–2346 (2012)
13. Ho, M., Chiang, C., Su, Y.: Object tracking by exploiting adaptive region-wise linear subspace representations and adaptive templates in an iterative particle filter. Pattern Recogn. Lett. 33(5), 500–512 (2012)
14. Wang, Z., Yang, X., Xu, Y., Yu, S.: CamShift guided particlefilter for visual tracking. Pattern Recogn. Lett. 30(4), 407–413 (2009)
15. Wang, W., Wang, B.: Feature-driven motion model-based particle-filter tracking method with abrupt motion handling. Opt. Eng. 51, 4 (2012)
16. Mallikarjuna Rao, G., Satyanarayana, C.: Visual object target tracking using particle filter: a survey. IJIGSP 5(6), 57–71 (2013)
17. Moon, S., Lee, J., Nam, D., Kim, H., Kim, W.: A comparative study on multi-object tracking methods for sports events. In: 19th International Conference on Advanced Communication Technology (ICACT) (2017)
18. Bewley, A., Ge, Z., Ott, L., Ramos, F., Upcroft, B.: Simple online and real time tracking. In: IEEE International Conference on Image Processing, Phoenix, AZ, pp. 3464–3468 (2016)
19. Mekonnen, A., Lerasle, F.: Comparative evaluations of selected tracking by detection approaches. IEEE Trans. Circuits Syst. Video Technol. 29(4), 996–1010 (2019)
20. Rezatofighi, H., Tsoi, N., Gwak, J., Sadeghian, A., Reid, I., Savarese, S.: Generalized intersection over union: a metric and a loss for bounding box regression. In: Proceedings of the IEEE Conference on Computer Vision and Pattern Recognition, pp. 658–666 (2019)

SDNStat-Sec: A Statistical Defense Mechanism Against DDoS Attacks in SDN-Based VANET

Faycal Bensalah, Najib Elkamoun, and Youssef Baddi

Abstract Software-defined networking (SDN) is an evolving trend that delivers an architecture that separates the control plane from the data plane. Through the central control, the network is more dynamic, and network capacity is administered more efficiently and cost-effectively. Other technology that is gaining industry attention and has tremendous potential for use worldwide is Vehicle Ad-hoc Networks (VANETs). VANETs are essential components of Intelligent Transport Systems (ITS), and the core objective of all ITS systems is to provide improved transport safety and a range of applications for both drivers and passengers. VANET networks can profit from using an SDN controller. Through the decoupling of control and data in VANET, network intelligence is centralized in a logical structure, and the underlying network backbone is decoupled from all applications. The primary goal of this paper is to monitor and detect DDoS attacks in the context of real-time applications, such as accident prevention, traffic congestion warning, or communication. In this article, we propose a new method to detect such an attack during communication in a VANET network. This method is based on a variable control chart to monitor the quality of a specific process. This method allows to identify malicious nodes in real time by controlling each node that is part of the communication within the network using graphical views.

Keywords SDN · SDN-VANET · DDoS · SPC · Security

F. Bensalah (✉) · N. Elkamoun
STIC Laboratory, Faculty of Sciences, Chouaib Doukkali University, El Jadida, Morocco
e-mail: f.bensalah@ucd.ac.ma

N. Elkamoun
e-mail: Elkamoun.n@ucd.ac.ma

Y. Baddi
STIC Laboratory, ESTSB, Chouaib Doukkali University, El Jadida, Morocco
e-mail: Baddi.y@ucd.ac.ma

© The Editor(s) (if applicable) and The Author(s), under exclusive license to Springer 527
Nature Singapore Pte Ltd. 2021
F. Saeed et al. (eds.), *Advances on Smart and Soft Computing*, Advances in Intelligent
Systems and Computing 1188, https://doi.org/10.1007/978-981-15-6048-4_46

1   Introduction

Currently, network security can be considered a major issue in any network application. That is why, when considering the SDN-VANET concept, we have targeted our activities on identifying the key security challenges associated with the SDN and VANET [1] technologies. After an thorough review of the challenges of security posed by these two different technologies, we decided to investigate the impact of DDoS attacks on the control layer. There are several threat vectors which can affect them. These kinds of threats can strongly affect the performance of the SDN-based VANET network, and because of the central control, the overall network can be compromised if the controller becoming inaccessible. For the reasons mentioned before, this paper designs and tests a DDoS detection process for an VANET network based on SDN. The testing case consists of executing a DDoS attack and implementing a DDoS detecting module on the SDN controller, for our work, we choose to use the statistical process control (SPC), and the primary reason is its capability to monitor the variability in a network. Our paper is composed of five sections, the introduction being the first. The second section presents a summary of the work done by researchers in this area of research. The third section describes the detection system that we propose to detect attack using a variable control chart. In the fourth section, the authors analyze the performance of the proposed approach using simulators. The last part concludes our presented work and gives some perspectives for the future.

1.1   Description of Architecture

To provide an SDN-based VANET system, the recommended platform architecture consists of the following SDN:

- **SDN controller**: The central logic intelligence of the VANET.
- **SDN wireless node**: The SDN controller controls the elements of the data plan.
- **SDN RSU**: The fixed elements of the data plan.

Figure 1 represents the software-defined VANET communication between the components.

1.2   Operation Overview

There are different software-defined VANET working modes which are based on the level of control of the SDN controller, although the main idea of the SDN [2], that is the decoupling of the control and the data plan. In this paper, we just focus on the centralized control mode [3]. This is the mode of operation in which all operations

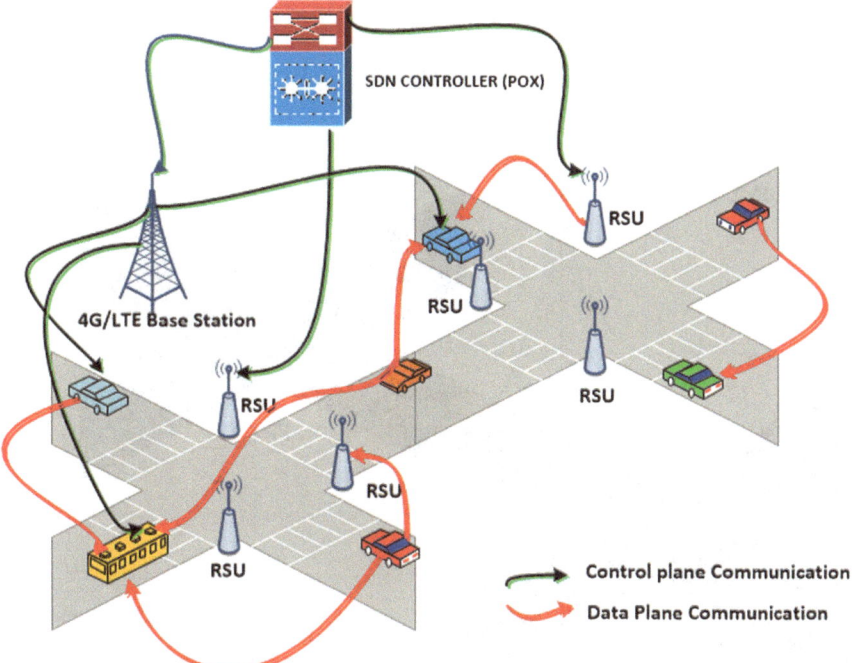

Fig. 1 Software-defined VANET communication network

of the SDN wireless nodes, and the underlying RSUs are controlled by the SDN controller. The SDN controller will define all the flow rules on how the traffic is to be managed, as shown in Fig. 2.

1.3 DoS and DDoS Attack

The denial of service (DoS) attack [4] is a malicious user action intended to affect the normal functioning of the network. If this is attempted by a host group, instead of a single host, it is referred to as a distributed denial of service (DDoS). A denial of service (DoS) attack is an attack that aims to prevent a computer application, server, or Web site from responding to requests from its legitimate users. In the case of DDoS [4], the attacker aggressively floods the target with massive traffic until it becomes unable to serve its users.

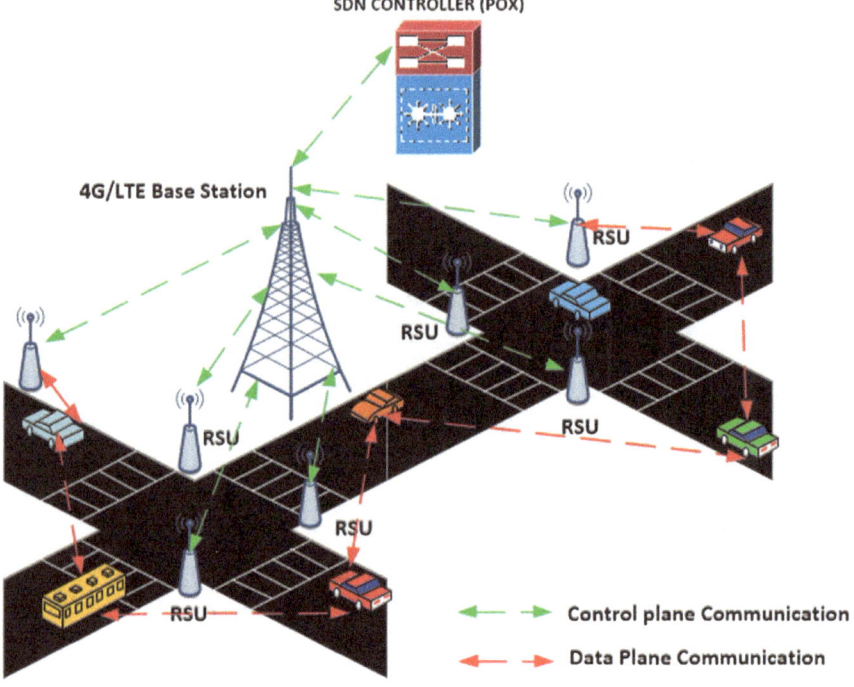

Fig. 2 Centralized control mode

2 Related Work

This section presents a review of the literature that summarizes the work already done and the existing solutions to mitigate the impact of DDoS attacks on the performance of the SDN network. These works propose solutions to properly manage the traffic to be analyzed by the IDS during DDos attacks, where the quantity of flows received is much greater than its processing capacity. A solution has been proposed in Dharmad-hikari et al. [5] to protect SDN networks against DDOS attacks and which is based on IP filtering technique. The mechanism consists in analyzing the user's behavior and activities in the network in order to calculate the number of connection attempts of his IP address. Another solution presented by Sahay et al. [6], the authors exploit the centralization and programmability offered by SDN technology and propose an autonomous management system in which the Internet Service Provider (ISP) and its customers collaborate to mitigate the impact of DDoS attacks.

FloodGuard [7] is a new platform to defend against DoS attacks in SDN networks proposed by Wang et al. [7]. The authors' objective is to avoid saturation of the switch tables of the transmission plan switches and controller overload. To achieve these objectives, they implement two supervision and flow management modules in the controller.

None of the existing solutions can simultaneously reduce the controller load, the bandwidth between the controller, and the switch (noted by BW of the P.C.). In our work, we aim to achieve these objectives simultaneously in order to maintain acceptable network performance during DDoS attacks.

3   The Proposed Detection System

3.1   Basic Concept

Our fundamental idea was inspired by the variations in end-to-end delay and throughput. In the normal case, the variations in throughput vary according to the continuously changing topology of the network and also depends on the speed of the vehicles. The end-to-end delay is the time it takes for a packet to reach its destination. The number of transit vehicles as well as their speed has an influence on the E2E delay. The attack results in a significant degradation of the throughput and the end-to-end delay as a result of the attack. Our detection method is based on the supervision of the two metrics that we mentioned earlier, with the double limit defined by the graph. These graphical representations are called control charts based on statistical process control.

3.2   Quality Control Chart

A quality control chart is a graphical representation used to study the variations of a process over time. The data collected are plotted in temporal order. In general, a control chart always has a center line (CL) for the mean, an upper line for the upper control limit (UCL) and a lower line for the lower control limit (LCL). The values of these lines are calculated from the historical data of the process being monitored. By comparing the current data with these lines, we can determine whether the process change is compliant (controlled) or unpredictable (uncontrolled). These control charts are mostly used in the industrial sector to control the quality of a given product before it reaches the end customer. In general, there are two types of charts: attributive control charts and variable control charts. These charts are considered powerful tools for the graphical interpretation of numerical data that facilitate the monitoring of anomalies in a controlled process. Among these charts, we have the Shewhart control chart for individual measurement [8].

3.3   The Shewhart Control Chart for Individual Measurement

The Shewhart control chart for individual measurement [9] is used when monitoring and controlling a process consisting of units that can be measured periodically. These measurements are made in quantity. An assumption of using moving range chart is that the data must be normally distributed. The difference between data, x_i and its predecessor, $i - 1$, is calculated as moving range

$$\text{MR}i = |xi - xi - 1| \tag{1}$$

For m individual values, there are $m - 1$ ranges. Next, the arithmetic mean of these values is calculated as,

$$\overline{\text{MR}} = \frac{\sum_{i=2}^{m} \text{MR}i}{m - 1} \tag{2}$$

To calculate the control limits for single values

$$\text{UCL} = \bar{x} + 3\frac{\overline{\text{MR}}}{d_2} \tag{3}$$

$$\text{CL} = \bar{x} \tag{4}$$

$$\text{LCL} = \bar{x} - 3\frac{\overline{\text{MR}}}{d_2} \tag{5}$$

Process variability can be monitored using the following equations:

$$\text{UCL} = D_4\overline{\text{MR}} \tag{6}$$

$$\text{CL} = \overline{\text{MR}} \tag{7}$$

$$\text{LCL} = D_3\overline{\text{MR}} \tag{8}$$

where $\overline{\text{MR}}$ is the average of moving ranges extracted from two successive observations x, and the UCL and the LCL are defined as the upper control limit and the lower control limit, respectively. The constants d_2, D_3, and D_4 are called factors for control limits. They are tabulated according to the size of the samples. The control chart for individual measurements consists of two graphs: the first is used to monitor the individual values for speed slip. The second is used to represent the moving range used for quality control. The individual measurement chart shows its corresponding moving range chart. Moving range chart displays moving ranges of two successive measurements.

3.4 The Proposed Monitoring Mechanism

In this method, we monitor the throughput as well as the end-to-end delay by indicating their averages using the control chart. Table 1 summarizes the process for interpreting the detection method.

To put our method to the test, each receiver node is monitored through the SDN controller. The objective is to identify the presence of attack. The method is therefore implemented in the SDN controller to detect the attack through the supervision of the two metrics mentioned before. The monitoring is automatically launched when the node starts receiving data.

Diagram 1 describes all the steps that we follow to detect the attack. First, we need to specify the parameters of the processing that we are going to monitor. In our case, we observe the communication mechanism by monitoring two metrics: throughput and delay E2E. Second, collect the data in the normal case, i.e., without the presence of the attack. Then, we calculate the chart parameters (CL, UCL, LCL) using the equations that we explained previously. Then, we combine the results of the two previous steps by transforming them into graphical representations. In order to

Parameter	Value
The observation x	Throughput of end-to-end delay
Average of observation x	Center line $= \bar{x}$
Average moving range of observation x	Center line $= \overline{\mathrm{MR}}$
Upper control limit for individual observations	$\mathrm{UCL} = \bar{x} + 3\frac{\overline{\mathrm{MR}}}{d_2}$
Lower control limit for individual observations	$\mathrm{LCL} = \bar{x} - 3\frac{\overline{\mathrm{MR}}}{d_2}$
Upper control limit for moving range observations	$\mathrm{UCL} = D_4\overline{\mathrm{MR}}$
Lower control limit for moving range observations	$\mathrm{LCL} = D_3\overline{\mathrm{MR}}$

Table 1 Process for interpreting the detection method

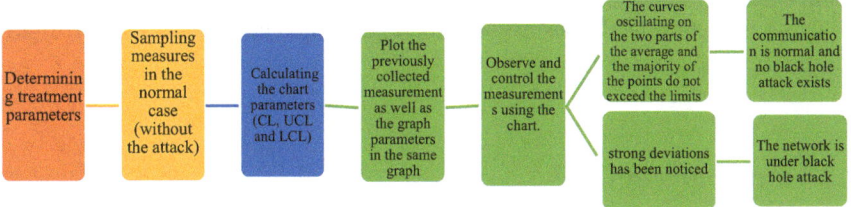

Diagram 1 Diagram follows to detect the attack

benefit from the advantages of the control chart, a correct and appropriate calculation is required. For this, we need a minimum of 20 values, which is a requirement defined by the chart itself. In our case, we plot each calculated value of the two selected metrics previously highlighted (throughput and E2E delay). This is a real-time monitoring. The following section presents our case study in which we test our method using the Mininet and NS-2 simulators, after which we will analyze the results obtained.

4 Simulation and Analysis

In this section, a short description of the experimental design is presented. Next, we study the feasibility of our DDoS detection methodology. According to the type of attack, volumetric attacks that deplete the device throughput can be identified by measuring the rate of change of the throughput before and during the attack to detect the presence of an attack.

4.1 Simulation Tools

Our solution is based on the DELTA framework [10]. We will realize a DDoS detection scheme, and we will take into account all the parameters defined in this method.

Moreover, we will implement the essence of this solution for the VANET network based on SDN. The main goal of our work is to study cases of DDoS attacks using UDP packets to provide a high quality of service that meets the requirements for real-time services.

As we will explain in the following sections for the requirements of our studies, we will consider the very small size of the packet windows and assume that the topology can be handled as static. We will consider instantaneous properties, which do not depend on mobility, and we are also interested in really small time intervals. Vehicle-to-vehicle communication (V2V) is a data plan communication, it is not part of our field of interest in this work, and we will direct our study in another way, the communication between the vehicles, the base-stations, and controller (Fig. 3).

To evaluate the effectiveness of our proposed method through NS-2 simulations, we specified the parameters and the tools in Table 2.

4.2 Environment Used

There are several current controllers available. The one used in our proposal is the POX. The POX is strongly used for studies, it is fast and light, and it is designed

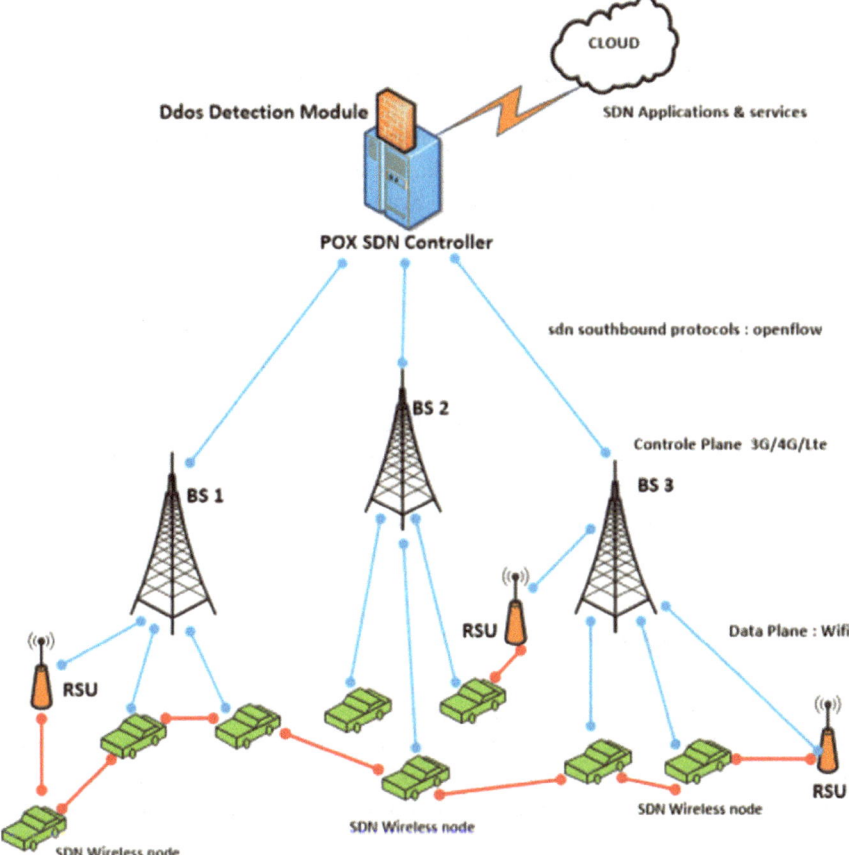

Fig. 3 Proposed topology

Table 2 Simulation parameters

Simulation	Descriptions
Mininet	For emulating the network topology
NS-2 simulator	Vanet simulation
Host	Machine Xeon, 64G RAM
Open daylight controller	Ubuntu server for release controller
VMware workstation	Simulation virtual environment
Other tools	DELTA framework Pentmenu for launching (Slowloris: using netcat; TCP ACK flood: using hping3 and TCP SYN flood: using hping3) iperf tool for monitoring network throughout before and during attack

as a real platform, which makes it possible to create a custom controller. POX [11] provides a framework to communicate with SDN switches device using the Open-Flow or OVSDB protocol. Mininet [12] is the network emulator that we used in our research to perform the simulations. It is the standard network emulation tools which can be used for the SDN. Packet generation is executed by Scapy [13]. This is efficient package generation tool, for package manipulation, scanning, sniffing, trace routing, attacking, and forging packages. Scapy is utilized in our research to assess UDP packets and to spoof the source IP address of the packets.

4.3 Network Setup

This experiment was carried out on an HP server equipped with an Intel(R) Xeon processor, 3.4 GHz power, 64 GB RAM, and a 10/100/1000 Mbit/s network interface. The OS is Linux Ubuntu 14.04, and Mininet version 2.2.x was run as standard on Windows 10. Mininet 2.2.1 supports OpenFlow version 1.3.

4.4 Network Topologies

For our test, we performed a tree type network of depth 2 with 8 switches and 32 hosts using Mininet. Figure 4 shows the topology network. We used Open Virtual Switch (OVS). The POX L3_learning module has been used for the controller. In this module, two functions are added to be used for statistics collection and calculation of parameters.

In the VANET case, only the RSUs and the base stations are static. The host is moving from one position to another and can be connected to various nodes in the network. We suggest that if the network gets a current state of nodes and switches connected to them, the controller can execute a dynamic algorithm to detect DDoS attacks.

4.5 Computation of Control Limits

The parameters of the chart used are calculated in the normal case according to the equations that we mentioned in the previous section. Therefore, the measurements are collected with Eqs. (2)–(7) without attack. The results are summarized and illustrated in Table 3.

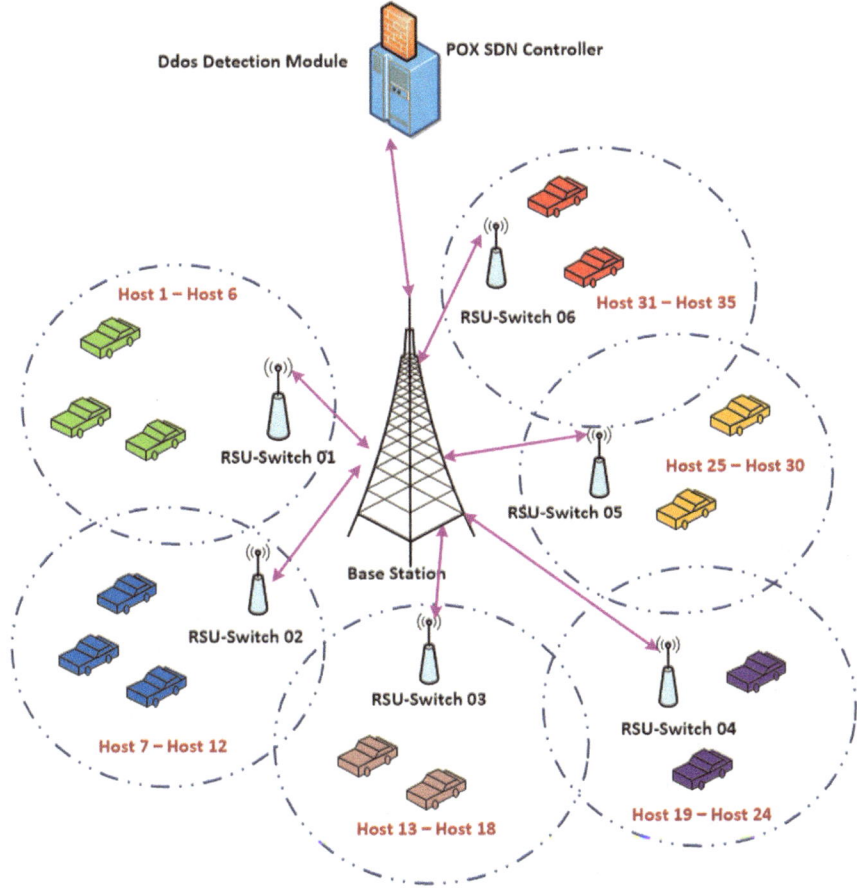

Fig. 4 Topology for the test bed

Table 3 Calculated parameters of the control chart

Parameter	Value	
The monitoring x	Throughput	E2E delay
Average of monitoring x	0.4200	0.0280
Average moving range of monitoring x	0.0643	0.0034
Upper control limit for individual monitoring	0.6087	0.0301
Lower control limit for individual monitoring	0.2521	0.0165
Upper control limit for moving range monitoring	0.2343	0.0187
Lower control limit for moving range monitoring	0.0	0.0

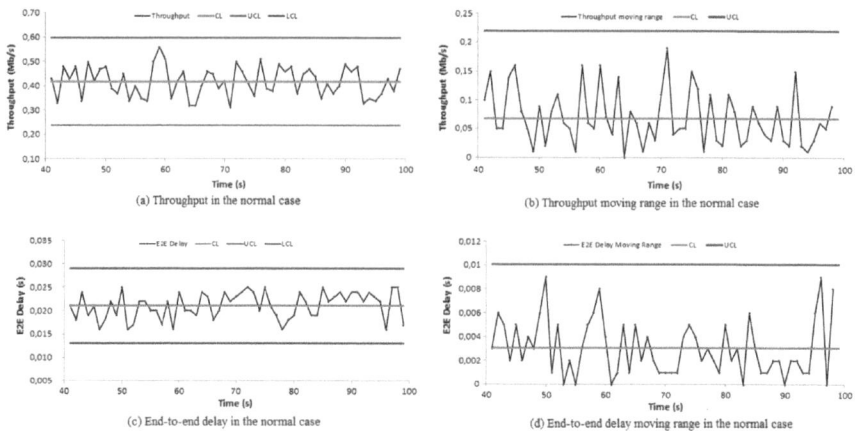

Fig. 5 Monitoring in the normal case (without the attack)

4.6 Simulations Results

To calculate the parameters, the table requires a minimum of 20 samples. The number of samples is required by the control chart to ensure efficient and flawless results. In our case, we take the values instantaneously, which means that the granularity of the control is fixed at one second. So, it is real-time monitoring, as we mentioned earlier. This detection system is deployed in the SDN controller, which gives it the right to check the data received.

In a first stage, we supervised the network in the normal case (without the attack). Both metrics (discussed previously) are monitored using the control card whose parameters are calculated in the previous section (Table 3). After plotting the graphs of the two metrics, we can notice in Fig. 5 that these two curves oscillate between the limits of the graph and on either side of the average. We can therefore say that the communication takes place in a normal environment without the presence of the attack.

In this case, it can be noted that the variations of the metric curves (throughput and end-to-end delay) are below the lower control limit. As we can see in Fig. 6a–d, the Shewhart statistic crossed the upper control limit in the presence of the greedy behavior. We note also that there was a strong deviation in the case of the attacked network. The Shewhart method presents constant upper and lower limits, but the calculated value depends on the normal case.

4.7 Generalization of the Detection System

To generalize our detection approach, we draw a tolerance interval based on the number of nodes. The tolerance interval is defined as the difference between the

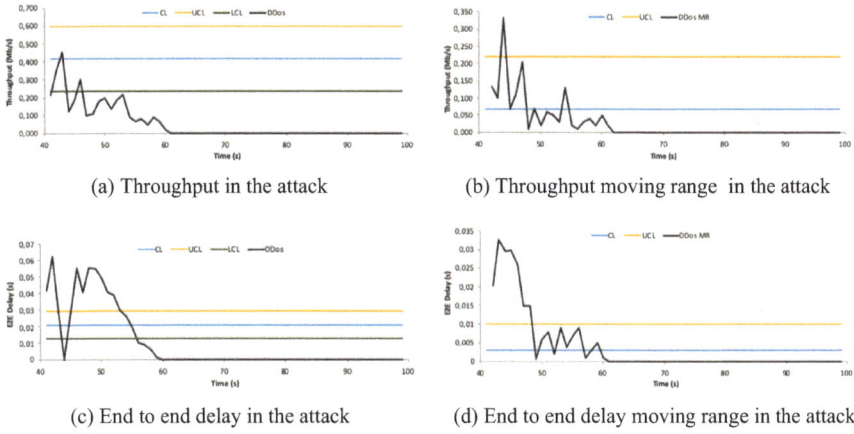

(a) Throughput in the attack

(b) Throughput moving range in the attack

(c) End to end delay in the attack

(d) End to end delay moving range in the attack

Fig. 6 Monitoring in the attack case

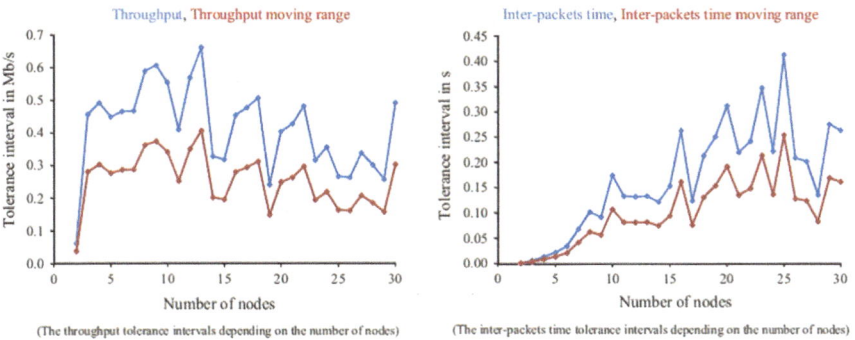

(The throughput tolerance intervals depending on the number of nodes)

(The inter-packets time tolerance intervals depending on the number of nodes)

Fig. 7 Tolerance ranges

UCL and the LCL. The curves shown in Fig. 7 indicate the detected variations which depend mainly on the number of nodes, i.e., the size of the network. Therefore, we need to calculate the chart parameters as a function of the network size to ensure better monitoring.

As mentioned above, the detection parameters and the tolerance interval depend on the number of nodes in the network.

5 Conclusion

In our paper, we presented a scheme for detecting DDos attacks in VANET environments. This scheme is based on statistical process control which is a very famous tool in the industrial domain in terms of supervision. Its strong point lies in its ability to detect abnormal variations and variations in all theoretical and real fields. This allows us to easily differentiate between normal and critical situations. This tool has obvious advantages that can be useful and interesting in the field of security. Also, the technique that we propose does not require any changes in the IEEE standards or in the routing protocol, which is considered a significant advantage. As we simulate our approach to obtaining the results that can be used as the basis for the developing a dynamic detection algorithm for DDoS attacks and its implementing it in the SDN controller, we consider this to be the main focus of our future research, with the next step being the mitigation of DDoS attacks in the SDN-based VANET scheme. The first part of the mitigation will be the identification of the source of the attack and then its elimination from the network.

References

1. Kalinin, M., Krundyshev, V., Zegzhda, P., Belenko, V.: Network security architectures for VANET. In: Proceedings of the 10th International Conference on Security of Information and Networks, pp. 73–79. ACM, Oct 2017
2. Xia, W., Wen, Y., Foh, C.H., Niyato, D., Xie, H.: A survey on software-defined networking. IEEE Commun. Surv. Tutor. 17(1), 27–51 (2014)
3. Todorova, M.S., Todorova, S.T.: DDoS attack detection in SDN-based VANET architectures, p. 175, June 2016.
4. Somani, G., Gaur, M.S., Sanghi, D., Conti, M., Buyya, R.: DDoS attacks in cloud computing: issues, taxonomy, and future directions. Comput. Commun. 107, 30–48 (2017)
5. Dharmadhikari, C., Kulkarni, S., Temkar, S., Bendale, S.: A study of DDoS attacks in software defined networks (2019)
6. Sahay, R., Blanc, G., Zhang, Z., Debar, H.: ArOMA: an SDN based autonomic DDoS mitigation framework. Comput. Secur. 70, 482–499 (2017)
7. Wang, H., Xu, L., Gu, G.: Floodguard: a DoS attack prevention extension in software-defined networks. In: 2015 45th Annual IEEE/IFIP International Conference on Dependable Systems and Networks, pp. 239–250. IEEE, June 2015
8. Montgomery, D.C.: Statistical Quality Control, vol. 7. Wiley, New York (2009)
9. Roes, K.C., Does, R.J., Schurink, Y.: Shewhart-type control charts for individual observations. J.Qual. Technol. 25(3), 188–198 (1993)
10. Lee, S., Yoon, C., Lee, C., Shin, S., Yegneswaran, V., Porras, P.A.: DELTA: a security assessment framework for software-defined networks. In: NDSS, Feb 2017
11. Kaur, S., Singh, J., Ghumman, N.S.: Network programmability using POX controller. In: ICCCS International Conference on Communication, Computing and Systems, vol. 138. IEEE (2014)
12. De Oliveira, R.L.S., Schweitzer, C.M., Shinoda, A.A., Prete, L.R.:Using mininet for emulation and prototyping software-defined networks. In: 2014 IEEE Colombian Conference on Communications and Computing (COLCOM), pp. 1–6. IEEE (2014)
13. Biondi, P.: Packet generation and network based attacks with Scapy. CanSecWest/core05 (2005)

Smart Expiry Food Tracking System

Haneen Almurashi, Bushra Sayed, Ma'ab Khalid, and Rahma Bouaziz

Abstract Food wastage in domestic environments represents a very serious issue common in the world because it has serious implications for wasted energy. People always forget to consume food they purchased before the expiration date, or sometimes, they over purchase food, and then throw them away without using. The aim of this research is to propose a system named Desired Before Expired (DBE) to prevent foods waste for consumers. The application helps grouping similar food types together in a fridge, keeps track of food expiration dates, and sends an automatic notification to the user before the purchased food expires or when the quantity of the product is low. In addition, the generation of an automated personalized shopping list. It can also perform other functions such as proposing receipt with items available in the fridge and eating habit analysis. We are confident that such an application may lead to a reduction in food waste in domestic environments.

Keywords Food waste (FLW) · Smart fridge · Smart home · Android application · Quick response (QR)

1 Introduction

Product wastage due to expiration is an issue that is evident in almost every household and needs to be addressed. People seem to have difficulty to sort their product when they are at risk of expiring. Not only it is very important to prevent products from expiring and avoid losing money by getting rid of expiring products. In addition, we need to keep track product's expired date, allowing us sorting products [1].

H. Almurashi · B. Sayed · M. Khalid · R. Bouaziz (✉)
Department of Computer Science, Taibah University, Medina 42353, Saudi Arabia
e-mail: rkammoun@taibahu.edu.sa

H. Almurashi
e-mail: haneenalmurashi1993@gmail.com

R. Bouaziz
ReDCAD, University of Sfax, Sfax, Tunisia

F. Saeed et al. (eds.), *Advances on Smart and Soft Computing*, Advances in Intelligent Systems and Computing 1188, https://doi.org/10.1007/978-981-15-6048-4_47

541

For all those reasons and in order to monitor the expiration date of basic consumer products in the fridge, we propose an android application named "Desired Before Expired." The application is created for people who are looking to avoid consequence of expired products like food poisoning. The application contains expiring notifications, definition of an expiration time to products and expired checker and other features. The proposed application can help people in sorting products in a fast and easy way.

The motivations behind developing such an android application are:

– Economic benefit is mainly to reduce over purchasing of items. We can save a significant amount of money if we use all products before their expiring dates.
– Social benefits are helping people by keeping track of when products are going to expire, it prevents wastage, and if we have surplus product, we can donate it to food rescue organizations.
– Environmental benefits to reduce resource use essential to product production, many resources are needed to generate product such as water and energy. By wasting those products and especially food, these resources are also wasted.

The rest of the article will be organized as follows. Section two will present the related works with a focus on smart refrigerator and similar applications. In section three, we will discusses the methodology used and the proposed application. Section four highlights the details of the proposed GUI of the application. Finally, Section five concludes the proposed application and gives the direction for future extension.

2   Related Works

On the topic of Reduction Food Loss and Waste Reduction. Authors in [2] system keeps track of products based on location data; the location data relating to a geographical location of the scanning device and included in the electronic scanning notification. End user can extend the estimated usage life by placing the product in a preferred storage environment and receiving notification from the scanning device that in product location. This system makes user update with their product by sending notification to the scanning device, including a recalculated usage expiration date. The system works over a network that contains a server computer including a processor and a client computing device.

Smart refrigerator is a refrigerator, which has been programmed to detect what kind of products are being stored inside it and keep a track of the stock through barcode or RFID scanning. This type of refrigerator is often equipped to automatically determine when a food item needs to be restocked [3].

An RFID tag application for transport has been investigated intensively for some time now. Recently, RFID applications for saving food from waste have become a new hot topic. The work in [4] proposes an RFID system when placing an item in the refrigerator or taking them out. The user must scan the item on an RFID reader to update the status of the refrigerator. For every item, scanning takes own RFID tag in the database. It acts as an expiration reminder of product, four days before the expiry date.

Internet of Things (IoT) is an extensive network of connected components working in unison by coordinating, acknowledging, and sharing the resources in the network. Authors on [5] propose a smart frameworks including components such as the centralized server, home security, smart phones, smart thermostat, centralized air condition, connected lights, windows and ventilation control, smart TV, and smart fridge. One of the framework component is smart fridge that keeps track of stored food items, monitors the measure of grocery items, and sends reminders to the occupants of smart home. Smart camera embedded in refrigerator enables the occupant to view items in smart phone.

In [6], prototype allows users to be notified proactively, when a food arrives to its expiration date. This system combines various resources in order to scan barcode, identify, and store data related to products, with a smartphone. Later, notifications are sent freely to consumers by pop-up or SMS, to avoid uneaten food loss. We believe that smartphones can be efficiently used in this fight against food waste. Indeed, they are natively equipped with hardware (camera, microphone) and software (barcode, automatic speech recognition, etc.) that can be combined and easily employed in order to collect data and also remind important deadlines to consumers. This system can be used in the supermarket, during shopping or, at home, during the storage of the products in cupboards and fridge.

Commercial solutions for food tracking have also become available in recent years; especially, the idea of using a smart phone application to save food item expiration dates has been studied before. An application which provides the manual entry of item names and expiring dates has been developed before in Android [7, 8] as well as in iOS [9]. In [10], user needs to manual entry, removing and assigning expiry date for products. However, this way of entry for each item's name and expiration date is difficult with many items stressful and time wasted. Even though these applications save the details of the food in a database and give notifications, it takes a long time to manually enter each item into the application and may be inconvenient to a typical user.

3 Methodology

The main idea is to propose a smart and sophisticated application that monitors the expiration date of consumer products and sending reminders of consuming the product before the expiration and sending the slogans at the end of the expiration date of the food product. We have made the assumption when using a handheld device like a smartphone or a tablet since the goal is to make the proposed application available for all. Contrary to previous works, the date or name of the product is not entered manually. Rather, the consumer scans the Quick Response QR code [11] after purchasing the product.

The application provides several other benefits, including inventory and reminder of food products after their expiration and add it to the shopping list to facilitate the process of remembering the amount of food that the user really (remove) needs without purchase large quantities that are not consumed on the expiry date.

Toward the further justification and development of the overall approach, it was important to gather the opinions of future users. Specifically, there is a need to know how many household members are interested on the subject and need smart system to reduce food waste. Therefore, a semi-structured interview was conducted with 643 household members. This interview included questions to reveal opinions about the overall proposed system, the associated concerns of households in using their fridge, and the amount of food waste. The result of such an interview was motivated to develop the DBE application.

4 The Proposed Desired Before Expired System

4.1 The Overall Design

The application is designed for Android platforms. Figure 1 illustrates how the system works as a whole. The user needs to scan the QR code of the product that contains the product details (name, Expiring date…). The system compares the expiring date of the product with the current date. If the product already expired, the system sends an error message. Otherwise, the database checks for the QR code id. If the id does not exist in the database, the system will ask the user to insert the count's number (number of items to be added to the fridge). After that, the product is stored and retrieved in the database through the application's connection to the Web server, so

Fig. 1 Desired Before Expired architecture

that the server updates the product data in the database. The system continuously checks when the quantity of the product is low or the item will going to expire, notifications will be generated and the product will be added to the user's shopping list.

The proposed Desired Before Expired application also allows users to print a shopping list and share it using social media applications. Using an Internet connection, the system can suggests recipes based on food stored in the fridge.

4.2   Main Features of DBE System

In this section, features offered by the proposed Desired Before Expired Android application will be presented:

– The item will be added to the system database as the users put it in refrigerator. The users have to scan the barcode before they put the item in the fridge. The user specify the number of items from each product. If the users want to put another item, they should scan the barcode again.
– The user can check product list of available and remaining items in the fridge through the application.
– The user can define the minimum quantity for every item in home server database; when reaching the specified quantity, the system will generate a notification to the user.
– The system records every items and transactions so user will be able to see transactions and taken items history by providing the counts number that shows how many number of pieces of this item from the proposed application.
– The expiry dates are determined by the system, but user is allowed to change every item's expiry date through the proposed application.
– The system can generate automatically the shopping list, which contains list of items that are consumed or the ones that have been expired. It is generated when the items quantity is less than the minimum quantity set by user and the time on its period. The user can add a product manually if it is not in the "shopping list."

The product must contain a QR code, and we had difficulty spreading the products containing the product with the rapid response code in the marketing centers, but according to our study of the extent of the spread of the QR code in the future and its wider use in marketing purposes [11]. There is widespread use of the QR code in the future for several reasons, including that QR code keeps information bigger well in a small space unlike barcode also the barcode has a space ten times that of a QR code loaded with the same amount of information in both [10]. The QR code can be read in a 360° direction to avoid interference. These features are attractive to the user, which saves time and effort.

Figure 2 presents the flowchart of the application. It shows that when the application is launched, it calls the main activity scan QR code. Initially, the user passes the product over the smartphone camera, system would check for QR code id and the

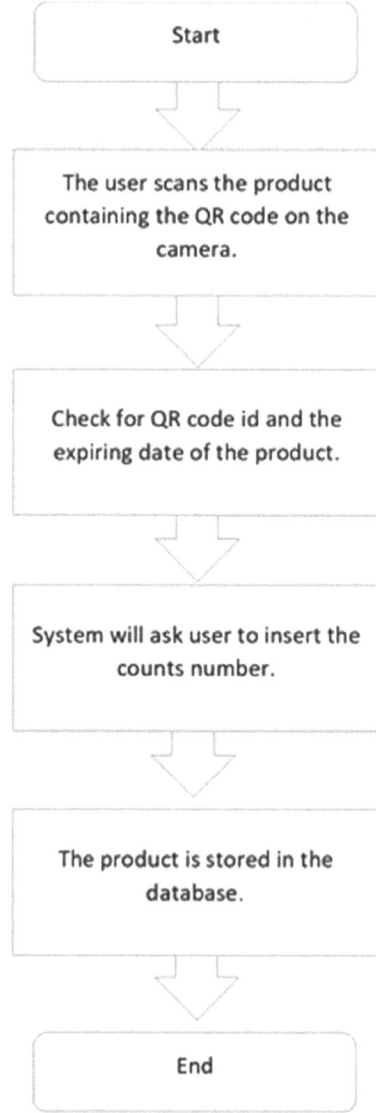

Fig. 2 Flowchart showing steps of product scanning

expiry date of the scanned product. The system checks for the group of products that apply constraints. When the expiry date of the product approaches, counts number is 1 or 0. Or when the quantity of the product is low. If one of these constraints at least applies, the system shows the products with satisfied constraints to the user's notifications button.

The system checks for the group of products that apply constraints. When the product expires, counts number is 0. Or when the quantity of the product is finished. If one of these constraints at least applies, the system moves the products with satisfied constraints to the user's shopping list.

4.3 Graphical User Interfaces

In this section, we present a prototype of how the application works. Initially, the QR code on the product will be scanned as shown in Fig. 3a, where the QR contains product details such as (production date, end date, product number, product weight, components). In addition, in this phase, the user will insert counts number as shown in Fig. 3b and press on "add" button so the product will be saved in the product list.

Product List
The stored list is to display various foods stored in the fridge without opening the door. This list displays the list of all the food items that are already added to the database. It also contains information about food name, expiry date, weight as shown by Fig. 4a. After scanning the QR code of the product, it will be added to the product list. The expiry date is shown for the user as a colored progress bar. The green color indicates that more than 7 days remain to the product to expire. The orange bar is between 7 and 3 days and the red is three or fewer days until expiration. The detail button is to show details about the product such counts number that the user already added

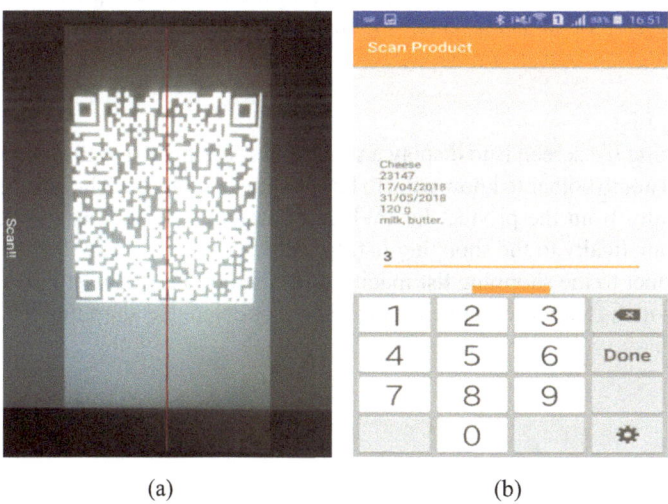

(a)   (b)

Fig. 3 **a** Scanning products QR code. **b** Add counts number of each new scanned item

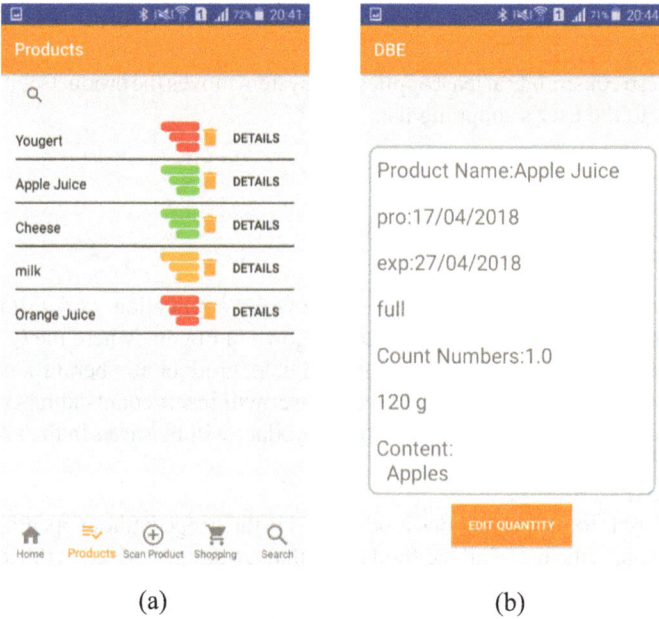

(a)   (b)

Fig. 4 **a** The product list, **b** detail of product and edit quantity

during the scanning phase, content, weight, production, and expiry date. In addition, from the detail button, the user can update the quantity. User can search or delete a product from the product list as shown by Fig. 4b.

This view has the items in the order of the closest day of expiry, so the item, which is going to expire soon, will show up on top of the list. In the options, the user has a choice to add a new item, search for an item, and change the preferences for the application and exit from the application.

Shopping List

The shopping list screen is to display a suggested shopping list to the user so that the users need not to bother to know what to buy as shown in Fig. 5a. This list is generated automatically from the product list. When a product is finished or expired, it will move automatically to the shopping list. In addition, the user have the possibility to add a product to the shopping list manually. In the options, the user has a choice to add a new item, search for an item, print, share or delete the shopping list as shown by Fig. 5b.

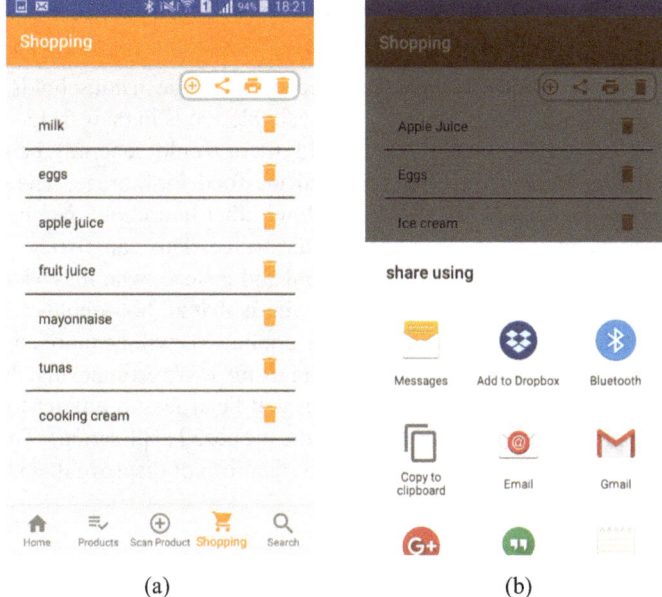

Fig. 5 **a** Shopping list. **b** Sharing the shopping list

5 Results and Discussion

Two challenges await the preliminary proposed application to prevent food lost and facilitate daily tasks.

5.1 The Use of Quick Response (QR) Code

An unexpected finding from the study showed how users interested by knowing details about food they buy. To store such details, we used the rapid response code that keeps information bigger well in a small space unlike barcode. The QR code can be read in a 360° direction to avoid interference. These features are attractive to the user, which saves time and effort. Therefore, the application is relied on products containing the QR code, which provides ease of use for the user. Given that this study employed a relatively small sample pool, this theme needs further exploration.

5.2 Impact on the Awareness of Available Food in the Fridge

The main dilemma currently leading to expired food wastage in households is that not all household members have knowledge of available foods in their fridge at a given moment. This is because, within a household, there would generally be a primary purchaser of food who often solely arranged the food for storage. The searching function used in the proposed application allows other household members to gain actionable knowledge of foods available in the fridge. This can also be applied to members who may not reside in the household and instead, who may visit and still require use of the refrigerator. However, our aim is that all households will show a noticeable increase in awareness. Users of the application will be more conscious of when particular foods would expire than before using it. We estimate that the amount of food waste using the proposed application will be at least a quarter to a half of what they had previously produced (without the proposed application). These initial assumptions will need to be evaluated to either confirm or disprove them.

5.3 Effectiveness in Reducing Expired Food Waste

This study's intention was to promote reductions of expired food waste by increasing the awareness of available foods for everyone residing in the household in question. We estimate that users will minimize the amount of food that would expire in the fridge. Further exploration and analysis are required in order to obtain the extent of the impact of the proposed application in raising the awareness of available foods for all. Members of a household and in return reducing the amount of expired food waste in the household for a significant period beyond their participation in the study.

6   Conclusion and Future Work

We have presented a novel android application that makes people more up to date on their products existing in the fridge. The proposed application enables customers to track their product and sort them in a product list with an easy way. Just they need to scan the QR code that appears on the product, which contains the necessary information about the product like expiry date; then, the application will remind customers before the expiry date, to avoid the waste of food.

We aim to add more feature to this application as to record frequent products of the user and make them as a favorite, to enable the user to customize the notifications to be sent in the time user prefers and to enable the user to sort the product in the product list by name or any other criteria. Displaying calories for various foods, calculating body mass index (BMI), Nutrition suggestion based on body mass index

(BMI) and for the various user with illnesses such as diabetes, high blood and/or pressure.

References

1. Guo, X., Broeze, J., Groot, J., Axmann, H., Vollebregt, M.: A global hotspot analysis on food loss and waste and associated greenhouse gas emissions. CCAFS Working Paper No. 290. Wageningen, The Netherlands: CGIAR Research Program on Climate Change, Agriculture and Food Security (CCAFS) (2019)
2. Haimi, S.U.: System and method for tracking shelf-life and after-opening usage life of medicaments, foods and other perishables. U.S. Patent No. 10,325,241, 18 June 2019
3. Krishna, M.B., Verma, A.: A framework of smart homes connected devices using Internet of Things. In: 2016 2nd International Conference on Contemporary Computing and Informatics, pp. 810–815. Noida (2016)
4. Sami, M., Atalla, S., Hashim, K.F.B.: Introducing innovative item management process towards providing smart fridges. In: 2019 2nd International Conference on Communication Engineering and Technology (ICCET). IEEE (2019)
5. Black, K.: What is a smart refrigerator? Conjecture Corporation [Online]. Available: https://www.wisegeek.com/what-is-asmart-refrigerator.htm (2017)
6. Aizawa, K.: FoodLog: Multimedia food recording platform and its application. In: Proceedings of the 5th International Workshop on Multimedia Assisted Dietary Management (MADiMa'19), p. 32. Association for Computing Machinery, New York, NY, USA (2019)
7. Pérez, J.: Before: don't let it expire [Online]. Available: https://itunes.apple.com/us/app/before-dont-let-it-expire/id971420683?mt=8 (2016)
8. Corgi, I.: Fridge, foods, expiration date [Online]. Available: https://play.google.com/store/apps/details?id=jp.gr.java_conf.indoorcorgi.mykura (2016)
9. Lindo, A.: Expiry control [Online]. Available: https://play.google.com/store/apps/details?id=com.expirycontrol (2017)
10. Keluskar, K.: Fridge-O-matic. Pramana Res. J. (2019)
11. Khan, T.: A cloud-based smart expiry system using QR code. In: 2018 IEEE International Conference on Electro/Information Technology (EIT). IEEE (2018)

Toward a Real-Time Personal Protective Equipment Compliance Control System Based on RFID Technology

El Mehdi Mandar, Wafaa Dachry, and Bahloul Bensassi

Abstract For many years, companies have been looking for ways to minimize workers' exposure to security hazards. One way of doing that is by ensuring that workers are at all working times wearing the necessary personal protective equipment (PPE) needed for the assigned task. To this end, in this paper, we present in this work, a real-time PPE compliance control system based on radio-frequency identification (RFID) technology (RTPPECCS). This system gives the possibility to ensure in real time that all workers present on the worksite are compliant with PPE safety policy. The system was developed while keeping in mind that safety systems are usually not stand-alone systems but come as add-ons that work with other management systems such as warehouse management systems (WMS) and enterprise resources planning (ERP) systems.

Keywords PPE · RFID · Personal protective equipment · RFID reader · RFID tags · Workplace safety · Worksite security · Compliance control

1 Introduction

Industrial environments are filled with heavy equipment and machinery. They are often intensely populated by workers that need to have a clear idea of how they can perform their task safely. These workers are often obliged to wear safety garments to lessen the effects of potential accidents. However, these safety garments feel unnatural to the workers, so they tend to take them off, which increases their exposure to

E. M. Mandar (✉) · B. Bensassi
Laboratory of Industrial Engineering, Information Processing and Logistics, Faculty of Science, Ain Chock, Casablanca, Morocco
e-mail: mandar.elmehdi@gmail.com

B. Bensassi
e-mail: Bahloul_bensassi@yahoo.fr

W. Dachry
Laboratory of Engineering, Industrial Management and Innovation, FST, Hassan I University, Settat, Morocco
e-mail: Wafaa.dachry@gmail.com

© The Editor(s) (if applicable) and The Author(s), under exclusive license to Springer Nature Singapore Pte Ltd. 2021
F. Saeed et al. (eds.), *Advances on Smart and Soft Computing*, Advances in Intelligent Systems and Computing 1188, https://doi.org/10.1007/978-981-15-6048-4_48

workplace risks. To solve this problem, companies started adopting the capabilities offered by information and communication technologies (ICT), namely the radio-frequency identification (RFID) technology, which is widely used in many industries and is known for its efficiency and cheap implementation cost. Many studies have covered the use of RFID technology in workplace safety management aimed to monitor and improve operators' safety. In this paper, we propose a system that presents higher PPE monitoring capabilities than preceding ones in a cost-effective way. Besides, the system's architecture allows its use as a stand-alone or add-on system. In other words, the proposed system can be used as an add-on, along with other systems.

This paper is organized as follows: in Sect. 2, we present the different systems conceived for the same purpose as the proposed system. In Sect. 3, we present the system architecture and modules. In Sect. 4, a PPE injury simulation is presented. Finally, in Sect. 5, we give a conclusion of the conducted work, and finally, we discuss the future perspectives of this work.

2 Literature Review

RFID is defined as a method of identifying unique items using radio waves. It consists of a reader (or an interrogator), which communicates with a transponder (radio-frequency identification, n.d.). This transfer of information happens without the devices making any actual physical contact and is used in items we use every day like car keys, employee identification cards, medical equipment, highway toll tags, and security access cards [1].

Although radio-frequency identification (RFID) technology has been around for some time, the use of the technology in supply chain management (SCM) and its associated operations is still being explored and adopted by many companies [2]. The RFID technology is a relatively cheap technology that can turn any object into a communicating object. This ability, coupled with systems processing capabilities, makes this technology even more appealing. In this work, we use these capabilities to help companies maintain their safety policies regarding the personal protective equipment (PPE). Many previous works discussed and proposed approaches and systems as solutions to personnel safety-related problems.

In [3], a system based on RFID technology was provided. The system assigned a unique identifier to each unique PPE. The identifiers were encoded in RFID tags embedded in PPEs. When scanned by an RFID reader, the user can get information about the PPE, the worker, and the state of the PPE by checking PPEs inspections history. This system, however, does not ensure PPE policy compliance.

In [4], a real-time RFID localization system was presented. The general purpose of the system is asset traceability. However, it also provides personnel safety capabilities. For instance, in the case of an accident, the rescue team would know on which floor employees are in with the help of RFID readers placed on each floor. The system cannot know the exact location of personnel wearing RFID tags; it can only know on which floor they are. Compared to the proposed system, this system provides less accuracy on the localization of employees. Furthermore, it does not consider the question of PPE.

In [5], a real-time RFID localization system was introduced. The purpose of the system is construction safety management. Each employee on the construction site wears a helmet with an embedded RFID tag. Multiple fixed readers on each floor determine the exact location of the employee using the least-squares method, which uses the signals exchanged between the tags and three readers to estimate the tag's localization. In case of accidents, this system gives greater intervention accuracy than the system presented in [4] as well as the proposed system. However, it is only useful after an accident occurs, unlike the proposed system, which ensures that all employees present in the workplace are well equipped against security hazards.

In [6], an RFID system for checking PPE compliance at the entrance of worksites was presented. The system makes it possible to read RFID tags embedded in employees' PPE at the entrance of worksites. The system aims to verify that each operator has the necessary PPE before granting or denying access to the construction site as well as time-stamping employees' entries and exits. However, this system solves only the problem of employees entering the site without their PPEs since they can take them off after entering with no reaction from the system, unlike the proposed system in this work, which ensures PPE compliance at all times and can also change compliance requirements depending on the area of the workplace or the employee's assigned task.

In [7], a system very similar to the one presented in this work was introduced. However, the system uses miniaturized Bluetooth Low Energy (BLE) as a means of communication between a BLE microsensor embedded on workers PPE and a wearable device capable of sensing if the worker is wearing the necessary PPEs. The BLE microsensor includes a microchip with a programmable processor and a power supply. In comparison with the RTPPECCS, this system provides almost the same possibilities; however, the hardware needed for it is more expensive as the proposed system works with PDAs mountable RFID readers. Also, the system needs a dedicated wearable device that does not integrate with other systems that use PDAs or other devices. In addition, this system does not provide SOS alerts or danger zones alerting capabilities.

In [8], an RFID-based PPE monitoring system was presented. It consists of RFID tags embedded in the PPEs in addition to fixed RFID readers. Each time a worker passes near an RFID reader, the systems check if the worker is wearing the necessary PPEs. In [9], a similar system consists of RFID readers fixed at predefined work areas entrances, a sensor embedded on the helmet, ensuring its proper wear and RFID readers embedded on other PPEs.

However, operators can learn RFID readers' locations and remove PPEs when they are out of range. In [10], an image processing PPE compliance system was described. The system takes pictures of operators at entrances of work areas. The system processes the images and checks the presence, the positioning, and the appropriateness of the PPEs. However, this system, like the ones presented in [5, 10], only solves the problem of PPE compliance at the entrance of the worksite and not inside the worksite.

In this work, we propose a real-time personal protective equipment compliance control system with the help of RFID technology. The system aims to continuously monitor PPE compliance with the help of RFID tags placed on PPEs and RFID readers mounted on workers' personal digital assistants (PDA).

3   A Real-Time Personal Protective Equipment Compliance Control System Based on RFID Technology

The protection of assets and personnel, as well as sustainability, is prime objectives of modern industrial enterprises [5]. To be able to reach higher performance levels, each process and each workspace must be safe. To this end, companies enforce workplace policies to ensure the safety of their human and non-human resources. An unsafe work environment can never reach its full performance as it is difficult for human resources to work to their fullest ability and concentration, knowing the existence of impending dangers. In this vision, the proposed system aims to achieve higher levels of workplace safety by enforcing safety PPE safety policies in real time. The system uses RFID technology to ensure that, at all times, every worker inside the workplace wears the necessary PPE. Each workplace environment can require a different set of PPE specific to the tasks and tools executed in that specific area. Each PPE is embedded with an RFID tag that contains an identifier unique to the operator wearing it to avoid problems of operators wearing each other's PPE. Each operator possesses a wearable personal digital assistant equipped with an RFID reader which reading range is adjusted to read only the tags present on the body and the personal space of the operator. Figure 1 outlines the architecture of the proposed system.

3.1   Data Collection Layer

This layer is composed of RFID readers mounted on operators' connected PDAs, RFID tags, and a Wi-Fi router.

The RFID reader mounted on the PDA reads all RFID tags on the body of the operator and sends this information to the system via Wi-Fi. The system stocks and then processes that information to formulate a reply.

The RFID tags are embedded in the PPE (helmets, goggles, gloves, boots, high visibility vests, and other protection equipment). At each moment, the operator must wear all the necessary equipment to perform the assigned tasks.

The decision to have portable handheld readers instead of fixed ones is backed by the fact that this kind of system is generally not a stand-alone system. It generally comes as a part of a more comprehensive system managing the whole workplace,

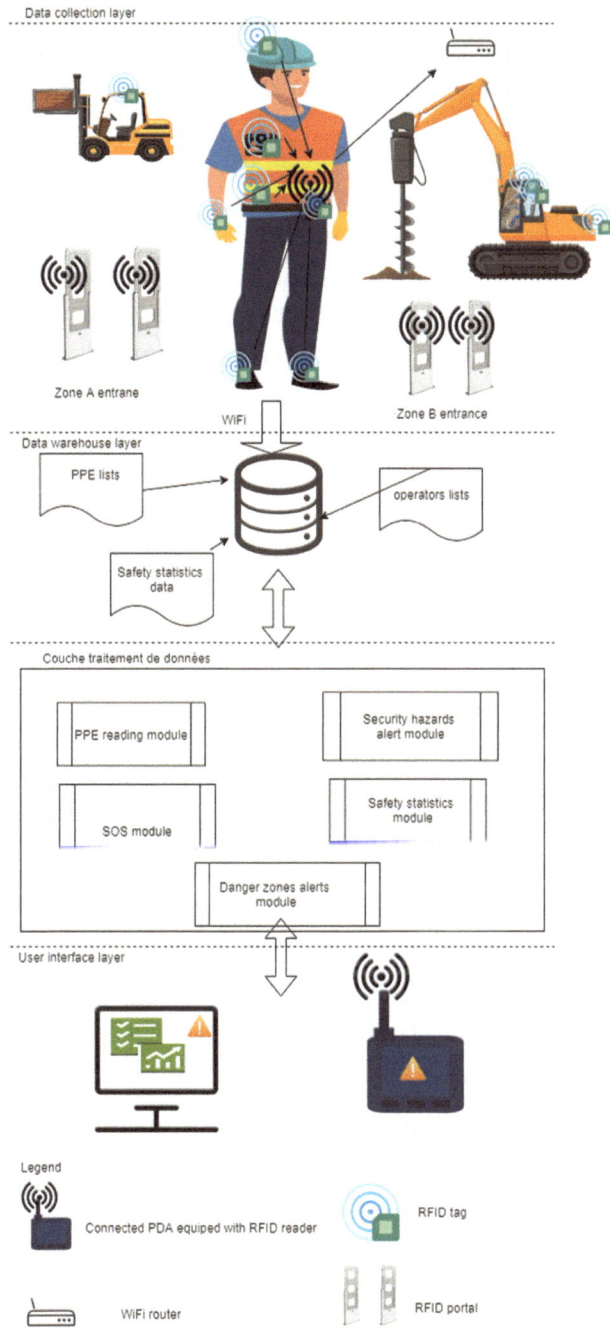

Figure 1 System's architecture

not just its safety aspect. That is why we chose RFID readers mounted on PDAs to give the possibility to the system to be used along with other management systems that require personnel to use PDA in their tasks, for instance, warehouse operators.

3.2 Data Warehouse Layer

This layer is composed of a database that holds all the information and data that flows within the system. The database contains two types of data:

– Static data: PPE lists, operators' lists, RFID tags, and RFID readers.
– Dynamic data: Safety policy infringements alerts' statistics (number of safety policy infringement and their gravity, by individuals, teams, and zones).

3.3 Data Processing Layer

This layer is responsible for processing data available in the data warehouse layer. It is the core of the system and contains multiple modules.

a. PPE reading module
   The objective of this module is to increase personnel safety in case of accidents. The RFID reader mounted on the PDAs must at all times read the entirety of the RFID tags present in the PPE and badge unique to the operator. After the operator enters specific workplaces, this module performs a series of checks to determine if the operator is wearing the necessary PPE. If it is not the case, an alert is displayed on the screen of the PDA, précising the missing element or elements of PPE. Each alert comes with a different sound for a different type of PPE missing in order to minimize the time checking what type of PPE the operator is missing.
   The system also uses signal strength between the tag and the reader to estimate the distance between them in order to check for the correct wearing of PPE. For instance, helmets and boots should not be too close to the RFID reader normally placed at the operator's hip.
   If the RFID reader reads other tags like SKU tags, stock placement tags, vehicle tags, machine tags, or other operators' tags momentarily in addition to the operators' own tags, this module displays no alerts considering those readings as unwanted or noise. However, this module keeps track of these unwanted readings to determine the approximate location of the operator. Figure 2 describes the PPE reading checks procedure.

b. SOS module
   This module is responsible for emitting SOS alerts, which can be triggered in two ways.

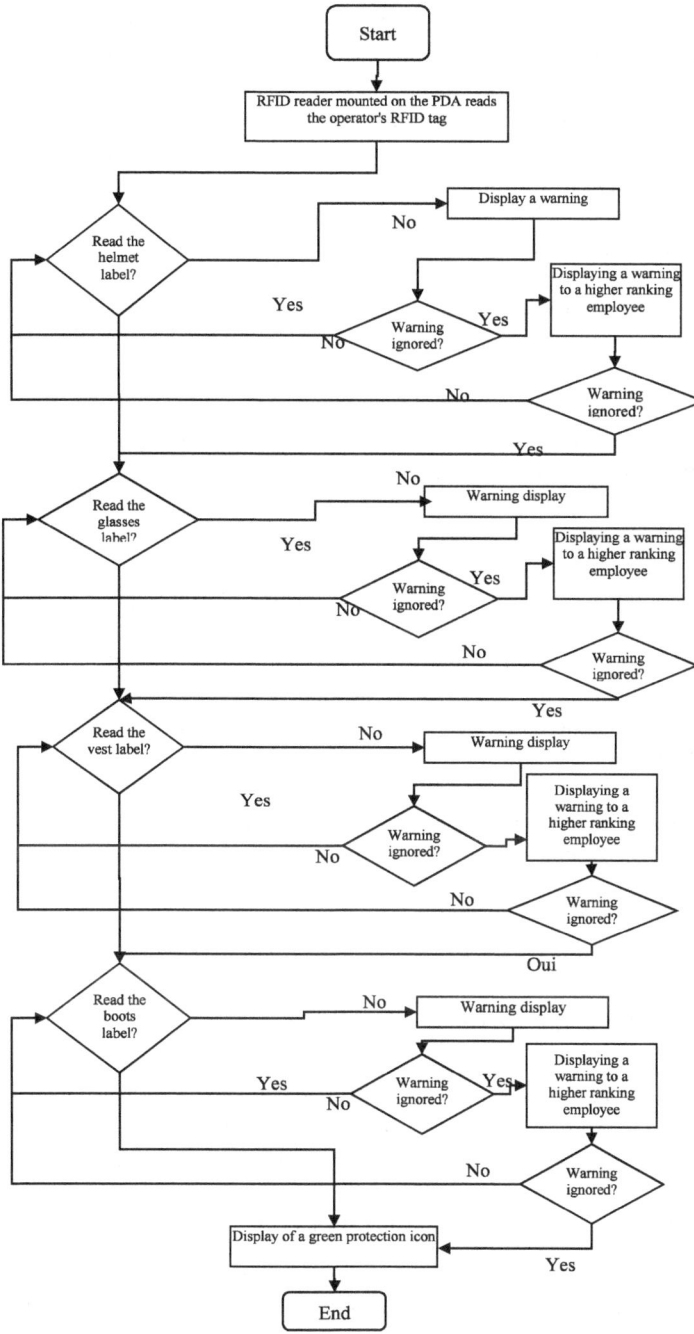

Fig. 2 PPE reading checks

The first way an SOS alert can be triggered is when the RFID reader no longer reads one or more of the operator's RFID tags, and the corresponding alert is ignored.

The second way is by pressing the SOS button present on the PDA of the operator. In case of an accident, other operators can help their colleagues by pressing the SOS button in addition to reading the RFID tag of static equipment (stock placement tag, machine tag, and other equipment tags) with their RFID reader to lead the intervention team to the location of the accident.

c. Safety statistics module

This module keeps track of all the data collected by the RFID readers mounted on the personnel PDAs in order to extract meaningful information about personnel safety.

For instance, know which PPE is the subject of the most safety alerts can help the organization understand the reasons behind it. Furthermore, the organization can be guided with this information to find solutions or better protective equipment. These statistics can also help to identify which part of the work environment it is harder for personnel to comply with safety policies.

d. Security hazards alert module

This module lets personnel inform on security hazards present in the work environment.

In these cases, operators can press a dedicated button present on their PDAs in addition to issuing the approximate location of the security hazard by reading the RFID tag of static equipment. Personnel can also describe the security hazard and its level of urgency.

e. Danger zones alert module

This module is responsible for displaying alerts to workers when their RFID readers pick one or more RFID tags associated with equipment, zones, or machinery that workers are not permitted to enter or work with. This procedure is presented in Fig. 3.

3.4   User Interface Layer

This layer is responsible for the interfaces that each personnel gets when accessing the system. The proposed system is developed in a Web environment and can be accessed, after authentication, by any computer, PDA, and smartphone connected to the Internet.

The system proposes two primary interfaces:

-   The first one: accessible to everyone on in the workplace with the duty to comply with safety policies. It keeps track of individual safety information, such as policy infringement alerts number. Besides, in the middle of the interface is a space reserved for alerts display.
-   The second one: accessible to team leaders and higher-ups, shows individual, as well as collective safety policy infringements, alerts statistics.

Fig. 3 Danger zones alert procedure

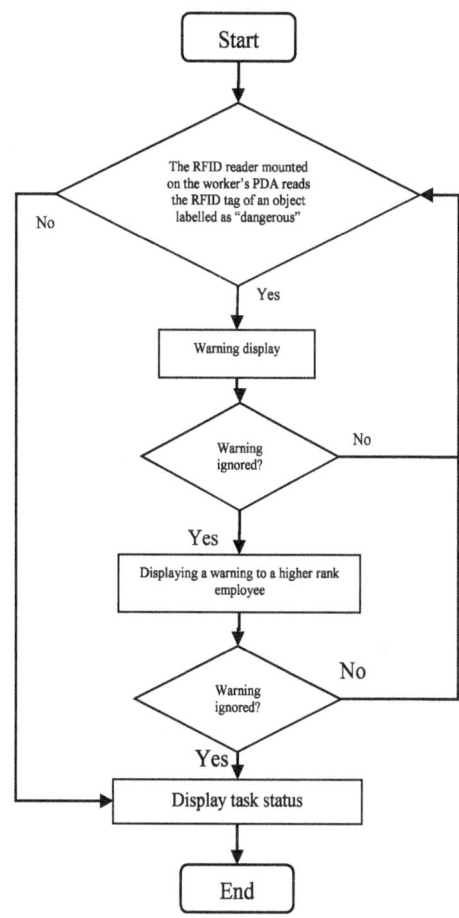

4 PPE Injury Simulation

In this section, a PPE injury scenario is given to demonstrate the proposed system potential to reduce accidents. In this simulation, only head injuries are considered. The scenario is based on the year 2012 in the USA, where 65,000 work head injury cases and 1020 head injury deaths were reported as in [11]. According to the Bureau of Labor Statistics, 84% of the workers who suffered head injuries were not wearing PPE. To better highlight the effectiveness of the compared systems, we consider in this simulation only 84% (54,600) of the 65,000 cases that were not wearing PPE. PPE compliance monitoring systems discussed in the literature review section can be categorized into four categories, as shown in Table 1. For each category, the accident reduction' rate is estimated.

Figure 4 shows the estimated head injury cases for each PPE compliance system category.

Table 1 Categorization of the safety systems discussed in Sect. 2

No compliance	Entrance compliance	Checkpoints compliance	All-time compliance
0% accident reduction	20% accident reduction	60% accident reduction	99% accident reduction
Method and system of managing the safety of a plurality of personal protection equipment items	Image recognition for personal protective equipment compliance enforcement in work areas	RFID and PPE: Concerning workers' safety solutions and cloud perspectives a reference to the Port of Bar	Worn personal protective equipment compliance system (BLE)
A passive RFID monitoring system	Mobile passive radio-frequency identification (RFID) portal for automated and rapid control of personal protective equipment (PPE) on construction sites		Real-time personal protective equipment compliance control system
RFID-based real-time locating system for construction safety management			

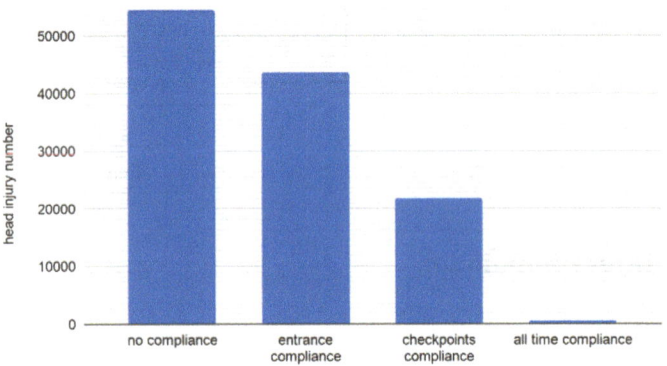

Fig. 4 Head injury number by system category

The graph above shows strong relativity between checking for compliance and compliance. The more compliance is verified; the higher the compliance is and the lesser the work injuries and deaths number. Figure 5 displays the number of deaths for each system category.

This simulation incorporates injuries statistics as well as their costs. In [11], head injury cost varies from $100,000 to $3,100,000 (3.1 million) while in [12], the average

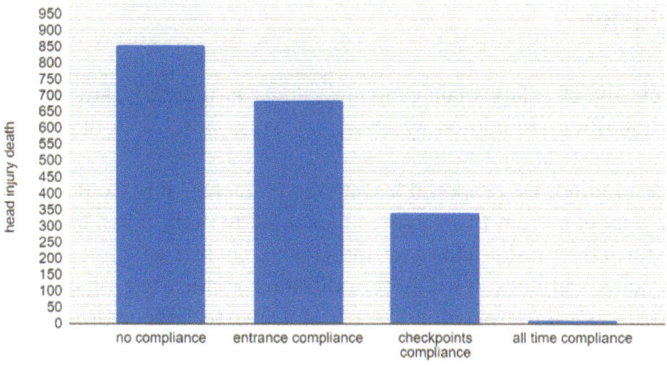

Fig. 5 Head injury deaths by system category

is estimated at $135,000. The graph in Fig. 6 shows head injury costs for each system category.

The graph above shows that even if all-time compliance systems may be more expensive than other systems categories, they can save millions of dollars for companies and show a great return on investment (ROI). The fact that all-time compliance systems head injury costs are just $73,710,000, compared to $2,948,400,000 for checkpoints compliance systems and $5,896,800,000 for entrance compliance systems, shows that companies willing to upgrade to their PPE compliance systems to all-time compliance systems can expect great ROI quickly considering that the cost of an all-time compliance system is estimated to range from $100,000 to $400,000 (based on other RFID systems costs that use the same components). Considering the average head injury cost of $135,000, companies can start seeing a return on investment when the system prevents one to three head injuries.

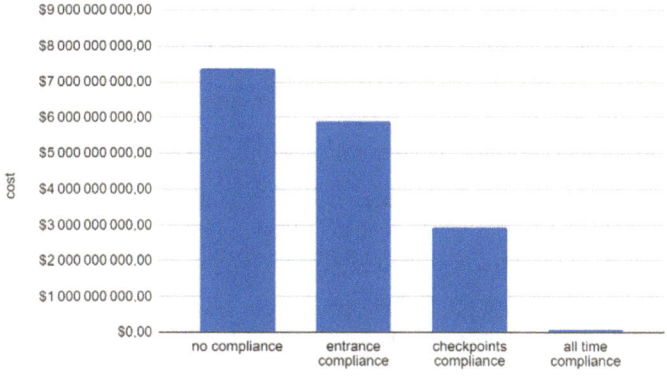

Fig. 6 Head injuries costs per system category

5 Conclusion

The particularity of safety management systems is that they are not stand-alone systems. They often are add-on systems that work along with other management systems.

With that in mind, we proposed a system architecture that would work with any other management system that requires personnel to carry a handheld or wearable PDA to perform their tasks. This way, the system can be easily integrated into other existing systems and does not require a different architecture of its own to work. The proposed system presents a new, economically feasible approach that ensures that personnel complies with safety policies not just at specific checkpoints of the workplace but at all times. Furthermore, it gives the possibility to monitor personnel safety in real time. The proposed system is in the prototyping phase. The future perspectives of this work will concern the system's implementation and test as well as some system improvements that aim to add more capabilities to the system as well as more versatility.

References

1. Radio Frequency Identification (RFID): What is it? U.S. Department of Homeland Security (2017). Retrieved from: https://www.dhs.gov/radio-frequency-identification-rfidwhat-it. Last accessed 07 Jan 2020
2. Smith, A., Manna, D.: RFID technologies and warehouse applications: case studies (2019)
3. Alan, N.A.: Method and system of managing the safety of a plurality of personal protection equipment items. U.S. Patent US10176347B2*, 8 Jan 2019
4. Costin, A., Pradhananga, N., Teizer, J.: Leveraging passive RFID technology for construction resource field mobility and status monitoring in a high-rise renovation project. Autom. Constr. 24, 1–15 (2012). https://doi.org/10.1016/j.autcon.2012.02.015
5. Lee, H.S., Lee, K.P., Park, M., Baek, Y., Lee, S.: RFID-based real-time locating system for construction safety management. J. Comput. Civ. Eng. 26(3), 366–377 (2012). https://doi.org/10.1061/(asce)cp.1943-5487.0000144
6. Kelm, A., Laußat, L., Meins-Becker, A., Platz, D., Khazaee, M.J., Costin, A.M., Teizer, J.: Mobile passive Radio Frequency Identification (RFID) portal for automated and rapid control of Personal Protective Equipment (PPE) on construction sites. Autom. Constr. 36, 38–52 (2013). https://doi.org/10.1016/j.autcon.2013.08.009
7. Obrien, J.: Worn personal protective equipment compliance system. U.S. Patent US2017/0206534A1, 20 July 2017
8. Bauk, S., Schmeink, A.: RFID and PPE: concerning workers' safety solutions and cloud perspectives a reference to the Port of Bar (Montenegro). In: 2016 5th Mediterranean Conference on Embedded Computing (MECO) (2016). https://doi.org/10.1109/meco.2016.7525756
9. Musu, C., Popescu, V., Giusto, D.: Workplace safety monitoring using RFID sensors. In: 2014 22nd Telecommunications Forum Telfor (TELFOR) (2014). https://doi.org/10.1109/telfor.2014.7034494
10. Au, K.W., et al.: Image recognition for personal protective equipment compliance enforcement in work areas. U.S. Patent US9695981B2, 4 July 2017

11. The Hard Truth about Safety Helmet Injuries and Statistics. https://www.hexarmor.com/posts/the-hard-truth-about-safety-helmet-injuries-and-statistics. Last accessed 06 Feb 2020.
12. PPE: Head, shoulders, knees and toes, Sandy Smith, 20 Nov 2015. https://www.ehstoday.com/ppe/article/21917224/ppe-head-shoulders-knees-and-toes-infographic. Last accessed 06 Feb 2020

Outer Weighted Graph Coloring Strategy to Mitigate the Problem of Pilot Contamination in Massive MIMO Systems

Abdelfettah Belhabib, Mohamed Boulouird, and Moha M'Rabet Hassani

Abstract Massive MIMO technology has emerged in the last decade as one of the most elected/popular keys proposed to enhance the quality of wireless communication within multicell massive MIMO systems. However, this technology suffers from a harmful constraint called the pilot contamination problem (PCP). Due to the scarcity of the available pilot resources, the same set of pilot sequences is reused across different cells, which directly leads to generate PCP upon users that reuse the same pilot resources. Thus, to overcome this constraint, the present paper proposes a novel decontaminating strategy, which is called the outer-weighted graph coloring strategy (OWGC). Specifically, users across each cell are firstly separated into two groups (i.e., inner and outer users) using a threshold of division, which is based on the large-scale fading coefficients. Therefore, inner users of all cells are obliged to reuse the same set of pilot sequences, while outer users are assigned with pilots based on a graph coloring strategy.

Keywords Massive MIMO systems · Pilot contamination problem · Weighted graph coloring

1 Introduction

Under the constraint of the scarcity of pilot resources, the requirement to improve the spectral efficiency in multicellular systems remains as a target to fulfill. Accordingly,

A. Belhabib (✉) · M. Boulouird · M. M. Hassani
Instrumentation, Signals and Physical Systems (I2SP) Group, Faculty of Sciences Semlalia, Cadi Ayyad University, Marrakesh, Morocco
e-mail: abdelfettah.belhabib@edu.uca.ac.ma

M. Boulouird
e-mail: m.boulouird@uca.ac.ma

M. M. Hassani
e-mail: hassani@ucam.ac.ma

M. Boulouird
National School of Applied Sciences, Cadi Ayyad University, Marrakesh, Morocco

F. Saeed et al. (eds.), *Advances on Smart and Soft Computing*, Advances in Intelligent Systems and Computing 1188, https://doi.org/10.1007/978-981-15-6048-4_49

it is necessary to find out some solutions to this issue without needs to either deploying more base stations (BS) or to scarify with many pilot resources. Hence, massive MIMO (MM) technology was proposed to achieve that goal, which is mainly based on increasing the number of antennas deployed in BSs (i.e., hundreds) [1]. Therefore, many users can be served simultaneously, which can enhance the per cell achievable rate [1]. Specifically, as higher the number of antennas deployed in BSs, as bigger the degrees of freedom that BSs get. In addition, the focus of the beam toward the desired user becomes narrow and can reach its target accurately. However, MM technology has some issues, e.g., pilot contamination problem (PCP), which is the main result of reusing the same set of pilots across several cells. This problem was considered as one of the biggest problems that limit the expected performance of MM technology [2].

2 Related Works

To mitigate this problem of PCP, potential works have been carried out since the appearance of MM [3]. Some techniques consider that this problem can be mitigated through shifting different phases of communication across adjacent cells; however, the estimated channels are contaminated by the data transmitted in adjacent cells instead of being contaminated by pilots [4, 5]. In the work of [6], a soft pilot reuse strategy was considered, which proposes to wisely share the available pilot resources between cells, and each cell can partially benefit from a set of orthogonal pilots; thus, the problem of PCP can be reduced; however, either fewer users can be served or more pilot resources are required to allocate orthogonal pilots to each cell, which can shrink the bandwidth left for data transmission. Regarding the constraint of the scarcity of pilots resources, Mochaourab et al. [7] propose a cooperative strategy based on some coalitions between BSs; hence, BSs of the same coalition can benefit from extra pilot resources and enhancing the per cell achievable rate. Nevertheless, this strategy assumes that users of different cells benefit from the same signal quality [i.e., signal-to-interference-plus-noise ratio (SINR)]. In [8], a graph coloring theory had been exploited to reduce the PCP upon users that reuse the same pilot sequences, and in other words, the strength of interference—between users that employ the same pilot sequences—is computed; afterward, users are allocated with the optimal pilots that would generate the tiniest strength of interferences upon them; however, as the number of served users becomes large compared to the available pilot resources, this decontaminating strategy tends to perform as a random pilot assignment strategy. Stand from that point, and regarding the fact that users within cells benefit from largely unequal performances, and to provide equally performances to all users. Specifically, users who are close to BSs benefit from high SINR compared to edge users, which are facing severe PCP. Hence, our proposed strategy first separate users of each cell into two groups labeled: inner and outer users. Therefore, an outer-weighted graph coloring (OWGC) strategy is proposed to enhance the SINR of outer users.

The remaining of this paper is organized as follows: Sect. 3 presents the system model adopted herein. Section 4 sheds light on the different steps of communication within multicell MM systems. The proposed OWGC is presented in Sect. 5. In Sect. 6, simulation results comparing the proposed OWGC, the conventional pilot assignment-based weighted graph coloring (WGC-PA) [8], and the conventional strategy (CS) [1] are presented. Finally, a conclusion is presented in Sect. 7.

Throughout the present paper, $(.)^H$, $\mathcal{C}$, $\mathcal{E}(.)$ denote, respectively, the transpose conjugate, the set of complex numbers, and the expectation symbol.

3   System Model

Let us adopt a multicell MM system, where each cell has an hexagonal shape, and it is equipped with a centered BS of M antennas serving K single-antenna users. The K users are randomly distributed within each cell. The independent identically distributed (i.i.d.) Rayleigh fading channel model [9] is considered herein. Therefore, the channel model between a user k in the jth cell to a BS of the lth cell is expressed as:

$$h_{l,k,j} = g_{l,k,j}\sqrt{\beta_{l,k,j}} \tag{1}$$

where $g_{l,k,j} \in \mathcal{C}^{M \times 1}$ denotes the small-scale (i.e., fast fading) fading (SSF) vector. Notice that the SSF vectors of different users are considered to be i.i.d., and each vector follows a complex normal Gaussian distribution of zero-mean and a unit variance, i.e., $g_{l,k,j} \sim \mathcal{CN}(0, I_M)$. $\beta_{l,k,j}$ denotes the large-scale (i.e., slow fading) fading (LSF) coefficient, which modelates both path loss and shadow fading. And it can be expressed [6] as:

$$\beta_{l,k,j} = \frac{z_{l,k,j}}{(r_{l,k,j}/R)^\alpha} \tag{2}$$

Here, $z_{l,k,j}$ accounts for shadowing, and it is a log-normal distribution between the user k of cell j to the BS of cell l. Moreover, $z_{l,k,j}$ obeys to a Gaussian distribution of zero-mean and a standard deviation σ_{shadow}. While $r_{l,k,j}$ denotes the distance between the considered pair user-BS and R is the cell radius, α denotes the path loss exponent.

4   Communication Scheme

Under the time division duplex (TDD), the communication between users and their BSs can be divided into frames [3] as it is shown in Fig. 1.

Hence, each frame is split into four specific subframes, which are (still referring to Fig. 1):

Fig. 1 Illustration of one
frame under TDD protocol

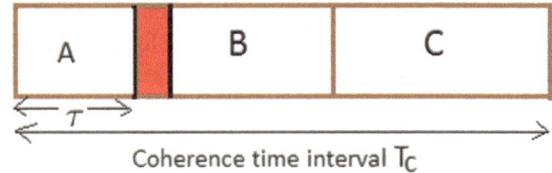

Coherence time interval T_C

1. Pilot transmission (i.e., A)
2. Channel estimation (i.e., between A and B)
3. Uplink detection (i.e., B)
4. Data precoding (i.e., C)

To reveal the main result of PCP, let us focus on the channel estimation phase. Hence, after the K users of all L cells uplink their pilot sequences of length τ (i.e., phase "A"), e.g., the pilot sequences of the kth user are $\Phi_k = [\phi_k^{[1]}, \phi_k^{[2]}, \ldots, \phi_k^{[\tau]}]$, where $||\Phi_k||^2 = 1$. Therefore, the received pilot signal $y_{l,k}^p \in C^{M \times \tau}$ from a user k at the BS of lth cell is expressed as:

$$y_{l,k}^P = \sqrt{\rho_p} \sum_{j=1}^{L} h_{l,k,j} \Phi_k + n_l \qquad (3)$$

where ρ_p denotes, respectively, the power used by users to transmit their pilots, while $n_l \in C^{M \times \tau}$ is an additive white Gaussian noise (AWGN) at the antennas BS of the lth cell.

Hence, BSs estimate the channel state information (CSI) through correlating their received pilot signals with the local pilot sequences. Thus, based on LS estimator, the estimated channel of the kth user of lth cell at the lth BS is:

$$\hat{h}_{l,k,l} = \frac{y_{l,k}^p}{\sqrt{\rho_p}} \Phi_k^H = h_{l,k,j} + \sum_{j=1, j \neq l}^{L} h_{l,k,l} + \frac{n_l \Phi_k^H}{\sqrt{\rho_p}} \qquad (4)$$

It is well recognized from (4) that the estimated channel is a sum of three components, which are, respectively, (from left to right) the desired channel, the contaminating channels, and the noise. Accordingly, the estimated channel in (4) is contaminated by the undesired channels of users that employ the same pilot sequence as the kth user of cell l. This problem is called PCP, and it is depicted in Fig. 2.

When BSs estimate the users' channels [i.e., (4)], BSs construct detectors and precoders, respectively, for data detection and data precoding. Thus, by considering the maximum ratio combining (MRC) detector, the $\text{SINR}_{l,k}$ of the kth user of the lth cell is given [6] as:

$$\text{SINR}_{l,k} = \frac{|h_{l,k,l} h_{l,k,l}^H|^2}{\sum_{j=1, j \neq l}^{L} |h_{l,k,j} h_{l,k,j}^H|^2 + |\epsilon_{1,k}|^2 / \rho_u} \xrightarrow{M \to \infty} \frac{\beta_{l,k,l}^2}{\sum_{j=1, j \neq l}^{L} \beta_{l,k,j}^2} \qquad (5)$$

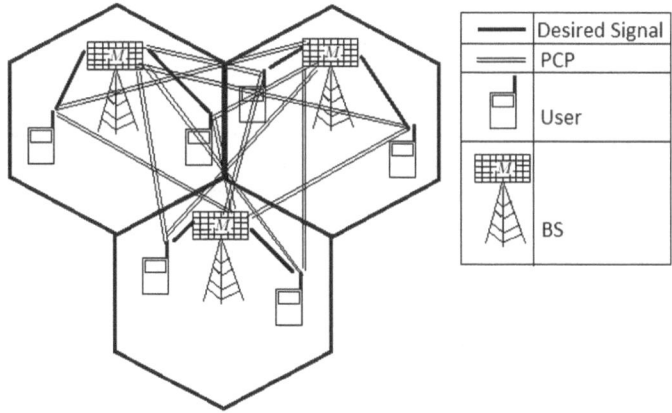

Fig. 2 Illustration of the PCP

where ρ_u denotes the uplink power transmission, while $\epsilon_{l,k}$ denotes the interference component. Therefore, the average uplink achievable rate in the lth cell is given as:

$$C_l^u = (1 - \mu) \sum_{k=1}^{K} \mathcal{E}(1 + \mathrm{SINR}_{l,k}) \tag{6}$$

Similarly to (5), and by adopting the maximum ratio transmitting (MRT), the per-cell downlink achievable rate can be expressed as follows:

$$C_l^d = (1 - \mu) \sum_{k=1}^{K} \mathcal{E}\left(1 + \frac{|h_{l,k,l}h_{l,k,l}^H|^2}{\sum_{j\neq l}^{L} |h_{j,k,l}h_{j,k,l}^H|^2 + |\epsilon_{l,k}|^2/\rho_u}\right) \tag{7}$$

where $0 < \mu = \tau/T_c < 1$ denotes the parameter that evaluates the loss of spectral efficiency.

5 Proposed Strategy

To enhance the quality of service for all users across cells, users are firstly split into inner and outer users based on their LSF coefficient [6] as follows:

$$\beta_{l,k,l}^2 > \rho_l : \begin{cases} \text{Yes} & k \text{ is an inner user} \\ \text{No} & k \text{ is an outer user} \end{cases} \tag{8}$$

where ρ_l denotes the threshold of the lth cell, and it can be expressed as:

$$\rho_l = \frac{\lambda}{K} \sum_{k=1}^{K} \beta_{l,k,l}^2 \tag{9}$$

Here, $0 < \lambda < 1$ is an adjusting parameter that depends on the system configuration. Therefore, inner users of all cells are randomly assigned with the same set of pilot sequences (i.e., full pilot reuse), and the number of pilot sequences K^c required for inner users is computed as follows:

$$K^c = \max\{K_i^c / i = 1, 2, \ldots, L\} \tag{10}$$

where K_i^c denotes the number of inner users in the ith cell. Hence, inner users are assigned with their pilots. For the outer users, the strength of PCP $\zeta_{\langle j,k \rangle \langle l,k' \rangle}$ upon two outer users of different cells (i.e., $\langle l, k' \rangle$ and $\langle j, k \rangle$) which are employing the same pilots can be computed as [8]:

$$\zeta_{\langle j,k \rangle \langle l,k' \rangle} = \frac{\beta_{\langle l,k' \rangle, j}^2}{\beta_{\langle j,k \rangle, j}^2} + \frac{\beta_{\langle j,k \rangle, l}^2}{\beta_{\langle l,k' \rangle, l}^2} \tag{11}$$

Thereafter, (11) computes the strength of interference generated between the user k of cell j (i.e., $\langle j, k \rangle$) and the user k' of cell l (i.e., $\langle l, k' \rangle$), and this can be expressed mathematically through considering a graph $G = (E, V)$, where V denotes the vertex of G and it denotes all outer users, while E is the edges of G and it denotes the strength of PCP across outer users Fig. 3. V and E can be expressed for $\{(k, k') = 1, \ldots, K\}$, $\{l = 1, \ldots, L\}$ as follows:

$$\begin{cases} V = \left\{ \langle j, k \rangle : \left\{ k : \beta_{j,k,j}^2 \leq \rho_j \right\} \right\} \\ E = \left\{ \zeta_{\langle j,k \rangle \langle l,k' \rangle} : k \neq k' \text{ and } j \neq l \right\} \end{cases} \tag{12}$$

Fig. 3 Illustration of $G = (E, V)$ in case of $K = 1$ and $L = 4$

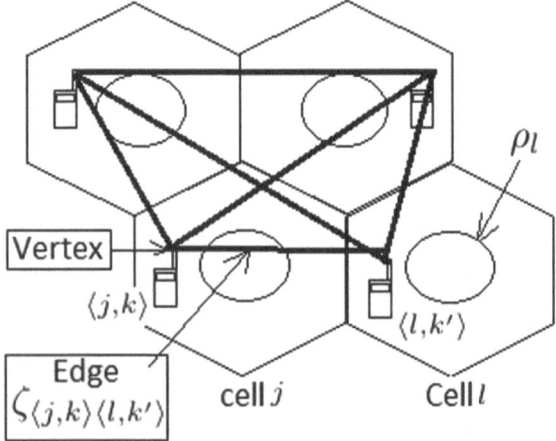

Therefore, the strength of PCP generated between outer users that reuse the same pilots is minimized through the algorithm given in [8] restricted only to outer users (i.e., $K_i^o = K - K_i^c$ for $i = 1, \ldots, L$). The complexity of our proposed strategy is on the order of $O(S(K^oL)^3)$, which is less complicated compared to the conventional WGC-PA that reaches $O(S(KL)^3)$.

6 Simulation Results

In this section, we adopt the system model described in Sect. 3. Therefore, a set of Monte-Carlo simulation is considered to evaluate the performance of our proposed strategy compared to both the CS and the WGC-PA strategies. The parameters are fixed according to [8] as it is depicted in Table 1.

On one hand, Fig. 4 shows the average uplink achievable rate against the parameter λ for: the CS, the WGC-PA, and the proposed OWGC-PA applied for a set of $K = 16$ users and $S = 16$. It is well seen that the performances of the proposed strategy outperform those of both CS and WGC-PA by about 1 bps/Hz and 1.5 Bps/Hz, respectively.

On the other hand, Fig. 5 depicts the average downlink achievable rate against the grouping parameter λ. It is well seen that the proposed strategy performs better than both other strategies (i.e., CS and WGC-PA) by about 8 bps/Hz and as we consider that users tend to be all in the center (i.e., $\lambda \to 0.9$), the proposed strategy loses about 5 bps/Hz. Nevertheless, it is still performing better than the two other strategies.

Figure 6 shows the average uplink achievable rate against the number of antennas deployed in BSs for $K = S = 16$. It is well recognized that the proposed strategy performs better than CS and WGC-PA until about $M = 256$ where WGC-PA takes back to perform better than the proposed strategy. Hence, WGC-PA needs a large number of antennas to win against our proposed strategy.

Table 1. System settings

Number of cells	$L = 7$
Number of antennas BS	$32 \leq M \leq 256$
Number of users per cell	$10 < K \leq 16$
Number of pilot symbols S	$K < S \ll LK$
Cell radius	$R = 500$ m
Inner radius	$r = 30$ m
Adjustement parameter	$0 < \lambda < 1$
Log normal shadow fading	$\sigma_{shadow} = 8$ dB
The transmited power between users-BSs	$\rho_p = \rho_u = \rho_d = 15$ dB m
Path loss exponent	$\alpha = 3$
Pilot overhead parameter μ	$\mu \to \mu \frac{S}{K}$, fixed $\mu = 0.05$

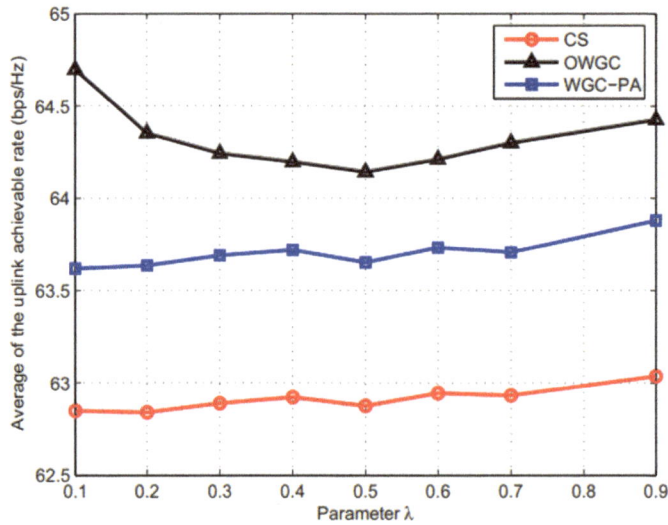

Fig. 4 Average uplink achievable rate against λ, $K = S = 16$, $\lambda = 0.1$

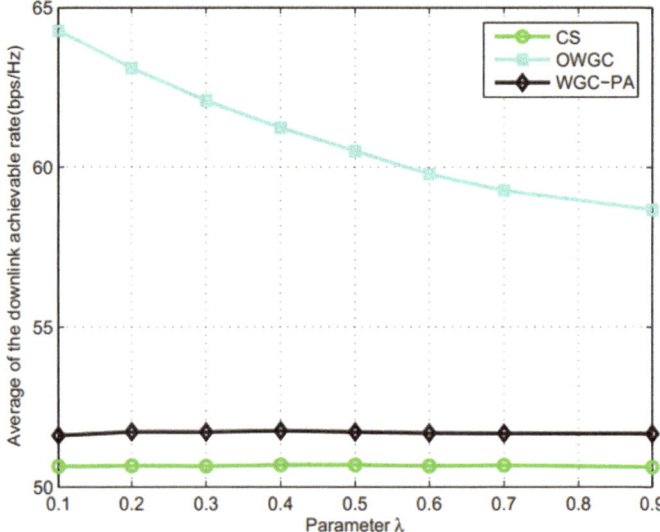

Fig. 5 Average downlink achievable rate against λ, $K = S = 16$, $\lambda = 0.1$

Figure 7 depicts the average downlink achievable rate against M. We observe that the proposed strategy outperforms both strategies (i.e., CS and WGC-PA) by about 8 bps/Hz for $M = 32$, and the gap increases to reach about 15 bps/Hz for $M = 256$.

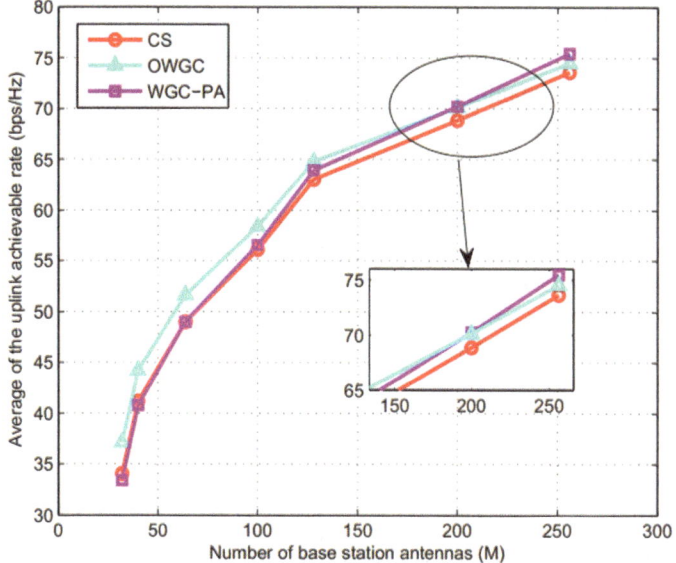

Fig. 6 Average uplink achievable rate where $K = S = 16$, $\rho_p = \rho_u = 15$ dB

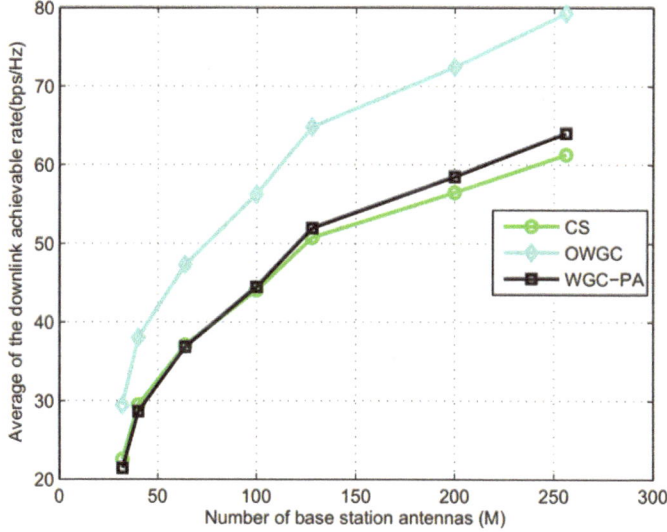

Fig. 7 Average downlink achievable rate where $K = S = 16$, $\rho_d = 15$ dB m

7 Conclusion

Regarding the scarcity of pilot resources and the problem of pilot contamination, this paper proposes an enhanced decontaminating strategy, which is called the OWGC strategy. OWGC assumes that users within cells must be considered separately. Specifically, inner users enjoy a higher quality of service compared to inner users, which are directly facing the problem of PCP. Hence, OWGC separates users within cells into two groups of users (i.e., inner and outer users), and afterward, inner users of all cells are assigned with the same set of pilots, while outer users are served based on the WGC-PA strategy. Through simulation results, the proposed OWGC strategy outperforms both CS and WGC-PA strategies.

References

1. Marzetta, T.L.: Noncooperative cellular wireless with unlimited numbers of base station antennas. IEEE Trans. Wirel. Commun. **9**(11), 3590–3600 (2010)
2. Rusek, F., et al.: Scaling up MIMO: opportunities and challenges with very large arrays. IEEE Signal Process. Mag. **30**(1), 40–60 (2013)
3. Elijah, O., Leow, C.Y., Rahman, T.A., Nunoo, S., Iliya, S.Z.: A comprehensive survey of pilot contamination in massive MIMO 5G system. IEEE Commun. Surv. Tutor. **18**(2), 905–923 (2016)
4. Mahyiddin, W.A.W.M., Martin, P.A., Smith, P.J.: Pilot contamination reduction using time-shifted pilots in finite massive MIMO systems. In: IEEE 80th Vehicular Technology Conference, pp. 1–5. VTC2014-Fall, Vancouver, BC (2014)
5. Jin, X., Wang, J., Wang, Y.: Improved soft pilot reuse combined with time-shifted pilots in massive MIMO systems. In: 2018 IEEE 87th Vehicular Technology Conference, pp. 1–5. VTC Spring, Porto (2018)
6. Zhu, X., et al.: Soft pilot reuse and multicell block diagonalization precoding for massive MIMO systems. IEEE Trans. Veh. Technol. **65**(5), 3285–3298 (2016)
7. Mochaourab, R., Bjornson, E., Bengtsson, M.: Adaptive pilot clustering in heterogeneous massive MIMO networks. IEEE Trans. Wirel. Commun. **15**(8), 5555–5568 (2016)
8. Zhu, X., Dai, L., Wang, Z., Wang, X.: Weighted-graph-coloring-based pilot decontamination for multicell massive MIMO systems. IEEE Trans. Veh. Technol. **66**(3), 2829–2834 (2017)
9. Zheng, K., Ou, S., Yin, X.: Massive MIMO channel models: a survey. Int. J. Antennas Propag. **2014** (2014). Article ID 848071

A Comprehensive Study of Dissemination and Data Retrieval in Secure VANET-Cloud Environment

M. A. Al-Shabi

Abstract Over the recent decades, incorporating vehicular ad hoc network (VANET) into cloud computing plays a vital role, since it provides a reliable safety journey to vehicular drivers, passengers, etc. However, attaining security and emergency message dissemination is still major bottleneck in VANET combined Cloud, due to the dynamic nature of vehicles and wireless communication. This paper presents the motivation with research gaps in the VANET-cloud environment. Integration of VANET into cloud environment provides ubiquitous advantages to the users in terms of the high level of infotainment information storage and faster response to the user request. Most of the works presented in the VANET-cloud environment provides security by authenticating vehicles only. However, there is also a chance to deploy malicious RSU on the roadsides. As yet, there is a research gap in authenticating roadside unit (RSU) in order to provide high level security to VANET-Cloud communication. And also, there is a lack in providing security to the user credentials stored in the cloud.

Keywords Vehicular ad hoc network-cloud · RSU · Hash-based authentication

1 Introduction

With the success of VANET technology among vehicular consumers focuses on the excellence of comfort and realism of safety provided by this technology [1]. With the progress of technology and sudden evolvement in vehicles lead to certain issues in VANET, due to flexibility, scalability and poor connectivity. To overcome these challenges, VANET is integrated into cloud computing. Cloud provides ubiquitous services and storage to vehicular users [2]. However, maintaining security in vehicular integrated cloud network becomes difficult.

In regard to conserving security cloud-based security and privacy information, dissemination scheme is introduced. Herein, identity-based signature (IBS) with

M. A. Al-Shabi (✉)
Department of Management Information System, College of Business Administration,
Taibah University, Medina, Saudi Arabia
e-mail: mshaby@taibahu.edu.sa

© The Editor(s) (if applicable) and The Author(s), under exclusive license to Springer 577
Nature Singapore Pte Ltd. 2021
F. Saeed et al. (eds.), *Advances on Smart and Soft Computing*, Advances in Intelligent
Systems and Computing 1188, https://doi.org/10.1007/978-981-15-6048-4_50

the pseudonym method is exploited to provide authentication to the vehicles [3]. Certificate-less pairing free encryption scheme (CCES) is introduced to provide authentication to vehicles. In this, signature-based information encryption and decryption are performed where the bilinear mapping method is used to generate keys for encryption and decryption [4]. A novel geo-location-based parking lock encryption mechanism is utilized to provide the security in VANET-cloud environment for location privacy and non-frameability [5]. In this, the Traffic Management Bureau (TMB) clustering of vehicles in the VANET environment reduces the high mobility of vehicles. The analytical model is utilized to cluster the vehicles in the network [6].

Link reliability-based clustering algorithm (LRCA) is exploited to provide efficient and reliable message transmission in VANET. LRCA-based clustering performs cluster head selection, cluster formation and cluster maintenance processes. Redundant unstable neighbors are filtered out using link lifetime-based sampling method [7]. Mobility prediction-based efficient clustering scheme (MPECS) is used in VANET where network regions are split through Voronoi diagram by using main intersections as centers of partitioned area. Using MPECS, each vehicle can predict its own longevity and cost for electing as cluster head [8]. Weighted k-medoid-based clustering approach is utilized to form clusters in vehicular network. In weighted k-medoid clustering, hybrid genetic algorithm is used to maintain the clusters where Tabu search is incorporated into genetic algorithm [9].

Sampling-based estimation scheme (SES) is introduced to route the packets from source to destination. SES method finds path using probabilistic contacts on overlapped road into multiple segments as samples. Each sample provides possible routing decision [10]. Bilateral Forwarder Determination method scheme is used to find best forwarder determination. Bilateral Forwarder Determination method (BFD) using in paper [11]. Hybrid relay selection scheme is used to disseminate emergency message among vehicles in VANET. Hybrid scheme considers spatial distribution of the next hop node with reference to the sending node. This relay selection scheme uses multiple criteria-based relay selection scheme to select group of relay nodes [12].

An adaptive data dissemination protocol-based emergency dissemination is proposed in vehicular network. This protocol proposes different mechanisms to dynamically adjust the beacon intervals [13]. Novel store and carry–forward scheme-based message dissemination is introduced in VANET. Here, SCF enabled vehicles only forward emergency message to the neighbor vehicles [14]. Time barrier-based emergency dissemination scheme is introduced in vehicular network. Herein, super node is selected initially in order to disseminate the emergency message. In this time barrier method, super node disseminates emergency message to the neighbor vehicles in timely manner [15]. Multi-channel coordination scheme is introduced for emergency dissemination in VANET. Multi-channel coordinator selection process is achieved using least average separation distance (LAD) to the vehicles where control channel is used for emergency dissemination [16].

From the above listed works, we have noticed that still there exist issues in VANET-cloud environment. The issues are summarized as follows:

- Providing security in VANET-cloud environment is difficult due to the frequent topology changes.
- Cluster formation in VANET-cloud environment is difficult due to the high mobility of the vehicle.
- Security is a major constraint in routing since most of the works does not concentrate on secure data transmission.
- Delay is high in emergency dissemination due to lack of parameter consideration in disseminator selection.

This paper presents the motivation with research gaps in the VANET-cloud environment. Integration of VANET into cloud environment provides ubiquitous advantages to the users in terms of the high level of infotainment information storage and faster response to the user request. The rest of this paper is structured as follows. Section 2 summarizes the related work regarding emergency message dissemination in VANET-cloud environment. Section 3 explains the problems that exist in the present VANET-cloud emergency dissemination. Section 4 concludes contribution of this work and comments with some future directions.

2 Related Work

In this section, we review works that are related in secure VANET-cloud, clustering, routing and emergency dissemination. A number of current studies explained and devised mechanism for secure combination of VANET and Cloud.

2.1 Secure VANET-Cloud

This section discusses the works that are related to the secure VANET-cloud environment. Kai et al. [17] have proposed secure and efficient privacy preserving cipher text retrieval in connected vehicular cloud computing. Herein, cipher text-based search system is implemented to exploit RSU as super peers for connected vehicular cloud computing. In this framework, all the computations and retrieval operations are handled by the super peers. In connected vehicular cloud computing, user documents are stored in the cloud to provide better security and high efficiency. Data retrieved vehicles get keys from source vehicles to decrypt the encrypted data. Herein, data are not stored securely, while data index is stored in hash form using SHA-1 algorithm that has less security compared to other SHA algorithm.

Yixian et al. [18] have proposed secure and efficient transmission method in connected vehicular cloud computing. The proposed connected vehicular cloud computing extends the fixed node of the traditional cloud to the mobile nodes. In this novel, computation approach is proposed to find the channel capacities of the multiple vehicular networks. Herein, secure and efficient transmission is achieved

Table 1 Merits and demerits of the secure VANET-cloud works

References	Merits	Demerits
Kai et al. [17]	High robustness in securing data	Security is provided only to the data index that tends to forging of data
Yixian et al. [18]	High reliability in securing cloud data	Security is not provided to the cloud data
Hamssa et al. [19]	Consumes less time in predicting malicious users	Absence of significant credentials consideration leads to poor security mechanism
Abderrahim et al. [20]	Introduces lightweight securing mechanism	Smart card-based authentication is used which is easily forgeable

through game theory and information theory. Nash equilibrium is executed in game theory in order to obtain optimal results. Cloud only sends emergency information to vehicles that are nearby accidental zone. This paper does not follow any security mechanism to ensure security of the stored information in cloud.

Hamssa et al. [19] have proposed trust model for secure group leader-based communication in vehicular environment. The proposed work consists of hybrid trust model (HTM) and misbehavior detection system (MDS). These systems are used to provide trust metric to each vehicle based on its behavior. Using this metric, this paper classifies the malicious from the normal nodes. The back end system is used to select the optimum group leader based on the higher trustworthiness. This work evaluates trust based on behavior that does not consider necessary metrics related to the misbehavior detection system.

Abderrahim et al. [20] have proposed security scheme in vehicular cloud network through lightweight encryption algorithm. Herein, each vehicular cloud users receives smart card by the vehicular cloud infrastructure, in order to verify the authentication of vehicle user. Message digest algorithm is used to generate hash values for user registered credentials such as ID, vehicle registration numbers. Authentication process execution depends on the random number generated. Since random numbers are generated at the time user registrations are integrated with the generated hash value by the vehicular cloud infrastructure, these credentials are verified in authentication process along with smart card. This paper uses smart card for authentication process that can be easily forged by the malicious users. Table 1 describes the merits and demerits of the secure VANET-cloud works.

2.2 Clustering

This portion discusses the works that are related to the clustering the vehicular cloud network. Lei et al. [21] have proposed clustering and probabilistic broadcasting using data dissemination scheme in VANET. In this, clusters are formed based on the direction of driving vehicles. Using clusters, each vehicle can transmit their data to the neighbor vehicle where cluster head is selected using link connectivity and packet successful transmission probability. Probabilistic forwarding approach is proposed to disseminate the data among the vehicles. Each cluster member forwards the received packet to its cluster head along with the computed probability that is associated with the number of times the same packet is received during one interval. Neighbors are discovered by broadcasting periodical beacons to its neighbors. Cluster head is selected based on two factors only which are not sufficient to form effective cluster in VANET environment.

Sarah et al. [22] have proposed secure and stable clustering algorithm based on hybrid mobility of vehicles and trust management scheme. New approach is proposed for trustworthy cluster head selection where stability and trust factors are considered. In trust management scheme, trust of data can be changed between vehicles and communication capabilities. In cluster formation step, each vehicle exchanges hello packets that contain ID, speed, acceleration and position. Cluster head is elected based on the computation of score where mobility metrics are considered. This work does not focus on security parameters in order to maintain security in vehicular environment.

Yasir et al. [23] have proposed Moth-Flame Optimization (MFO)-based clustering algorithm in VANET. In this work, clustering is performed using efficient clustering algorithm MFO which forms the clusters effectively. At first, network of the autonomous vehicles is generated by the randomly initializing their position with its respective region. The fitness function is calculated using the position of the moths in searching space. Based on the highest fitness function, cluster head is selected in the network. Cluster head selection is not effective, because it considers two metrics only such as distance and speed.

Yassine et al. [24] have proposed new communication model for VANET environment. New communication model considers road as a ring segment. New communication model forms clusters in order to maintain the stable routes during inter-cluster communication. In cluster head selection, candidate vehicle broadcasts hello message that contains ideal position and mobility metrics. Each vehicle provides vote to the neighbor vehicles based on probability value and ideal location. If there is no vehicle present in the ideal position, then vehicle near to the ideal position is selected for aggregation. Cluster head selection is based on the voting, and there are chances to select non-optimal vehicle as cluster head. Table 2 describes the merits and demerits of the clustering works.

2.3 Routing

This part deals with works that are associated with the routing in vehicular cloud network. Hui et al. [25] have proposed novel trust-based multicast routing for VANET. The proposed model computes two trust factors that include direct and

Table 2 Merits and demerits of the clustering works

References	Merits	Demerits
Lei et al. [21]	Effective data transmission	Cluster head selection is not effective due to lack of parameter consideration
Sarah et al. [22]	Considers QoS parameters while forming clusters that enhance the system performance	Consumes more bandwidth while selecting cluster head
Yasir et al. [23]	Consumes less time to form clusters	MFO-based cluster head selection cannot able to select an optimal head due to poor convergence
Yassine et al. [24]	Less process to form clusters in the network	Frequent formation of cluster due to ineffective head selection

indirect trust. The direct trust is computed based on the Bayesian theory and indirected trust is computed based on the evaluation credibility and activity. The fuzzy logic approach is used to discover the direct and indirect trust values. The total trust values of the nodes are obtained using defuzzification process. Based on the computed trust value, efficient routes are selected to transmit the data delivery. This paper finds route based on the trust factor only that leads to loss in transmitted packet.

Sha et al. [26] have proposed delay-aware relay selection in vehicular network. This paper proposes Greedy Parameter Stateless Routing (GPSR)-based routing protocol for data dissemination in the heterogeneous communication range. The neighbor table with asymmetric wirelength is established with the help of acknowledgement list and intermediate nodes. The vehicle with minimum expected transmission delay is selected as the next hop for the transmission. The expected transmission delay is combination of the expect delay and communication range. This factor is used to estimate the delay from the current packet carrying vehicle to destination. Relay node is selected based on the delay factor; hence, the routing is not effective in terms of transmission loss. Table 3 describes the merits and demerits of the routing works.

Table 3 Merits and demerits of the routing

References	Merits	Demerits
Hui et al. [25]	Increases scalability in data transmission	Transmission loss occurs while routing packet to the destination
Sha et al. [26]	High reliable data transmission	Lack of parameter consideration in routing induces packet losses

3 Emergency Data Dissemination

This section describes the works that are related to the emergency dissemination in vehicular network. Li et al. [27] have proposed reliable emergency message dissemination protocol for vehicular network. The proposed protocol consists of layout-aware ready to broadcast and clear to broadcast message handoff mechanism and redundant relay node adaptation mechanism. The emergency message broadcast is based on the two phases that are intersection broadcast process and road segment broadcast process. The back-off timers is set to the relay nodes to avoid overlap in time so that replies do not collide. By receiving the last clear to broadcast message, the data message transmission is commenced. Broadcast message-based dissemination leads to more delay in emergency message dissemination.

Daxin et al. [28] have proposed distributed position-based protocol for emergency message broadcasting in VANET. Herein, improved positio- based protocol is proposed to disseminate the emergency message among large-scale vehicular networks. Using the proposed protocol, emergency messages are only broadcasted to the interested region and rebroadcast of the messages depends on the information included in the message. Yusor et al. [29] have proposed accident management system based on vehicular network for intelligent transport system. Accident management system consists of three modules that are sensor module, speed monitor module and message and alert module. Herein, sensor module is used to monitor and control the sensors deployed in the network. Message and alert module is used to maintain the communication between RSU unit, ambulance and central server. Speed monitor module is used to measure the speed of each vehicle. The authors discussed in [28, 29] have more delay in emergency message dissemination, since it considered delay-oriented factors in emergency message dissemination. Table 4 describes the merits and demerits of the clustering works.

Table 4 Merits and demerits of the emergency dissemination works

References	Merits	Demerits
Li et al. [27]	Improves reliability in emergency data transmission	Broadcasting the emergency message induces broadcast storm
Daxin et al. [28]	Consumes less time to select disseminator	Delay is high while disseminating emergency message
Yusor et al. [29]	Increase scalability of the system	Emergency message dissemination is not effective due to lack of delay oriented percolation

4 Conclusion

The vehicular ad hoc network (VANET) into cloud computing plays a vital role, since it provides a reliable safety journey to vehicular drivers, passengers, etc. However, attaining security and emergency message dissemination is still major bottleneck in VANET combined cloud, due to the dynamic nature of vehicles and wireless communication. This paper presents a review of security in VANET-cloud environment which is difficult due to the frequent topology changes and presents cluster formation in VANET-cloud environment which is difficult due to the high mobility of the vehicle. The paper also explained that security is a major constraint in routing since most of the works did not concentrate on secure data transmission. We noted that the number of current studies explained and devised mechanism for secure combination of VANET and cloud.

References

1. Rasheed H., Zeinab R., Junggab S., Md Z. Alam B., Sangjin K., Heekuck O.: PB-MII: replacing static RSUs with public buses-based mobile intermediary infrastructure in urban VANET-based clouds. Cluster Comput. 20(3), 2231–2252 (2017)
2. Rakesh, S., Rojeena, B., Seung, Y.: Challenges of future VANET and cloud-based approaches. Wirel. Commun. Mobile Comput. 2018, 1–15 (2018)
3. Qamas, G., Senlin, L., Chao, W., Limin, P.G.Y.: Cloud-based security and privacy-aware information dissemination over ubiquitous VANETs. Comput. Stand. Interfaces 56, 107–115 (2018)
4. Qamas, G.K.S., Senlin, L., Limin, P.W.L., Guangluo, Y.: Secure authentication framework for cloud-based toll payment message dissemination over ubiquitous VANETs. Pervasive Mobile Comput. 48, 43–58 (2018)
5. Qamas, G.K.S., Senlin, L., Chao, W., Limin, P., Qianrou, C.: PIaaS: cloud-oriented secure and privacy-conscious parking information as a service using VANETs. Comput. Netw. 124, 33–45 (2017)
6. Raghavendra P., Arun P., Rajeev T., Dhananjay S.: Analytical model for clustered vehicular ad hoc network analysis. ICT Exp. 4, 160–164 (2018)
7. Xiang, J., Huiqun, Y., Guisheng, F., Huaiying, S., Liqiong, C.: Efficient and reliable cluster-based data transmission for vehicular ad hoc networks. Mobile Inf. Syst. 2018, 1–15 (2018)
8. Islam, T.A., Hossam, M.F., Ayman, M.B.: Mobility prediction-based efficient clustering scheme for connected and automated vehicles in VANETs. Comput. Netw. 150, 217–233 (2019)
9. Rejab, H., Eesa, A., Tarek, M., Hervé, G.: Construction of a stable vehicular ad hoc network based on hybrid genetic algorithm. Telecommun. Syst. 1–13 (2018)
10. Chao, S., Jie, W., Ming, L., Huanyang, Z.: Efficient routing through discretization of overlapped road segments in VANETs. J. Parallel Distrib. Computi. 102, 57–70 (2017)
11. Junling, S., Xingwei, W., Min, H., Keqin, L., Sajal, K.D.: Social-based routing scheme for fixed-line VANET. Comput. Netw. 113, 230–243 (2017)
12. Osama, R., Mohamed, Q.: A hybrid relay node selection scheme for message dissemination in VANETs. Fut. Gener. Comput. Syst. 93, 1–17 (2019)
13. Renöe, O., Carlos, M., Azzedine, B., Michelle, S.W.: Reliable data dissemination protocol for VANET traffic safety applications. Ad Hoc Netw. 63, 30–44 (2017)
14. Truc, D.T.N., Thanhm, V. L., Hoang-Anh, P.: Novel store–carry–forward scheme for message dissemination in vehicular ad-hoc networks. ICT Exp. 3, 193–198 (2017)

15. Syed, S.S., Asad, W.M., Rahman, A.U., Sohail, I., Samee, U.K.: Time BARRIER-based emergency message dissemination in vehicular ad-hoc networks. IEEE Access **7**, 16494–16503 (2019)
16. Odongo, S.E., Jhihoon, J., Dong, S.H.: CMD: A multichannel coordination scheme for emergency message dissemination in ieee 1609.4. Mobile Inf. Syst. **2018**, 1–13 (2018)
17. Kai, F., Xin, W., Katsuya, S., Hui, L., Yintang, Y.: Secure and Efficient Privacy-Preserving Ciphertext Retrieval in Connected Vehicular Cloud Computing. IEEE Netw. **32**(3), 52–57 (2018)
18. Yixian, Y., Xinxin, N., Lixiang, L., Haipeng, P.: A secure and efficient transmission method in connected vehicular cloud computing. IEEE Netw. **32**(3), 14–19 (2018)
19. Hamssa H., Abed El. S., Carole B., Anis L.: Trust model for secure group leader-based communications in VANET. Wirel. Netw. 1–23 (2018)
20. Abderrahim, A., Oussama, A., Habiba, C., Hicham, E., Nabil, H.: xxTEA-VCLOUD: a security scheme for the vehicular cloud network using a lightweight encryption algorithm. In; CCIOT 2018 Proceedings of the 2018 International Conference on Cloud Computing and Internet of Things, pp. 67–72 (2018)
21. Lei, L., Chen, C., Tie, Q., Mengyuan, Z., Siyu, L., Bin, Z.: A data dissemination scheme based on clustering and probabilistic broadcasting in VANETs. Veh. Commun. **13**, 78–88 (2018)
22. Sarah, O., Rachida, A., Joel, J.P.C.R., Said, T.: Secure and stable Vehicular Ad Hoc Network clustering algorithm based on hybrid mobility similarities and trust management scheme. Veh. Commun. **13**, 128–138 (2018)
23. Yasir, A.S., Hafiz, A.H., Farhan, A., Muhammad, F.K., Muazzam, M., Tabassam, N.: CAMONET: Moth-flame optimization (MFO) based clustering algorithm for VANETs. IEEE Access **6**, 48611–48624 (2018)
24. Yassine, H., Mohamed, B.A., Mohammed, B.: ACO and PSO algorithms for developing a new communication model for VANET applications in smart cities. Wirel. Personal Commun. **96**(2), 2039–2075 (2017)
25. Hui, X., San-shun, Z., Ben-xia, L., Li, L., Xiang-guo, C.: Towards a novel trust-based multicast routing for VANETs. Secur. Commun. Netw. **2018**, 12 (2018)
26. Sha, W., Chuanhe, H., Danxin, W.: Delay-aware relay selection with heterogeneous communication range in VANETs. Wirel. Netw. 1–10 (2018)
27. Li, T., Xiong, N.N., Gao, J., Song, H., Liu, A., Cai, Z.: Reliable code disseminations through opportunistic communication in vehicular wireless networks. IEEE Access (2018)
28. Daxin, T., Chao, L., Xuting, D., Zhengguo, S., Qiang, N., Min, C., Victor, C.M.L.: A Distributed Position-Based Protocol for Emergency Messages Broadcasting in Vehicular Ad Hoc Networks. IEEE Internet of Things J. **5**(2), 1218–1227 (2018)
29. Yusor, R.B.A., Omar, A.M., Namar, A.T., Nor, F.A., Suleman, K., Muhammad, A.: Accident management system based on vehicular network for an intelligent transportation system in urban environments. J. Adv. Transp. **2018**, 11 (2018)

Toward to Autonomous Broker for Virtual Network Provisioning and Monitoring

Mohammed Errais, Mostafa Bellafkih, and Mohammed Al Sarem

Abstract Network virtualization is an emerging concept that aims to facilitate the integration of new technologies regardless physical layer. To do this, the new concept proposes to evolve the current Internet business model. The new business model divides the Internet access provider to several actors. This actors diversity poses several challenges, especially when allocating resources and monitoring. For this, we propose in this work a new approach based on the enhanced telecom operation management business process framework. An approach that aims at ensuring an automatic supply of resources and a real-time monitoring based on the SLA defines between different actors.

Keywords Network virtualization environment · Network administration · Quality of service · Enhanced telecom operation management eTOM · Frameworks · SOA architecture · Next-generation operations support system NGOSS

1 Introduction

Network virtualization [1] is the perfect solution to the phenomenon of Internet ossification [2]. Indeed, the multitude of equipment and technologies deployed in the global network has become a major obstacle to its evolution. This is explained by the difficulty of changing the architecture and thus the migration to added value services.

M. Errais (✉)
Research and Computer Innovation Laboratory, Hassan II University of Casablanca, Casablanca, Morocco
e-mail: mahammed.errais@gmail.com

M. Bellafkih
Network Laboratory, INPT, Rabat, Morocco

M. Al Sarem
College of Computer Science and Engineering, Taibah University, Medina, Saudi Arabia
e-mail: mohsarem@gmail.com

© The Editor(s) (if applicable) and The Author(s), under exclusive license to Springer Nature Singapore Pte Ltd. 2021
F. Saeed et al. (eds.), *Advances on Smart and Soft Computing*, Advances in Intelligent Systems and Computing 1188, https://doi.org/10.1007/978-981-15-6048-4_51

Network virtualization allows telecom operators to migrate to new technologies without worrying about physical media. The principle is to add an abstraction layer between physical devices and deployed networks. This layer is provided by virtualization techniques that have made their success in virtualization of data centers. The principle of network virtualization is to deploy independent virtual nodes on existing physical nodes. The connection between these virtual nodes creates logical networks that are logically independent of the underlying physical network.

The implementation of the concept of network virtualization requires the evolution of the current Internet business model [3, 4]. This later is based on a main operator who is the Internet service provider. In the new business model, this provider is divided into three actors. The infrastructure provider (FI) is responsible for deploying the physical equipment. The virtual network provider (LIF) is responsible for deploying virtual networks on physical devices. The third actor is the service provider that deploys the value-added services on the virtual links. This new business model poses several challenges to the actual deployment of the network virtualization concept [4]. These challenges can be summarized in the dynamic supply of resources, the supervision of the quality of service, and the security of the exchanges between the different actors.

In the literature, several research works have focused on the study of network virtualization [1–7]. These research works have made it possible to focus on the issues and propose solutions for the supply of resources and communication between the various actors. However, these works do not provide a complete solution for resource provisioning and monitoring of SLA-defined [8] virtual networks between partners.

In this work, we propose a new approach based on the eTOM framework [9] for dynamic resource provisioning and monitoring of SLA requirements in real time. This will make it possible to migrate to an autonomous system capable of meeting the needs of the various actors in terms of supply and supervision in real time.

This document is organized as follows. In the first part, we introduce the concept of network virtualization and the organization of the eTOM framework. In the second part, we present the autonomous system for the supply and monitoring of virtual networks. In the third part, we discuss the obtained results during the experimental phase.

2 Related Works

2.1 Network Virtualization Concept

Network virtualization is essentially the segmentation of traditional networks with multiple logical segments. These segments are built via the connection between virtual nodes that are deployed on physical nodes. The virtual links are independent of the underlying physical network and the other virtual links. The new architecture allows the deployment of new value-added services without making any changes to the architecture of the Internet. This will facilitate the migration to new communication technologies and the integration of services.

Setting up network virtualization is possible by migrating to a new business model. Indeed, the existing model is based on the Internet service provider (ISP), whose main income is the sale of connectivity. In this model, value-added service providers, such as video-on-demand (VoD) and Internet-TV (IPTV) are forced to adapt to technologies deployed by the ISP. This poses significant challenges in procurement, given the wide variation in resource use. Several business models of network virtualization are presented in the literature. However, the most used model has four main actors [4].

- Infrastructure provider (IP): The entity that holds the physical devices and virtualization tools for creating virtual entities. This actor is responsible for ensuring the deployment of access and control tools for virtual nodes.
- Virtual network provider (VNP): The entity responsible for deploying and creating virtual networks. To do this, it must interact with the different providers of partner infrastructures to select the best offer before creating the virtual nodes.
- Service provider (SP): It is responsible for providing added value services. To do this, the service provider interacts with the VNP to deploy virtual networks that meet the technological requirements of these services.
- End user (EU): End consumer of value-added services.

The multitude of stakeholders and the complexity of operations pose significant challenges to the implementation of the new business model. For this purpose, the use of a model to unify trade is a necessity. A model that must be able to unify the exchanges between the various actors according to a logical and transparent sequence for the essential operations, in particular the supply of the resources and the assurance of the service continuity.

2.2   eTOM Framework

The eTOM framework as proposed by the TMFORUM [10] aims to implement a powerful tool for trade modeling in the telecommunications industry. It is a set of business processes structured according to several levels of abstraction (Fig. 1) each of which focuses on a particular type of operation. The eTOM framework is mainly aimed at organizing exchanges between telecom operators and their partners. An organization based on a sequence of business processes defined by a group of experts from several countries and horizons [10, 11].

The eTOM provides a standard for the acquisition and/or implementation of software and hardware necessary for the implementation of unified support systems. It presents a common language for business process modeling, which allows easy integration of solutions from different partners.

The eTOM business processes provide an overview of all exchanges during the course of an operation. Thus, the use of these business processes for standardization of exchanges in network virtualization will overcome several challenges. However, the

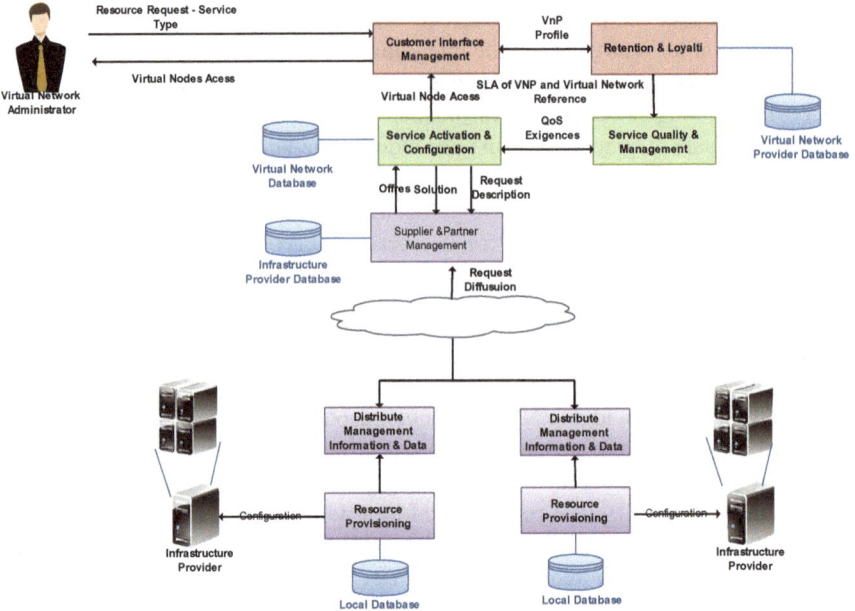

Fig. 1 Resource allocation operation according to the eTOM framework

abstract nature of business processes makes the projection of the eTOM framework in a context as complex as network virtualization a difficult operation.

3 Autonomous Broker

The proposed work consists mainly of establishing a transparent exchange between the different partners for the completion of the two essential operations, supply and real-time monitoring. For this purpose, resource provision requires the virtual network provider to discover the resources through the different infrastructure providers. Similarly, monitoring the QoS of the virtual networks created requires the exchange with all the FIs responsible for the physical resources. In order to regroup the exchanges, we propose at first the establishment of an autonomous broker between these two partners. The implementation of the broker-based business processes requires a clear and effective approach to have all the entities necessary for the structuring of information and implementation of processes. This process can be summarized in four main steps. The first step is to describe the nominal flow of the allocations and monitoring operations. The second is the division of operations according to a sequence of eTOM business processes. The last step is the implementation of different components of the broker.

3.1 Nominal Flow of Supply and Monitoring Operations

Dynamic resource allocation: During a request for a virtual network by the service provider, the virtual network provider must establish a mapping of the virtual network. To do this, the VnP establishes the needs of the network according to the type of service that will be deployed. The mapping includes the minimum resource requirements of the virtual nodes, as well as the desired geographic location. Then, the VnP transmits the mapping of the virtual network to the broker. The broker then proceeds to discover the resources available through the partner infrastructure providers. After receiving offers from these different partners, the broker selects according to the following criteria:

- Geographical location of the physical node
- Availability of resources (RAM, CPU, storage, number of network map)
- The current state of the physical node (number of virtual nodes deployed, CPU consumption, available storage ...)
- Current statistics on link status (average rate, burst size, average delay ...)
- The rental price of the resources.

The virtual network can be created on equipment owned by different infrastructure providers. The choice is made by virtual nodes taking into consideration the different criteria on the basis of a score calculated according to the TOPSIS method [12]. After validation of the choice, a request for supply is made to the InP which owns the selected resource. The InP then proceeds to the creation of the nodes according to the requested configuration, before transmitting the access identifiers to the VnP. An SLA is then established and defined the minimum requirements to be met for each virtual node. Then the VnP starts the configuration procedure based on the network protocols to establish the logical connection between the different virtual nodes and thus build the virtual network.

The nominal conduct of the surveillance operation: Real-time monitoring is essentially based on the SLA verification operation. That is, the system checks the indicators collected from the nodes in regular intervals. The purpose of this operation is to evaluate compliance with the requirements defined during creation in the SLA. To do this, the first step is to retrieve the performance indicators from the resources. These indicators are then mapped to calculate the quality indicators. These are compared with the requirements in the contract in order to come out with a conclusion. The difficulty of the monitoring operation in the field of network virtualization is expressed in two essential points. The first is the unification of the process of collecting indicators regardless of the technologies deployed. The proposed solution is the sequence of eTOM business processes that ensure a uniform process. The second point is the choice of performance indicators that are able to describe the real state of the virtual resources. For this purpose, we propose two types of performance indicators.

- System indicators that are inspired by virtual machine administration systems including the percentage of CPU used, RAM memory, available storage, and physical node state.
- Network indicators inspired by NGN monitoring systems [13] such as actual throughput, jitter, delay, and percentage of lost packets.

3.2 Modeling Provisioning Operation According to the eTOM Framework

For the modeling of the resource allocation operation, two eTOM process groupings are required. The first modelizes the sequence of steps during the negotiation between the broker and the VnP. While the second models the exchanges between the broker and the infrastructure provider (Fig. 1).

Figure 1 illustrates the sequence of processes during the supply operation. This operation is initiated by a request from the administrator of the VnP actor. This request includes the nature of the service, the type of resource and the location. This information is received and formalized via the customer and management interface (CIMP) process. The latter transmits this information and the VnP reference to the retention and loyalty (RIP) process. The RIP process loads the profile before checking its creditworthiness. If the VnP actor is solvent the RIP creates the request report which includes the VnP profile, the nature of service, and the location of the requested device. The report is then passed to the service activation and configuration (ACP) process which is responsible for finding the most optimal solution. To do this, the ACP process identifies the QoS requirements needed for the requested service and issues a call for tenders to the different InP partners via the supplier and partner management (SPMP) process.

After receipt of the RFP, the distribute management and information data (DMIDP) process creates an offer report and initiates a collection request to the resource data collection and proceeding (RDCP) process. The latter loads the information available from the hypervisors and the local database. This information is grouped by the DMIDP process that formats it to finalize the offer report. This report is then passed to the SPMP process.

The last step is to select the best offer. The service activation and configuration process are based on several indicators grouped in the global report (see Sect. II-1-A). At this stage, the selection problem becomes a multi-criteria problem. Thus, the decision is made on the basis of a score calculated by the TOPSIS method.

Once the best offer is selected the SACP process creates a supply report, before updating the local database. The report is sent to the InP concerned to proceed with the actual creation of the virtual node. Once the operation is performed, an SLA report of the node is established.

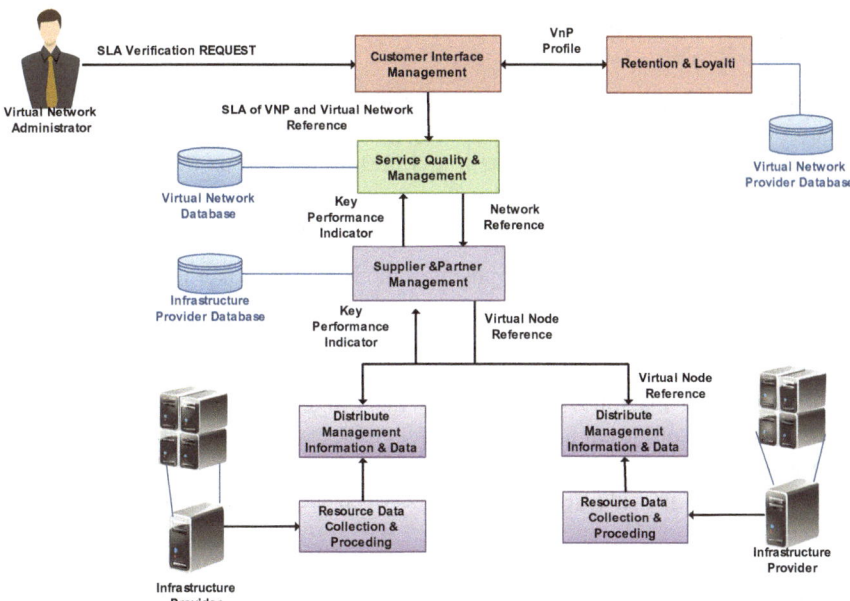

Fig. 2 Monitoring operation according to the eTOM framework

3.3 Modeling Monitoring Operation According to the eTOM Framework

The SLA check operation is based on the VnP profile and the SLA generated during the provisioning phase. The operation (Fig. 2) can be initiated by the VnP via the customer interface management process, or periodically by the service quality management (CQMP) process.

After initiating the SLA [14] check operation, the SQMP process loads the SLA defined during the allocation operation. Then, the SQMP opens a verification report that includes the node reference, the VnP identity, and the performance indicator thresholds. Once the report is open, the SQMP initiates a collection request by specifying the identity of the node to the SPMP process. The latter loads the requested information from the local InP database that holds the physical resources. Then, the listening operation is initiated via the DMID process. The values of the indicators collected in real time are transmitted to the SPM process that formats them before sending them to the SQM process. The service quality management process performs the check operation before completing the global check operation.

4 Experimentation

The objective of the experimentation phase is to evaluate the broker in real-time resource allocation and real-time monitoring cases. To do this, the test bench deploys the various components of the system architecture as well as the main actors of network virtualization [15] (Fig. 3).

Figure 3 illustrates the test bench that includes the following entities:

- Broker server: Enables the deployment of the EJB [16] broker module as defined in the system architecture.
- InP server: Includes multiple virtual machines. Each machine has the EJB infrastructure provider module as defined in the system architecture.
- InP terminals: machines connected to the broker via a switch. They allow administrators of virtual network providers to access interfaces published by the Broker.
- Linux routers: provide the connection between the different servers of the test bench.
- Hypervisors: the hypervisors offered by the infrastructures providers. They are responsible for providing the runtime environment for virtual machines. In our case, we used KVM [17] hypervisors. The deployed servers have a computing capacity of i7, 4 core of 3.6 Ghz each

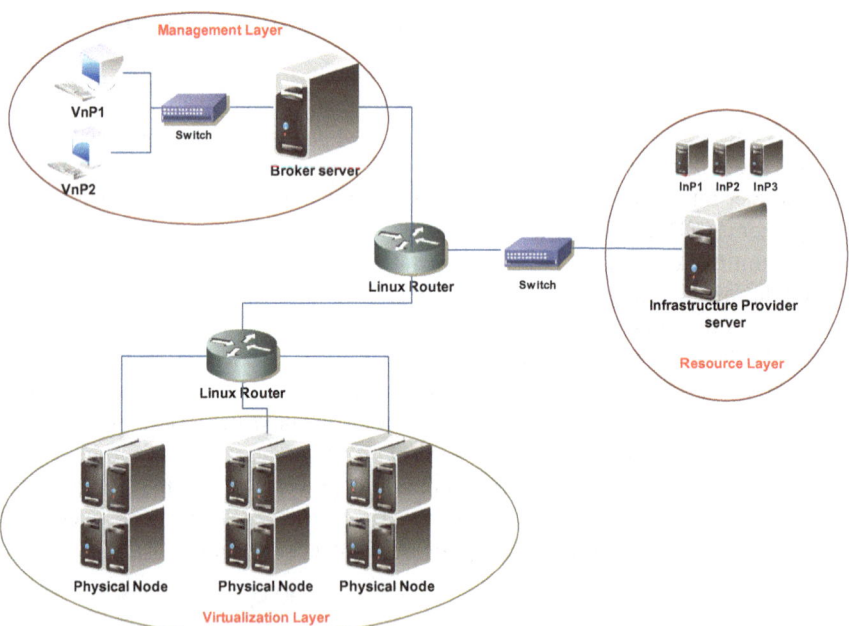

Fig. 3 Test bench

The autonomous broker evaluation is based mainly on three criteria. The first is the time of supply of resources that must be optimal to ensure the reliability of the broker. The second criterion is the negotiation time during the service provision phase. While the third criterion is the SLA verification time during the real-time monitoring operation. During the first experimental phase, a single infrastructure provider is used. Then, we increase the number of InP to evaluate the behavior of the system.

Figure 4 shows the variation in resource provisioning time by the number of infrastructure providers deployed in the test bench. The figure shows that the allocation time increases significantly with the evolution of InP involved in the operation. This is due to the time needed to negotiate offers and the collection of real-time data (CPU, state of the hypervisor, RAM). However, the allocation time remains acceptable (150 s for 60 InP). This is explained by the choice of technical tools including the use of Web services for the exchange of messages.

The provisioning stage can be broken down into two stages: the negotiation and the creation and installation of the virtual node. For the second step, the time does not depend on the number of InP. However, the negotiation phase absolutely depends on the exchanges established between the different InP and the autonomous broker.

Figure 5 illustrates the variation of negotiation time according to the number of provider infrastructure offering offers. It is clear that the negotiation time depends on the number of infrastructure providers involved. This is explained by the exchanges between the InP and the broker and the collection of data necessary for decision-making.

The recovery time varies considerably depending on the number of nodes in the virtual network. This is understandable given the difficulty of collecting and processing these indicators in real time. However, the use of eTOM business processes and the use of libvirt functions have reduced the cost of this operation. This is clearly reflected on the calculated time (35 s for 60 virtual nodes) (Fig. 6).

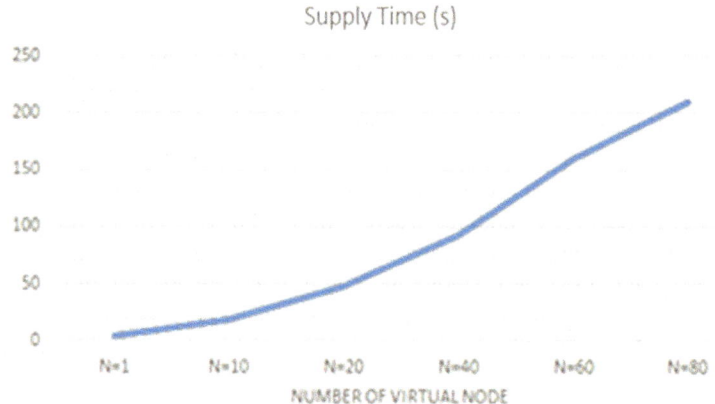

Fig. 4 Evolution of resource allocation time according to the number of infrastructure providers deployed

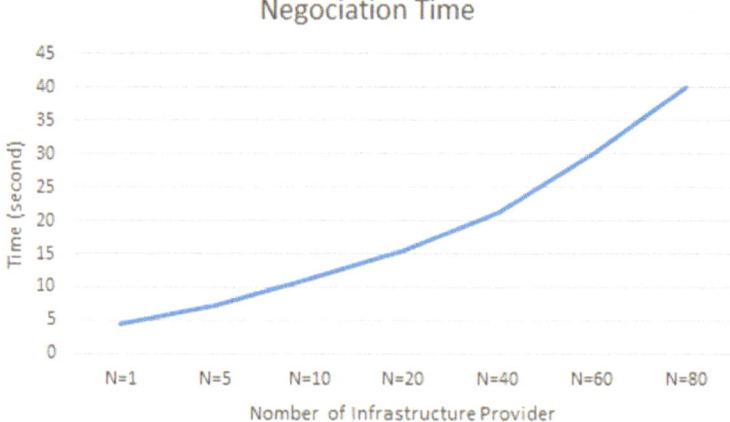

Fig. 5 Negotiation time according to the infrastructure provider number

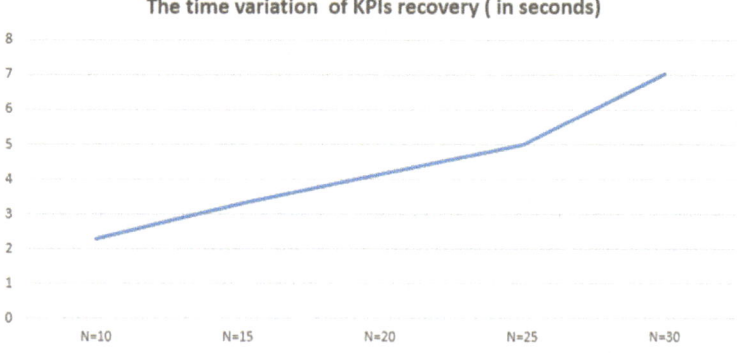

Fig. 6 Variation of indicator collection time by number of virtual nodes

The establishment of the stand-alone broker has helped address two key issues of network virtualization, provisioning, and monitoring. To this end, the adaptation of the eTOM framework has helped to unify and structure exchanges between the various factors involved in network virtualization operations. The evaluation of the broker in real experimental cases has allowed focusing on these strengths and limitations. Indeed, the use of EJB modules to deploy the modules of the system architecture made it possible to minimize the time of the exchanges and thus reduce the time of the monitoring and allocation operations. The virtual networks monitoring makes it possible to detect any anomaly on the bases of the SLAs established during the creation phase. However, without the correction integration, the monitoring operation will not have a great impact on virtual network assurance.

5 Conclusion

Network virtualization is an emerging concept that will undoubtedly solve the problem of the ossification of the Internet. However, the implementation of this concept faces several difficulties. To this end, several research projects have dealt with the issue of resource allocation and real nood monitoring. However, much of this work has not incorporated important concepts such as the SLA and the standardization of exchanges between actors.

The adaptation of the eTOM business processes helped to unify the exchanges between the actors and to structure the essential operations notably the allocation of the resources and the monitoring of the virtual nodes. For this purpose, the autonomous broker presents a solution for the two essential issues of network virtualization. The experimental phase tested the system in real deployment cases. This made it possible to highlight the strengths and limitations of such an approach.

Virtual network monitoring is a critical step in the successful completion of an effective network virtualization architecture. However, monitoring is not enough. Thus, the migration to an autonomous system for the correction of anomalies is a necessity. Moreover, the optimization of the cost of resource allocation will be a considerable asset to facilitate the integration of this concept in the telecommunications industry.

References

1. Ferguson, B.: The official Vcp5 certification guide. Vmware Press, Aug 2015. ISBN: 0789749319
2. Amarasinghe, H., Belbekkouche, A., Karmouch, A.: Aggregation based discovery for virtual network environments. In: Proceedings of the IEEE International Conference on Communications (ICC 2012), pp. 1276–1280. IEEE Press, Ottawa (2012)
3. Xu, Y., Han, Y., Niu, W., Li, Y., Lin, T., Ci, S.: A reference model for virtual resource description and discovery in virtual networks. In: Proceedings of ICCSA. Springer, Brazil, pp 297–310 (2012)
4. Rabah, S., El Barachi, M., Kara, N., Dssouli, R., Paquet, J.: A service oriented broker-based approach for dynamic resource discovery in virtual networks. J. Cloud Comput. Adv. Syst. Appl. 4, 3 (2015). https://doi.org/10.1186/s13677-015-0029-5
5. Chowdhury, N.M.M.K., Boutaba, R.: Network virtualization: state of the art and research challenges. IEEE Commun. Mag. 47(7), 20–26 (2009)
6. Seddiki, M.S., Frikha, M., Nefzi, B., Song, Y.: Automated controllers for bandwidth allocation in network virtualization. In: The 32nd IEEE International Performance Computing and Communications Conference (IPCCC 2013), San Diego, California, USA, pp. 1–7 (2013)
7. Errais, M., SEkkaki, A.: Implementation of new broker based on the eTOM framework for dynamic supply of resources during the composition of virtual networks. In: International Symposium on Networks, Computer and Communications, Morocco, 16–17 May 2017
8. Fajjari, I., Ayari, M., Pujolle, G.: VN-SLA: a virtual network specification schema for virtual network provisioning. In: Ninth International Conference on Networks, France, pp. 11–16, Apr 2010
9. Business Process Framework (eTOM), Enhanced Telecom Operation management, GB921, version 7.2.

10. The TM FORUM : http ://www.tmforum.org/
11. The NGOSS Real World use case , Version 1.2, GB921, T
12. Gong, S., Chen, J., Zhao, S., Zhu, Q.: Virtual network embedding with multi-attribute node ranking based on TOPSIS. In: KSII Trans. Int. Inf. Syst. **10**(2) (2016)
13. Errais, M., Bellafkih, M., Ranc, D.: Autonomous system for network monitoring and service correction in IMS Architecture. Int. J. Comput. Sci. Appl. **12**(1) (2015). ISSN 0972-9038
14. Errais, M., Bellafkih, M., Ranc, D.: Cost optimization of monitoring and supervision of ip multimedia subsystem networks. In: 4th international conference on next generation networks and services (NGNs), 2–4 Dec 2012, Portugal
15. Panda, D., Rahman, R., Lane, D.: 'EJB3 In Action . Manning Publications Co., Greenwich (2007)
16. Tutschku, K., Zinner, T., Nakao, A., Tran-Gia, P.: Network virtualization: implementation steps towards the future internet. In: Proceedings of the Workshop on Overlay and Network Virtualization at KiVS (2009)
17. Jie, R.Y., Zhu, D.H.T.: KVM virtualization technology: real and analytical principles. Intel Virtualization Technology Department Senior Virtualization, 1 Oct 2013, ISBN-10: 7111439007

Comparative Study on Random Traveling Wave Pulse-Coupled Oscillator Algorithm of Energy-Efficient Wireless Sensor Networks

Zeyad Ghaleb Al-Mekhlafi, Jalawi Alshudukhi, and Khalil Almekhlafi

Abstract Wireless sensor networks (WSNs) cannot manage the cost of concurrent transmission and get-together of information and for most conditions, the battery substitution is stunning upon the weariness of a center points battery vitality. At this moment, vitality productive conventions comprise crucial plan prerequisites for the WSN so as to build the lifetime and guarantee fruitful transmission of information from sensor hub (SN) source to target, it gets critical to keep up SNs openness. Random traveling wave pulse-coupled oscillator (RTWPCO) has exhibited to be healthy, effective, and impenetrable to check for deafness and bundle crash under various WSN including insightful models. In any case how much vitality effective is eaten up in SNs, which sends RTW-PCO as its self-association has never been referenced. To this limitation, we conducted a relative report on vitality productive in RTWPCO for WSNs. Using self-organizing scheme energy efficient WSNs by adopting a traveling wave biologically inspired network systems based on phase locking of Pulse Coupled Oscillator (PCO) model regards sensor nodes as observed in the flashing synchronization behaviors of fireflies and secretion of radio signals as firing. The amusement work was carried out using Java programming language. Energy proficiency in the self-assertive variety of the two plans (RTWPCO and PCO) was in like manner seen to be higher than the need variety of the plans.

Keywords RTWPCO · Energy efficient · WSN · PCO

Z. G. Al-Mekhlafi (✉) · J. Alshudukhi
University of Hail, Hail 81481, Saudi Arabia
e-mail: ziadgh2003@hotmail.com

J. Alshudukhi
e-mail: j.alshudukhi@uoh.edu.sa

K. Almekhlafi
Taibah University, CBA-Yanbu, Medina 42353, Saudi Arabia
e-mail: drkhalilalmekhlafi@gmail.com

F. Saeed et al. (eds.), *Advances on Smart and Soft Computing*, Advances in Intelligent
Systems and Computing 1188, https://doi.org/10.1007/978-981-15-6048-4_52

1   Introduction

In the last couple of years, smaller-scale small-scale electrical mechanical systems and wireless sensor networks (WSNs) have pulled in impressive consideration from analysts. WSNs are pretty much nothing and sensible devices with identifying, taking care of, and transmitting limits of normal wonders of interest [13, 15]. They have different application possibilities, including military, mechanical, and horticultural checking frameworks [4]. Furthermore, the constrained preparing capacity and correspondence sweep are two significant highlights describing the WSNs innovation [11]. Since these impediments are crucial to the general lifetime of the WSN, they ought to be seen while organizing a coordinating show [6]. Since the disappointment of individual nodes legitimately speaks to the entirety of coordinates with time, it is almost certain for bundle transferring also, standard identifying to the base station (BS) to be displayed to veritable jeopardization.

This is increasingly conceivable, particularly in light of the fact that an ever-increasing number of SNs will stop activities because of their depleted energy [3, 19]. Each steering convention estimation transmits data from the sources to the objectives and it is required to grow the framework introduction, whether or not the inciting regard decreases. Be that as it may, one of the most critical issues looked at by the WSN development has been the issue of incorporation holes. In this way, the principle remaining conceivable course of action decision or choice in medium and tremendous associations in hostile regions is discretionary dropping of (SNs) by unmanned vehicles or low-flying helicopters. In any occasion, when there is a credibility of deterministic sending, the issue of consideration openings will come up, especially in perspective on the SNs missing the mark on battery vitality. Such an issue gets even more plainly testing, particularly for those centers masterminded close to the BS. Such nodes generally speak to the framework bottleneck in light of their high information handing-off undertaking. Furthermore, SNs are left to work autonomously after they are at first conveyed, along these lines confounding the issue of inclusion. As portrayed by Wang et al. [18], WSN speaks to a gathering of thousands of minor SNs which are fit for performing remote commandoes, which constrained figuring and detecting. The commands issue is connected with the RTS or CTS pack in the IEEE 802.11 Macintosh show [8] and happens at the moment that a transmitter sends a control group to start a transmission while the goal is tuned to another channel. In the wake of sending diverse mentioning, if the transmitter does not get any reaction, it may reason that the beneficiary is never again reachable [19].

Figure 1 delineates the vitality productive logical arrangement that encompasses two rule instruments, namely organic enlivened system framework and non-natural motivated system framework.

In this investigation, we conducted a relative report on vitality proficient in RTWPCO for WSNs. Using self-arranging plan, vitality proficient WSNs by getting a voyaging wave organically enlivened system frameworks dependent on stage locking of PCO model considers SNs to be found in the blasting synchronization practices of fireflies and release of radio banner as ending [2, 3]. The re-DESYNC enactment

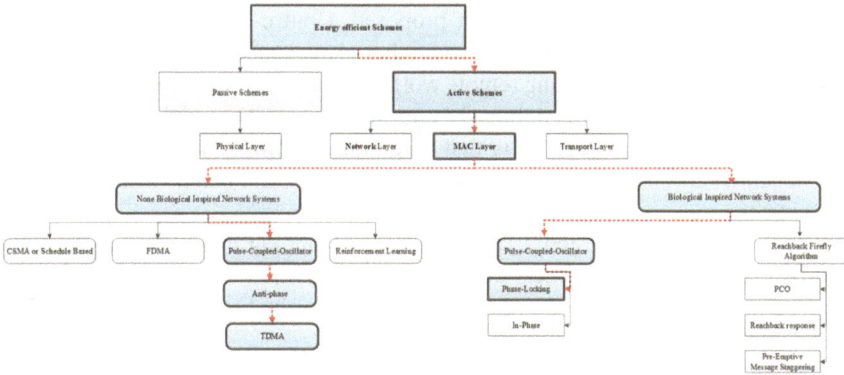

Fig. 1 Grouping of the vitality productive component in WSNs

work was finished utilizing Java programming language. Vitality productivity in the irregular variation of the two plans (RTWPCO and PCO) was additionally seen to be giant than the need variation of the plans.

2 Related Works

So far, the establishment overview of the vitality productivity of PCO on communication booking reflects on the bundle crash bypassing using a self-sorting out procedure in WSNs. This transmission booking is as often as possible applied for WSNs. Vitality proficiency is ensured in WSNs by express embedding vitality minimization shows into the fundamental identifying model of the SNs, (for instance, lessening the per bundle vitality usage) or bypassing the height vitality utilization of any solitary hub inside the system [4, 6, 15].

Various mechanisms of energy-efficient PCO for transmission booking have been proposed by utilizing self-organizing technique in WSNs [2–4, 14, 17, 18]. These components are arranged into non-naturally enlivened system frameworks and organically propelled system frameworks. This assessment focuses on naturally roused system frameworks that depend on stage locking traveling wave pulse coupled oscillator (TWPCO) and non-organically motivated system frameworks that depend on against stage time division multiple access (TDMA) in the PCO model.

Wireless communication scientists of the direct of the PCO model have proposed a variety of vitality proficiency approaches on WSNs. The PCO model could be copied to make WSNs. Such as with SNs, vitality effectiveness techniques for PCO models have decentralized direct and limited individual getting ready limit. Furthermore, they are principally limited in the manner that they convey. Subsequently, biologically inspired network system models and non-organically motivated system framework models can be gathered either as natural or scientific models.

In [9], the creators from the start proposed a pulse-coupled synchronization and scheduling protocol, PulseSS, a show that gives sort out synchronization and moderately sensible anticipating remote work systems with a gathered structure. For this condition, the show was dubiously moved by the PCO model from numerical science. Significantly intriguing is that of giving organizing and synchronization functionalities by abusing crucial physical layer hailing and neighborhood structure resuscitates.

The work in [7] suggested a DESYNC: self-arranging de-synchronization and TDMA on WSN. The examiners utilized the TDMA show and found that the new proposed model demonstrated updated execution in evaluation with existing TDMA models for WSNs.

The creators in [16] proposed a desynchronization issue which was attended to in the organization of an enhancement issue. The hypothetical outcomes and recreations displayed that further research is expected to enhance the assembly pace of the considerable number of strategies and to perhaps arrange under which presumptions it is useful to utilize one in disservice of another.

Disclosures are distinct, a self-sifting through framework that depends upon arrange locking PCO was first made. The purpose for this particular framework was to stream the SN of data from the edge of the framework to the BS to ricochet breaker to foresee deafness. Following this, a prompt sporadic-based strategy and a de-synchronization-based approach were made by focusing on the counter stage given in the PCO portray. The explanation for making the two arrangements was to settle the issue of an insignificant contact among SNs and to meanwhile proceed and a commensurate ricochet tally that is polished by strategies for the essential powerful extent and the data gathering extent. Regardless, the desynchronization-based instrument that was envisioned by those experts is legitimate to be connected with WSN, as it requires an unprecedented condition of data-grouping and limit and clearness of plan as opposed to the data gathering extent.

According to [3] proposed a RTWPCO scheme that is a self-sorting out strategy for vitality effective SNs balanced on fire beetle synchronization. They have shown an unrivaled execution than the TWPCO and PCO, which lessened the use of power inside the framework coming to fruition to expand the life expectancy of the whole WSN. Likewise, this instrument decreases the amount of dropped data provoked growing the information gathering extent, subsequently engaging the proposed plan to demonstrate gain in trustworthiness and vitality proficient in WSN. This evaluation does not stand out from reach-back with show the favored introduction over the PCO and TWPCO.

In perspective on the over audit, the past works have helped this appraisal to accomplish a superior seeing correspondingly as than connect with correspondence stages, procurement, and framework synchronization to be disconnected. In any case, the proposed instruments that have been researched above are not authentic to be applied to WSNs because of the way that it must hold brisk to the topology.

3 Proposed Mathematical Module

Radio correspondences of SNs add to the rising vitality costs on the SN, particularly in the dynamic mode (get and transmit). Truly, the vitality devoured in transmitting and accepting modes is bigger contrasted with rest and inert modes that they are thought to be unimportant right now. Various diverse SN organize applications will bring about different requests and particulars, which cause it to be progressively tangled in having undefined necessities for all applications. For instance, the apportionment of the TWPCO in the imperativeness capability process is essentialness eating up, thus making it inapplicable for various applications [5]. On the other hand, the RTWPCO is favored on the grounds that it does not assimilate a ton of vitality, along these lines making it conceivable to augment the life expectancy of the entire system. As indicated by Al-Mekhlafi [1], Hyun et al. [10], Molisch et al. [12], alluding to the customary component of gathering information, a tree topology that is established at the BS is not changed due to many-to-one highlights of the checking applications. In addition, the information parcel transmissions, just as the rest timings are being isolated to stay away from radio impedance and high-power, are now required to relegate control parcels to continue the system topology and incorporate themselves with the encompassing SNs, or stay away from any impedance and high-power among the encompassing SNs. On account of a system topology change, SNs need to ship additional control bundles to reinforce the topology since it requires more vitality and time.

This causes it to be progressively tangled in having undefined necessities for all applications. For instance, the apportionment of the TWPCO in the imperativeness capability process is very essentialness eating up, thus making it inapplicable for various applications [5].

Be that as it may, the progression of time has caused the ϕ_i to move toward T (which is the most extreme). At T, the oscillator ϕ_i fires before ϕ_i returns to zero. So also, the oscillator ϕ_j which is multiplied with the terminating oscillator $\emptyset_i$ is incited, along these lines causing the relating stage ϕ_j to be moved by a tiny sum $\Delta(\phi_j)$, where: $\phi_j = \Delta(\phi_j) + \phi_j$.

Right now, stage clock ϕ_i, level 1 li including counterbalance T are available in every SN n_i (where $1 \leq i \leq N$). As indicated by the recipe, the level field can connote the SN's jump check from the BS. Toward the start of the procedure, each degree of the SN is designated to a high worth. The counterbalance T alludes to the interim between the hubs of level 1 and $l-1$ which happens in the midst of the correspondence of information. In this way, a transmission which contains controlled data and SN information is conveyed by the SN n_i when the stage ϕ_i comes to T. The stage will return to 0 after the past advance. In this manner, when the message by hub n_j from n_i at level $l_i < l_j$ is gathered, the hub l_j will be changed to its level as $l_j = l_i + 1$.

$$\phi_j = \phi_j + \text{PRC}_a * \cos \frac{\pi}{2 * T} * \phi_j + \text{PRC}_b * (T - \phi_j) \tag{1}$$

Figure 2 shows the evaluation of PRC ($\Delta_s(\phi)$) which furthermore satisfies the

Fig. 2 Assessment of PRC $\Delta_s(\phi)$ accompanied by RIC mock-up and QIF mock-up

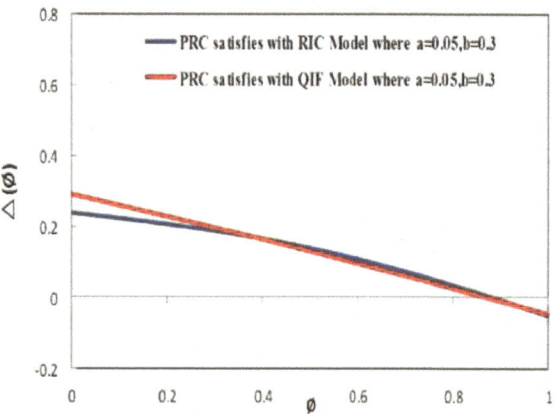

RIC model and the QIF model where $PRC_a = 0.05$ and $PRC_b = 0.3$ ought to be arranged in the two lines when $T = 0.2$.

The TWPCO system proposes that the stage will be actuated and balanced by the hub as exhibited in Condition 1, where PRC_a and PRC_b are the antigenic determinant that impact the pace of array. Unaccompanied by thinking about the initial period of hubs, the voyaging wave marvel that is in agreement to stage securing in the PCO is executed in the midst of the ordinary transmission among SNs by applying Condition 1 in the PRC work.

Present of randomization-based recipe, balance T_i is changeably picked among zero and T^{max} as displayed beneath:

$$\tau_i^{prev} = \tau \varepsilon T_{i,t}^{max} < \tau_i^{trans'} \tag{2}$$

$$\tau_i^{next} = \tau \varepsilon T_{i,t}^{min} < \tau_i^{trans'} \tag{3}$$

whereby T_i suggests the game plan of time assessed for transmission of messages in $\in_i$. Be that as it may, they will be considered as zero if T_i^{Prev} or T_j^{next} is not accomplished. Tailing it, the association is gotten as a guide for the center point n_i to change the offset T_i:

$$P_i = (1 - \alpha) * P_i + \alpha * P_i^{mid} \tag{4}$$

It is basic to fathom that in the randomization-based condition, the SN n_i neither stores message correspondence nor includes any information of message transmission timing F_i in the messages. At any rate, this condition is not commonly helpful in understanding the covered center point issue despite it is considered as uncomplicated, and what it simply needs is a base control overhead diverged from various segments. From Fig. 3, it will in general be seen that both the message transmission timing similarly as offset T_i with SNs are relentless when the randomization-based

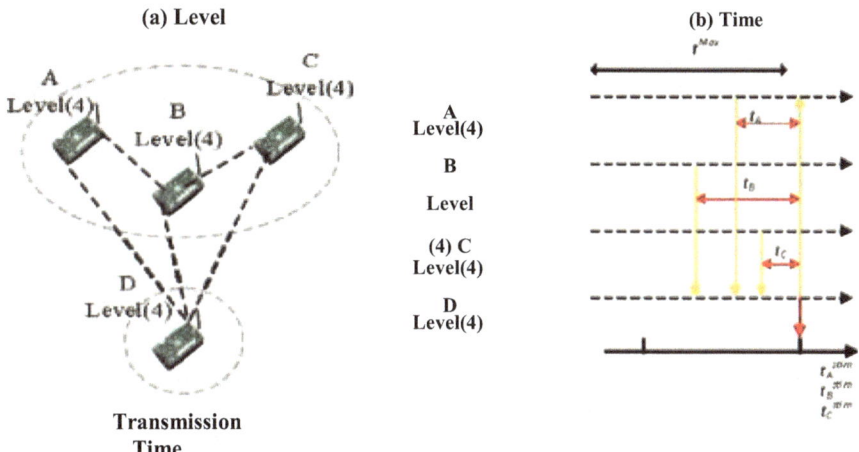

(a) Level

(b) Time

Transmission
Time

Fig. 3 Message communicate timing and counterbalance in the randomization-based calculation

formula is applied at the same time. Implying a comparable figure, it might be seen that SNs *A*, *B*, and *C* share the equivalent bounce tallies from the SN. Right now, correspondence communicate timings are flowed alongside the counterbalance T. It was seen that the two SNs *A* and *C* went about as two-jump neighbor SNs, yet no brief transmission happened between them. With respect to RTWPCO instrument, data must be accomplished from a solitary jump upstream neighbor SN, explicitly SN *D* as appeared in Fig. 3. This plans to forestall gigantic usage between the two-bounce neighbor SNs. The sorted out data of SN has a time t_i and holds stage ϕ_i (with the end goal that ϕ_i $e[0, P]$), level l_i, counterbalance T_i ($0 < T_i < T^{max}$), correspondence communicate timing table e_i, and SN information D_i. P alludes to information gathering period, while level l_i connotes the quantity of bounces from the BS that was at first set to a big incentive by remembering that a BS consistently contain the degree of 0. Counterbalance T_i sets the correspondence communicate period among SN n_i and a SN, subsequently activating SN n_i that was initially fixed to T^{max}. Acknowledge that SN thickness exceptionally impacts the way toward setting the best balance T^{max}, correspondence communicate timing table e_i, and SN information D_i. In addition, e_i is used to keep data on correspondence communicate timing of SN n_j inside two jumps from SN n_i.

4   Simulation

In this investigation, we completed perceptions and appraisal of the reenactment, test results with the goal that we could assess the presentation under an assortment of WSN situations. In light of the exploratory outcomes, we plot assumes a significant job in improving the energy efficiency and expanding the SNs relentless states. Such improvement accomplished by the RTWPCO plot is ascribed to the way of modifying the quality assessment of the capacities, which thus adds to upgrade the exactness of the fire beetle dependent based on the PCO-model calculations. To approve the RTWPCO plot, the consequences of the trials were contrasted with the PCO model. The following segment tends to clarify the trial arrangement and evaluate the exactness and after effects of the RTWPCO plot.

4.1 System Configuration

The proposed RTWPCO system was reproduced by setting up different SNs. The propagation of the analyses was completed through a PC (Intel CoreTM $i5$ 2410 M, CPU @ 2.30 GHz and 2.30 GHz, 4G Smash, Windows 7) by utilizing a test framework executed in Java. As such, at whatever point, a hub gets transmission messages from two distinct hubs simultaneously, the beneficiary hub would not have the option to get the two messages. The information gathering degree cycle, T was set to 1 s, while the most basic substitute, t^{max} was set to 0.1 s. According to Conditions 1 and 4, PRC_a, PRC_b, and a were set to 0.01, 0.5, and 0.5, respectively. The directly off the bat compose readings were un-systematically planned. The scopes of the message header and the individual SN information were set to 2 B, while message transmission control data was fixed to 1 B. The CSMA/CA values consolidated a back-off accessibility of 1 ms with a back-off of 4 ms. Tailing it, the aftereffects of the RTWPCO component were contrasted with those of PCO and TWPCO. In addition, so as to keep up the objectivity, the PCO and TWPCO were re-actualized. This included utilizing programming language which is Java so the two strategies could work or run utilizing comparative test systems together with steady programming and equipment stages. As displayed in Table 1, a succinct examination of different parameters utilized in the reproduction is additionally clarified.

4.2 Conversation of Experimental Results

The TWPCO model in Fig. 4, when the hubs 100 and 90 ate up 5.27089 and 4.13808 (mJ) of vitality independently, while the bespoke RTWPCO framework simply governed around 2.30856 and 1.78087 (mJ). Thusly, it is displayed which proposed scheme diminished the vitality capacity up to 48% stood out from the PCO and TWPCO models for every hub. Thus, the proportion of devoured power accomplished through the RTWPCO system is roughly not exactly the PCO and TWPCO instruments. This specific result goes about as one of the responsibilities to the proposed portion barely essential at the present time. In the meantime, the improvement is the consequence of different portrayals of the degrees of centers.

As introduced in Fig. 5, the aftereffects of the recommended RTWPCO instrument are contrasted with the PCO and TWPCO components, especially on the absolute

Table 1 Values arrangement

Parameters	Values	
	Scenario I	Scenario II
Channel frequency	2.4 GHz	2.4 GHz
Number of nodes 10.20, 30, 40, 50, 60, 70, 80, 90, 100		30
MIN_TIME_STEP	0.00001	0.00001
Packet data Size	16 bits	8.16, 40, 80, 160, 400, 800 bites
Energy model	MICAz	MICAz
Data rate	250 kbps	250 kbps
Transmit power	52.2 pW	52.2 pW
Receive power	59.1 pW	59.1 pW
Idle power	60 pW	60 pW
Sleep power	3 pW	3 pW
Initial energy	100 J	100 J

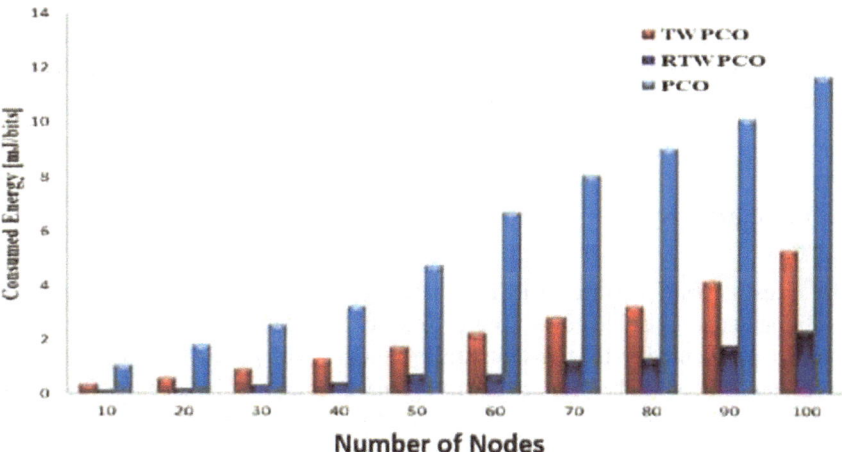

Fig.4 Vitality effectiveness proportion dependent on number of hubs in the transmission territory of RTWPCO component

power utilization of the hubs as indicated by the information bundle size when the quantity of SN is equivalent to 30. Unmistakably the information parcel size of SN 100 B appeared to expend a colossal measure of energy contrasted with the rest, for example, fifty B through the transmission, therefore causing quicker power channel. In the meantime, the proposed RTWPCO component spread the enormous information bundle size to the big level hubs rather than the basic hubs. It makes the measure of energy utilized diminished by the information parcel size in SNs fifty B and 100 B up to 16 and 56%, separately. Along these lines, it very well may be inferred that the devoured power proportion of the RTWPCO component is around

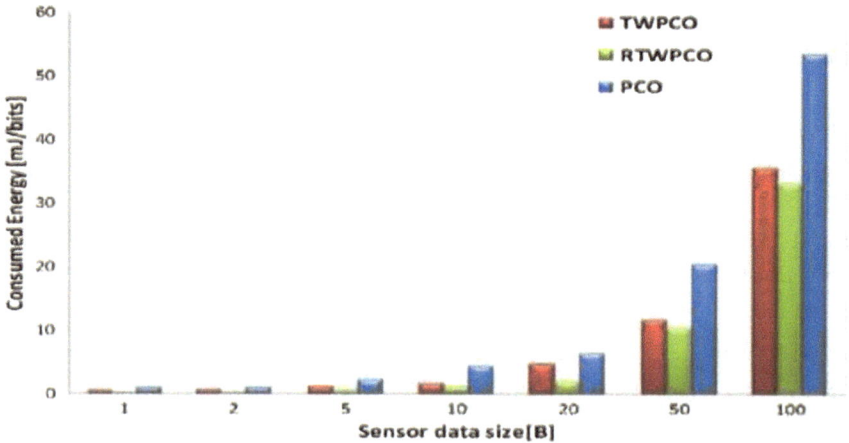

Fig. 5 Vitality proficiency proportion dependent on information bundle size in the transmission province of RTWPCO instrument

not exactly the PCO and TWPCO for all information parcel measures because of the nearness of more noteworthy measure of got bundles in the PCO and TWPCO systems that makes it increasingly successful.

5   Conclusion

The investigation is restricted to PCO and RTWPCO just think about the single condition of energy efficiency in view of WSN correspondence; the most noteworthy powers are devoured by the transmit-control together with get control procedures of a correspondence in a hub. The need elective plan whose activity has been recognized as being erased while picking an afterward jump hub to is identified with any semblance of an immediate correspondence which is additionally reached out to land at a BS. These need elective plans (RTWPCO and PCO) devoured less energy contrasted with the irregular elective plans whose activity has been portrayed as whimsical when picking a next bounce node. Be that as it may, future research ought to affirm and improve the presentation of these systems by utilizing desynchronization in the trial assessment just as the genuine detecting condition to stay away from a parcel impact.

References

1. Al-Mekhlafi, Z.G., Hanapi, Z.M., Othman, M., Zukarnain, Z.A.: A firefly-inspired scheme for energy-efficient transmission scheduling using a self-organizing method in a wireless sensor

networks. JCS **12**(10), 482–494 (2016)

2. Al-Mekhlafi, Z.G., Hanapi, Z.M., Othman, M., Zukarnain, Z.A.: Travelling wave pulse coupled oscillator (twpco) using a self-organizing scheme for energy-efficient wireless sensor networks. PLOS ONE **12**(1) (01 2017) 1–27

3. Al-Mekhlafi, Z.G., Hanapi, Z.M., Othman, M., Zukarnain, Z.A., Shamsan Saleh, A.M.: Random traveling wave pulse-coupled oscillator algorithm of energy-efficient wireless sensor networks. Int. J. Distrib. Sens. Netw. **14**(4), 1550147718768991 (2018)

4. Al-Mekhlafi, Z.G., Hanapi, Z.M., Saleh, A.M.S.: Firefly-inspired time synchronization mechanism for self-organizing energy-efficient wireless sensor networks: a survey. IEEE Access **7**, 115229–115248 (2019)

5. Chipara, O., Lu, C., Stankovic, J.A., Roman, G.C.: Dynamic conflict-free transmission scheduling for sensor network queries. IEEE Trans. Mob. Comput. **10**(5), 734–748 (2011)

6. Darabkh, K.A., Odetallah, S.M., Al-qudah, Z., AlaF, K., Shurman, M.M.: Energy- aware and density-based clustering and relaying protocol (ea-db-crp) for gathering data in wireless sensor networks. Appl. Soft Comput. **80**, 154–166 (2019)

7. Degesys, J., Rose, I., Patel, A., Nagpal, R.: Desync: self-organizing desynchronization and tdma on wireless sensor networks. In: Proceedings of the 6th International Conference on Information Processing in Sensor Networks. IPSN '07, New York, NY, USA, ACM, pp 11–20 (2007)

8. Fruth, M.: Probabilistic model checking of contention resolution in the iEEE 802.15. 4 low-rate wireless personal area network protocol. In: Second International Symposium on Leveraging Applications of Formal Methods, Verification and Validation (isola 2006), IEEE, pp. 290–297 (2006)

9. Gentz, R., Scaglione, A., Ferrari, L., Hong, Y.P.: Pulsess: a pulse-coupled synchronization and scheduling protocol for clustered wireless sensor networks. IEEE Internet of Things J. **3**(6), 1222–1234 (2016)

10. Hyun, S.H., Kim, G., Yang, D.: Pd-desync: practical and deterministic desynchronization in wireless sensor networks. KSII Trans. Internet Inf. Syst. **13**(8) (2019)

11. Mann, P.S., Singh, S.: Energy-efficient hierarchical routing for wireless sensor networks: a swarm intelligence approach. Wireless Pers. Commun. **92**(2), 785–805 (2017)

12. Molisch, A.F., Balakrishnan, K., Chong, C.C., Emami, S., Fort, A., Karedal, J., Kunisch, J., Schantz, H., Schuster, U., Siwiak, K.: IEEE 802.15. 4a channel model-final report. IEEE P802 **15**(04), 0662 (2004)

13. Nunez, F., Wang, Y., Doyle, F.J.: Bio-inspired hybrid control of pulse-coupled oscillators and application to synchronization of a wireless network. In: 2012 American Control Conference (ACC), IEEE, 2818–2823 (2012)

14. Phung, K.H., Lemmens, B., Mihaylov, M., Tran, L., Steenhaut, K.: Adaptive learning based scheduling in multichannel protocol for energy-efficient data-gathering wireless sensor networks. Int. J. Distrib. Sens. Netw. **9**(2), 345821 (2013)

15. Raji, K.A., Gbolagade, K.A.: A survey of different techniques for energy-efficient, reliability and fault tolerant in wireless sensor networks. Int. J. Wirel. Commun. Mobile Comput. **7**(1), 19 (2019)

16. Silvestre, D., Hespanha, J., Silvestre, C.: Desynchronization for decentralized medium access control based on gauss-seidel iterations. In: 2019 American Control Conference (ACC), July 2019, pp. 4049–4054

17. Taniguchi, Y., Hasegawa, G., Nakano, H.: Self-organizing transmission scheduling considering collision avoidance for data gathering in wireless sensor networks. J. Commun. **8**(6), 389–397 (2013)

18. Wang, Y., Nunez, F., Doyle, F.J.: Energy-efficient pulse-coupled synchronization strategy design for wireless sensor networks through reduced idle listening. IEEE Trans. Signal Process. **60**(10), 5293–5306 (2012)

19. Youn, S.: A comparison of clock synchronization in wireless sensor networks. Int. J. Distrib. Sens. Netw. **9**(12), 532986 (2013)

Computational Informatics

Impact of Inspirational Motivation on Organizational Innovation (Administrative Innovation, Process Innovation, and Product Innovation)

Ali Ameen, Sultan Alshamsi, Osama Isaac, Nadhmi A. Gazem, and Fathey Mohammed

Abstract Inspiration is constantly cited as one of the most important managers who have to improve employee productivity. It can play a vital role in the administrative innovation, process innovation, and product innovation. Current research utilized smart PLS as an suitable instrument of structural equations modeling (SEM) for analyzing the 389 valid survey questionnaires for assessing the proposed model that is based on the transformational leadership characteristic (inspirational motivation) to identify its effect on organizational innovation in the government sector in the United Arab Emirates. The main independent construct is inspirational motivation as. The dependent construct is represented by product innovation, process innovation, and administrative innovation. The researchers have described the relationship between the different constructs. This study could improve insight into the significance of TL and organizational innovation. The results showed that the independent variables could help in predicting innovation. Furthermore, the proposed model could explain 45.7% of the variance that was seen in the administrative innovation, 42.8% of product innovation, and 41.9% in process innovation.

Keywords Inspirational motivation · Organizational innovation · Product innovation · Process innovation · Administrative innovation

A. Ameen (✉) · S. Alshamsi · O. Isaac
Lincoln University College, Bharu, Malaysia
e-mail: ali.ameen@aol.com

S. Alshamsi
e-mail: Ali_ukm@yahoo.com

O. Isaac
e-mail: Osama2isaac@gmail.com

N. A. Gazem
Taibah University, Medina, Kingdom of Saudi Arabia
e-mail: nadhmigazem@gmail.com

F. Mohammed
University Utara Malaysia (UUM), Changlun, Malaysia
e-mail: fathey.m.ye@gmail.com

© The Editor(s) (if applicable) and The Author(s), under exclusive license to Springer Nature Singapore Pte Ltd. 2021
F. Saeed et al. (eds.), *Advances on Smart and Soft Computing*, Advances in Intelligent Systems and Computing 1188, https://doi.org/10.1007/978-981-15-6048-4_53

1 Introduction

Inspirational motivation was one of the major transformational leadership (TL) components which played a vital role in the changing and turbulent business environment by redefining and embracing the business processes and organizational performance. In their studies, Avolio et al. [1] and García-Morales et al. [2] stated that the TL was an important technique that determined the effect of the leaders on the organizational performance and business activities. Additionally, they observed that in the past few years, TL has shown a close relationship with the various determinants of the organizational performance.

In the last 20–25 years, the public sector in UAE has undergone a substantial change with respect to the increasing globalization [3, 4]. UAE is one of the fastest-growing countries in the Middle-East due to the implementation of the innovative governance and infrastructural systems, since the UAE has strived to position itself and its public sector as the service-oriented economy [5]. It has led to a significant restructuring of the traditional public administrative sectors into more customer-centric and competitive platforms. Finally, the UAE public sector has changed its scope in the past few years and has started implementing changes which were similar to those being implemented in the private sector. Due to this pressure to change, all public-sector organizations in the UAE have shown a higher interest in implementing the TL, which could help them acquire better results.

2 Literature Review

2.1 Organizational Innovation (OI)

Innovation can be defined in multiple ways; a significant amount of research indicates that it is about creative things. For instance, Rogers [6] suggests that innovation is about creating new object, practice, or idea based on an individual's view. The same author illustrated a second concept named as diffusion of innovation that focuses on the dispersion of innovation over time through specific channels within social system members. On the other hand, some researcher think that there is a distinction between innovation and creativity, and the latter is defined as the process to produce new ideas, while innovation creates and implements new products, processes, and ideas [7–11].

Thus, it is concluded that creativity is an element of innovation [12].

While services and goods are different in nature and in terms of their characteristics, numerous researchers claim that the theories and concepts that are applicable to the manufacturing sector can actually be transferred and used in the service context using the assimilation approach [13, 14]. For the purpose of examining the service sectors' innovation, Droege et al. [14] mention that the same models of innovation were used in these studies in the service context as in the sector of manufacturing.

There is a need to examine the factors of the study that affecting the organizational innovation.

2.2 Inspirational Motivation (IM)

Inspirational motivation (IM) was seen to be one of the four important dimensions related to TL, which indicated the degree to which the TL inspired the employees, instilled an encouraging, confident, optimistic environment along with a clear vision. TL has been considered as an ideal leadership style in various organizations. TL has helped in achieving the different organizational outcomes like organizational performance [15] and employee. TL could instigate a high order need [16]. In their study, stated that the TL motivated the employees and generated some positive emotions, created an inspirational vision, and directed the followers for achieving their objectives. Based on the proposed dimensions of the TL by the earlier studies, [17] derived four TL dimensions. This study will utilize inspirational motivation, the above arguments, and other supporting results that led to the following hypothesis:

H1: Inspirational motivation has a positive effect on product innovation.
H2: Inspirational motivation has a positive effect on process innovation.
H3: Inspirational motivation has a positive effect on administrative innovation.

3 Research Method

3.1 Overview of the Proposed Conceptual Framework

The correlations between the parameters conjectured in the conceptual model have been obtained from available literature. Figure 1 displays independent variable which was inspirational motivation, and the dependent variables which were the product innovation, process innovation, and administrative innovation. The proposed model has three hypotheses to be tested.

3.2 Development of Instrument and Data Collection

In this study, the researchers developed the questionnaire tool which consisted of 16 questions. Variables were measured using a Likert scale which recommended in the previous studies [4, 18–22]. This information was collected by delivering the self-managed questionnaire "in-person" to the employees in the Abu Dhabi Executive Council Authority, UAE, in the period between March 2018 and April 2019. Out of the 500 questionnaires that were distributed, 389 responses were seen to be suitable

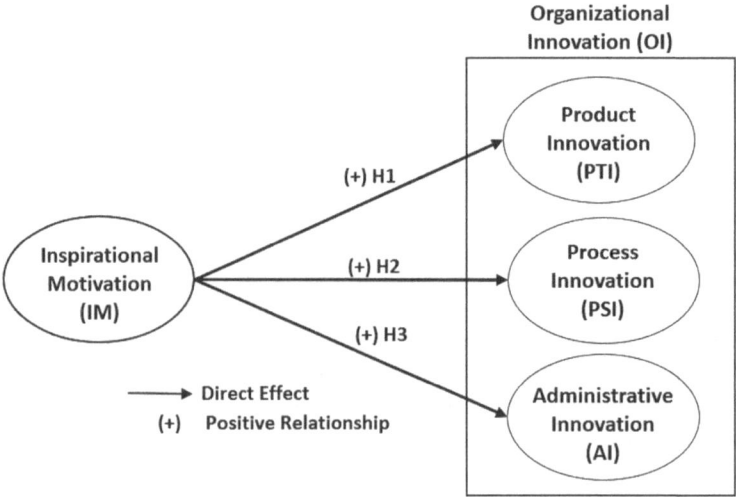

Fig. 1 Conceptual framework

for analysis. This sample size was sufficient as stated by Krejcie and Morgan [23] and Tabachnick and Fidell [24].

4 Data Analysis and Results

The researchers used the SmartPLS 3.0 software for examining their model, with the help of the partial least squares (PLS) variance-based structural equation model (VB-SEM) [25, 26]. They used a two-stage analytical [27] for (a) assessing the measurement model (i.e., reliability and validity), and (b) assessing the structural model (i.e., hypothesized relationship analysis). The main reason for choosing SEM as a statistical method for this study is that SEM offers a simultaneous analysis which leads to more accurate estimates [28].

4.1 Measurement Model Assessment

The results of the data analysis indicated that the different parameters like the composite reliability (CR), Cronbach's alpha, average variance extended (AVE) along with the factor loadings were higher than the suggested values [29, 30] as described in Table 1.

The discriminant validity refers to the degree to which the different articles differentiate among the concepts and measure the constructs. The researchers used the Fornell–Larcker factor for analyzing the discriminant value in the measurement

Table 1 Measurement model assessment

Constructs	Item	Loading (>0.7)	M	SD	α (>0.7)	CR (>0.7)	AVE (>0.5)
Inspirational motivation (IM)	IM1	0.944	3.724	1.068	0.962	0.973	0.899
	IM2	0.952					
	IM3	0.941					
	IM4	0.956					
Product innovation (PTI)	PTI1	0.939	3.722	1.096	0.940	0.961	0.893
	PTI2	0.949					
	PTI3	0.946					
Process innovation (PSI)	PSI1	0.964	3.632	1.069	0.955	0.971	0.917
	PSI2	0.960					
	PSI3	0.949					
Administrative innovation (AI)	AI1	0.913	3.352	1.041	0.944	0.957	0.818
	AI2	0.918					
	AI3	0.873					
	AI4	0.897					
	AI5	0.921					
	AI6	Deleted					

Key: *IM* inspirational motivation, *PTI* product innovation, *PSI* process innovation, *AI* administrative innovation

model, and the results have been presented in Table 2. The results indicated that the square root of the AVE on the diagonals (presented in bold) was higher than the correlation occurring between the constructs (corresponding row and column values), which indicated a higher relationship between the different concepts and their markers compared to the other concepts suggested in the model [31, 32]. Hair et al. [25] described this as satisfactory discriminant validity. It was noted that the exogenous constructs showed a correlation of <0.85. Thus, the discriminant validity of all the constructs was satisfactory.

Table 2 Fornell–Larcker criterion

	AI	IM	PSI	PTI
AI	**0.904**			
IM	0.676	**0.948**		
PSI	0.744	0.647	**0.958**	
PTI	0.782	0.655	0.780	**0.945**

Key: *IM* inspirational motivation, *PTI* product innovation, *PSI* process innovation, *AI* administrative innovation

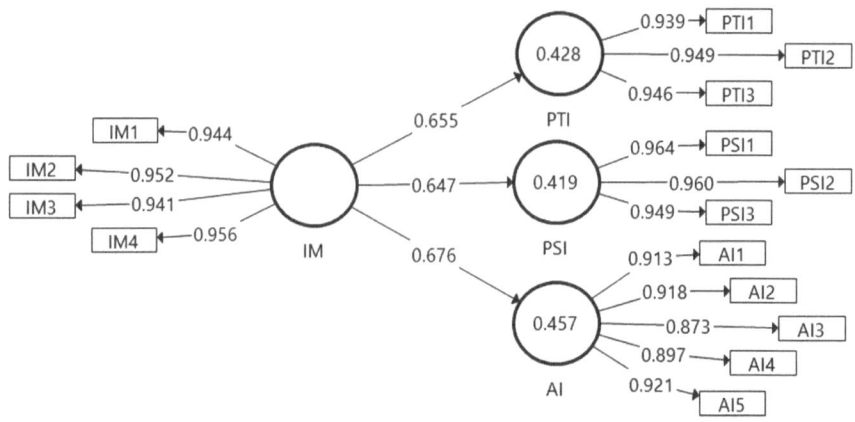

Fig. 2 PLS algorithm results

Table 3 Result of direct effect hypotheses

Hypothesis	Relationship	Std beta	Std error	t-value	p-value	Decision	R^2
H1	IM → PTI	0.655	0.036	18.310	0.000	Supported	0.43
H2	IM → PSI	0.647	0.036	18.078	0.000	Supported	0.42
H3	IM → AI	0.676	0.031	21.483	0.000	Supported	0.46

Key: *IM* inspirational motivation, *PTI* product innovation, *PSI* process innovation, *AI* administrative innovation

4.2 Structural Model Assessment

The researchers tested the structural model after computing the beta (β), R^2, and corresponding t-values using a bootstrapping process using a resample size of 5000 [25].

Figure 2 and Table 3 presented the results of the hypothesis tests and structural model assessment. Inspirational motivation explains forty-three percent of the variance in product innovation, 42% of the variance in process innovation, and 46% of the variance in administrative innovation. Furthermore, the R^2 values showed an acceptable explanatory power, which indicated that it was a substantial model [33, 34].

5 Discussion and Implications

The researchers noted that the IM positively affected the product innovation among the employees of the Abu Dhabi Executive Council Authority, UAE, as shown in earlier studies [17]. This was based on the fact that when the leaders spread a higher sense of optimism, were enthusiastic about the objectives to be fulfilled, presented a vision for the future, and were confident that these goals could be achieved, then the organizations developed novel services, diversified the services, and implemented many new ideas and technologies in the organization.

The results indicated that the IM positively affected the process innovation among the employees of the Abu Dhabi Executive Council Authority, UAE, as shown in the earlier studies [17, 35–39]. This was based on the fact that when the leaders spread a higher sense of optimism, were enthusiastic about the objectives to be fulfilled, presented a vision for the future, and were confident that these goals could be achieved, then the organization adapted new technologies during the work processes, showed a quick response to the changes, and developed novel processes and services.

Furthermore, IM also positively affected the administrative innovation among the employees of the Abu Dhabi Executive Council Authority, UAE, as shown in the earlier studies [17, 18, 22, 40, 41]. This was based on the fact that when the leaders spread a higher sense of optimism, were enthusiastic about the objectives to be fulfilled, presented a vision for the future, and were confident that these goals could be achieved, then the organization offered a higher administrative support to their employees, implemented a higher performance evaluation system, provided an openly communicative environment, considered the creativity of the employees during the hiring process, and also provided a more participative work environment.

Thus, it was seen that the TL was a new and emerging concept, which has not been completely understood by the majority of the organizations in the UAE and the Arabic world. The researchers presented this concept in the public sector of UAE organizations. They studied the role played by the internal leadership activities on the individual, group, or organizational performance and determined how it could be used for improving the general performance [3, 35, 41–43].

One of the limitations of this research was that the data collected were cross-sectional and not of a longitudinal nature. The longitudinal method may improve the understanding of correlations and causation between variables. Future research should be undertaken to find out the relationship between variables by conducting cross-cultural studies as recommended in previous studies.

6 Conclusion

In this study, the researchers aimed to investigate the effect of the IM on the organizational innovation among the employees of the Abu Dhabi Executive Council Authority, UAE. They presented the results related to the concept of TL, which were described in earlier studies. The researchers proposed a new model which included IM as an independent variable, while organizational innovation (i.e., product, process, and administrative innovation) was used as a dependent variable. According to the results, all the hypotheses were significant. The independent variable could explain 42.8% of the variation noted in the product innovation, 41.9% of the variance noted in the process innovation, and 45.7% of the variance noted in the administrative innovation.

7 Appendix A

See Table 4.

Table 4 Instrument for variables

Variable	Measure	Source
Inspirational motivation (IM)	"IM1: Leaders talk optimistically about the future" "IM2: Leaders talk enthusiastically about what needs to be accomplished" "IM3: Leaders articulate a compelling vision of the future" "IM4: Leaders express confidence that goals will be achieved"	[44]
Product innovation (PTI)	PTI1: We always develop new product and services PTI2: We try to introduce and diversify our product to suit customer needs PTI3: We always try applying new idea/technology at our organization	[45]
Process innovation (PSI)	PSI1: In my organization, new technology is adapted for improving the work processes (computers, wireless networking, etc.) PSI2: In our organization, we try new methods for improving processes (paperless environment, online learning, etc.) PSI3: Our organization is quick to respond to changing needs of its customer	[45]
Administrative innovation (AI)	AI1: Administrative support is always there for employees AI2: Employees compensation system is linked to performance AI3: Our institution has a new and improved performance evaluation system AI4: At our organization, we believe in the open communication environment AI5: In our organization, employees are hired on their creativity AI6: In our institution, there is a participative working environment	[45]

References

1. Avolio, B., Bass, B.M., Jung, D.I.: Re-examining the components of transformational and transactional leadership using the multifactor leadership questionnaire. J. Occup. Organ. Psychol. **72**, 441–462 (1999)
2. García-Morales, V.J., Jiménez-Barrionuevo, M.M., Gutiérrez-Gutiérrez, L.: Transformational leadership influence on organizational performance through organizational learning and innovation. J. Bus. Res. **65**(7), 1040–1050 (2012)
3. Alameri, M., Ali Ameen, G. S., Khalifa, A., Alrajawy, I., Bhaumik, A.: The mediating effect of creative self-efficacy on the relation between empowering leadership and organizational innovation. Test Eng. Manag. **81**, 1938–1946 (2019)
4. Alameri, M., Ameen, A., Isaac, O., Khalifa, G.S.A., Bhaumik, A.: Examining the moderating influence of job complexity on the relationships between empowering leadership and organizational innovation. Test Eng. Manag. **81**, 1930–1937 (2019)
5. Suliman, A., Al Kathairi, M.: Organizational justice, commitment and performance in developing countries: The case of the UAE. Empl. Relat. **35**(1), 98–115 (2012)
6. Rogers, E.M.: Diffusion of Innovations, 4th edn. The Free Press, New York (1995)
7. Trott, P.: Innovation Management and New Product Development. Pearson Education Limited (2005)
8. Ameen, A., Ahmad, K.: The role of finance information systems in anti financial corruptions: a theoretical review. In: 11 International Conference on Research and Innovation in Information Systems (ICRIIS'11), 267–272 (2011)
9. Baharuden, A.F., Isaac, O., Ameen, A.: Factors influencing big data & analytics (BD&A) learning intentions with transformational leadership as moderator variable: Malaysian SME perspective. Int. J. Manag. Hum. Sci. **3**(1), 10–20 (2019)
10. Ameen, A., Ahmad, K.: Information systems strategies to reduce financial corruption. In: Benlamri, R., Sparer, M, (ed.) Springer Proceedings in Business and Economics, Dubai, UAE, vol. 1, pp. 731–740. Springer, Berlin (2017)
11. Ameen, A., Ahmad, K.: A conceptual framework of financial information systems to reduce corruption. J. Theor. Appl. Inf. Technol. **54**(1), 59–72 (2013)
12. West, M.A., Farr, J.L.: Innovation and Creativity at Work: Psychological and Organizational Strategies. Wiley, New York (1990)
13. de Vries, E.J.: Innovation in services in networks of organizations and in the distribution of services. Res. Policy **35**(7), 1037–1051 (2006)
14. Droege, H., Hildebrand, D., Forcada, M.A.H.: Innovation in services: present findings, and future pathways. J. Serv. Manag. **20**(2), 131–155 (2009)
15. Klamo, L., Huang, W.W., Wang, K.L., Le, T.: Successfully Implementing E-Government: Fundamental Issues and a Case Study in the USA. IEEE (2006)
16. Rowold, J., Schlotz, W.: Transformational and transactional leadership and followers' chronic stress. Leadersh. Rev. **9**(1), 35–48 (2009)
17. Aydogdu, S., Asikgil, B.: The effect of transformational leadership behavior on organizational culture: an application in pharmaceutical industry. Int. Rev. Manag. Mark. **1**(4), 65–73 (2011)
18. Alneyadi, B.A., Al-shibami, A.H., Ameen, A., Bhaumik, A.: Effect of transformational leadership on human capital among public sector employees in Abu Dhabi. Int. J. Emerg. Technol. **10**(1a), 16–22 (2019)
19. Al-Obthani, F., Ameen, A.: Influence of overall quality and innovativeness on actual usage of smart government: an empirical study on the UAE public sector. Int. J. Emerg. Technol. **10**(1a), 141–146 (2019a)
20. Al-Obthani, F., Ameen, A.: Association between transformational leadership and smart government among employees in UAE public organizations. Int. J. Emerg. Technol. **10**(1a), 98–104 (2019b)
21. Al-Ali, W., Ameen, A., Issac, O., Nusari, M., Ibrhim Alrajawi, M.: Investigate the influence of underlying happiness factors on the job performance on the oil and gas industry in UAE. Int. J. Manag. Hum. Sci. **2**(4), 32 (2018)

22. A. Al-Mulla, Asma; Ameen, Ali; Isaac, Osama; Nusari, Mohammed; Hamoud Al-Shibami, "The Effect of Organizational Tensions, Merge Policy and Knowledge Sharing on Managing Organizational Change: The Context of Abu Dhabi National Oil Organizations," *J. Eng. Appl. Sci.*, vol. 14, no. 8, pp. 2517–2531, 2019
23. Krejcie, R.V., Morgan, D.W.: Determining sample size for research activities. Educ. Psychol. Meas. **38**, 607–610 (1970)
24. Tabachnick, B.G., Fidell, L.S.: Using multivariate statistics. PsycCRITIQUES **28**, 980 (2007)
25. Hair, J.F., Hult, G.T.M., Ringle, C., Sarstedt, M.: A Primer on Partial Least Squares Structural Equation Modeling (PLS-SEM), 2nd edn. Sage, London (2017)
26. Ringle, C.M., Wende, S., Becker, J.-M.: SmartPLS 3. Bonningstedt: SmartPLS (2015)
27. Hair, J.F., Black, W.C., Babin, B.J., Anderson, R.E.: Multivariate data analysis: a global perspective (2010)
28. Isaac, O., Abdullah, Z., Aldholay, A.H., Ameen, A.: Antecedents and outcomes of internet usage within organisations in Yemen: an extension of the Unified theory of acceptance and use of technology (UTAUT) model. Asia Pacific Manag. Rev. **1**(1), 72–92 (2019)
29. Kline, R.B.: Principles and Practice of Structural Equation Modeling, 3rd edn. The Guilford Press, New York (2010)
30. Hair, J.F., Black, W.C., Babin, B.J., Anderson, R.E.: Multivariate Data Analysis. New Jersey (2010)
31. Fornell, C., Larcker, D.F.: Evaluating structural equation models with unobservable variables and measurement error. J. Mark. Res. **18**(1), 39–50 (1981)
32. Chin, W.W.: The partial least squares approach to structural equation modeling. In Marcoulides, G.A. (ed.) Modern Methods for Business Research, pp. 295–358. Lawrence Erlbaum Associates, New Jersey (1998)
33. Cohen, J.: Statistical Power Analysis for the Behavioral Sciences, 2nd edn. Lawrence Associates, Erlbaum (1988)
34. Chin, W.W.: Issues and opinion on structural equation modeling. MIS Q. **22**(1), 7–16 (1998)
35. Alhefiti, S., Ameen, A., Bhaumik, A.: The impact of the leadership and strategy management on organizational excellence: moderating role of organizational culture. J. Adv. Res. Dyn. Control Syst. **06**(Special Issue), 748–759 (2019)
36. Mohamed, M.S., Khalifa, G.S.A., Nusari, M., Ameen, A., Al-Shibami, A.H., El-Shazly, A.-E.: Effect of organizational excellence and employee performance on organizational productivity within healthcare sector in the UAE. J. Eng. Appl. Sci. **13**(15), 6199–6210 (2018)
37. Alkhateri, A.S., Abuelhassan, A.E., Khalifa, G.S.A., Nusari, M.: The Impact of perceived supervisor support on employees turnover intention: the mediating role of job satisfaction and affective organizational commitment. Int. Bus. Manag. **12**(7), 477–492 (2018)
38. Alshamsi, O., Ameen, A., Nusari, M., Abuelhassan, A.E., Bhumic, A.: Towards a better understanding of relationship between Dubai Smart Government characteristics and organizational performance. Int. J. Recent Technol. Eng. **8**(2S10), 310–318 (2019)
39. Haddad, A., Ameen, A., Isaac, O., Alrajawy, I., Al-Shbami, A., Midhun Chakkaravarthy, D.: The impact of technology readiness on the big data adoption among UAE organisations. In: Sharma, N., Chakrabarti, A., Balas, V.E. (eds.) Data Management, Analytics and Innovation, pp. 249–264. Springer Singapore, Singapore (2020)
40. Mohamed, F., Nusari, M., Ameen, A., Raju, V., Bhaumik, A.: Impact of strategy implementation (strategy, structure, and human resources) on organizational performance with in Abu Dhabi Police, UAE. Test Eng. Manag. **81**, 1914–1920 (2019)
41. Alkatheeri, Y., Ameen, A., Isaac, O., Duraisamy, B., Nusari, M., Gamal, A., Khalifa, S.: The effect of big data on the quality of decision-making in Abu Dhabi Government Organisations. In: 3rd International Conference on Data Management, Analytics and Innovation, 2019, p. 50 (2020)
42. Isaac, O., Abdullah, Z., Ramayah, T.: Antecedents and outcomes of internet usage within organizations in Yemen: an extension of the unified theory of acceptance and use of technology (UTAUT) model. Bus. Process Manag. J. Emerald (2019)

43. Alshamsi, O., Ameen, A., Isaac, O., Khalifa, G.S.A., Bhumic, A.: Examining The Impact Of Dubai Smart Government Characteristics On User Satisfaction. Int. J. Recent Technol. Eng. **8**(2S10), 319–327 (2019)
44. Wang, Y.S., Li, H.T., Li, C.R., Wang, C.: A model for assessing blog-based learning systems success. Online Inf. Rev. **38**(7), 969 (2014)
45. Hussain, H.K.: Assessing the moderating effects of organizational innovation on the relationship between transformational leadership and job satisfaction (2015)

Managing the Smart Grid in Free Market

Rim Marah, Inssaf El Guabassi, Sanae Larioui, and Mohammed Abakkali

Abstract It is confident that renewable energies in smart grid promise to meet the growing demand of energy while reducing carbon emissions. But because of their intermittent and stochastic nature, the need to find a compromise between the use of two sources of energy, one renewable and the other non-renewable, which would allow consumers to have electricity at an affordable cost while reducing carbon emissions is a necessity. However, there is not yet an efficient mechanism for distributing electricity in free market of smart grid where multiple suppliers coexist. This paper proposes a distribution algorithm of electricity prioritizing the use of green resources and minimizing the use of polluting conventional resource.

Keywords Smart grid · Optimization · Microgrid · Free electricity market

1 Introduction

Recently, a vision of modern electrical network commonly known as the smart grid (SG) is shaping up in several countries. The emergence of these visions around the world gives birth to several definitions of what could be the tomorrow's electrical network.

The smart grids [1, 2] combine between the classical electrical distribution networks (e.g., these technologies collect and consolidate information as close as producers, consumers or during the delivery of energy). Moreover, they process more intelligently the information.

The energy challenges that the world faces in the coming years will change the way we produce, transport and consume our electricity. Fundamentals changes are essential to meet the growing demand for electricity while minimizing our greenhouse gas emissions and maximizing the use of green resources.

R. Marah (✉) · S. Larioui · M. Abakkali
System Control and Decision Laboratory, ENSIT Engineering School, Tanger, Morocco
e-mail: ensate.rim@gmail.com

I. El Guabassi
FS, Abdelmalek Essaadi University, P.O. BOX 2121, 93000 Tetuan, Morocco

© The Editor(s) (if applicable) and The Author(s), under exclusive license to Springer
Nature Singapore Pte Ltd. 2021
F. Saeed et al. (eds.), *Advances on Smart and Soft Computing*, Advances in Intelligent
Systems and Computing 1188, https://doi.org/10.1007/978-981-15-6048-4_54

The use of renewable sources only to satisfy all the demand of electricity at the same cost of network parity is not yet possible because of the high investment cost and their intermittent and stochastic nature.

Hence, the need to find a compromise between the use of two sources of energy, one renewable and the other non-renewable, would allow consumers to have electricity at an affordable cost while reducing carbon emissions.

In free electricity market where multiple suppliers coexist, the electrical network is a combination of several interconnected sources, combining solar energy, wind energy, as well as others renewable energy sources and the conventional source of fuel. The use of different sources becomes a necessity not only to meet demand in electricity, but also to reduce carbon emissions. However, there is not yet an efficient mechanism for distributing electricity.

This paper proposes a distribution algorithm of electricity assuming the existence of a free market where too sources can supply several consumers.

In this paper, we will focus on the microgrid level of the smart grid and propose an algorithm of distribution of energy based on normal distribution in order to optimize the use of green energy. This algorithm regulates the demand of the microgrid depending on the values of consumption of each local level.

The paper is organized as follows: After this introduction, we will present in section smart grid architecture and modelization of the smart grid architecture, and we will give a model of it. Indeed, in section microgrid level managing, we will describe in detail what is happening in the microgrid level. In mathematical prediction section, we will use the normal distribution to predict the consumption of each local level.

After, in section distribution algorithm, we will present an algorithm based on standard normal distribution to optimize the use of green energy. Finally, we will close our paper by a conclusion and some perspectives.

2 Smart Grid Architecture

Smart grid technologies develop the existing infrastructure [3] (power lines, substations, control rooms) and improve the real-time evaluation of the system status. New equipment and digital devices can be deployed strategically to complement the existing ones.

The electricity companies from around the world have understood the need to set up new systems smart grid, in order to add a new layer of equipment to interconnect all resources (super high voltage grids to the micro and pico networks ultra-low voltage) to buildings and houses.

Using a combination of centralized systems and distributed intelligence in strategic control nodes. For example, thermal plant orders and sites of renewable energy to the transmission and distribution control centers, via commercial and industrial infrastructure and individual residences.

**Grid & Market Management Systems
(EMS-DMS-MMS)**

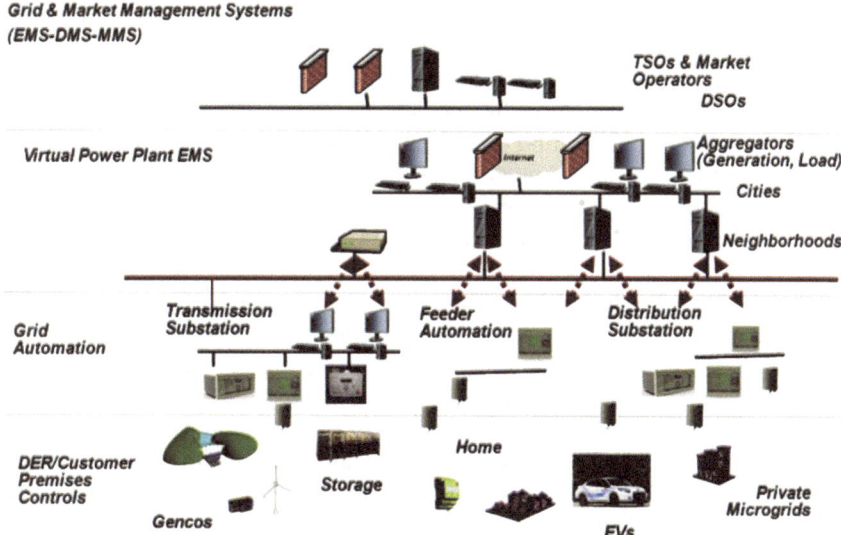

Virtual Power Plant EMS

TSOs & Market
Operators
DSOs

Aggregators
(Generation, Load)
Cities
Neighborhoods

Grid
Automation

Transmission
Substation

Feeder
Automation

Distribution
Substation

DER/Customer
Premises
Controls

Gencos

Storage

Home

EVs

Private
Microgrids

Fig. 1 Smart grid architecture

Smart grid layers require a system of systems approach, with different needs in terms of security. It relies on a multitude of actors and stakeholders, each with a specific role and activities in each field.

The architecture of the smart grid is the architecture of a combination of systems interconnected in a distributed way.

Each of these systems collaborates individually in the construction of the global complex system which is smart grid. Each sub-system, with its associated resources, requires specific functions and security solutions. For example, the solution secures the erase and management systems for domestic energy.

At each level, security measures should be enough to reduce the risks. To effectively protect the entire network, the various sub-systems are not necessarily aligned with the sub-system with the highest safety requirements, each with a specific role to play in the ecosystem.

Players must analyze smart grid security levels, from the perspective of a comprehensive risk assessment each type of use of the smart grid and for the sub-systems considered within the overall architecture [3, 4].

Figure 1 shows the layered architecture of the smart grid.

2.1 Smart Grid Modelization

The smart grids combine between the classical electrical distribution networks (e.g., these technologies collect and consolidate information as close as producers, consumers or during the delivery of energy). Moreover, they process more intelligently the information. Smart grids are intended to smooth the consumption curve, to reduce overall consumption, to balance supply and demand and to integrate new technologies. Constraints are sending commands or energy requirements and optimize energy flow, to perform preventive maintenance, control and minimize costs permutations to optimize investments, and in the same time, to deal with the generation of renewable energy (in erratic production) and storage. Smart grids combine information and communication technology (ICT)to the electrical networks. The two systems communicate in parallel with distribution networks.

That embedded [5, 6] intelligence must allow better adjustment between electricity production and consumption as well as the integration of renewable energies.

Therefore, these critical systems are large-scale and potentially impacting many people, it is important to ensure that they function properly before they are implemented. In such cases, the simulation proves to be very useful to carry out verifications and to evaluate different hypotheses and scenarios of the architecture. However, because they involve many technical areas including electronics, information processing and telecommunications, smart grids [7, 8] are a typical example of complex systems, difficult to design.

This architecture is designed using different tools and languages that can manipulate state machines, systems equational, discrete events than statistical models.

This justifies the establishment of a cosimulation environment in order to conserve and take advantage of this diversity. In short, a general approach is giving for smart grid architecture [5, 6].

The overlay of the electrical network infrastructure and modern ICT technologies increases the need of make this system self-stabilizing [9, 10] of smart grid and allow it to self-correct without any human intervention.

3   Microgrid Level Managing

A microgrid is a grouping of related houses or buildings, commonly called an Ecodistrict. The microgrid is the most important echelon in the smart grid, and it is the bridge between the distributors' offer and the consumers's demands.

Hence, it must manage both users and suppliers. The microgrid must therefore match the inputs and outputs of the other two levels, while managing energy according to priorities.

The current electrical network is evolving to become a combination of several interconnected sources. However, the intermittent nature of renewable energy and the unpredictability of electricity consumption imply that a microsource cannot guarantee permanent nutrition to its consumers.

Typically, the needs of the consumer will be satisfied by the nearest microsources. This introduces a new paradigm for the electricity market. A new architecture is installed, in which several electricity suppliers can coexist. This new electricity market is a free market.

Given the benefits of having a free market for electricity, in this work, we suggest a real-time model with several energy suppliers connected to several consumers, using game theory. This approach emerges as a mathematical framework to design and model the smart grid. The proposed objective function does not only consider line losses but also production and cost.

4   Mathematical Prediction

In this section, we will predict the Demand of each consumer based on his consumption history communicated to the microgrid and the decision support tool the normal distribution N (conso$_{moy}$, σ^2).

Let us first analyze the characteristics of the normal distribution.

Definition 4.1 A random real variable X follows a normal distribution noted.

$X \sim N$ (μ, σ^2) of standard deviation σ if this real random variable admits as probability density function $p(x)$ defined for any real number x by (Fig. 2)

$$P(x) = \left(\frac{1}{\sigma\left(\sqrt{2\pi}\right)} \right) \times e^{1/2\left(\frac{x-\mu}{\sigma}\right)^2} \tag{1}$$

Definition 4.2 If random real variable X follows a normal distribution noted.

$X \sim N$ (μ, σ^2) and if a and b are two reals such as a, then the probability of being between a and b is noted

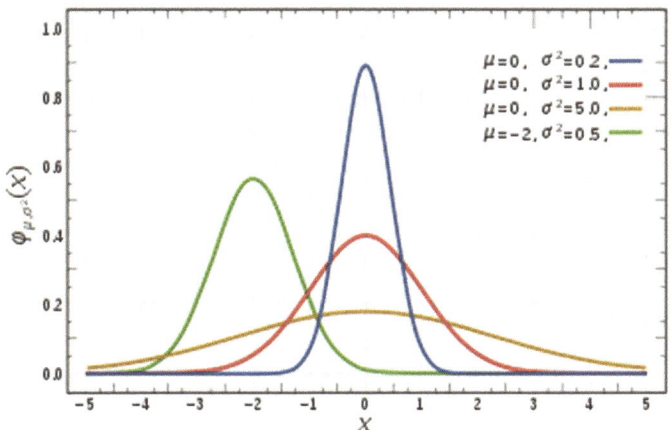

Fig. 2 Probability density function

$$P(\alpha \le X) = F(b) - F(c) = \varnothing\left(\frac{b - \mu}{\sigma}\right) - \varnothing\left(\frac{a - \mu}{\sigma}\right) = \alpha \tag{2}$$

$\varnothing$ is the standard normal distribution function N (0, 1)

$$\varnothing(x) = \int \frac{1}{\sqrt{2\pi}} e^{-(t)^2/2} dt \tag{3}$$

The random variable $X* = (X - \mu)/\sigma$ follows the standard normal distribution. The interval $[a, b] = [\mu - t\sigma, \mu + t\sigma]$ is called the normality rang with confidence level α, this means that α percent of the population lie in the interval $[a;b]$, in our case, we look for configuration of normal distribution according to the consumption capacities of each local level i, here are the characteristics:

1. $\mu = \text{conso}_{\text{moyi}}$
2. $[\mu - 3\sigma, \mu + 3\sigma]$ is the normality range of the confidence level 99.7%. A margin of ε around $\text{conso}_{\text{moyi}}$ so $3\sigma = \varepsilon$.

Once the microgrid has made a consumption prediction for each local level, it sums up the results, according to which it makes its energy demand from the TD level.

5   Distribution Algorithm

We consider the following mathematical model; we have two key elements:

1. Consumers: They are divided into several distributed groups. So, the overall demand will be cumulated for a neighborhood, an industrial zone or for a database area.
2. Sources: The many sources can be based on solar energy, wind or coal energy.

   To model our network, we assume that we have:
   $S = \{r, nr\}$ is suppliers set, renewable sources and non-renewable one.
   $C = \{1, 2, \ldots m\}$ is the set of customers.
   At each time t, each node $s \in S$ produces P_s^t respecting with maximum capacity (available energy) $P_{s-\text{max}}^t$ respecting

$$P_s^t \le P_{s-\text{max}}^t, \quad \forall s \in S \tag{4}$$

P_{s-c}^t indicates the energy received by the consumer c from the source s at a time t.

D_c^t the request of the consumer c

$$P_{r,c}^t + P_{nr,c=}^t D_c^t, \forall c \in C \tag{5}$$

We define C_S as a set of all consumers connected to the same source s, and we assume that $C_S \ne \varnothing$

The line losses are represented by $P_{s,c}^{t,\text{loss}}$

$$\sum P_{s,c}^{t} + P_{s,c}^{t,\text{loss}} \leq P_{s,\text{max}}^{t}. \tag{6}$$

After the modelization of our network, let us move to our distribution algorithm. The algorithm executes the following instructions:

1. Each local level sends its consumption to the linked microgrid, which constitute a study sample
2. Based on the values received, the microgrid makes a future prediction for each local level by using the normal distribution
3. The microgrid adds up the consumption of all the local levels related
4. The microgrid sends the request to the renewable source first, if it is unable to cover all the requested capacity, it sends the order to the non-renewable one
5. It waits for a return from TD level.

where

$\varnothing(x*)$ is the standard normal distribution function.

$P_{c-\text{max}}^{t}$ is the maximal consumption of local level c.

$P_{c-\text{min}}^{t}$ is the minimal consumption of local level c.

X is the table of values of normal distribution of every local level.

Algorithm 1: Distribution Algorithm

```
begin
    foreach c ∈ C do
        receive (msg ==<"consumption, P_c^t", c >);
        ∅(x *) = rund[0 ... 1];
        X* = approximation ∅(x *);
        if (X_c < P_{c,min}^t)   then
            X_c = P_{c,min}^t;
        end
        if (X_c < P_{c,max}^t)   then
            X_c = P_{c,max}^t;
            P_c^t = P_c^t + X_c ;
            Stock X_c ;
        end
    end
    return ∑ P_c^t = D^{total};
        if P_{r,max}^t ≥ D^{total} then
            serve all levels c;
            sent (msg == "green energy use");

        end
        if P_{r,max}^t < D^{total} then
            D^{total} = P_{r,max}^t + P_{nr}^t ;
            sent (msg == "mixed energy use");
        end
end
```

6   Conclusion

In this work, we have presented a modelization of the smart grid, we have seen that smart grid is considered as a complex system that has three distinct levels: the local level, the microgrid and the T&D network. We focused on the microgrid level of the smart grid, and we have proposed an algorithm to optimize the distribution of energy by privileging the use of green one. The algorithm used normal distribution as a mathematical prediction. This work constitutes one of others of our global attempt to manage all smart grid. We will propose, therefore, other algorithms to pilot other smart grid levels.

References

1. Rim, M., Hibaoui, A.: Algorithms for Smart Grid Management. Elsevier Sustainable Cities and Society (2018)
2. Rim, M., Hibaoui, A.: Modelization of smart grid managing the local level. In: IRSEC16 (2016)

3. Ruj, S., Pal, A.: Analyzing cascading failures in smart grid under random and targeted attacks (2014)
4. Ashok, A., Hahn, A., Govindarasu, M.: Cyber physical security of wide area monitoring, protection and control in a smart grid environment. J. Adv. Res. (2014)
5. Wenye, W., Zhuo, L.: Cyber security in smart grid: survey and challenges. Elsevier Computer Network Journal (2013)
6. Yan, Y., Qian, Y., Sharif, H., Tipper, D.: A survey on smart grid communication infrastructures. IEEE Commun. J. (2013)
7. Wang, W., Xu, Y., Khanna, M.: A survey on the communication architectures in smart grid. Elsevier Comput. Netw. J. (2011)
8. Wenpeng, L., Wenyuan, L.: Smart metering and infrastructure in smart grid. In: Clouds, Communications, Open Source, and Automation, pp. 925–930 (2014)
9. Rim, M., Hibaoui, A.: Model checking self-stabilising in embedded systems with linear temporal logic. Int. J. Adv. Comput. Sci. Appl. (2015)
10. Rim, M., Hibaoui, A.: Formalism of self-stabilization with linear temporal logic and its verification. In: Proceedings of 2015 IEEE World Conference on Complex Systems, WCCS (2015)

An Haptic Display for Visually Impaired Pedestrian Navigation

Sara Alzighaibi, Rahma Bouaziz, and Slim Kammoun

Abstract Sight is one of the major senses that allows people to travel safely, freely, and independently of others. In their daily life, visually impaired people are confronting challenging tasks such as reading or writing. Navigation and direction finding, particularly in the unfamiliar environment, are remaining as one of the main issues for visually impaired (VI) individuals. Several ways have been emerged to help VI orientation and mobility such as electronic orientation aids which can be further studied and developed. This work aims to improve VI navigation by proposing a new tactile interface that will be designed and implemented with the use of vibrators and Arduino board. The proposed interface targets to increase the quality of life and independence for VI. Also, evaluation performed with VI pedestrian will give us further information and knowledge regarding haptic perceptions and effectiveness of such interface and interaction technique.

1 Introduction

In the last few decades, visually impaired (VI) have received a lot of attention aimed at supporting them in their daily tasks and improving the quality of their life and

S. Alzighaibi · S. Kammoun (✉)
Information System Department, College of Computer Science and Engineering, Taibah University, Medina, Saudi Arabia
e-mail: skammoun@taibahu.edu.sa

S. Alzighaibi
e-mail: sara.alnuzha@gmail.com

R. Bouaziz
Computer Science Department, College of Computer Science and Engineering, Taibah University, Medina, Saudi Arabia

ReDCAD, University of Sfax, Sfax, Tunisia

S. Kammoun
Research Laboratory of Technologies of Information and Communication & Electrical Engineering, University of Tunis, Tunis, Tunisia

F. Saeed et al. (eds.), *Advances on Smart and Soft Computing*, Advances in Intelligent Systems and Computing 1188, https://doi.org/10.1007/978-981-15-6048-4_55

635

allowing them to practice any of the activities of daily living (ADL). Braille printer, screen reader application, or adapted smartphones were proposed to help visually impaired user in their daily activities. Till it developed, in addition, devices that talk or make other sounds became available in everyone's hands today, extending from bank ATMs and TVs. Despite all of these successful innovations, one of the main challenges facing this group is navigation that considered as core issue are still happening. In their daily displacements, VI people use conventional mobility aids such as white cane and guide dog to perform their day-to-day activities, and however, these assistive devices remain unusable in unknown environments. Hence, the importance of electronic orientation aids (EOA) appears, and the central role of GPS-based systems is to complement and go beyond what is impossible with the use of traditional assistive devices (e.g., white cane or guide dog) and not to replace them. As a result, numerous studies have emerged focusing on designing sophisticated systems based on new technologies (e.g., GPS, geographic information system (GIS), virtual reality (VR), adapted user interfaces, etc.) aiming to help VI in their daily M&O tasks [1–4].

In the absence of vision, audio-based interfaces have been proposed as an intuitive solution to indicate directional instructions to VI pedestrian, due to not specific learning required. Synthesized text or text-to-speech (TTS) was used in commercialized devices such as Kapten (Kapsys Inc.) and Trekker (HumanWare Inc.).

3D sound has been also investigated providing binaural rendering over headphones using a specialization engine [5, 6]. The main idea here is to generate informational auditory content at the spatial position, which directly coincides with that of a specific target (i.e., direction to take). In this case, the use of stereo headphone is required to produce binaural 3D sound. The major inconvenient when using audio outputs is that instructions could have a high level of intrusion and may not be well received in noisy environments. Further, the imposed cognitive load may be inadequate as for a VI person environmental sounds are primordial to understand his surroundings. Although the colossal work was done previously in proposing navigational aids, we have noted that commercialized EOA is a simple adaptation of cars-based guidance systems.

Trying to take into consideration pedestrian specificity and especially in the absence of vision, we will present a new haptic user interface that can be plugged with any EOA. Our goal is not to present a whole way-finding system, but only to focus on information rendering how we can provide EOA users with the best directional instruction helping them to navigate in a safe manner.

The rest of the paper is presented as follow. Section 2 describes related works with a focus on haptic interfaces proposed for VI pedestrian. In Sect. 3, we will present our proposed interface, detail our design choices, and present a preliminary user evaluation of the proposed interface. Section 4 will conclude the paper, discuss results, and present future works.

2 Related Work

Over the last ten years, tactile and haptic feedback has been investigated as sensory substitution and has proved useful in supporting users in navigating daily tasks. Numerous works demonstrate that vibration base tactile feedback able to provide comfortable and effective interactions during human–computer interaction [7–10]. Ghiani et al. [11] propose a system that use tactile feedback to provide the user with directional instructions using two actuators located on the user's two fingers. The study found that tactile feedback offered to the users was extremely useful and beneficial when providing the user with dynamic information such as the proximity to an obstacle and the distance veered away from the right orientation. Srikulwong and O'Neill [12] present two wearable haptic devices for navigation purposes used on the waist. User evaluation demonstrates that using waist for indicating the direction to take can be just as effective as map direction.

Kammoun et al. and Brock et al. [8, 13] presented a study aimed to reduce the cognitive load required for blind people in following navigation instructions within unknown environments. An Arduino wearable prototype was developed by using two wrist bracelets. Results proposed that vibrations can assist users in the maintenance of their path; however, patterns were quite confusing. This strengthened that walking in unfamiliar path is stressful and a demanding task and cognitive load must be lowered to a minimum. Pocket-based navigation using an Android maps application was also explored [14, 15]. Users could receive non-visual guidance using vibration cues as the device is in the pocket. This model was advantageous since it gave the user constant feedback and did not require custom-built tactile applications. The study recorded that pedestrians could use this mobile-based compass without requiring individual turn instructions. The application could be distributed in the form of an Android application. In the same philosophy, [16] propose to use smartphone vibrator to indicate direction by coding them through different patterns. Rümelin et al. [16] explore using vibration on a smartphone to provide turn-by-turn walking instructions to people with visual impairments. A first prototype was built, and a user study was conducted where eight participants walked along a pre-programmed route using the vibration feedback methods and no audio output. The result showed that a single vibrator could provide clear instructions. Variation in detection and response times for stimuli applied across various parts of the body has been also investigated. Up to 13 body, locations were proposed and evaluated. Results found that the wrist, feet, and spine were the worst body parts in terms of detection. The interesting finding regarding this study was that visual workload did not affect the detection efficiency of the wearable technology. This asserts that such technology could be used to navigate visually challenged individual since visual workload did not affect vibrotactile device efficiency.

Shoes-based tactile interfaces for VI navigation purpose has been also investigated. Kammoun et al. [17] assess the ability for a pedestrian to keep right decision when receiving directional instructions on their shoes. A preliminary field trial with two blindfolded participants shown that such configuration can guide pedestrian to take the right direction. Velázquez et al. [18] present a new wearable way-finding system that combines GPS positioning for outdoor localization and tactile-foot stimulation for information display. Based on works described below, we will design a new interface. Also, changing certain parameters including frequency, duration, and intensity of vibration are possible to improve further outcomes and information provided to the visually impaired user. All of the previous initiatives indicate that there is no one approved accurate output interface out for EOA, and the field still allows for more creativity and innovation.

3   The Proposed Interface

This project provides an essential opportunity to make an advance investigation on the best method to use the haptic techniques so as to improve the mobility and orientation for VI individuals. Our work will focus only on interface design. We think that interface can be plugged with any electronic orientation aid used by VI pedestrian or way-finding system used with smartphone. The primary objective of this research is to propose and design a new tactile interface that can be combined with a GPS-based system to help VI users when moving within unknown spaces.

3.1   Hardware and Software

An Arduino Uno board and four vibrating actuators are the main components of the proposed interface (see Fig. 1). The Arduino is a microcontroller, which contains integrated circuits that are tiny computers designed to carry out lightweight operations

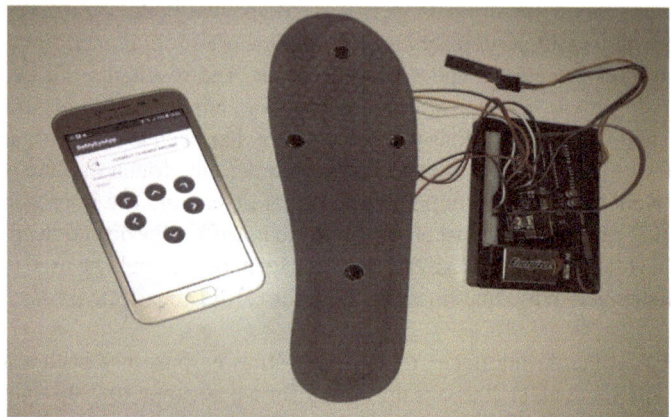

Fig. 1 First prototype composed by an Arduino board connected to four vibrators

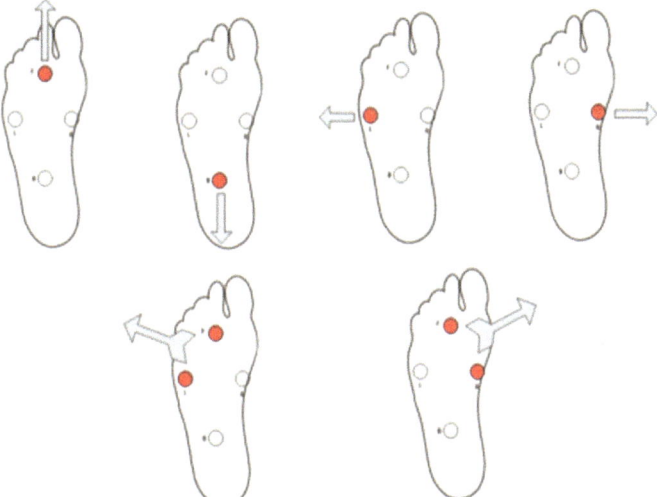

Fig. 2 Six different directions are defined. The arrow indicates the direction indicated by the red vibrator

with open source software. Each vibrator is connected to the Arduino board and can be independently controlled with a specific vibrating frequency command. The tactile display can be used on the right or left foot according to user preferences (e.g., dominant foot), and it is completely wearable by putting the Arduino board, the battery, and the Bluetooth transmission module in the pocket or attached in the user's ankle.

One of the aims of our study is to reduce the time and effort required of blind user to understand the directional instruction. To do that, we define a simple pattern for direction encoding based on activating the actuator according to direction to take.

Six directions have been defined as presented in Fig. 2, four simple directions (e.g., forward, right, left, and backward) and two composed direction (e.g., forward-left and forward-right). Simple directions are indicated by one vibration from the corresponding actuator. However, composed one will be indicated by activating the two close actuators.

3.2 Indoor Prelimenary Evaluation

To evaluate the proposed system, we implement a framework as presented by Fig. 3.

15 blindfolded voluntaries users were involved in this first experiment (13 female and two male), and none of them has complained of any problem in sight and has no previous experience with any haptic devices. The average age of all participant was

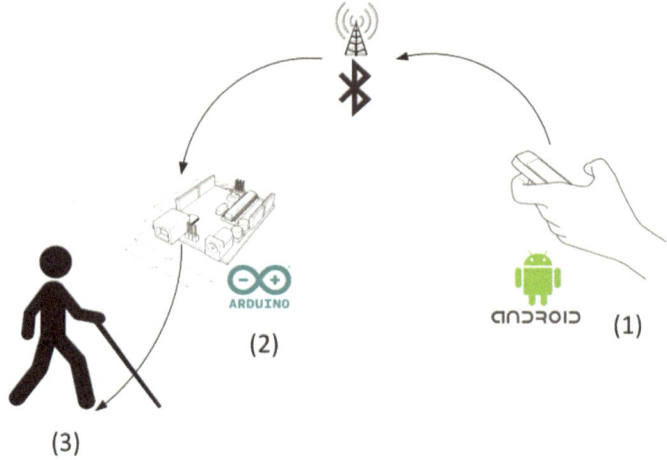

Fig. 3 Evaluation framework. The usr sends turning instructions through an Android application connected via a Bluetooth connection with the haptic shoes

33 years old. Participants were invited individually, and each session takes between 15 or 20 min.

The experimenter explains aims and goals of this evaluation and prototype composition in the beginning of each session. Before starting the experiment, users were asked to sign an agreement form. Each participant was asked to wear the prototype on their foot while sitting. During the test, all users have the full freedom to choose preferred foot in which the test will be implemented. The experimenter plays each direction five times (e.g., forward, backward, right, left, up-right, and up-left). In total for each session, 30 randomly direction was played, and for each participant, 75 stimulus will be used. After one played direction, the user fill a checkbox table to indicate the filled direction.

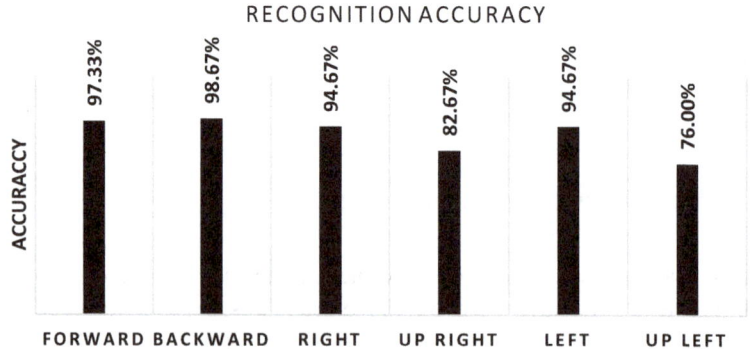

Fig. 4 Direction recognition accracy

Table 1 Number of correct answers for each directions played 75 times for all 15 users		Correct answers	Correct (%)
	Forward	73	97.33
	Backward	74	98.67
	Right	71	94.67
	Up-right	62	82.67
	Left	71	94.67
	Up-left	57	76.00

As presented by Fig. 4 and Table 1, the four main directions (e.g., up, forward, right, left) was well recognized and can be clearly determined with an accuracy rate more than 94%. However, for the composite direction, up-right and up-left, the recognition accuracy was, respectively, 82.67 and 76%.

4 Conclusion and Discussion

We present a haptic interface prototype to be used by VI in way-finding scenarios. In this paper, we constringed our focus on how directional information will be displayed to the user paying no attention to localization techniques, which is still a problematic to be considered due to the lack of positioning accuracy. This shoes-based haptic interface prototype enables the deployment of different feedback approaches. Not only the number of actuators can vary, as the type of vibrations patterns and dimensions therein (frequency, duration, interval). To better understand how these dimensions can be used, a thorough evaluation needs to take place. Our plans include validating different designs within the haptic design space and deriving a set of guidelines for the usage of such haptic feedback for way-finding purposes. In future work, we plan to perform a set of user experiences using the proposed evaluation framework presented in Sect. 4. This framework enables us to eliminate guidance and localization errors and help us to understand the effectiveness of such a design.

References

1. Kammoun, S., Parseihian, G., Gutierrez, O., Brilhault, A., Serpa, A., Raynal, M., Oriola, B., MacÉ, M.J.-M., Auvray, M., Denis, M., Thorpe, S.J., Truillet, P., Katz, B.F.G., Jouffrais, C.: Navigation and space perception assistance for the visually impaired: the NAVIG project. IRBM **33** (2012). https://doi.org/10.1016/j.irbm.2012.01.009
2. Loomis, J.M., Golledge, R.G., Klatzky, R.L., Speigle, J.M., Tietz, J.: Personal Guidance System for the Visually İmpaired. ACM Press, New York (1994). https://doi.org/10.1145/191028.191051
3. Ran, L., Helal, S., Moore, S.: Drishti: an integrated indoor/outdoor blind navigation system and service. In: Second IEEE Annual Conference on Pervasive Computing and Communications,

2004. Proceedings, pp. 23–30 (2004). https://doi.org/10.1109/PERCOM.2004.1276842
4. Helal, A., Moore, S.E., Ramachandran, B.: Drishti: an integrated navigation system for visu- ally impaired and disabled. In: Proceedings of Fifth International Symposium on Wearable Computing, pp. 149–156 (2001). https://doi.org/10.1109/ISWC.2001.962119
5. Katz, B.F.G., Kammoun, S., Parseihian, G., Gutierrez, O., Brilhault, A., Auvray, M., Truillet, P., Denis, M., Thorpe, S., Jouffrais, C.: NAVIG: augmented reality guidance system for the visually impaired: combining object localization, GNSS, and spatial audio. Virtual Real. **16** (2012). https://doi.org/10.1007/s10055-012-0213-6
6. Walker, B.N., Lindsay, J.: Navigation performance with a virtual auditory display: effects of beacon sound, capture radius, and practice. Hum. Factors **48**, 265–278 (2006)
7. Srikulwong, M.: A comparison of two wearable tactile interfaces with a complementary display in two orientations. In: Haptic and Audio Interaction Design (2010)
8. Kammoun, S., Jouffrais, C., Guerreiro, T., Nicolau, H., Jorge, J.: Guiding blind people with haptic feedback. In: Frontiers in Accessibility for Pervasive Computing (Pervasive 2012). Newcastle, UK (2012)
9. Pielot, M., Henze, N., Heuten, W., Boll, S.: Evaluation of continuous direction encoding with tactile belts. In: Lecture Notes in Computer Science (including subseries Lecture Notes in Artificial Intelligence and Lecture Notes in Bioinformatics), pp. 1–10. Springer (2008). https:// doi.org/10.1007/978-3-540-87883-4_1
10. Spiers, A.: Explorations of Shape-Changing Haptic Interfaces for Blind and Sighted Pedestrian Navigation. Max Planck Institute for Intelligent Systems (2019)
11. Ghiani, G., Leporini, B., Paternò, F.: Vibrotactile feedback as an orientation aid for blind users of mobile guides. In: Proceedings of the 10th International Conference on Human Computer Interaction with Mobile Devices and Services—MobileHCI'08, p. 431 (2008). https://doi.org/ 10.1145/1409240.1409306.
12. Srikulwong, M., O'Neill, E.: A comparative study of tactile representation techniques for landmarks on a wearable device. In: Proceedings of the 2011 Annual Conference on Human Factors in Computing Systems, pp. 2029–2038. ACM (2011)
13. Brock, A., Kammoun, S., Macé, M., Jouffrais, C.: Using wrist vibrations to guide hand move- ment and whole body navigation. I-Com **13**, 19–28 (2014). https://doi.org/10.1515/icom.2014. 0026
14. Pielot, M., Poppinga, B., Boll, S.: PocketNavigator: vibro-tactile waypoint navigation for everyday mobile devices. In: 11th International Conference International Conference on Human-Computer Interaction with Mobile Devices and Services, pp. 423–426 (2010)
15. Pielot, M., Poppinga, B., Heuten, W., Boll, S.: A tactile compass for eyes-free pedestrian navigation. In: Lecture Notes in Computer Science (LNCS) (including its subseries Lecture Notes in Artificial Intelligence (LNAI) and Lecture Notes in Bioinformatics), vol. 694, pp. 640– 656 (2011). https://doi.org/10.1007/978-3-642-23771-3_47
16. Rümelin, S., Rukzio, E., Hardy, R.: NaviRadar: a novel tactile information display for pedestrian navigation. In: Proceedings of the 24th Annual ACM Symposium on User Interface Software and Technology, pp. 293–302. ACM (2011)
17. Kammoun, S., Bouhani, W., Jemni, M.: Sole based tactile information display for visually impaired pedestrian navigation. In: 8th ACM International Conference on PErvasive Tech- nologies Related to Assistive Environments, PETRA 2015—Proceedings (2015). https://doi. org/10.1145/2769493.2769561
18. Velázquez, R., Pissaloux, E., Rodrigo, P., Carrasco, M., Giannoccaro, N., Lay-Ekuakille, A.: An outdoor navigation system for blind pedestrians using GPS and tactile-foot feedback. Appl. Sci. **8**, 578 (2018). https://doi.org/10.3390/app8040578

GPU Parallelization for Accelerating 3D Primitive Equations of Ocean Modeling

Abdullah Aysh Dahawi, Norma Binti Alias, and Amidora Idris

Abstract Graphics processing unit (GPU) has become a powerful computation platform not only for graphic rendering purposes, but also for multi-purpose computations. Using various software, such as NVIDIA's Compute Unified Device Architecture (CUDA) programming model, the developers can use the GPU without a graphics programming background. In this paper, we describe the implementation of 3D primitive equations solver for incompressible and inviscid fluid flow in rotating frame with hydrostatic balance using desktop platform equipped with a GPU. The governing equations for this study consist of six dependent variables, three velocity components, temperature, salinity, and pressure. The finite difference method (FDM) is used to discretize the mathematical model based on forward-time backward-space (FTBS) scheme. It is realized that using a single Tesla K20c GPU card, the CUDA implementation of the ocean circulation model within two days simulation runs 216 times faster than a serial C++ code running on a single core of an Intel(R) Xeon(R) CPU E5-2620 2.10 GHz processor. The results reveal that the ocean circulation is feasible on this type of platform and that model can be run within minutes.

Keywords GPU parallel computing · CUDA · Ocean circulation · FDM

A. A. Dahawi · A. Idris
Department of Mathematics, Faculty of Science, Universiti Teknologi Malaysia, Iskandar Puteri, Johor, Malaysia
e-mail: abdullah_aysh@yahoo.com

A. Idris
e-mail: amidora@utm.my

N. B. Alias (✉)
Center for Sustainable Nanomaterials, Ibnu Sina Institute for Scientific and Industrial Research, Universiti Teknologi Malaysia, Iskandar Puteri, Johor, Malaysia
e-mail: norma@ibnusina.utm.my

© The Editor(s) (if applicable) and The Author(s), under exclusive license to Springer Nature Singapore Pte Ltd. 2021
F. Saeed et al. (eds.), *Advances on Smart and Soft Computing*, Advances in Intelligent Systems and Computing 1188, https://doi.org/10.1007/978-981-15-6048-4_56

1 Introduction

Nowadays, graphics processing unit (GPU) has become a very efficient and powerful programmable unit not only for graphic rendering purposes but also for general-purpose computations [1]. It has been considered as a massive parallel co-processor to the central processing unit (CPU) because its design depends on the stream processing architecture that is suitable for computing intensive parallel tasks [2]. This device has turned into an attractive alternative for high-performance scientific computing due to the relative low cost and high arithmetic computation power [3]. Many advantages can be acquired using the GPU parallelization such as accelerating the computations and minimizing the power demand [4]. In addition, many general-purpose applications targeting the GPUs have been developed such as OpenGL, DirectX and Compute Unified Device Architecture (CUDA) [5]. Unlike OpenGL and DirectX, CUDA, produced by NVIDIA in 2006, is a programmable platform extending the C programming language for efficient parallel execution [6]. Using CUDA, programmers and researchers can directly execute their data parallel computations on GPU without having a graphics programming background.

High-performance computing (HPC) using GPU has attracted the interest of many researchers in computational science and other scientific fields, such as fuzzy theory [7], computational biology and systems biology [8–11], molecular dynamics [12, 13], artificial intelligence [14], and weather forecasting [15, 16]. In ocean circulation modeling, Mak et al. have implemented a new parallelization of the Regional Ocean Modeling System using CUDA Fortran [17]. They obtained $8\times$ speedup compared with a serial code which had been performed in Athlon II processor clocked at 2.9 GHz. Compared to other parallel platforms, they have also achieved a $2.5\times$ speedup over OpenMP. They conclude that GPUs are close to MPI cluster in spite of the less cost of GPU. Chen et al. have ported three-dimensional ocean model to the GPU using CUDA programming models [18]. Without using computation-reduction techniques, they could accelerate the running of the model directly and explicitly. Their NVIDIA GTX460 GPU achieved a speedup of 20 times relative to CPU code running on an Intel Core i7-920. Bleichrodt et al. introduced a numerical simulation for the barotropic ocean model [19]. They utilized the central finite difference for spatial direction and Adams–Bashforth for time integration to discretize the two-dimensional barotropic vorticity equation of ocean model. The computations were conducted on NVIDIA Tesla M1060 GPU and an Intel Xeon L5420 2.50 GHz CPU. The results showed a speedup of up to $48\times$ for a grid size of 2049.

In this paper, we use a GPU to implement a new parallelization of a 3D ocean model which is represented by the hydrostatic Boussinesq primitive equations. Physically relevant references can be found in [20]. This model describes the momentum, pressure, salinity concentration changes, and heat equation that investigate the temperature behavior of the ocean. Using fully explicit forward-time backward-space (FTBS) scheme, the mathematical model is discretized. However, finite difference grid resolution limits the simulation's accuracy and performance. In other words, increasing the grid resolution leads to increase the computational cost. That is due to the requirements of the numerical stability, which need finer grid with small time step [17].

2 Governing Equations

Physically, ocean model consists of six basic fields of interest. Each field can be represented as a time-dependent and space-dependent function in which the ocean resides. These fields are three-component vector velocity $\mathbf{v}$, pressure p, thermodynamic energy T, and salinity S, which form a complete set of dependent variables representing the ocean circulation [20].

2.1 The Primitive Equations

The mathematical model of the ocean circulation is represented by the Navier–Stokes equations under the hydrostatic and Boussinesq approximations known as the primitive equations. These equations correspond, respectively, to the conservation of momentum in the x- and y-directions, the hydrostatic equation, continuity equation, and time evolution of temperature and salinity (concentration) fields.

$$\frac{\partial u}{\partial t} + u\frac{\partial u}{\partial x} + v\frac{\partial u}{\partial y} + w\frac{\partial u}{\partial z} + \frac{1}{\rho_0}\frac{\partial p}{\partial x} - fv = 0 \tag{1}$$

$$\frac{\partial v}{\partial t} + u\frac{\partial v}{\partial x} + v\frac{\partial v}{\partial y} + w\frac{\partial v}{\partial z} + \frac{1}{\rho_0}\frac{\partial p}{\partial y} + fu = 0 \tag{2}$$

$$\frac{\partial p}{\partial z} + g\rho_0\left[1 - \alpha(T - T_0) + \gamma(S - S_0)\right] = 0 \tag{3}$$

$$\frac{\partial u}{\partial x} + \frac{\partial v}{\partial y} + \frac{\partial w}{\partial z} = 0 \tag{4}$$

$$\frac{\partial T}{\partial t} + u\frac{\partial T}{\partial x} + v\frac{\partial T}{\partial y} + w\frac{\partial T}{\partial z} = 0 \tag{5}$$

$$\frac{\partial S}{\partial t} + u\frac{\partial S}{\partial x} + v\frac{\partial S}{\partial y} + w\frac{\partial S}{\partial z} = 0 \tag{6}$$

where f is the Coriolis parameter given by the following relation called beta-plane approximation:

$$f = f_0 + \beta y. \tag{7}$$

where $\beta = 2.2 \times 10^{-11}$ m^{-1}s^{-1} at mid-latitudes, and y is the distance in meters. The constants ρ_0, T_0 and S_0 represent, respectively, the density reference, temperature reference, and salinity reference while α is the thermal expansion coefficient and γ is contraction of saline. Typical seawater values are $\rho_0 = 1028$ kg/m^3, $T_0 = 10°$, $C = 283\ K$, $S_0 = 35, \alpha = 1.7 \times 10^{-4} K^{-1}$ and $\gamma = 7.6 \times 10^{-4}$ where g represents the effective gravitational acceleration which is equal to 9.81 m s^{-2}.

In this study, we used the one-degree annual temperature and salinity statistical mean of Southern Ocean as initial condition of the model. The study area lies between $-60°$ and $-70°$ latitude and between $5°$ and $45°$ longitude with 100 m of depth. These data are collected by the National Oceanic and Atmospheric Administration (NOAA) [21]. Besides, the boundary conditions of prognostic nonlinear problem of ocean general circulation have been chosen for the model [22].

2.2 Discretization of the Model

In the GPU programming, the performance can be impacted by the chosen numerical scheme. In other words, memory access needs to be coalesced to achieve maximal bandwidth by making adjacent points in the spatial dimension for the stencil. Therefore, the time space, backward finite difference scheme in the space direction is used, while forward finite difference is used for time integration.

The domain of this study is $\Omega = [0, L] \times [0, W] \times [0, D]$, where L, W and D are the length, width, and depth, respectively. This domain can be divided in a grid with N_x, N_y and N_z grid points:

$$(x_i, y_j, z_k), \quad i = 0, 1, \ldots, N_x - 1, j = 0, 1, \ldots, N_y - 1, k = 0, 1, \ldots, N_z - 1.$$

In the x-direction, the step size is $\Delta x = L/N_x - 1$, in the y-direction it is $\Delta x = W/N_y - 1$ and in z-direction is $\Delta x = D/N_z - 1$.

The equations of (1–6) are then written in difference equations form as:

$$u_{i,j,k}^{n+1} = F_{i,j,k}^n + \frac{\Delta t}{\rho_0 \Delta x}\left(p_{i-1,j,k}^n - p_{i,j,k}^n\right) \tag{8}$$

$$v_{i,j,k}^{n+1} = G_{i,j,k}^n + \frac{\Delta t}{\rho_0 \Delta y}\left(p_{ij-1,k}^n - p_{ij,k}^n\right) \tag{9}$$

$$p_{i,j,k}^n = p_{i,j,k-1}^n + \rho_0 g \Delta z\left[\alpha\left(T_{i,j,k}^n - T_0\right) - \gamma\left(S_{i,j,k}^n - S_0\right) - 1\right] \tag{10}$$

$$w_{i,j,k}^n = w_{i,j,k-1}^n + \frac{\Delta z}{\Delta x}\left(u_{i-1,j,k}^n - u_{i,j,k}^n\right) + \frac{\Delta z}{\Delta y}\left(v_{i,j-1,k}^n - v_{i,j,k}^n\right) \tag{11}$$

$$T^{n+1} = H^n \tag{12}$$

$$S^{n+1} = I^n \tag{13}$$

where

$$F_{i,j,k}^n = u_{i,j,k}^n + \frac{\Delta t}{\Delta x} u_{i,j,k}^n \left(u_{i-1,j,k}^n - u_{i,j,k}^n\right) + \frac{\Delta t}{\Delta y} v_{i,j,k}^n \left(u_{i,j-1,k}^n - u_{i,j,k}^n\right)$$
$$+ \frac{\Delta t}{\Delta z} w_{i,j,k}^n \left(u_{i,j,k-1}^n - u_{i,j,k}^n\right) + \Delta t (f_0 + \beta j \Delta y) v_{i,j,k}^n \tag{14}$$

$$G_{i,j,k}^n = v_{i,j,k}^n + \frac{\Delta t}{\Delta x} u_{i,j,k}^n \left(v_{i-1,j,k}^n - v_{i,j,k}^n\right) + \frac{\Delta t}{\Delta y} v_{i,j,k}^n \left(v_{i,j-1,k}^n - v_{i,j,k}^n\right)$$
$$+ \frac{\Delta t}{\Delta z} w_{i,j,k}^n \left(v_{i,j,k-1}^n - v_{i,j,k}^n\right) - \Delta t (f_0 + \beta j \Delta y) u_{i,j,k}^n \tag{15}$$

$$H_{i,j,k}^n = T_{i,j,k}^n + \frac{\Delta t}{\Delta x} u_{i,j,k}^n \left(T_{i-1,j,k}^n - T_{i,j,k}^n\right) + \frac{\Delta t}{\Delta y} v_{i,j,k}^n \left(T_{i,j-1,k}^n - T_{i,j,k}^n\right)$$
$$+ \frac{\Delta t}{\Delta z} w_{i,j,k}^n (T_{i,j,k-1}^n - T_{i,j,k}^n) \tag{16}$$

$$I_{i,j,k}^n = S_{ij,k}^n + \frac{\Delta t}{\Delta x} u_{i,j,k}^n \left(S_{i-1,j,k}^n - S_{i,j,k}^n\right) + \frac{\Delta t}{\Delta y} v_{i,j,k}^n \left(S_{i,j-1,k}^n - S_{i,j,k}^n\right)$$
$$+ \frac{\Delta t}{\Delta z} w_{i,j,k}^n (S_{ij,k-1}^n - S_{ij,k}^n) \tag{17}$$

3 Hardware and Software

CUDA is a parallel computing platform and programming model developed by NVIDIA for its own GPU device. The hardware layout makes some restrictions which leads to the best performance, but it undermines portability. Therefore, a brief discertion is given here to understand the hardware layout and memory organization.

The CUDA architecture depends on the single instruction, multiple data (SIMD) pattern. This implies that the same instructions on different data are executed by multiple threads. A GPU has several multiprocessors (MPs, 13 for the Tesla K20c), each containing 192 cores. The fine-grained data parallelism, therefore, can be carried out by running several hundreds of threads simultaneously on the GPU.

The GPU code is usually called CUDA "kernel" which is the computation core of the CUDA programming model. A CUDA kernel typically generates a very large number of threads to exploit data parallelism. Hundreds or maybe thousands of threads, depending on the device model, are grouped into (1D, 2D, or 3D) "blocks".

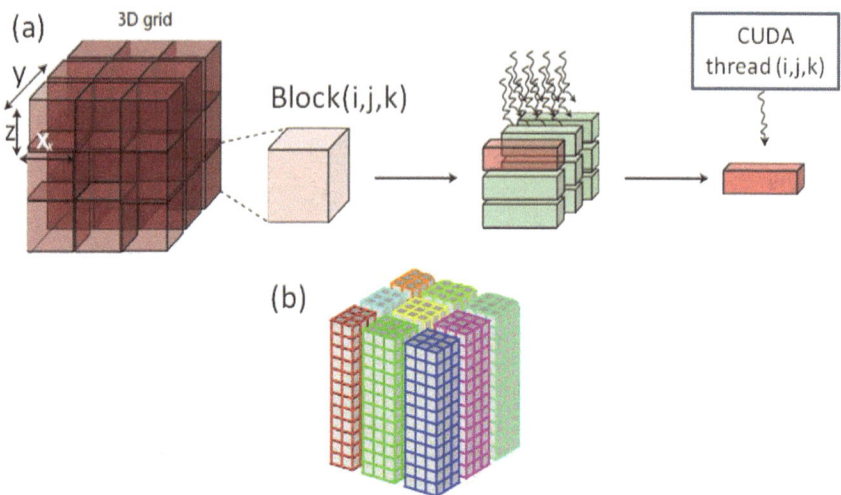

Fig. 1 3D grid **a** represents 3D decomposition. **b** Representation of a 2D Z-pencil decomposition

These blocks form a (1D, 2D, or 3D) grid, as shown in Fig. 1. The thread blocks are scheduled for execution in the MPs. The MP registers determine the number of threads in the block that can run concurrently. The threads per block can be defined to be as much as possible to reduce the idle time of threads, but appropriate number of blocks per grid should be set that all MPs remain busy.

3.1 Decomposition of 3D Ocean Domain

This study was restricted to ocean basins with vertical sidewalls and flat ocean bottom. The number of the domain's computational nodes can be represented by Nx, Ny and Nz in x, y and z directions, respectively. Also, because the vertical relation between pressure and vertical velocity component of the domain mesh points are separate, we can apply the domain decomposition technique in 2D Z-pencil decomposition, which decompose the domain vertically in two dimensions as shown in Fig. 1b. On the other hand, because the threads block synchronization, where threads per the same block can be synchronized, the computational nodes in the vertical block can share the updating results at the same time step. Therefore, the host code was defined as two-dimensional grid of three-dimensional blocks.

On the other hand, device memory stores the data in a 1D array that can represent a 3D domain. The corresponding domain cell indices are shown in Fig. 2, while the translation map from 3 to 1D is given by:

$$A[P] = B[i][j][k]$$

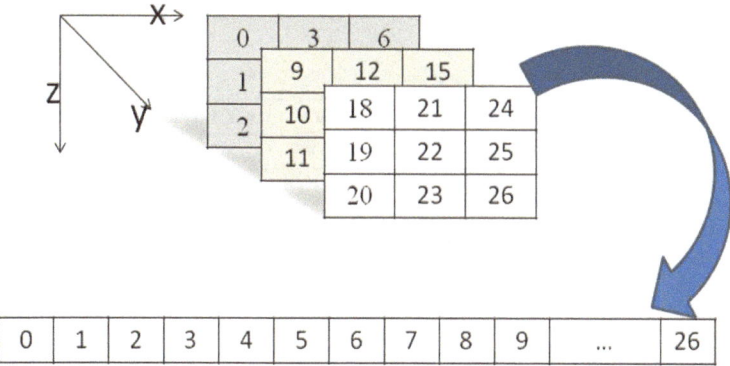

Fig. 2 Representation of 3D domain by 1D array

where

$$P = k + i*Nx + j*Nx*Nz$$

$$i = blockIdx.x*blockDim.x + threadIdx.x$$

$$j = blockIdx.y*blockDim.y + threadIdx.y$$

$$k = blockIdx.z*blockDim.z + threadIdx.z$$

while, A is a 1D array with $Nx*Ny*Nz$ elements and B is a 3D matrix of the same number of elements (see Fig. 2) and P is the index that accesses the GPU DRAM while $(blockDim.x, blockDim.y, blockDim.z)$ represents single block dimensions and $(blockIdx.x, blockIdx.y, blockIdx.z)$ is the identification ID of the block with 3D coordinates. The thread ID in a block is given by $(threadIdx.x, threadIdx.y, threadIdx.z)$.

Figure 3 shows the algorithm of GPU implementation. The code consists of two main parts. The first part includes the host-side code that involves the declaration of all variables and parameters as well as the main loop for time marching to advance the calculation in time. Second is the device code that consists of three kernels to execute algorithm of the numerical methods that were performed in sequential approach. As mentioned above, the variables of the problem should be updated every time step by the results in the previous time step, and hence, we need to force a global synchronization by using separate kernels for calculations which leads to update the threads in every CUDA blocks before going to the next time step of the computations.

In the current implementation, the new values of the variables u,v,w,T,S,p are calculated at every time step by calling the kernel "Main_Cul" which solves the

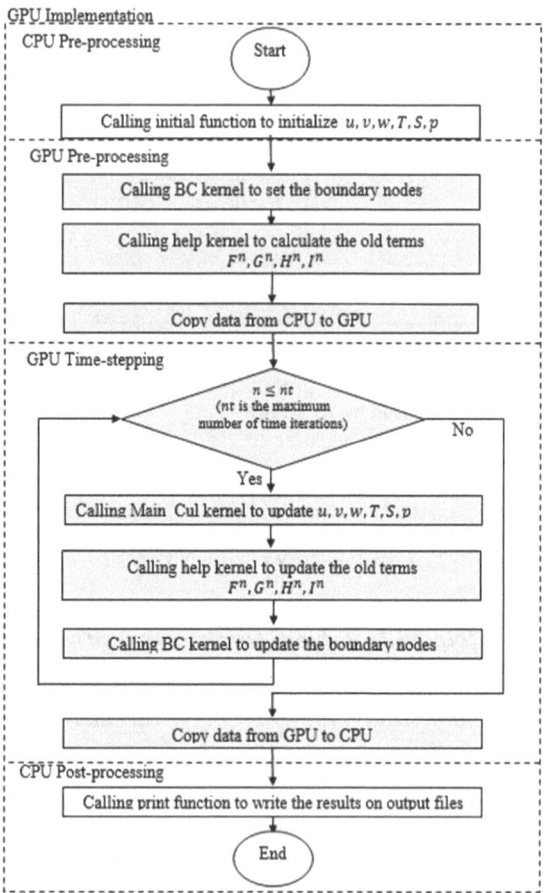

Fig. 3 Flowchart of the parallel algorithm

linear algebraic equation directly. Then, the boundary nodes are updated using the kernel "BC". In addition, some auxiliary matrices are needed to temporarily store the previous time step values and then use it to update the variables. These matrices are included in the third kernel "help_f".

4 Results and Discussion

4.1 GPU Performance Evaluation

Ocean model running in serial code was compared with CUDA running one single GPU. The CPU is used for all experiments in an Intel Xeon E5-2620 v2 processor with 32 GB of RAM. The CUDA implementation uses NVIDIA Quadro K4000 as shown in Table 1. The ocean domain for this study is 4360 km × 778 km × 100 m with a computational grid of 128 × 128 × 32 and 2 days simulation with 2880 time steps. The serial code takes 10,906.6 s to simulate the mathematical model. The execution time for the GPU is only 67.62 s, which means the GPU implementation is 161.3 times faster than CPU implantation.

Table 2 and Fig. 4 show the runtime of the parallel algorithm with respect to the mesh size. The results (shown in Fig. 5) demonstrate that the speedup increases when the domain size increases. That is because the serial code must wait until the previous instruction ends, as opposed to the GPU execution where the data are distributed to the threads blocks and executed simultaneously.

Moreover, the threads of blocks in GPU influence the speedup. This is probably due to the overhead caused by CUDA. At some point, the speedup should stagnate, when the total number of active threads is reached. At that point, the further increasing of number of threads does not result in a larger speedup, but it will only add more overhead [19]. In Table 3, the effect of difference in the number of threads per block has been presented by giving the run time and the speedup in all cases. The optimal number of threads per block of this study is 64, where the mesh size is 128 × 128 × 32, which yields a speedup of up to 216× over the serial implementation. This means that when the block dimension is defined to be greater than 64, as shown in Figs. 6 and 7, the speedup will decrease.

Table 1 Computer specifications

GPU	NVIDIA Quadro K4000
CPU	Intel(R) Xeon(R) CPU E5-2620 v2@2.10 GHz
Memory	32 GB RAM
OS	Windows7 Professional
CUDA	CUDA 7.0
Host compiler	Visual C++ 10

Table 2 Run time and speedup of CPU and GPU in different mesh size

Size	No. of time step	CPU run time (s)	GPU run time (s)	Speedup
16 × 16 × 4	2880	0.46	0.71	0.64
32 × 32 × 8		3.24	1.933	1.67
64 × 64 × 16		68.4	11.76	5.81
128 × 128 × 32		10,906.6	67.62	161.3

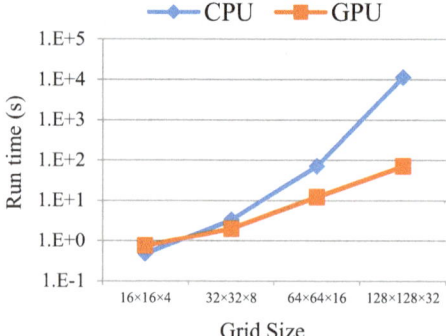

Fig. 4 Run time of CPU and GPU in different mesh size

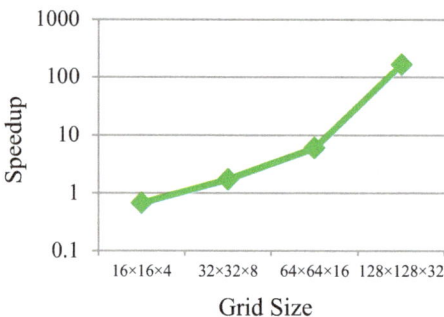

Fig. 5 Speedup of GPU compared to CPU in different mesh size

	No. of threads	CPU run time (s)	GPU run time (s)	Speedup
Table 3 Run time and speedup of CPU and GPU in different number threads per block	1	10,906.6	142.39	76.59667111
	4		98.5	110.7269036
	8		79.56	137.0864756
	16		69.5	156.9294964
	32		68.4	159.4532164
	64		50.49	216.0150525
	128		59.4	183.6127946
	256		76.54	142.4954272
	512		99.7	109.3941825

5   Conclusion

This study presented the implementation of the primitive equations for the ocean circulation model on desktop platform with GPU. The model profile consists of six dependent variables, three velocity components, temperature, salinity, and pressure.

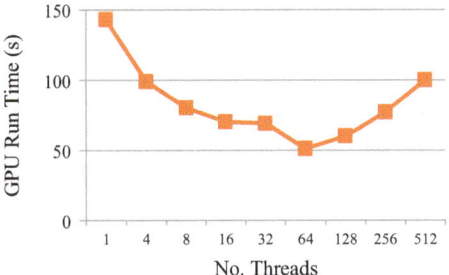

Fig. 6 Run time of the parallel code in different number of threads per block

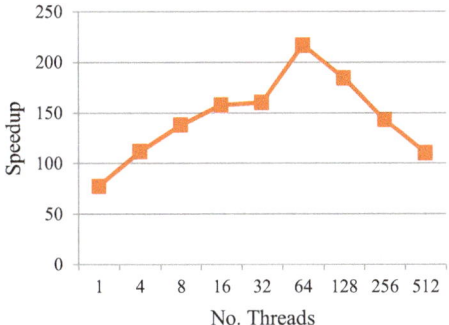

Fig. 7 Speedup of the parallel code in different number of threads per block

The relations between these variables form a coupled of nonlinear PDEs. FDM was used based on FTBS scheme to discretize and solve the model numerically. The simulation of the model was executed sequentially using C++ programming language, while NVIDIA's CUDA programming model was utilized to accelerate the computations of the discretized form of the governing equations. The main contribution of this paper is on accelerating the simulation of the ocean circulation model. CUDA C++ yields a speedup of up to 216× over a serial implementation of 2 days simulation. Results showed that GPU desktop platforms have substantial potential to compute intensive parallel tasks.

Acknowledgements We would like to thank the Research Management Centre (RMC) at Universiti Teknologi Malaysia (UTM) for funding this project under grant number PY/2019/01171/RJ130000.7854.5F253. We also would like to thank the Ibnu Sina Institute for Scientific and Industrial Research (ISI-SIR), Faculty of Science, Universiti Teknologi Malaysia for supporting this research.

References

1. Li, D., Wu, H., Becchi, M.: Nested parallelism on GPU: exploring parallelization templates for irregular loops and recursive computations. In: 44th International Conference on Parallel Processing 2015, pp. 979–988. IEEE, Beijing (2015)
2. García-Feal, O., et al.: IberWQ: A GPU accelerated tool for 2D water quality modeling in rivers and estuaries. Water 12(2), 413 (2020)
3. Hadi, N.A., Alias, N.: 3-dimensional human head reconstruction using cubic spline surface on CPU-GPU platform. In: Proceedings of the 2019 4th International Conference on Intelligent Information Technology, pp. 16–20. Association for Computing Machinery, Da, Nang, Viet Nam (2019)
4. Panzer, I., et al.: High performance regional ocean modeling with GPU acceleration. In: Oceans-San Diego. IEEE (2013)
5. Aijia, O., et al.: Parallel hybrid pso with cuda for ld heat conduction equation. Comput. Fluids 110, 198–210 (2015)
6. Nvidia, CUDA C Programming Guide. Nvidia (2015)
7. Mukaram, M.Z., Norma, A., Tahir, A.: High performance analysis in generating FTTM graph of pseudo degree zero using CPU-GPU. In: AIP Conference Proceedings, p. 020005. AIP Publishing LLC (2018)
8. Manconi, A., et al.: G-CNV: A GPU-Based tool for preparing data to detect CNVs with read-depth methods. Front. Bioeng. Biotechnol. 3(28) (2015)
9. Hanyu, J., Narayan, G.: CUDAMPF: a multi-tiered parallel framework for accelerating protein sequence search in HMMER on CUDA-enabled GPU. BMC Bioinform. 106(1), 106 (2016)
10. 3-Dimensional human head reconstruction using cubic spline surface on CPU-GPU Platform. In: Proceedings of the 4th International Conference on Intelligent Information Technology
11. Vouzis, P.D., Sahinidis, N.V.: GPU-BLAST: using graphics processors to accelerate protein sequence alignment. Bioinformatics 27(2), 182–188 (2011)
12. Anderson, J.A., Lorenz, C.D., Travesset, A.: General purpose molecular dynamics simulations fully implemented on graphics processing units. J. Comput. Phys. 227(10), 5342–5359 (2008)
13. Voelz, V.A., et al.: Molecular simulation of ab initio protein folding for a millisecond folder NTL9 (1–39). J. Am. Chem. Soc. 132(5), 1526–1528 (2010)
14. Zhou, Y., Zeng, J.: Massively Parallel A* Search on a GPU (2015)
15. Govett, M., et al.: Parallelization and performance of the NIM weather model on CPU, GPU, and MIC processors. Bull. Am. Meteor. Soc. 98(10), 2201–2213 (2017)
16. Schalkwijk, J., et al.: Weather forecasting using GPU-based large-Eddy simulations. Bull. Am. Meteor. Soc. 96(5), 715–723 (2015)
17. Mak, J., Choboter, P., Lupo, C.: Numerical ocean modeling and simulation with CUDA. In: OCEANS 2011. IEEE (2011)
18. Chen, B., Zhu, J., Li, L.: Accelerating 3d ocean model development by using gpu computing. In: Future Control and Automation, pp. 37–43. Springer (2012)
19. Bleichrodt, F., Bisseling, R.H., Dijkstra, H.A.: Accelerating a barotropic ocean model using a GPU. Ocean Model. 41, 16–21 (2012)
20. Samelson, R.M.: The Theory of Large-scale Ocean Circulation. Cambridge University Press, Cambridge (2011)
21. NOAA, World Ocean Atlas 2013 V2. 2015, National Oceanographic Data Center: USA.
22. Sarkisyan, A.S., Sündermann, J.: Modelling Ocean Climate Variability. Springer Science & Business Media. Springer (2009)

Author Index

A

Abakkali, Mohammed, 625
Abdulgabar, Abdulnasser Abdulgaleel, 295
Abdullah, Nibras, 341
Abouelmehdi, Karim, 211
Achtaich, Khadija, 261
Adoni, Wilfried Yves Hamilton, 273
Adoui El Ouadrhiri, Abderrahmane, 3
Ait Abdelali, Hamd, 517
Ait idar, Hafsa, 421
Alaidaroos, Alawi S., 383
Aldowah, Hanan, 329
Al-Hadhrami, Tawfik, 341
Al-Hameli, Bassam Abdo, 223
Alharbe, Mahmood Abdulghani, 403
Al-Hetar, Abdulaziz Mohammed, 295
Alias, Norma Binti, 643
Almekhlafi, Khalil, 599
Al-Mekhlafi, Zeyad Ghaleb, 599
Al-Mohaimeed, Muhannad, 189
Al-Moslmi, Tareq, 431
Almurashi, Haneen, 541
Alromema, Waseem, 71
Al-Sarem, Mohammed, 71, 189
Alsarem, Mohammed, 223
Alsewari, AbdulRahman A., 223
Al-Shabi, M. A., 577
Alshadadi, Ahmed, 61
Alshami, Maged, 61
Alshamsi, Sultan, 613
Alshudukhi, Jalawi, 599
Al-Wosabi, Abdo Ali, 431
Al-Zekri, Burhan T., 61
Alzighaibi, Sara, 635
Amazal, Houda, 107
Ameen, Ali, 613
Amine, Aouatif, 153

Anbar, Mohammed, 341
Azbeg, Kebira, 357
Azizi, Nabiha, 37

B

Ba Alawi, Abdulfattah E., 61
Bachaoui, ELMostafa, 467
Baddi, Youssef, 249, 443, 527
Bahashwan, Abdullah Ahmed, 341
Bahri, Nisrine, 505
Bamatraf, Mohammed A., 93, 383
Basir, Mohammad Aizat, 167
Behja, Hicham, 235
Belhabib, Abdelfettah, 567
Belhadaoui, Hicham, 421
Belkacem, Soundes, 119
Bellafkih, Mostafa, 587
Beni-Hssane, Abederrahim, 211
Ben Lahmar, El Habib, 261
Benlahmer, El Habib, 285
Bensalah, Faycal, 527
Bensassi, Bahloul, 303, 553
Ben Souda, Souad, 201
Benzebouchi, Nacer Eddine, 37
Bin Merdhah, Mansoor H., 383
Bin Shamlan, Mohammed H., 383
Bouaziz, Rahma, 541, 635
Boudhane, Mohcine, 143
Boukhamla, Assia, 37
Boukhdir, Khalid, 319
Boulila, Wadii, 189
Boulouird, Mohamed, 567
Bourja, Omar, 517
Bourzex, François, 517

C

Cakula, Sarma, 143
Chbihi Louhdi, Mohammed Reda, 235
Chemlal, Yman, 481
Cheriguene, Soraya, 37
Cherradi, Bouchaib, 15

D

Daanouni, Othmane, 15
Dachry, Wafaa, 303, 553
Dahaoui, Fatimaez-Zahra, 235
Dahawi, Abdullah Aysh, 643
Demraoui, Lamiae, 235
Dendani, Najdette, 37
Derrouz, Hatim, 517

E

El Alaoui, Imane, 153
El Bennani, Kaoutar, 467
El Filali, Sanaa, 261
Elfilali, Sanaa, 285
El Guabassi, Inssaf, 625
El Haddadi, Anass, 481
Elhaloui, Loubna, 285
Elkamoun, Najib, 495, 527
Elmoufidi, Abdelali, 177
Emara, Abdel Hamid, 189
Errais, Mohammed, 189, 587
Ezziri, Salma, 27

F

Fagroud, Fatima Zahra, 261
Faroukhi, Abou Zakaria, 153
Fauzi, Muhammad Ali, 391
Fetjah, Laila, 357
Filali, Reda, 421

G

Gahi, Youssef, 153
Gallofré Ocaña, Marc, 431
Gazem, Nadhmi A., 613
Guezouli, Larbi, 119

H

Hammami, Nacer Eddine, 37
Hanshi, Sabri M., 341
Haouari, Rajae, 517
Hasbi, Abderrahim, 249, 443
Hassani, Moha M'Rabet, 567

Hussin, Mohamed Saifullah, 167

I

Idalene, Asmaa, 319
Idris, Amidora, 643
Isaac, Osama, 613
Isnin, Ismail Fauzi Bin, 455

J

Jai Andaloussi, Said, 3, 357
Jai-Andaloussi, Said, 177

K

Kaleem, Muhammad, 455
Kammoun, Slim, 635
Kechid, Samir, 131
Khadir, Omar, 27, 411
Khalid, Ma'ab, 541
Khaloufi, Hayat, 211
Kissi, Mohamed, 107
Ko'adan, Mohammed A., 93
Koulali, Rim, 51
Krichen, Moez, 273

L

Lakrami, Fatima, 495
Lamari, Mouna, 37
Larioui, Sanae, 625

M

Mahmud, Mufti, 79
Mandar, El Mehdi, 303, 553
Marah, Rim, 625
Martiri, Edlira, 391
Mechti, Seifeddine, 273
Medromi, Hicham, 319, 505
Miah, Yunus, 79
Mohamad, Kamaruddin Malik, 371
Mohammad, Muneef A., 61
Mohammed, Ala'a M., 61
Mohammed, Fathey, 613
Mouncif, Hicham, 467
Moussaid El Idrissi, Sohaib, 411

N

Nahhal, Tarik, 273
Noureddine, Dhouha Ben, 273

O
Ochetto, Ouail, 177
Ouchetto, Ouail, 3, 357
Oulad Haj Thami, Rachid, 517

P
Pratt, Madara, 143
Prima, Chowdhury Nazia Enam, 79

R
Radman, Borhan M. N., 61
Ramdani, Mohammed, 107
Rifqi, Hanane, 201

S
Saadaoui, Safa, 505
Sabtu, Saiful Bahari Mohd, 371
Sadik, Mohamed, 505
Sadouki, Fatiha, 131
Saeed, Adeeb M., 61
Saeed, Ahmed Y. A., 61
Saeed, Faisal, 71, 189
Saeed, Osama Y. A., 61
Saif, Alaa A., 61
Saoudi, El Mehdi, 3
Sarem Al, Mohammed, 587
Sayed, Bushra, 541

Seema, Sharmeen Jahan, 79
Shafal, Khalid Q., 93
Shamim Kaiser, M., 79
Skouta, Ayoub, 177
Sounni, Hind, 495

T
Tabaa, Mohamed, 285, 505
Tmiri, Amal, 15
Toumi, Hicham, 261

U
Ul Rehman, Shafiq, 329
Umar, Irfan, 329

Y
Yang, Bian, 391
Yusof, Yuhanis, 167

Z
Zaidani, Hajar, 51
Zaim, Maryeme, 51
Zain, Adnan A., 383
Zamma, Abdellah, 201
Zemrane, Hamza, 249, 443
Zennayi, Yahya, 517